Generalized Boltzmann Physical Kinetics

Generalized Boltzmann Physical Kinetics

BORIS V. ALEXEEV

Moscow Fine Chemical Technology Institute
Moscow 119571, Russia

2004

ELSEVIER

Amsterdam – Boston – Heidelberg – London – New York – Oxford
Paris – San Diego – San Francisco – Singapore – Sydney – Tokyo

ELSEVIER B.V.
Sara Burgerhartstraat 25
P.O. Box 211, 1000 AE
Amsterdam, The Netherlands

ELSEVIER Inc.
525 B Street, Suite 1900
San Diego, CA 92101-4495
USA

ELSEVIER Ltd
The Boulevard, Langford Lane
Kidlington, Oxford OX5 1GB
UK

ELSEVIER Ltd
84 Theobalds Road
London WC1X 8RR
UK

First edition 2004

Library of Congress Cataloging in Publication Data
A catalog record is available from the Library of Congress.

British Library Cataloguing in Publication Data
A catalogue record is available from the British Library.

ISBN: 0-444-51582-8

♾ The paper used in this publication meets the requirements of ANSI/NISO Z39.48-1992 (Permanence of Paper).
Printed in The Netherlands.

Contents

Preface vii

Historical Introduction and the Problem Formulation 1

Chapter 1. Generalized Boltzmann Equation 21

1.1. Mathematical introduction. Method of many scales 21
1.2. Hierarchy of Bogolubov kinetic equations 33
1.3. Derivation of the generalized Boltzmann equation 39
1.4. Generalized Boltzmann ***H***-theorem and the problem of irreversibility of time 61
1.5. Generalized Boltzmann equation and iterative construction of higher-order equations in the Boltzmann kinetic theory 79
1.6. Generalized Boltzmann equation and the theory of non-local kinetic equations with time delay 83

Chapter 2. Theory of Generalized Hydrodynamic Equations 91

2.1. Transport of molecular characteristics 91
2.2. Hydrodynamic Enskog equations 94
2.3. Transformations of the generalized Boltzmann equation 96
2.4. Generalized continuity equation 98
2.5. Generalized momentum equation for component 101
2.6. Generalized energy equation for component 106
2.7. Generalized hydrodynamic Euler equations 112
2.8. Boundary conditions in the theory of the generalized hydrodynamic equations 124

Chapter 3. Strict Theory of Turbulence and Some Applications of the Generalized Hydrodynamic Theory 133

3.1. About principles of classical theory of turbulent flows 133
3.2. Theory of turbulence and generalized Euler equations 136
3.3. Theory of turbulence and the generalized Enskog equations 149
3.4. Generalized hydrodynamic equations and quantum mechanics 153

Chapter 4. Physics of a Weakly Ionized Gas 161

4.1. Relaxation of charged particles in "Maxwellian" gas and the hydrodynamic aspects of the theory 161
4.2. Distribution function of the charged particles in the "Lorentz" gas 168
4.3. Charged particles in alternating electric field 176
4.4. Conductivity of a weakly ionized gas in crossed electric and magnetic fields 179

Chapter 5. Kinetic Coefficients in the Theory of the Generalized Kinetic Equations 187

5.1. Linearization of the generalized Boltzmann equation 187
5.2. Approximate modified Chapman–Enskog method 195
5.3. Kinetic coefficient calculation with taking into account the statistical fluctuations 208

Chapter 6. Some Applications of the Generalized Boltzmann Physical Kinetics 215

6.1. Investigation of the generalized Boltzmann equation for electron energy distribution in a constant electric field with due regard for inelastic collisions 215
6.2. Sound propagation studied with the generalized equations of fluid dynamics 226
6.3. Shock wave structure examined with the generalized equations of fluid dynamics 238

Chapter 7. Numerical Simulation of Vortex Gas Flow Using the Generalized Euler Equations 241

7.1. Unsteady flow of a compressible gas in a cavity 241
7.2. Application of the generalized hydrodynamic equations: to the investigation of gas flows in channels with a step 254
7.3. Vortex and turbulent flow of viscous gas in channel with flat plate 266

Chapter 8. Generalized Boltzmann Physical Kinetics in Physics of Plasma and Liquids 287

8.1. Extension of generalized Boltzmann physical kinetics for the transport processes description in plasma 287
8.2. Dispersion equations of plasma in generalized Boltzmann theory 297
8.3. Generalized dispersion relations for plasma: theory and experiment 313
8.4. To the kinetic and hydrodynamic theory of liquids 323

Appendices

A1. Derivation of energy equation for invariant $E_\alpha = m_\alpha V_\alpha^2/2 + \varepsilon_\alpha$ 339
A2. Three-diagonal method of Gauss elimination technics for the differential third-order equation 347
A3. Some integral calculations in the generalized Navier–Stokes approximation 352
A4. Three-diagonal method of Gauss elimination technique for the differential second-order equation 354
A5. Characteristic scales in plasma physics 356
A6. Dispersion relations in the generalized Boltzmann kinetic theory neglecting the integral collision term 357

References 361

Subject Index 367

Preface

For about 130 years the Boltzmann equation has belonged to the fundamental equations of physics. The destiny of this equation is as dramatic as that of its great creator. Even in Boltzmann's days there was a complete awareness that his equation acquires a fundamental importance for physics and that its range of validity stretches from transport processes and hydrodynamics all the way to cosmology – thus fully justifying the keen attention it attracted and debates it provoked. Both sides of the dispute have exhausted their arguments. Thus, the development of Boltzmann kinetic theory has turned out to be typical of any revolutionary physical theory – from rejection to recognition and further to a kind of "canonization".

About twenty years ago it was shown by the author of this book that taking into account the variation of the distribution function over times of the order of the collision time led to additional terms in the Boltzmann equation, which were proportional to mean time *between* collisions of particles and therefore to the Knudsen number and viscosity in the hydrodynamic limit of the theory. Moreover, it turns out that these terms – whose influence grows with an increase in the Knudsen number – cannot be omitted in the case of small Knudsen numbers because these terms contain small parameters in front of senior derivatives. Then these terms should be conserved in the theory in the whole diapason of evolution of Knudsen numbers. I have been working in the kinetic theory for more than 40 years and this conclusion was dramatic first and foremost for myself.

Therefore, the case in point is of unprecedented situation in physics, when the fundamental physical equation is revised. During my stay in Marseille as an Invited Professor A.J.A. Favre reminded me Henri Poincaré's phrase after the death of a great Austrian physicist: "Boltzmann was wrong, but his mistake is equal to zero". It's a pity, but the situation in the kinetic theory is more serious. Obviously, changing the fundamental equation leads – to some extent – to possible changes of the known results in the modern transport theory of physics. This book reflects the scale of these alterations. It is safe to say – as the main result of the generalized Boltzmann kinetic theory – that this theory has showed it to be a highly effective tool for solving many physical problems in the areas where the classical theory runs into difficulties.

The author is deeply indebted to V.L. Ginzburg and F. Uhlig for their interest in this work and in the subject in general. I am thankful to V. Mikhailov and A. Fedoseyev for their cooperation.

October, 2003

"Alles Vergängliche
ist nur ein Gleichniss!"
Boltzmann's epigraph
for his "Vorlesungen über Gastheorie"

Historical Introduction and the Problem Formulation

In 1872 L. Boltzmann, then a mere 28 years old, published his famous kinetic equation for the one-particle distribution function $f(\mathbf{r}, \mathbf{v}, t)$ (Boltzmann, 1872). He expressed the equation in the form

$$\frac{Df}{Dt} = J^{\mathrm{st}}(f), \tag{I.1}$$

where J^{st} is a collision ("stoß") integral, and

$$\frac{D}{Dt} = \frac{\partial}{\partial t} + \mathbf{v} \cdot \frac{\partial}{\partial \mathbf{r}} + \mathbf{F} \cdot \frac{\partial}{\partial \mathbf{v}} \tag{I.2}$$

is a substantial (particle) derivative, $\mathbf{v}$ and $\mathbf{r}$ being the velocity and the radius-vector of the particle, respectively.

Eq. (I.1) governs the transport processes in a one-component gas which is sufficiently rarefied so that only binary collisions between particles are of importance. While we are not concerned here with the explicit form of the collision integral (which determines the change of the distribution function f in binary collisions), note that it should satisfy the conservation laws. For the simplest case of elastic collisions in a one-component gas, we have

$$\int J^{\mathrm{st}} \psi_i \, \mathrm{d}\mathbf{v} = 0 \quad (i = 1, 2, 3), \; \mathrm{d}\mathbf{v} = \mathrm{d}v_x \, \mathrm{d}v_y \, \mathrm{d}v_z, \tag{I.3}$$

where ψ_i are collisional invariants ($\psi_1 = m$, $\psi_2 = m\mathbf{v}$, $\psi_3 = mv^2/2$, m is the mass of the particle) related to the laws of conservation of mass, momentum, and energy.

Integrals of the distribution function (i.e., its moments) determine the macroscopic hydrodynamic characteristics of the system, in particular, the number density of particles

$$n = \int f \, \mathrm{d}\mathbf{v} \tag{I.4}$$

and the temperature T:

$$\frac{3}{2} k_B T n = \frac{1}{2} m \int f (\mathbf{v} - \mathbf{v}_0)^2 \, \mathrm{d}\mathbf{v}. \tag{I.5}$$

Here k_B is the Boltzmann constant, and $\mathbf{v}_0$ is the hydrodynamic flow velocity. It follows then that, multiplying the Boltzmann integro-differential equation term by term by the collisional invariants ψ_i, integrating over all particle velocities, and using the conservation laws (I.3), we arrive at the differential equations of fluid dynamics, whose general form is known as the hydrodynamic Enskog equations.

The Boltzmann equation (BE) is not of course as simple as its symbolic form above might suggest, and it is in only a few special cases that it is amenable to solution. One example is that of a Maxwellian distribution in a locally, thermodynamically equilibrium gas in the event, where no external forces are present. In this case, the equality

$$J^{\mathrm{st}} = 0 \tag{I.6}$$

is met, giving the Maxwellian distribution function

$$f^{(0)} = n\left(\frac{m}{2\pi k_B T}\right)^{3/2} \exp\left(-\frac{mV^2}{2k_B T}\right), \tag{I.7}$$

where $\mathbf{V} = \mathbf{v} - \mathbf{v}_0$ is the thermal velocity.

It was much later, years after Boltzmann's death in 1906, that an analytic method for solving the Boltzmann equation was developed for the purpose of calculating transport coefficients. This method, developed in 1916–1917 by Chapman (1916, 1917) and Enskog (1917, 1921a), led to explicit expressions for the coefficients of viscosity, thermal conductivity, diffusion, and later thermal diffusion in a system with a small parameter (which, for Chapman and Enskog's particular problem of a nonreacting gas, was the Knudsen number, the ratio of the particle's mean free path to a characteristic hydrodynamic dimension).

However, even in Boltzmann's days there was a complete awareness that his equation acquires a fundamental importance for physics and that its range of validity stretches from transport processes and hydrodynamics all the way to cosmology – thus fully justifying the keen attention it attracted and debates it provoked.

Of the many results L. Boltzmann derived from his kinetic equation, one of the most impressive is the molecular-kinetic interpretation of the second principle of thermodynamics and, in particular, of the statistical meaning of the concept of entropy. It turned out that it was possible to define the function

$$H = \int f \ln f \, \mathrm{d}\mathbf{v} \tag{I.8}$$

(H is the first letter in the English word *heat* and German word *Heizung*) which behaves monotonically in a closed system.

If the relation between S, the entropy per unit volume of an ideal gas, and the H-function is written in the form

$$S = -kH + const, \tag{I.9}$$

then one can prove the inequality

$$\frac{\partial S}{\partial t} \geqslant 0. \tag{I.10}$$

The laconic formula

$$S = k \ln W, \tag{I.11}$$

connecting the entropy S and the thermodynamic probability W, is inscribed on Boltzmann's tombstone.

Ever since their creation, Boltzmann's physical kinetics and the Boltzmann equation have received severe criticism, much of which remains of interest even today. Let us elaborate on this.

To begin with, Boltzmann's contemporaries were very much in the dark regarding the relation between the Boltzmann equation and classical mechanics – in particular, with the Newton equation. The Boltzmann equation was obtained in a phenomenological manner, based on convincing physical arguments, and reflects the fact that the distribution function does not change along the particle's trajectory between collisions but rather changes as a result of an "instantaneous" interaction between colliding particles.

J. Loschmidt noted in 1876 that the Boltzmann equation underlying the H-theorem includes only the first time derivative whereas the Newton equation contains the second one ("square of time") and hence the equations of motion are reversible in time. This means that if a system of hard-sphere particles starts a "backward" motion due to the particles reversing their direction of motion at some instant of time, it passes through all its preceding states up to the initial one, and this will increase the H-function whose variation is originally governed by reversible equations of motion. The essential point to be made here is that the observer cannot prefer one of the situations under study, the "forward" motion of the system in time, in favor of the second situation, its "backward" motion. In other words, the problem of the reversibility of time arises here.

Although somewhat differently formulated, essentially the same objection was made in 1896 by Planck's student E. Zermelo, who noted that the H-theorem was inconsistent with the Poincare's "recurrence" theorem proved in 1890, and stated that any physical system, even with irreversible thermodynamic processes operating in it, had a nonzero probability of returning to its original state. Boltzmann himself fully aware of this possibility, wrote in the second part of his Lectures on the Theory of Gases (see Boltzmann, 1912, p. 251): "As a result of the motion of gas molecules, the H-function always decreases. The unidirectional nature of this process does not follow from the equations of motion, which the molecules obey. Indeed, these equations do not change if time changes sign."

There is a well-known example from the probability theory, which Boltzmann employed as an argument in his discussions – sometimes very heated ones – with Zermelo, Planck, and Ostwald. If a six-sided die is thrown 6000 times, one expects each side to turn up about 1000 times. The probability of, say, a six turning up 6000 times in succession has a vanishing small value of $(1/6)^{6000}$. This example does not clear up the matter, however. Nor do the two papers which Boltzmann's student P. Ehrenfest wrote in co-authorship with T. Afanas'eva-Ehrenfest after the death of the great Austrian physicist.

Their first model, reported by Afanas'eva-Ehrenfest at the February 12, 1908 meeting of the Russian Physico-Chemical Society, involved the application of the H-theorem to the "plane" motion of gas (Ehrenfest, 1972). Suppose P-molecules, nontransparent to one another, start moving normally to the axis y and travel with the same velocity in the direction of the axis x. Suppose further that in doing so they undergo elastic collisions with Q-particles, squares with sides at an angle of 45° to the axis y, which are nontransparent to the molecules and are all at rest.

It is readily shown that shortly after, all the molecules will divide themselves into four groups, and it is a simple matter to write down the change in the number of molecules P in each group in a certain time Δt and then to define a "planar-gas" H-function

$$H = \sum_{i=1}^{4} f_i \ln f_i, \tag{I.12}$$

where f_i is the number of molecules of the ith kind, i.e., of those moving in one of the four possible directions. If all the velocities reverse their direction, the H-function starts to increase and reverts to the value it had when the P-molecules started their motion from the y axis. While this simple model confirms the Poincare–Zermelo theorem, it does not at all guarantee that the H-function will decrease when the far more complicated Boltzmann model is used.

P. and T. Ehrenfest's second model (Ehrenfest, 1979), known as the lottery's model, features two boxes, A and B, and N numbered balls to which there correspond "lottery tickets" placed in a certain box and which are all in box A initially. The balls are then taken one by one from A and transferred to B according to the number of a lottery ticket, drawn randomly. Importantly, the ticket is not eliminated after that but rather is returned to the box. In the event that the newly drawn ticket corresponds to a ball contained in B, the ball is returned to A. As a result, there will be approximately $N/2$ balls in either box.

Now suppose one of the boxes contains n balls – and the other accordingly $N - n$ balls – at a certain step s in the drawing process. We can then define Δ, a function, which determines the difference in the number of balls between the two boxes: $\Delta = n - (N - n) = 2n - N$. In "statistical" equilibrium, $\Delta = 0$ and $n = N/2$, the dependence $\Delta(s)$ will imitate the behavior of the H-function in a Boltzmann gas.

This example is also not convincing enough because this "lottery" game will necessarily lead to the fluctuation in the Δ function, whereas the Boltzmann kinetic theory completely excludes fluctuations in the H-function. By the end of his life, Boltzmann went over to the fluctuation theory, in which the decrease of the H-function in time was only treated as the process the system is most likely to follow. This interpretation, however, is not substantiated by his kinetic theory since the origin of the primary fluctuation remains unclear (the galactic scale of such fluctuation included).

One of the first physicists to see that the Boltzmann equation must be modified in order to remove the existing contradictions was J. Maxwell. Maxwell thought highly of the results of Boltzmann, who, in his turn, did much to promote Maxwell electrodynamics and its experimental verification.

We may summarize Maxwell's ideas as follows. The equations of fluid dynamics are a consequence of the Boltzmann equation. From the energy equation, limiting ourselves to one dimension for the sake of simplicity and neglecting some energy transfer mechanisms (in particular, convective heat transfer), we obtain the well-known heat conduction equation

$$\frac{\partial T}{\partial t} = a^2 \frac{\partial^2 T}{\partial x^2}. \tag{I.13}$$

The fundamental solution of Eq. (I.13) up to the dimensional constant is

$$T(x,t) = \frac{1}{2\sqrt{\pi a^2 t}} \exp\left(-\frac{x^2}{4a^2 t}\right) \tag{I.14}$$

and represents the temperature at the point x at the instant t provided at time $t = 0$, an amount of heat $c\rho$, with ρ the density and a the thermal diffusivity of the medium, evolved at the origin of coordinates. Defining an argument of the function T as $\theta = a^2 t$ with the dimension of a coordinate squared, we obtain

$$T = \frac{1}{2\sqrt{\pi\theta}} \exp\left(-\frac{x^2}{4\theta}\right). \tag{I.15}$$

The temperature distribution given by this equation is unsatisfactory physically. For small values of θ, the temperature at the heat evolution point $x = 0$ is indefinitely large. On the other hand, at any arbitrarily distant point x the temperature produced by an instantaneous heat source will be different from zero for arbitrarily small times. While this difference may be small, it is a point of principal importance that it has a finite value.

As Landau and Lifshitz noted in their classical *Course of Theoretical Physics* (Landau and Lifshitz, 1988, p. 283), "The heat conduction process described by the equations obtained here has the property that any thermal perturbation becomes instantaneously felt over all space". This implies an infinitely fast propagation of heat, which is absurd from the point of view of molecular-kinetic theory. In the courses of mathematical physics, this result is usually attributed to the fact that the heat conduction equation is derived phenomenologically, neglecting the molecular-kinetic mechanism of heat propagation. However, as already noted, the parabolic equation (I.13) follows from the Boltzmann equation. Some of Maxwell's ideas, phenomenological in nature and aimed at the generalization of the Boltzmann equation, are discussed in Woods' monograph (Woods, 1993).

Although the examples above are purely illustrative and the exhaustive list of difficulties faced by the Boltzmann kinetic theory would, of course, be much longer, it should be recognized that after the intense debates of the early 20th century, the search for an alternative kinetic equation for a one-particle distribution function has gradually levelled off or, perhaps to be more precise, has become of marginal physical importance. Both sides of the dispute have exhausted their arguments. On the other hand, the Boltzmann equation has proven to be successful in solving a variety of problems, particularly in the calculation of kinetic coefficients. Thus, the development of Boltzmann kinetic theory has turned out to be typical of any revolutionary physical theory – from rejection to recognition and further to a kind of "canonization".

Work on the hyperbolic equation of heat conduction was no longer directly related to the Boltzmann equation, but rather was of a phenomenological nature. Without expanding into details of this approach, we only point out that the idea of the improvement of Eq. (I.13) was to introduce the second derivative with respect to time thus turning Eq. (I.13) into the hyperbolic form

$$\tau_{\mathrm{rel}} \frac{\partial^2 T}{\partial t^2} + \frac{\partial T}{\partial t} = a^2 \frac{\partial^2 T}{\partial x^2}, \tag{I.16}$$

where τ_{rel} is treated as a certain relaxation kinetic parameter with the dimensions of time. For the first time in modern physics this idea was formulated by Davydov (1935) (see also an interesting discussion on the priority between Cattaneo (1948, 1958) and Vernotte (1958b, 1958a)). The wave equation (I.16) leads to final propagation velocities for a thermal perturbation – although it should be remarked parenthetically that the quasi-linear parabolic equations may also produce wave solutions.

A breakthrough period in the history of kinetic theory occurred in the late 1930s and early 1940s, when it was shown through efforts of many scientists – of which Bogolyubov certainly tops the list – how, based on the Liouville equation for the multiparticle distribution function f_N of a system of N interacting particles, one could obtain a one-particle representation by introducing a small parameter $\varepsilon = n v_b$, where n is the number of particles per unit volume and v_b is the interaction volume (Bogolyubov, 1946; Born and Green, 1946; Green, 1952; Kirkwood, 1947; Yvon, 1935). This hierarchy of equations is usually referred to as the Bogolyubov or BBGKY (Bogolyubov–Born–Green–Kirkwood–Yvon) chain.

We do not present the technical details in Introduction but refer the reader to the classical works cited above or, for example, to Alekseev (1982). Some fundamental points of the problem are worth mentioning here, however.

(1) Integrating the Liouville equation

$$\frac{\partial f_N}{\partial t} + \sum_{i=1}^{N} \mathbf{v}_i \cdot \frac{\partial f_N}{\partial \mathbf{r}_i} + \sum_{i=1}^{N} \mathbf{F}_i \cdot \frac{\partial f_N}{\partial \mathbf{v}_i} = 0 \tag{I.17}$$

subsequently over phase volumes $\mathrm{d}\Omega_{s+1}, \ldots, \mathrm{d}\Omega_N$ ($\mathrm{d}\Omega_j \equiv \mathrm{d}\mathbf{r}_j\, \mathrm{d}\mathbf{v}_j$), one obtains a kinetic equation for the s-particle distribution function, with the distribution function f_{s+1} in the integral part of the corresponding equation.

In other words, the set of integro-differential equations turns out to be a linked one, so that in the lowest-order approximation the distribution function f_1 depends on f_2. This means formally that, strictly speaking, the solution procedure for such a set should be as follows. First find the distribution function f_N and then solve the set of BBGKY equations subsequently for decreasingly lower-order distributions. But if we know the function f_N, there is no need at all to solve the equations for f_s and it actually suffices to employ the definition of the function

$$f_s = \int f_N(t, \Omega_1, \ldots, \Omega_N)\, d\Omega_{s+1} \cdots d\Omega_N. \tag{I.18}$$

We thus conclude that the rigorous solution to the set of BBGKY equations is again equivalent to solving Liouville equations. On the other hand, the seemingly illogical solution procedure involving a search for the distribution function f_1 is of great significance in the kinetic theory and in nonequilibrium statistical mechanics. This approach involves breaking the BBGKY chain by introducing certain additional assumptions (which have a clear physical meaning, though). These assumptions are discussed in detail below.

(2) For a nonreacting gas, the Boltzmann equation is valid for two time scales of distribution functions: one of the order of the mean free time of the particles, and the other the hydrodynamic flow time. The Boltzmann equation is invalid for time lengths of the order of the collision times. Notice that a change from the time scale to the length scale can of course be made if desired.

(3) After the BBGKY chain has been broken and f_2 represented as a product of one-particle distribution functions (which is quite reasonable for a rarefied gas), the Boltzmann equation cannot be written in the classical form with only one small parameter ε and it reduces instead to the Vlasov equation in a self-consistent field.

(4) Since the Boltzmann equation does not work at distances of the order of the particle interaction radius (or at the r_b scale), Boltzmann particles are pointlike and structureless, and it is one of the inconsistencies of the Boltzmann theory that the resulting collision cross sections of the particles enter the theory by the collision integral.

(5) Usually the one-particle distribution function is normalized to the number of particles per unit volume. For Boltzmann particles the distribution function is "automatically" normalized to an integer, because a point-like particle may only be either inside or outside a trial contour in a gas – unlike finite-diameter particles which of course may overlap the boundary of the contour at some instant of time. Another noteworthy point is that the mean free path in the Boltzmann kinetic theory is only meaningful for particles modelled by hard elastic spheres. Other models face difficulties related, though, to the level of one-particle description employed. The requirement for the transition to a one-particle model is that molecular chaos should exist prior to a particle collision.

The advent of the BBGKY chain led to the recognition that whatever generalization of the Boltzmann kinetic theory is to be made, the logic to be followed should involve all the elements of the chain, i.e., the Liouville equation, the kinetic equations for s-particle distribution functions f_s, and the hydrodynamic equations. This logical construction was not generally adhered to.

In 1951, N. Slezkin published two papers (Slezkin, 1951b, 1951a) on the derivation of alternative equations for describing the motion of gas. The idea was to employ Meshcherskii's variable-mass point dynamics theory (Meshcherskii, 1897), well known for its jet propulsion applications.

The assumption of a variable-mass particle implies that, at each point, a liquid particle, close to this point and moving with a velocity **v**, adds or loses a certain mass, whose absolute velocity vector **U** differs, as Slezkin puts it, by a certain appreciable amount from the velocity vector **v** of the particle itself. Since there are different directions for this mass to come or go off, an associated mass flux density vector **Q** is introduced.

By applying the laws of conservation of mass, momentum, and energy in the usual way, Slezkin then proceeds to formulate a set of hydrodynamical equations, of which we will here rewrite the continuity equation for a one-component nonreacting gas:

$$\frac{\partial \rho}{\partial t} + \frac{\partial}{\partial \mathbf{r}} \cdot (\rho \mathbf{v} + \mathbf{Q}) = 0. \tag{I.19}$$

The mass flux density **Q** is written phenomenologically in terms of the density and temperature gradients.

Thus, the continuity equation is intuitively modified to incorporate a source term giving

$$\frac{\partial \rho}{\partial t} + \frac{\partial}{\partial \mathbf{r}} \cdot \rho \mathbf{v} = \frac{\partial}{\partial \mathbf{r}} \cdot \left(D \frac{\partial \rho}{\partial \mathbf{r}} + \beta \frac{\partial T}{\partial \mathbf{r}} \right), \tag{I.20}$$

where the coefficient D is that of self-diffusion, and β is related to thermal diffusion. Thus, we now have fluctuation terms on the right-hand side of Eq. (I.20), which are generally proportional to the mean free time τ and, hence, after Eq. (I.20) has been made dimensionless, to the Knudsen number which is small in the hydrodynamic limit.

At very nearly the time of publication of Slezkin's first paper (Slezkin, 1951a), Vallander (1951) argued that the standard equations of motion were ill grounded physically and should therefore be replaced by other equations, based on the introduction of additional mass Q_i and energy t_i fluxes ($i = 1, 2, 3$):

$$Q_i = D_1 \frac{\partial \rho}{\partial r_i} + D_2 \frac{\partial T}{\partial r_i}, \qquad t_i = k_1 \frac{\partial \rho}{\partial r_i} + k_2 \frac{\partial T}{\partial r_i},$$

where, to quote, "D_1 is the density self-diffusion coefficient, D_2 is the thermal self-diffusion coefficient, k_1 is the density heat conductivity, and k_2, the temperature heat conductivity".

Heuristic and inconsistent with Boltzmann's theory, the work of Slezkin and Vallander came under sufficiently severe criticism. Shaposhnikov (1951) noted that in these papers, "which are almost identical in content... the essential point is that instead of the conventional expression $\rho \mathbf{v}_0$, additional effects – "concentration self-diffusion" and "thermal self-diffusion" – are introduced into the mass flux density which, in addition to the macroscopic mass transfer, cause a molecular mass transfer, much as the macroscopic energy and momentum transfer in a moving fluid goes in parallel with analogous molecular transport (heat conduction and viscosity)". Shaposhnikov then proceeds to derive the equation of continuity from the Boltzmann equation for a one-component gas and shows that the hydrodynamic equations of Slezkin and Vallander are in conflict with the Boltzmann kinetic theory.

Note that Slezkin and Vallander also modified the equations of motion and energy for a one-component gas in a similar way (by including self-diffusion effects). Possible consequences of additional mass transfer mechanisms for the Boltzmann kinetic theory were not analyzed by these authors.

Boltzmann's "fluctuation hypothesis" was repeatedly addressed by Ya. Terletskii (see, for instance, Terletskii, 1952, 1994) whose idea was to estimate fluctuations by using the expression of the general theorems of Gibbs (see, for example, Gibbs, 1934, pp. 85–88) yield for the mean-square deviation of an arbitrary generalized coordinate. To secure that fluctuations in statistical equilibrium be noticeable, Terletskii modifies the equation of the perfect gas state by introducing a gravitational term, which immediately extends his analysis beyond the Boltzmann kinetic theory, leaving the question about the irreversible change of the Boltzmann H-function unanswered.

In recent years, possible generalizations of the Boltzmann equation have been widely discussed in the scientific literature. Since the term "generalized Boltzmann equation" (GBE) has usually been given to any new modification published, we will only apply this term to the particular kinetic equation derived in Alexeev (1994, 1995c, 2000b) to avoid confusion.

L. Woods (see, e.g., Woods, 1990), following the ideas dating back to Maxwell (1860), introduces in his theory a phenomenological correction to the substantial first derivative on the left-hand side of the Boltzmann equation to take account of the further influence of pressure on transport processes. It is argued that the equation of motion of a liquid particle may be written as $\dot{\mathbf{v}} = \mathbf{F} + \mathbf{P}$, where $\mathbf{P}$ is a certain additional force, proportional to the pressure gradient: $\mathbf{P} = -\rho^{-1}\partial p/\partial \mathbf{r}$, with the result that the left-hand side of the Boltzmann equation becomes

$$\frac{Df}{Dt} \equiv \frac{\partial f}{\partial t} + \mathbf{v} \cdot \frac{\partial f}{\partial \mathbf{r}} + \left(\mathbf{F} - \frac{1}{\rho}\frac{\partial p}{\partial \mathbf{r}}\right) \cdot \frac{\partial f}{\partial \mathbf{v}}, \tag{I.21}$$

whereas the collisional term remains unchanged. The phenomenological equation (I.21) has no solid foundation and does not fall into the hierarchy of Bogolyubov kinetic equations.

At the heart of the kinetic theory of neutral and ionized gases is the Boltzmann equation (BE) which describes how the one-particle distribution function f_1 changes over times in the order of the mean time between collisions and in the order of the gas dynamic flow time. Despite certain difficulties in the theory long-positioned as classical, the Boltzmann equation, now 130 years old (Boltzmann, 1872, 1912), has had no alternatives until recently as the basis for physical kinetics.

A weak point of the classical Boltzmann kinetic theory is the way it treats the dynamic properties of interacting particles. On the one hand, as the so-called "physical" derivation of the BE suggests (Boltzmann, 1872, 1912; Chapman and Cowling, 1952; Hirschfelder, Curtiss and Bird, 1954), Boltzmann particles are treated as material points;

on the other hand, the collision integral in the BE brings into existence the cross sections for collisions between particles. A rigorous approach to the derivation of the kinetic equation for f_1 (KE_{f_1}) is based on the hierarchy of the Bogolyubov–Born–Green–Kirkwood–Yvon (BBGKY) equations. A KE_{f_1} obtained by the multi-scale method turns into the BE if one ignores the change of the distribution function (DF) over a time in the order of the collision time (or, equivalently, over a length in the order of the particle interaction radius). It is important to note (Alexeev, 1994, 1995c, 2000b) that accounting for the third of the scales mentioned above has the consequence that, prior to introducing any approximations destined to break the Bogolyubov chain, additional terms, generally of the same order of magnitude, appear in the BE. If the method of correlation functions is used to derive KE_{f_1} from the BBGKY equations, then a passage to the BE implies the neglect of nonlocal and time delay effects. Given the above difficulties of the Boltzmann kinetic theory (BKT), the following clearly interrelated questions arise.

First, what is a physically infinitesimal volume and how does its introduction (and, as a consequence, the unavoidable smoothing out of the DF) affect the kinetic equation (Alexeev, 1994)?

And second, how does a systematic account for the proper diameter of a particle in the derivation of the KE_{f_1} affect the Boltzmann equation? In the theory we develop here, we will refer to the corresponding KE_{f_1} as the generalized Boltzmann equation, or GBE.

Accordingly, our purpose in this introduction is first to explain the essence of physical generalization of the BE and then to take a look at the specifics of derivation of the GBE, when (as is the case in plasma physics) the self-consistent field of forces must of necessity be introduced. As the Boltzmann equation is the centerpiece of the theory of transport processes (TTP), the introduction of an alternative KE_{f_1} leads in fact to an overhaul of the entire theory, including its macroscopic (for example, hydrodynamic) aspects. Conversely, a change in the macroscopic description will inevitably affect the kinetic level of description. Because of the complexity of the problem, this interrelation is not always easy to trace when solving a particular TTP problem. The important point to emphasize is that at issue here is not how to modify the classical equations of physical kinetics and hydrodynamics to include additional transport mechanisms (in reacting media, for example), rather we face a situation in which, those involved believe, we must go beyond the classical picture if we wish the revised theory to describe the experiment adequately. The alternative TTPs can be grouped conventionally into the following categories:

(1) theories that modify the macroscopic (hydrodynamic) description and neglect the possible changes of the kinetic description;
(2) those changing the kinetic description at the KE_{f_1} level without bothering much whether these changes are consistent with the structure of the entire BBGKY chain, and
(3) kinetic and hydrodynamic alternative theories consistent with the BBGKY hierarchy.

One of the pioneering efforts in the first line of research was a paper by Davydov (1935), which stimulated a variety of studies (see, for instance, Sobolev, 1993) on the hyperbolic equation of thermal conductivity. Introducing the second derivative of temperature with respect to time permitted a passage from the parabolic to the hyperbolic heat conduction equation, thus allowing for a finite heat propagation velocity. However, already in his 1935 paper B.I. Davydov points out that his method "cannot be extended to the three-dimensional case" and that "here the assumption that all the particles move at the same velocity would separate out a five-dimensional manifold from the six-dimensional phase space, suggesting that the problem cannot be limited to the coordinate space alone". We note, however, that quasi-linear parabolic equations can also produce wave solutions.

Therefore, to hyperbolize the heat conduction equation phenomenologically (Cattaneo, 1958) is not valid unless a rigorous kinetic justification is given. The hyperbolic heat conduction equation appears when the BE is solved by the Grad method (Zhdanov and Roldugin, 2002), retaining a term which involves a derivative of the heat flow with respect to time and to which, in the context of the Chapman–Enskog method, no particular order of approximation can be ascribed. Following its introduction, stable and high-precision computational schemes were developed for the hyperbolic equation of heat conduction (Jarzebski and Thulli, 1986), whose applications included, for example, two-temperature nonlocal heat conduction models and the study of the telegraph equation as a paradigm for possible generalized hydrodynamics (Sobolev, 1993; Rosenau, 1993).

Major difficulties arose when the issue of existence and uniqueness of solutions to the Navier–Stokes equations was addressed. O.A. Ladyzhenskaya has shown for three-dimensional flows that, under smooth initial conditions, a unique solution is only possible over a finite time interval. Ladyzhenskaya even introduced a "correction" into the Navier–Stokes equations in order that its unique solvability could be proved (see the discussion in Klimontovich, 1995). It turned out that, in this case, the viscosity coefficient should be dependent on transverse flow-velocity gradients – with the result that the very idea of introducing kinetic coefficients should be overhauled.

G. Uhlenbeck, in his review of the fundamental problems of statistical mechanics (Uhlenbeck, 1971), examines in particular the Kramers equation (Kramers, 1940) derived as a consequence of the Fokker–Planck equation

$$\frac{\partial f}{\partial t} + \mathbf{v} \cdot \frac{\partial f}{\partial \mathbf{r}} + \mathbf{a} \cdot \frac{\partial f}{\partial \mathbf{v}} = \beta \left[\frac{\partial}{\partial \mathbf{v}} \cdot (\mathbf{v} f) + \frac{k_B T}{m} \frac{\partial}{\partial \mathbf{v}} \cdot \frac{\partial f}{\partial \mathbf{v}} \right], \tag{I.22}$$

where $f(\mathbf{r}, \mathbf{v}, t)$ is the distribution function of Brownian particles, $\mathbf{a}$ is the acceleration due to an external field of forces, and $m\beta$ is the coefficient of friction for the motion of a colloid particle in the medium. What intrigues Uhlenbeck is how Kramers goes over from the Fokker–Planck equation (I.22) to the Einstein–Smoluchowski equation

$$\frac{\partial \rho}{\partial t} + \frac{\partial}{\partial \mathbf{r}} \cdot \left(\frac{\mathbf{a}}{\beta} \rho - \frac{k_B T}{m\beta} \frac{\partial \rho}{\partial \mathbf{r}} \right) = 0 \tag{I.23}$$

(ρ is the density) which has the character of the hydrodynamic continuity equation. In Uhlenbeck's words, "the proof of this change-over is very interesting, it is a typical

Kramers-style proof. It is in fact very simple, but at the same time some tricks and subtleties it involves make it very hard to discuss". The velocity distribution of colloid particles is assumed to be Maxwellian. The "trick", however, is that Kramers integrated along the line

$$\mathbf{r} + \frac{\mathbf{v}}{\beta} = \mathbf{r}_0, \tag{I.24}$$

and the number density of particles turned out to be given by the formula

$$n(\mathbf{r}_0, t) = \int f\left(\mathbf{r}_0 - \frac{\mathbf{v}}{\beta}, \mathbf{v}, t\right) \mathrm{d}\mathbf{v}. \tag{I.25}$$

So what exactly did H. Kramers do? Let us consider this change from the point of view of the generalized Boltzmann kinetic theory (GBKT) using, wherever possible, qualitative arguments to see things more clearly.

The structure of the KE_{f_1} is generally as follows

$$\frac{Df_1}{Dt} = J^B + J^{\mathrm{td}}, \tag{I.26}$$

where D/Dt is the substantial (particle) derivative, J^B is the (local) Boltzmann collision integral, and J^{td} is a nonlocal integral term incorporating the time delay effect. The generalized Boltzmann physical kinetics, in essence, involves a local approximation

$$J^{\mathrm{td}} = \frac{D}{Dt}\left(\tau \frac{Df_1}{Dt}\right) \tag{I.27}$$

for the second collision integral, here τ being the mean time *between* the particle collisions. We can draw here an analogy with the Bhatnager–Gross–Krook (BGK) approximation for J^B:

$$J^B = \frac{f_1^{(0)} - f_1}{\tau}, \tag{I.28}$$

whose popularity as a means to represent the Boltzmann collision integral is due to the huge simplifications it offers.

The ratio of the second to the first term on the right-hand side of Eq. (I.20) is given in the order of magnitude as

$$\frac{J^{\mathrm{td}}}{J^B} \approx \mathrm{O}\left(Kn^2\right) \tag{I.29}$$

and at large Knudsen numbers these terms become of the same order of magnitude. It would seem that, at small Knudsen numbers answering the hydrodynamic description, the contribution from the second term on the right-hand side of Eq. (I.26) is negligible.

This is not the case, however. When one goes over to the hydrodynamic approximation (by multiplying the kinetic equation by collision invariants and then integrating over velocities), the Boltzmann integral part vanishes, and the second term on the right-hand side of Eq. (I.26) gives a single-order contribution in the generalized Navier–Stokes description. Mathematically, we cannot neglect a term with a small parameter in front of the higher derivative. Physically, the appearing additional terms are due to viscosity and they correspond to the small-scale Kolmogorov turbulence (Alexeev, 1994, 2000b). The integral term J^{td}, thus, turns out to be important both at small and large Knudsen numbers in the theory of transport processes.

The important methodical issue to be considered is how classical conservation laws fit into the GBE picture. Continuum mechanics conservation laws are derived on the macroscopic level by considering a certain reference volume within the medium, which is enclosed by an infinitesimally thin surface. Moving material points (gas particles) can be either within or outside the volume, and it is by writing down the corresponding balance equations for mass, momentum flux, and energy that the classical equations of continuity, motion, and energy are obtained. In particular, we obtain the continuity equation in the form

$$\frac{\partial \rho^a}{\partial t} + \frac{\partial}{\partial \mathbf{r}} \cdot (\rho \mathbf{v}_0)^a = 0, \tag{I.30}$$

where ρ^a is the gas density, $\mathbf{v}_0^a$ is the hydrodynamic flow velocity, and $(\rho \mathbf{v}_0)^a$ is the momentum flux density obtained by neglecting fluctuations.

Thus, Boltzmann particles are fully "packed" in the reference volume. It would appear that in continuum mechanics the idea of discreteness can be abandoned altogether and the medium under study be considered as a continuum in the literal sense of the word. Such an approach is of course possible and indeed leads to Euler equations in hydrodynamics. But when the viscosity and thermal conductivity effects are to be included, a totally different situation arises. As it is well known, the dynamical viscosity is proportional to the mean time τ between the particle collisions, and a continuum medium in the Euler model with $\tau = 0$ implies that neither viscosity nor thermal conductivity are possible. The appearance of finite size particles in the reference contour leads to new effects.

Let a particle of finite radius be characterized, as before, by the position vector $\mathbf{r}$ and velocity $\mathbf{v}$ of its center of mass at some instant of time t. Then the fact that its center of mass is in the reference volume does not mean that all of the particles are there. In other words, at any given point in time there are always particles which are partly inside and partly outside of the reference surface, unavoidably leading to fluctuations in mass and hence in other hydrodynamic quantities.

There are two important points to be made here. First, the fluctuations will be proportional to the mean time between the collisions (rather than the collision time). This fact is rigorously established in Alexeev (1994, 1995c, 2000b), but it can also be made evident by means of quite simple arguments. Suppose we have a gas of hard spheres kept in a hard-wall cavity as shown in Figure I.1. Consider a reference contour drawn at a distance of the order of a particle diameter from the cavity wall. The mathematical expectation of the number of particles moving through the reference surface strictly

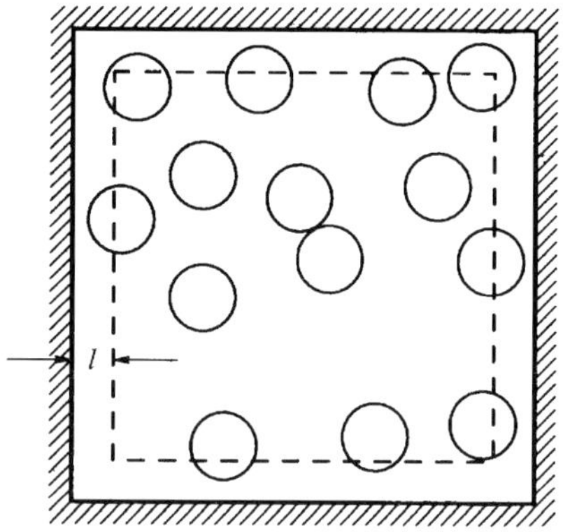

Fig. I.1. Closed cavity and the reference contour containing particles of a finite diameter.

perpendicular to the hard wall is zero. Therefore, in the first approximation, fluctuations will be proportional to the mean free path (or, equivalently, to the mean time between the collisions). As a result, the hydrodynamic equations will explicitly involve fluctuations proportional to τ. For example, the continuity equation changes its form and will contain terms proportional to viscosity (Alexeev, 1994). On the other hand – and this is the second point to be made – if the reference volume extends over the whole of the cavity, then the classical conservation laws should be obeyed, and this is exactly what the paper (Alexeev, 1994) proves. However, we will attempt here to "guess" the structure of the generalized continuity equation, using the arguments outlined above.

Neglecting fluctuations, the continuity equation should have the classical form (I.25) with

$$\rho^a = \rho - \tau A, \tag{I.31}$$

$$(\rho \mathbf{v}_0)^a = \rho \mathbf{v}_0 - \tau \mathbf{B}. \tag{I.32}$$

Strictly speaking, the factors A and $\mathbf{B}$ can be obtained from the generalized kinetic equation, in our case, from the GBE. Still, we can guess their form without appeal to the KE_{f_1}.

Indeed, let us write the generalized continuity equation

$$\frac{\partial}{\partial t}(\rho - \tau A) + \frac{\partial}{\partial \mathbf{r}} \cdot (\rho \mathbf{v}_0 - \tau \mathbf{B}) = 0 \tag{I.33}$$

in the dimensionless form using l, the distance from the reference contour to the hard wall (see Figure I.1) as a length scale. Then, instead of τ, the (already dimensionless) quantities A and $\mathbf{B}$ will have the Knudsen number $Kn_l = \lambda / l$ as a coefficient. In the limit

$$l \to 0, \quad Kn_l \to \infty$$

the contour embraces the entire cavity contained within hard walls, and there are no fluctuations on the walls. In other words, the classical equations of continuity and motion must be satisfied at the wall. Using hydrodynamic terminology, we note that the conditions

$$A = 0, \qquad \mathbf{B} = 0 \tag{I.34}$$

correspond to a laminar sublayer in a turbulent flow. Now, if a local Maxwellian distribution is assumed, then the generalized equation of continuity in the Euler approximation is written as

$$\frac{\partial}{\partial t}\left\{\rho - \tau\left[\frac{\partial \rho}{\partial t} + \frac{\partial}{\partial \mathbf{r}} \cdot (\rho \mathbf{v}_0)\right]\right\}$$
$$+ \frac{\partial}{\partial \mathbf{r}} \cdot \left\{\rho \mathbf{v}_0 - \tau\left[\frac{\partial}{\partial t}(\rho \mathbf{v}_0) + \frac{\partial}{\partial \mathbf{r}} \cdot \rho \mathbf{v}_0 \mathbf{v}_0 + \overleftrightarrow{I} \cdot \frac{\partial p}{\partial \mathbf{r}} - \rho \mathbf{a}\right]\right\} = 0, \tag{I.35}$$

where $\overleftrightarrow{I}$ is the unit tensor.

In the hydrodynamic approximation, the mean time τ between the collisions is related to the dynamic viscosity μ by

$$\tau p = \Pi \mu, \tag{I.36}$$

where the factor Π depends on the choice of a collision model and is $\Pi = 0.8$ for the particular case of neutral gas comprising hard spheres (Chapman and Cowling, 1952). The generalized equations of energy and motion are much more difficult to guess in this way, making the GBE indispensable. It is worthwhile though to say a few words about the treatment of the GBE (I.20) in the spirit of the fluctuation theory:

$$\frac{Df^a}{Dt} = J^B(f), \tag{I.37}$$

where J^B is the Boltzmann collision integral, and

$$f^a = f - \tau \frac{Df}{Dt}. \tag{I.38}$$

Thus, $\tau Df/Dt$ is the distribution function fluctuation, and writing Eq. (I.37) without taking into account Eq. (I.38) makes the BE nonclosed. From the viewpoint of the fluctuation theory, Boltzmann employed the simplest possible closure procedure:

$$f^a = f. \tag{I.39}$$

Now, having in mind the Kramers method, let us compare the generalized continuity equation (I.35) and the Einstein–Smoluchowski equation (I.23). Eq. (I.35) reduces to Eq. (I.23) if

(a) the convective transfer corresponding to the hydrodynamical velocity $\mathbf{v}_0$ is neglected;

(b) the temperature gradient is less important than the gradient of the number density of particles, $n\partial T/\partial \mathbf{r} \ll T\partial n/\partial \mathbf{r}$, and
(c) the temporal part of the density fluctuations is left out of account.

By integrating with respect to velocity v from $-\infty$ to $+\infty$ along the line

$$\mathbf{r} + \frac{\mathbf{v}}{\beta} = \mathbf{r}_0. \tag{I.40}$$

Kramers (see also Eq. (I.25)) introduced nonlocal collisions without accounting for the time delay effect. In our theory, the coefficient of friction $\beta = \tau^{-1}$, which corresponds to the binary collision approximation. If the simultaneous interaction with many particles is important and must be accounted for, additional difficulties associated with the definition of the coefficient of friction β arise, and the Einstein–Smoluchowski theory becomes semi-phenomenological. Overcoming these difficulties may require the use of the theory of non-Markov processes for describing Brownian motion (Morozov, 1996).

Note that the application of the above principles also leads to the modification of the system of Maxwell equations. While the traditional formulation of this system does not involve the continuity equation, its derivation explicitly employs the equation

$$\frac{\partial \rho^a}{\partial t} + \frac{\partial}{\partial \mathbf{r}} \cdot \mathbf{j}^a = 0, \tag{I.41}$$

where ρ^a is the charge per unit volume, and $\mathbf{j}^a$ is the current density, both calculated without accounting for the fluctuations. As a result, the system of Maxwell equations written in the standard notation, namely,

$$\begin{aligned} &\frac{\partial}{\partial \mathbf{r}} \cdot \mathbf{B} = 0, &&\frac{\partial}{\partial \mathbf{r}} \cdot \mathbf{D} = \rho^a, \\ &\frac{\partial}{\partial \mathbf{r}} \times \mathbf{E} = -\frac{\partial \mathbf{B}}{\partial t}, &&\frac{\partial}{\partial \mathbf{r}} \times \mathbf{H} = \mathbf{j}^a + \frac{\partial \mathbf{D}}{\partial t} \end{aligned} \tag{I.42}$$

contains

$$\rho^a = \rho - \rho^{\mathrm{fl}}, \qquad \mathbf{j}^a = \mathbf{j} - \mathbf{j}^{\mathrm{fl}}. \tag{I.43}$$

The ρ^{fl}, $\mathbf{j}^{\mathrm{fl}}$ fluctuations calculated using the GBE are given, for example, in Alexeev (1994, 2000b).

We shall now turn to approaches in which the KE_{f_1} can be changed in a way that is generally inconsistent with the BBGKY hierarchy. It has been repeatedly pointed out that using a wrong distribution function for charged particles may have a catastrophic effect on the macro-parameters of a weakly ionized gas.

Let us have a look at some examples of this. As it is well known, the temperature dependence of the density of atoms ionized in plasma to various degrees was first studied by Saha (1920) and Eggert (1919). For a system in thermodynamic equilibrium they obtained the equation

$$\frac{n_{j+1} n_e}{n_j} = \frac{s_{j+1}}{s_j} \frac{(2\pi m_e k_B T)^{3/2}}{h^3} \mathrm{e}^{-\varepsilon_j/(k_B T)}, \tag{I.44}$$

where n_j is the density number of j-fold ionized atoms, n_e is the number density of free electrons, m_e is the electron mass, k_B is the Boltzmann constant, h is the Planck constant, s_j is the statistical weight for a j-fold ionized atom (Fowler and Guggenheim, 1939), and ε_j is the jth ionization potential. The Saha equation (I.44) is derived for the Maxwellian distribution and should necessarily be modified if another velocity distribution of particles exists in the plasma. This problem was studied in Dewan (1961), in which, for illustrative purposes, the values of $n_{j+1}n_e/n_j$ calculated by the Maxwell distribution function are compared with those obtained by the Druyvesteyn distribution function, the average energies for both distributions being assumed equal. Let $T = 10^4$ K, $n_e = 10^{14}$ cm^{-3}, $\varepsilon_j = 10$ eV, the charge number $Z = 1$, $s_{j+1}/s_j = 1$. Then one arrives at Dewan (1961)

$$\frac{n_{j+1}n_e}{n_j} = 6 \times 10^2 \qquad \text{(calculation using the Druyvesteyn distribution)},$$

$$\frac{n_{j+1}n_e}{n_j} = 4.53 \times 10^{16} \qquad \text{(calculation using the Maxwellian distribution function; by the Saha formula)}.$$

As E. Dewan explained, “the discrepancy in fourteen orders of magnitude obtained above is clearly due to the fact that, unlike Maxwellian distribution, the Druyvesteyn distribution does not have a ‘tail’”.

In our second example, two quantities – the ionization rate constant and the ionization cross section – were calculated by Gryzinski, Kunc and Zgorzelski (1973), using the two above-mentioned distributions. The ionization cross section σ_i is defined by the following interpolation formula known to match satisfactorily the experimental data

$$\sigma_i = \frac{\sigma_0}{\varepsilon_i^2} G_i(\xi, \varsigma), \tag{I.45}$$

where $\sigma_0 = 6.56 \times 10^{-14}$ cm^2 (eV)^2, ε_i is the ionization potential of the atom, and ξ is a dimensionless parameter characterizing the atomic electron shell:

$$\xi = \frac{W}{\varepsilon_i}, \tag{I.46}$$

where W is the average kinetic energy of the atomic electrons, given by the formula

$$W = \frac{1}{N_e} \sum_{j=1}^{N_e} \varepsilon_j, \tag{I.47}$$

in which N_e is the number of electrons in the atom, and ε_j are the ionization potentials for the atom successively stripped of its electrons. The parameter ς is defined by the expression

$$\varsigma = \frac{U_e}{\varepsilon_i}, \tag{I.48}$$

where U_e is the energy of the electrons bombarding the atom. The neutral particle velocities are assumed to be much lower than the average electron velocity, and the plasma is taken to be uniform. The average value of the ionization cross section is then given by

$$\bar{\sigma}_i = \int_0^\infty \sigma_i(v_e) f(v_e)\,\mathrm{d}v_e, \tag{I.49}$$

and the ionization rates are evaluated by the formula

$$\overline{\sigma_i v_e} = \int_0^\infty \sigma_i(v_e) v_e f(v_e)\,\mathrm{d}v_e, \tag{I.50}$$

provided the function $G_i(\xi, \varsigma)$, defined as in Gryzinski, Kunc and Zgorzelski (1973),

$$G_i(\xi, \varsigma) = \frac{(\varsigma - 1)(1 + \frac{2}{3}\xi)}{(\varsigma + 1)(1 + \xi + \varsigma)} \tag{I.51}$$

is known.

Table I.1 illustrates in the dimensionless form the calculated values of $\bar{\sigma}_i$ and $\overline{\sigma_i v_e}$ for $\xi = 1$ and various $\widehat{T} = k_B T_e/\varepsilon_i$. It can be seen that the results obtained with different DFs can differ considerably, indeed, catastrophically so even for relatively small values of $\widehat{T}$. Thus, the reliable computation of DFs remains a topic of intense current interest in plasma physics problems, the weak effect of the DF form on its moments being rather an exception than the rule.

The use of collision cross sections which are "self-consistent" with kinetic equations is also suggested by the well-known Enskog theory of moderately dense gases (Enskog, 1921b). Enskog's idea was to describe the properties of such gases by separating the nonlocal part out of the essentially local Boltzmann collision integral. The transport coefficients, obtained in this way for the hard-sphere model, yielded an incorrect temperature dependence on the system's kinetic coefficients. To remedy this situation, the model of "soft" spheres was introduced to fit the experimental data (see, for instance, Koremans and Beenaker, 1960).

In the theory of the so-called kinetically consistent difference schemes (Chetverushkin, 1999), the DF is expanded in a power series of time, which corresponds to using an incomplete second approximation in the "physical" derivation of the Boltzmann equation (see the discussion in Klimontovich, 1995, 1997). The result is that the difference schemes obtained contain only an artificial ad hoc viscosity chosen specially for the problem at hand. Some workers followed the steps of Davydov by adding the term $\partial^2 f/\partial t^2$ to the kinetic equations for fast processes.

Bakai and Sigov (1996) suggest using such a term in the equation for describing DF fluctuations in a turbulent plasma. The so-called "ordering parameter" they introduce alters the very type of equation. To describe spatial nonlocality, Bakai and Sigov complement the kinetic equation with the $\partial^2 f/\partial x^2$ term and higher derivatives, including mixed time-coordinate partial derivatives – a modification which can possibly describe

Table I.1
Comparison of ionization cross sections $\bar{\sigma}_i$ and ionization rates $\overline{\sigma_i v_e}$ calculated with the Maxwellian and Druyvesteyn DF

$\widehat{T}$	Maxwellian DF		Druyvesteyn DF	
	$\bar{\sigma}_i$	$\overline{\sigma_i v_e}$	$\bar{\sigma}_i$	$\overline{\sigma_i v_e}$
0.1	4.206×10^{-6}	1.184×10^{-5}	1.278×10^{-27}	4.077×10^{-27}
0.2	8.262×10^{-4}	1.184×10^{-3}	4.382×10^{-9}	1.011×10^{-8}
0.3	5.029×10^{-3}	9.251×10^{-3}	2.128×10^{-5}	4.135×10^{-5}
0.4	1.259×10^{-2}	2.103×10^{-2}	5.403×10^{-4}	9.405×10^{-4}
0.5	2.194×10^{-2}	3.415×10^{-2}	2.773×10^{-3}	4.466×10^{-3}
0.6	3.180×10^{-2}	4.687×10^{-2}	7.305×10^{-3}	1.110×10^{-2}
0.7	4.143×10^{-2}	5.842×10^{-2}	1.376×10^{-2}	1.998×10^{-2}
0.8	5.047×10^{-2}	6.857×10^{-2}	2.145×10^{-2}	3.001×10^{-2}
0.9	5.875×10^{-2}	7.733×10^{-2}	2.973×10^{-2}	4.033×10^{-2}
1	6.624×10^{-2}	8.482×10^{-2}	3.813×10^{-2}	5.039×10^{-2}
2	1.079×10^{-1}	1.171×10^{-1}	9.918×10^{-2}	1.132×10^{-1}
3	1.195×10^{-1}	1.190×10^{-1}	1.233×10^{-1}	1.312×10^{-1}
4	1.209×10^{-1}	1.137×10^{-1}	1.311×10^{-1}	1.717×10^{-1}
5	1.185×10^{-1}	1.069×10^{-1}	1.320×10^{-1}	1.298×10^{-1}
6	1.146×10^{-1}	9.992×10^{-2}	1.299×10^{-1}	1.243×10^{-1}
7	1.102×10^{-1}	9.326×10^{-2}	1.263×10^{-1}	1.184×10^{-1}
8	1.056×10^{-1}	8.704×10^{-2}	1.222×10^{-1}	1.125×10^{-1}
9	1.010×10^{-1}	8.123×10^{-2}	1.179×10^{-1}	1.069×10^{-1}
10	9.662×10^{-2}	7.589×10^{-2}	1.137×10^{-1}	1.017×10^{-1}

non-Gaussian random sources in the Langevin equations (Reshetnyak and Shelepin, 1996). It is interesting to note that the GBE also makes it possible to include higher derivatives of the DF (see the approximation (5.8) in Alexeev, 2000b).

Clearly, approaches to the modification of the KE$_{f_1}$ must be based on certain principles, and it is appropriate to outline these in brief here. Of the approaches we have mentioned above, the most consistent one is the third, which clearly reveals the relation between alternative KE$_{f_1}$'s and the BBGKY hierarchy. There are general requirements which the generalized KE$_{f_1}$ must satisfy.

(1) Since the artificial breaking of the BBGKY hierarchy is unavoidable in changing to a one-particle description, the generalized KE$_{f_1}$ should be obtainable by the known methods of the theory of kinetic equations, such as the multiscale approach, the correlation function method, iterative methods, and so forth, or their combinations. In each of these, some specific features of the particular alternative KE$_{f_1}$ are highlighted.

(2) There must be an explicit link between the KE$_{f_1}$ and the way we introduce the physically infinitesimal volume – and hence with the way the moments in the reference contour with transparent boundaries fluctuate due to the finite size of the particles.

(3) In the nonrelativistic case, the KE$_{f_1}$ must satisfy the Galileo transformation.

(4) The KE_{f_1} must ensure a connection with the classical H-theorem and its generalizations.
(5) The KE_{f_1} should not lead to unreasonable complexities in the theory.

The last requirement needs some commentary. The integral collision terms – in particular, the Boltzmann local integral and especially the nonlocal integral with time delay – have a very complex structure.

The "caricature" the BGK approximation makes of the Boltzmann collision integral (to use Yu.L. Klimontovich's expressive word) has turned out to be a very successful approach, and this algebraically approximated Boltzmann collision integral is widely used in the kinetic theory of neutral and ionized gases.

The generalized Boltzmann equation introduces a local differential approximation for the nonlocal collision integral with time delay. Here, we are faced in fact with the "price–quality" problem familiar from economics. That is, what price – in terms of the increased complexity of the kinetic equation – are we ready to pay for the improved quality of the theory? An answer to this question is possible only through experience with practical problems.

A consistent theory meeting the above requirements is being developed, in particular, by Klimontovich (1995, 1997) and, the present author believes, within the GBE framework. One can recognize points of common ideology in the two approaches. However, whereas in Klimontovich's work, the treatment of the physically infinitesimal volume is transferred to the "upper echelon" of the BBGKY hierarchy and leads to a change in the Liouville equation, in the GBE theory, it turns out that approximated nonlocal terms can even be introduced at the level of a one-particle description. The essential point to be made here is that the GBE theory does well without specifying the smoothing procedure, whereas in Klimontovich's theory altering this procedure unavoidably modifies the alternative KE_{f_1}.

Vlasov (1978) attempted to eliminate the inconsistencies of the Boltzmann theory through the inclusion of additional dynamical variables (derivatives of the velocity) in the one-particle distribution function $f(\mathbf{r}, \mathbf{v}, \dot{\mathbf{v}}, \ddot{\mathbf{v}}, \dddot{\mathbf{v}}, \ldots, t)$. However, due primarily to the reasonable complexity requirement which should be met for a theory to be useful in practice, this approach is, in our view, too early to try until all the traditional resources for describing the DF are exhausted.

From this perspective, fluctuation terms in the GBE are due to the fact that the reference volume as a measuring element for a system of finite-sized particles is introduced without changing the DF form, used for describing point structureless particles.

The reader is referred to review (Alekseev, 1997) of some other theories of transport properties. To conclude, it remains only to note that the effects listed above will always be relevant to a kinetic theory, using a one-particle description including, in particular, applications to liquids or plasmas, where self-consistent forces with appropriately cut-off radius of their action are introduced to expand the capabilities of the GBE.

Other suggestions on possible generalizations of the Boltzmann equation are best described by leaning upon the basic principles of the generalized Boltzmann physical kinetics, developed by the present author. We now proceed to discuss these principles.

CHAPTER 1

Generalized Boltzmann Equation

1.1. Mathematical introduction. Method of many scales

In the sequel asymptotic methods will be used, but first of all, we will look at the method of many scales. The method of many scales is so popular that Nayfeh in his book (Nayfeh, 1972) written more than thirty years ago said that the method of many scales (MMS) is discovered by different authors every half a year. As a result, there exist many different variants of MMS. As a minimum four variants of MMS are considered in Nayfeh (1972). We are interested only in the main ideas of MMS, which are used further in the theory of kinetic equations.

From this standpoint we demonstrate MMS capabilities, using a typical example of solving a linear differential equation which also has the exact solution for comparing the results (Nayfeh, 1972). But in contrast to the usual consideration, which can be found in literature, we intend to bring this example up to a table and a graph.

Therefore, let us consider the linear differential equation

$$\delta \ddot{x} + \varepsilon \dot{x} + x = 0. \tag{1.1.1}$$

We begin with the special case when $\delta = 1$ and ε is a small parameter. Eq. (1.1.1) has the exact solution

$$x = a\,\mathrm{e}^{-\varepsilon t/2} \cos\left[t\sqrt{1 - \frac{1}{4}\varepsilon^2} + \varphi\right], \tag{1.1.2}$$

where a and φ are arbitrary constants of integrating. In the typical case of a small parameter in front of a senior derivative – in this case, it would be $\ddot{x}$ – the effects of a boundary layer can be observed. Using the derivatives

$$\dot{x} = -\frac{1}{2}\varepsilon x - a\,\mathrm{e}^{-\varepsilon t/2}\sqrt{1 - \frac{1}{4}\varepsilon^2}\,\sin\left[t\sqrt{1 - \frac{1}{4}\varepsilon^2} + \varphi\right],$$

$$\ddot{x} = -\frac{1}{2}\varepsilon \dot{x} - x\left(1 - \frac{1}{4}\varepsilon^2\right) + \frac{1}{2}a\varepsilon\,\mathrm{e}^{-\varepsilon t/2}\sqrt{1 - \frac{1}{4}\varepsilon^2}\,\sin\left[t\sqrt{1 - \frac{1}{4}\varepsilon^2} + \varphi\right],$$

for substitution in (1.1.1), we find the identical satisfaction of Eq. (1.1.1).

We begin with a direct expansion in small ε, using the series

$$x = x_0 + \varepsilon x_1 + \varepsilon^2 x_2 + \cdots, \tag{1.1.3}$$

and after differentiating

$$\dot{x} = \dot{x}_0 + \varepsilon \dot{x}_1 + \varepsilon^2 \dot{x}_2 + \cdots,$$
$$\ddot{x} = \ddot{x}_0 + \varepsilon \ddot{x}_1 + \varepsilon^2 \ddot{x}_2 + \cdots.$$

Substitute series (1.1.3) into (1.1.1) and equalize the coefficients in front of equal powers of ε, having

$$\ddot{x}_0 + x_0 = 0, \tag{1.1.4}$$
$$\ddot{x}_1 + x_1 = -\dot{x}_0, \tag{1.1.5}$$
$$\ddot{x}_2 + x_2 = -\dot{x}_1, \tag{1.1.6}$$
$$\ddot{x}_3 + x_3 = -\dot{x}_2 \tag{1.1.7}$$

and so on. The general solution of homogeneous equation (1.1.4) has the form

$$x_0 = a\cos(t + \varphi). \tag{1.1.8}$$

Substitute (1.1.8) in (1.1.5):

$$\ddot{x}_1 + x_1 = a\sin(t + \varphi). \tag{1.1.9}$$

General solution (1.1.1) should contain only two arbitrary constants. In this case, both constants a and φ are contained in the main term of expansion defined by relation (1.1.8). Then we need to find only the particular solution of Eq. (1.1.9); which can be found as follows

$$x_1 = -\frac{1}{2}at\cos(t + \varphi). \tag{1.1.10}$$

Really,

$$\dot{x}_1 = -\frac{1}{2}a\cos(t + \varphi) + \frac{1}{2}at\sin(t + \varphi),$$
$$\ddot{x}_1 = \frac{1}{2}a\sin(t + \varphi) - x_1 + \frac{1}{2}a\sin(t + \varphi).$$

After substituting in (1.1.9), we find the identity

$$a\sin(t + \varphi) - x_1 + x_1 = a\sin(t + \varphi).$$

Eq. (1.1.6) can be rewritten as

$$\ddot{x}_2 + x_2 = \frac{1}{2} a \cos(t+\varphi) - \frac{1}{2} a t \sin(t+\varphi), \tag{1.1.11}$$

and its solution is

$$x_2 = \frac{1}{8} a t^2 \cos(t+\varphi) + \frac{1}{8} a t \sin(t+\varphi). \tag{1.1.12}$$

Really,

$$\dot{x}_2 = \frac{1}{4} a t \cos(t+\varphi) - \frac{1}{8} a t^2 \sin(t+\varphi) + \frac{1}{8} a \sin(t+\varphi) + \frac{1}{8} a t \cos(t+\varphi),$$
$$\ddot{x}_2 = \frac{1}{2} a \cos(t+\varphi) - \frac{5}{8} a t \sin(t+\varphi) - \frac{1}{8} a t^2 \cos(t+\varphi).$$

The substitution into the left-hand side of Eq. (1.1.11) leads to the result

$$\begin{aligned} &\frac{1}{2} a \cos(t+\varphi) - \frac{5}{8} a t \sin(t+\varphi) - \frac{1}{8} a t^2 \cos(t+\varphi) \\ &\quad + \frac{1}{8} a t^2 \cos(t+\varphi) + \frac{1}{8} a t \sin(t+\varphi) \\ &= \frac{1}{2} a \cos(t+\varphi) - \frac{1}{2} a t \sin(t+\varphi). \end{aligned}$$

Then we state the identical satisfaction of Eq. (1.1.11) by solution (1.1.12). In an analogous way the solution of Eq. (1.1.7) is written as a cubic polynomial in t. For the first three terms of (1.1.3) series the solution is

$$\begin{aligned} x &= a \cos(t+\varphi) - \frac{1}{2} \varepsilon a t \cos(t+\varphi) \\ &\quad + \frac{1}{8} \varepsilon^2 a \left[t^2 \cos(t+\varphi) + t \sin(t+\varphi) \right] + \mathrm{O}\left(\varepsilon^3\right). \end{aligned} \tag{1.1.13}$$

At our desire the variable t can be considered as a dimensionless time. Suppose, of course, that we wish to have a solution for arbitrary time moments. But it is not possible in the developed procedure, because the series construction regards the successive terms of the series to be smaller than the foregoing terms; in other case, it is impossible to speak about the series convergence. But for fixed ε, the time moment can be found when a successive term of expansion is no smaller than the foregoing term. Figure 1.1 contains the comparison of the exact solution (1.1.2) with concrete parameters of calculations

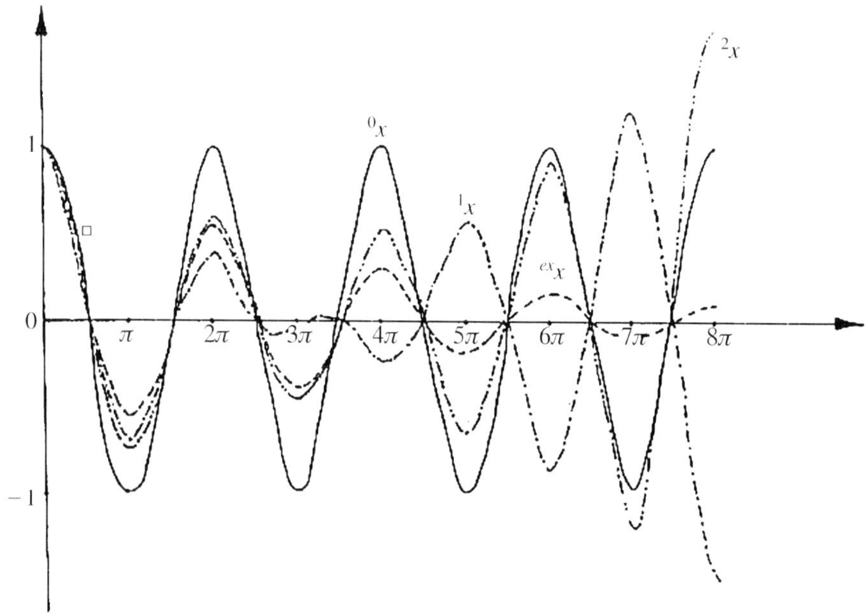

Fig. 1.1. Comparison of the solutions ^{0}x, ^{1}x, ^{2}x, obtained by the perturbation method, with the many scales solution $^{ms}x_2$ and the exact solution ^{ex}x for the case $a = 1$, $\varphi = 0$, $\varepsilon = 0.1$.

$\delta = 1$, $a = 1$, $\varphi = 0$, $\varepsilon = 0.2$ with the approximate solutions

$$^{0}x = \cos t,$$

$$^{1}x = \cos t - 0.1t\cos t,$$

$$^{2}x = \cos t - 0.1t\cos t + 0.005\left[t^2\cos t + t\sin t\right],$$

$$^{ex}x = e^{-0.1t}\cos\left(t\sqrt{0.99}\right).$$

As we could expect, the divergence of solutions ^{2}x and exact solution ^{ex}x appear when t is of order 10; or, in the common case, if $t \sim \varepsilon^{-1}$. But as it follows from Figure 1.1, the situation is much worse, because, for example, for $t = 4\pi$ the approximate solution ^{1}x gives a wrong sign in comparison with the exact solution ^{ex}x. For the mathematical model of an oscillator with damping, that is reflected by Eq. (1.1.1), it means that the approximate solution ^{1}x forecasts a deviation in the opposite direction for the mentioned oscillator. By the way, the solution ^{1}x is also worse in comparison with the solution ^{0}x, in this sense, the minor approximation is better than the senior ones.

This poses the question how to improve the situation while remaining in the frame of asymptotic methods. To answer this question, let us consider the exact solution (1.1.2). Exponential and cosine terms contained in this solution, can be expanded in the following series for small ε and fixed t:

$$\mathrm{e}^{-\varepsilon t/2} = 1 - \frac{1}{2}\varepsilon t + \frac{1}{8}(\varepsilon t)^2 - \frac{1}{64}(\varepsilon t)^3 + \cdots, \tag{1.1.14}$$

$$\begin{aligned}
\cos\left(t\sqrt{1-\frac{1}{4}\varepsilon^2} + \varphi\right) &= \cos\left[t\left(1 - \frac{1}{8}\varepsilon^2 - \frac{1}{128}\varepsilon^4 - \frac{1}{1024}\varepsilon^6 - \cdots\right) + \varphi\right] \\
&= \cos\left[t + \varphi - \frac{1}{8}t\varepsilon^2 - \frac{1}{128}t\varepsilon^4 - \frac{1}{1024}t\varepsilon^6 - \cdots\right] \\
&\cong \cos(t+\varphi) \\
&\quad + \left(\frac{1}{8}t\varepsilon^2 + \frac{1}{128}t\varepsilon^4 + \frac{1}{1024}t\varepsilon^6 + \cdots\right)\sin(t+\varphi) \\
&= \cos(t+\varphi) + \frac{1}{8}t\varepsilon^2\sin(t+\varphi) \\
&\quad + \frac{1}{128}t\varepsilon^4\sin(t+\varphi) + \cdots.
\end{aligned} \tag{1.1.15}$$

Obviously, the product of the first terms in expansions (1.1.14), (1.1.15) gives ${}^0x$, and retaining of terms of $\mathrm{O}(\varepsilon^3)$ leads to the result

$$\begin{aligned}
x &\cong a\left(1 - \frac{1}{2}\varepsilon t + \frac{1}{8}\varepsilon^2 t^2\right)\left(\cos(t+\varphi) + \frac{1}{8}t\varepsilon^2\sin(t+\varphi)\right) \\
&\cong a\cos(t+\varphi) - \frac{1}{2}a\varepsilon t\cos(t+\varphi) + \frac{1}{8}at\varepsilon^2\sin(t+\varphi) \\
&\quad + \frac{1}{8}a\varepsilon^2 t^2\cos(t+\varphi).
\end{aligned}$$

Then we state that the used construction of asymptotic solution is based indeed on the assumption that the combination εt is small. If it is not so (for t having the order ε^{-1}), then expansions (1.1.14), (1.1.15) are wrong or need to take into account all the terms of expansions. But the asymptotic expansion can be organized in another way, using additional variables:

$$T_1 = \varepsilon t \tag{1.1.16}$$

and

$$T_2 = \varepsilon^2 t. \tag{1.1.17}$$

In this case,

$$\mathrm{e}^{-\varepsilon t/2} = \mathrm{e}^{-T_1/2}, \tag{1.1.18}$$

and expansion (1.1.15) is replaced by other ones:

$$\begin{aligned}\cos\left(t\sqrt{1-\frac{1}{4}\varepsilon^2}+\varphi\right) &= \cos\left[t+\varphi-\frac{1}{8}t\varepsilon^2-\frac{1}{128}t\varepsilon^4-\frac{1}{1024}t\varepsilon^6-\cdots\right]\\ &= \cos\left[t+\varphi-\frac{1}{8}T_2\right]\\ &\quad+\left(\frac{1}{128}t\varepsilon^4+\frac{1}{1024}t\varepsilon^6+\cdots\right)\sin\left(t+\varphi-\frac{1}{8}T_2\right)\\ &= \cos\left(t+\varphi-\frac{1}{8}T_2\right)+\frac{1}{128}t\varepsilon^4\sin\left(t+\varphi-\frac{1}{8}T_2\right)\\ &\quad+\frac{1}{1024}t\varepsilon^6\sin\left(t+\varphi-\frac{1}{8}T_2\right)+\cdots. \qquad (1.1.19)\end{aligned}$$

Expansion (1.1.19) "is working" with the accuracy $\mathrm{O}(t\varepsilon^4)$ and therefore the movement in the direction of achieving a higher and higher accuracy leads to the appearance of new variables of the type

$$T_m=\varepsilon^m t \quad (m=0,1,2,\ldots) \qquad (1.1.20)$$

or, which is the same, to new time scales $T_0, T_1, T_2, \ldots$. As a result, we obtain the asymptotic solution of a new type

$$\begin{aligned}x(t,\varepsilon) &= x(T_0,T_1,T_2,\ldots,T_M;\varepsilon)\\ &= \sum_{m=0}^{M}\varepsilon^m x_m(T_0,T_1,\ldots,T_M)+\mathrm{O}(\varepsilon T_M). \qquad (1.1.21)\end{aligned}$$

The application of expansion (1.1.21) inevitably leads to the system of equations in partial derivatives, and the time derivative should be calculated, using the rule for differentiating a composite function:

$$\frac{\mathrm{d}}{\mathrm{d}t}=\frac{\partial}{\partial T_0}+\varepsilon\frac{\partial}{\partial T_1}+\varepsilon^2\frac{\partial}{\partial T_2}+\cdots. \qquad (1.1.22)$$

Then, with the accuracy $\mathrm{O}(\varepsilon^2)$, we have:

$$\begin{aligned}\frac{\mathrm{d}x}{\mathrm{d}t} &\cong \frac{\partial}{\partial T_0}(x_0+\varepsilon x_1+\varepsilon^2x_2)+\varepsilon\frac{\partial}{\partial T_1}(x_0+\varepsilon x_1)+\varepsilon^2\frac{\partial x_0}{\partial T_2}\\ &= \frac{\partial x_0}{\partial T_0}+\varepsilon\left(\frac{\partial x_1}{\partial T_0}+\frac{\partial x_0}{\partial T_1}\right)+\varepsilon^2\left(\frac{\partial x_2}{\partial T_0}+\frac{\partial x_1}{\partial T_1}\right). \qquad (1.1.23)\end{aligned}$$

Using series (1.1.21) and the procedure of decomposition of the derivative (1.1.22) (see also Nayfeh, 1972), we obtain the system of equations in partial derivatives defining x_0,

x_1, and x_2 in series (1.1.21). Obviously,

$$\frac{d^2x}{dt^2} \cong \left(\frac{\partial}{\partial T_0} + \varepsilon\frac{\partial}{\partial T_1} + \varepsilon^2\frac{\partial}{\partial T_2}\right)\left(\frac{\partial x}{\partial T_0} + \varepsilon\frac{\partial x}{\partial T_1} + \varepsilon^2\frac{\partial x}{\partial T_2}\right)$$
$$\cong \frac{\partial^2 x}{\partial T_0^2} + \varepsilon\frac{\partial^2 x}{\partial T_0 \partial T_1} + \varepsilon^2\frac{\partial^2 x}{\partial T_0 \partial T_2} + \varepsilon\frac{\partial^2 x}{\partial T_0 \partial T_1} + \varepsilon^2\frac{\partial^2 x}{\partial T_1^2} + \varepsilon^2\frac{\partial^2 x}{\partial T_0 \partial T_2}$$
$$= \frac{\partial^2 x}{\partial T_0^2} + 2\varepsilon\frac{\partial^2 x}{\partial T_0 \partial T_1} + \varepsilon^2\left(2\frac{\partial^2 x}{\partial T_0 \partial T_2} + \frac{\partial^2 x}{\partial T_1^2}\right)$$
$$\cong \frac{\partial^2 x_0}{\partial T_0^2} + \varepsilon\frac{\partial^2 x_1}{\partial T_0^2} + \varepsilon^2\frac{\partial^2 x_2}{\partial T_0^2} + 2\varepsilon\frac{\partial^2 x_0}{\partial T_0 \partial T_1} + 2\varepsilon^2\frac{\partial^2 x_1}{\partial T_0 \partial T_1}$$
$$+ \varepsilon^2\left(2\frac{\partial^2 x_0}{\partial T_0 \partial T_2} + \frac{\partial^2 x_0}{\partial T_1^2}\right)$$
$$= \frac{\partial^2 x_0}{\partial T_0^2} + \varepsilon\left(\frac{\partial^2 x_1}{\partial T_0^2} + 2\frac{\partial^2 x_0}{\partial T_0 \partial T_1}\right)$$
$$+ \varepsilon^2\left(\frac{\partial^2 x_2}{\partial T_0^2} + 2\frac{\partial^2 x_1}{\partial T_0 \partial T_1} + 2\frac{\partial^2 x_0}{\partial T_0 \partial T_2} + \frac{\partial^2 x_0}{\partial T_1^2}\right). \quad (1.1.24)$$

Equating now the coefficients at the same powers of ε in Eqs. (1.1.1) and (1.1.21), (1.1.24), we obtain a set of equations, written below, for terms containing ε^0, ε^1, ε^2.

$$\frac{\partial^2 x_0}{\partial T_0^2} + x_0 = 0, \quad (1.1.25)$$

$$\frac{\partial^2 x_1}{\partial T_0^2} + 2\frac{\partial^2 x_0}{\partial T_0 \partial T_1} + x_1 = -\frac{\partial x_0}{\partial T_0}, \quad (1.1.26)$$

$$\frac{\partial^2 x_2}{\partial T_0^2} + 2\frac{\partial^2 x_1}{\partial T_0 \partial T_1} + 2\frac{\partial^2 x_0}{\partial T_0 \partial T_2} + \frac{\partial^2 x_0}{\partial T_1^2} + x_2 = -\left(\frac{\partial x_1}{\partial T_0} + \frac{\partial x_0}{\partial T_1}\right). \quad (1.1.27)$$

The set of Eqs. (1.1.25)–(1.1.27) is not linked, it means, that equations served for obtaining the minor coefficients of expansion (1.1.21) do not contain the senior coefficients of the mentioned expansion. To underline this fact, the next system of equations contains on the right-hand sides only the terms known from the previous approximations:

$$\frac{\partial^2 x_0}{\partial T_0^2} + x_0 = 0,$$

$$\frac{\partial^2 x_1}{\partial T_0^2} + x_1 = -\frac{\partial x_0}{\partial T_0} - 2\frac{\partial^2 x_0}{\partial T_0 \partial T_1}, \quad (1.1.26')$$

$$\frac{\partial^2 x_2}{\partial T_0^2} + x_2 = -\frac{\partial x_0}{\partial T_1} - 2\frac{\partial^2 x_0}{\partial T_0 \partial T_2} - \frac{\partial^2 x_0}{\partial T_1^2} - \frac{\partial x_1}{\partial T_0} - 2\frac{\partial^2 x_1}{\partial T_0 \partial T_1}. \tag{1.1.27$'$}$$

The following step consists in a successive solution of Eqs. (1.1.25), (1.1.26′), (1.1.27′) and the corresponding next equations of this type. It is just this type of procedure, but only for kinetic equations, that will be realized in Section 1.3.

For linear equation (1.1.1) chosen as an example, these is no reason to use this rather complicated procedure and it is better to avoid expansion (1.1.21). For this aim it suffices to suppose that

$$x = x(T_0, T_1, T_2) \tag{1.1.28}$$

and calculate the first and second derivatives in (1.1.1) as in (1.1.23), (1.1.24), i.e.,

$$\frac{\mathrm{d}x}{\mathrm{d}t} = \frac{\partial x}{\partial T_0} + \varepsilon\frac{\partial x}{\partial T_1} + \varepsilon^2\frac{\partial x}{\partial T_2}, \tag{1.1.29}$$

$$\begin{aligned}\frac{\mathrm{d}^2 x}{\mathrm{d}t^2} &= \left(\frac{\partial}{\partial T_0} + \varepsilon\frac{\partial}{\partial T_1} + \varepsilon^2\frac{\partial}{\partial T_2}\right)\left(\frac{\partial x}{\partial T_0} + \varepsilon\frac{\partial x}{\partial T_1} + \varepsilon^2\frac{\partial x}{\partial T_2}\right) \\ &\cong \frac{\partial^2 x}{\partial T_0^2} + 2\varepsilon\frac{\partial^2 x}{\partial T_0 \partial T_1} + \varepsilon^2\left(\frac{\partial^2 x}{\partial T_1^2} + 2\frac{\partial^2 x}{\partial T_0 \partial T_2}\right).\end{aligned} \tag{1.1.30}$$

As a result, we have

$$\begin{aligned}&\frac{\partial^2 x}{\partial T_0^2} + 2\varepsilon\frac{\partial^2 x}{\partial T_0 \partial T_1} + \varepsilon^2\left(\frac{\partial^2 x}{\partial T_1^2} + 2\frac{\partial^2 x}{\partial T_0 \partial T_2}\right) + x \\ &\quad = -\varepsilon\left(\frac{\partial x}{\partial T_0} + \varepsilon\frac{\partial x}{\partial T_1}\right).\end{aligned} \tag{1.1.31}$$

After the procedure of term equating described above, we obtain the system of equations

$$\frac{\partial^2 x}{\partial T_0^2} + x = 0, \tag{1.1.32}$$

$$2\frac{\partial^2 x}{\partial T_0 \partial T_1} = -\frac{\partial x}{\partial T_0}, \tag{1.1.33}$$

$$2\frac{\partial^2 x}{\partial T_2 \partial T_0} = -\frac{\partial^2 x}{\partial T_1^2} - \frac{\partial x}{\partial T_1}. \tag{1.1.34}$$

A general solution of (1.1.32) has the form

$$x = A(T_1, T_2)\,\mathrm{e}^{\mathrm{i}T_0} + A^*(T_1, T_2)\,\mathrm{e}^{-\mathrm{i}T_0}. \tag{1.1.35}$$

Obviously,

$$\frac{\partial x}{\partial T_0} = \mathrm{i}A\,\mathrm{e}^{\mathrm{i}T_0} - \mathrm{i}A^*\,\mathrm{e}^{-\mathrm{i}T_0}, \tag{1.1.36}$$

$$\frac{\partial^2 x}{\partial T_0 \partial T_1} = \mathrm{i}\frac{\partial A}{\partial T_1}\,\mathrm{e}^{\mathrm{i}T_0} - \mathrm{i}\frac{\partial A^*}{\partial T_1}\,\mathrm{e}^{-\mathrm{i}T_0}. \tag{1.1.37}$$

Substitute (1.1.36), (1.1.37) in (1.1.33):

$$\left(2\frac{\partial A}{\partial T_1} + A\right)\mathrm{e}^{\mathrm{i}T_0} - \left(2\frac{\partial A^*}{\partial T_1} + A^*\right)\mathrm{e}^{-\mathrm{i}T_0} = 0. \tag{1.1.38}$$

Equality (1.1.38) should be fulfilled for arbitrary values of the independent variable T_0, but A and A^* are functions of another variables (T_1 and T_2). It can be realized only if the following relations are identities:

$$2\frac{\partial A}{\partial T_1} + A = 0, \tag{1.1.39}$$

$$2\frac{\partial A^*}{\partial T_1} + A^* = 0 \tag{1.1.40}$$

whence it follows that

$$A = b(T_2)\,\mathrm{e}^{-T_1/2}, \qquad A^* = b^*(T_2)\,\mathrm{e}^{-T_1/2}. \tag{1.1.41}$$

We use (1.1.34) for the definition of $b(T_2)$ and $b^*(T_2)$. Since now

$$x = b(T_2)\,\mathrm{e}^{-T_1/2+\mathrm{i}T_0} + b^*(T_2)\,\mathrm{e}^{-T_1/2-\mathrm{i}T_0}, \tag{1.1.42}$$

we have

$$\frac{\partial x}{\partial T_1} = -\frac{1}{2}x, \qquad \frac{\partial^2 x}{\partial T_1^2} = \frac{1}{4}x, \tag{1.1.43}$$

$$\begin{aligned}\frac{\partial^2 x}{\partial T_0 \partial T_2} &= \mathrm{i}\frac{\partial}{\partial T_2}\left[b(T_2)\,\mathrm{e}^{-T_1/2+\mathrm{i}T_0} - b^*(T_2)\,\mathrm{e}^{-T_1/2-\mathrm{i}T_0}\right] \\ &= \mathrm{i}\left[\frac{\partial b}{\partial T_2}\,\mathrm{e}^{-T_1/2+\mathrm{i}T_0} - \frac{\partial b^*}{\partial T_2}\,\mathrm{e}^{-T_1/2-\mathrm{i}T_0}\right]\end{aligned} \tag{1.1.44}$$

and (1.1.34) after substituting (1.1.42)–(1.1.44) takes the form

$$\left(2\mathrm{i}\frac{\partial b}{\partial T_2} - \frac{1}{4}b\right)\mathrm{e}^{-T_1/2+\mathrm{i}T_0} - \left(2\mathrm{i}\frac{\partial b^*}{\partial T_2} + \frac{1}{4}b^*\right)\mathrm{e}^{-T_1/2-\mathrm{i}T_0} = 0. \tag{1.1.45}$$

Relation (1.1.45) should be fulfilled for all T_1 and T_0 (with a depending on another variable T_2). Then the terms in round brackets are equal to zero. Thus,

$$
\begin{aligned}
&2\mathrm{i}\frac{\partial b}{\partial T_2} - \frac{1}{4}b = 0, \\
&2\mathrm{i}\frac{\partial b^*}{\partial T_2} + \frac{1}{4}b^* = 0,
\end{aligned} \tag{1.1.46}
$$

and

$$
b = b_0\,\mathrm{e}^{-\mathrm{i}T_2/8}, \qquad b^* = b_0^*\,\mathrm{e}^{+\mathrm{i}T_2/8}. \tag{1.1.47}
$$

As a result, we have the solution

$$
x = b_0\,\mathrm{e}^{-T_1/2+\mathrm{i}(T_0-T_2/8)} + b_0^*\,\mathrm{e}^{-T_1/2-\mathrm{i}(T_0-T_2/8)}. \tag{1.1.48}
$$

For convenience b_0 will be written as

$$
b_0 = \frac{1}{2}a\,\mathrm{e}^{\mathrm{i}\varphi}. \tag{1.1.49}
$$

Then

$$
x = \frac{1}{2}a\,\mathrm{e}^{-T_1/2}\,\mathrm{e}^{\mathrm{i}(T_0-T_2/8+\varphi)} + \frac{1}{2}a\,\mathrm{e}^{-T_1/2}\,\mathrm{e}^{-\mathrm{i}(T_0-T_2/8+\varphi)}. \tag{1.1.50}
$$

Another form of this solution is:

$$
x = a\,\mathrm{e}^{-T_1/2}\cos\left(T_0 - \frac{T_2}{8} + \varphi\right), \tag{1.1.51}
$$

or, using the variable t and the parameter ε, we get

$$
x = a\,\mathrm{e}^{-\varepsilon t/2}\cos\left(t - \frac{1}{8}\varepsilon^2 t + \varphi\right). \tag{1.1.52}
$$

In the previous variant of MMS, which leads to system (1.1.25), (1.1.26′), (1.1.27′), the solution ${}^{\mathrm{ms}}x_2$ obtained with the accuracy of $\mathrm{O}(\varepsilon^2 t)$, can be written in the same form.

For the above example ($\varepsilon = 0.2$, $a = 1$, $\varphi = 0$)

$$
{}^{\mathrm{ms}}x_2 = \mathrm{e}^{-0.1t}\cos(0.995t). \tag{1.1.53}
$$

The comparison of the exact solution ${}^{\mathrm{ex}}x$ with the approximate solution obtained by MMS ${}^{\mathrm{ms}}x_2$ in the region of a "claimed" accuracy $\mathrm{O}(\varepsilon^2 t)$ shows (see Table 1.1 and Figure 1.1) that in the considered range of t $(0 \div 8\pi)$ the MMS-solution is coincidental with the exact solution with the accuracy as the rule of three ciphers in spite of the

Table 1.1
Comparison of the solutions ${}^0x$, ${}^1x$, ${}^2x$, obtained by the perturbation method, with the many scales solution ${}^{ms}x_2$ and the exact solution ${}^{ex}x$ for the case $a = 1$, $\varphi = 0$, $\varepsilon = 0.2$, $\delta = 1$

t	${}^0x$	${}^1x$	${}^2x$	${}^{ms}x_2$	${}^{ex}x$
0	1	1	1	1	1
$\pi/4$	0.707	0.652	0.656	0.656	0.656
$\pi/2$	0	0	0.00785	0.00671	0.00673
$3/4\pi$	−0.707	−0.540	−0.552	−0.552	−0.552
π	−1	−0.686	−0.735	−0.730	−0.730
1.25π	−0.707	−0.429	−0.498	−0.487	−0.487
1.5π	0	0	−0.0236	−0.0147	−0.0147
1.75π	0.707	0.318	0.406	0.397	0.397
2π	1	0.372	0.569	0.533	0.533
2.25π	0.707	0.207	0.409	0.361	0.361
2.5π	0	0	0.0393	0.0179	0.0179
2.75π	−0.707	−0.0962	−0.329	−0.285	−0.285
3π	−1	−0.0575	−0.502	−0.389	−0.389
3.25π	−0.707	0.0149	−0.390	−0.267	−0.267
3.5π	0	0	−0.0550	−0.0183	−0.0183
3.75π	0.707	−0.126	0.323	0.204	0.204
4π	1	−0.257	0.533	0.284	0.284
4.25π	0.707	−0.237	0.440	0.198	0.198
4.5π	0	0	0.0707	0.0172	0.0172
4.75π	−0.707	0.348	−0.386	−0.147	−0.147
5π	−1	0.571	−0.663	−0.207	−0.207
5.25π	−0.707	0.459	−0.561	−0.147	−0.147
5.5π	0	0	−0.0864	−0.0153	−0.0154
5.75π	0.707	−0.570	0.520	0.105	0.105
6π	1	−0.885	0.892	0.151	0.151
6.25π	0.707	−0.681	0.751	0.108	0.108
6.5π	0	0	0.102	0.0133	0.0133
6.75π	−0.707	0.792	−0.722	−0.0754	−0.0753
7π	−1	1.199	−1.219	−0.110	−0.110
7.25π	−0.707	0.903	−1.011	−0.0803	−0.0803
7.5π	0	0	−0.118	−0.0111	−0.0112
7.75π	0.707	1.015	0.995	0.0540	0.0539
8π	1	−1.513	1.645	0.0804	0.0804

${}^0x = \cos t$, ${}^1x = \cos t(1 - 0.1t)$, ${}^2x = \cos t(1 - 0.1t) + 0.005t(t\cos t + \sin t)$,
${}^{ex}x = \mathrm{e}^{-0.1t}\cos(t\sqrt{0.99}) = \mathrm{e}^{-0.1t}\cos(t \cdot 0.99498744)$, ${}^{ms}x_2 = \mathrm{e}^{-0.1t}\cos(t \cdot 0.995)$.

fact that $\varepsilon = 0.2$ is not really very small parameter. This result significantly exceeds the accuracy of the traditional asymptotic method (Table 1.1 contains also solutions ${}^0x$, ${}^1x$, ${}^2x$). Since

$$\cos(t\sqrt{0.99}) - \cos(t\sqrt{0.995}) = 2\sin(t \cdot 0.994993719)\sin(t \cdot 0.000006281),$$

the divergence O(1) could appear only $t \approx 10^5$, but, in this case, the difference of cosines is completely suppressed by the exponential term e^{-10^4}. Then, the growth of relative error $({}^{ex}x - {}^{ms}x_2)/{}^{ex}x$ is of no significance. The evolution of the solution ${}^{ms}x_2$

is not shown in Figure 1.1 due to actual coincidence of the exact and many scales solutions.

We can recollect another analogy. In the kinetic theory of reacting gases, the calculation of kinetic coefficients (by explicit taking into account inelastic collisions) is connected with the investigation of convergence of Sonine polynomial expansions. In the region, where we can wait for a bad convergence, the perturbation effects are small, because they are suppressed by the exponential function $e^{-\overline{E}}$ with large $\overline{E}$, where $\overline{E}$ is dimensionless activation energy. Only in the theory of reactions with high energy barriers it is not unreasonable to consider the mentioned effects, but of course it is better on the basis of the generalized Boltzmann physical kinetics because, as it will be shown, the traditional Boltzmann physical kinetics leads to significant errors in calculations of high energetic "tails" of distribution functions.

Let us consider now Eq. (1.1.1) for the case $0 < \delta \ll 1$, $\varepsilon \approx 1$. After substitution

$$x = u(t)\,e^{-(\varepsilon/(2\delta))t} \tag{1.1.54}$$

one obtains

$$\ddot{u} + \left(\frac{1}{\delta} - \frac{\varepsilon^2}{4\delta^2}\right)u = 0 \tag{1.1.55}$$

or in the mentioned case

$$\ddot{u} = \frac{\varepsilon^2}{4\delta^2}u. \tag{1.1.56}$$

For conditions $u(0) = 1$, $u(\infty) = 0$ we have

$$u = e^{-(\varepsilon/(2\delta))t} \tag{1.1.57}$$

and

$$x = e^{-(\varepsilon/\delta)t}. \tag{1.1.58}$$

Let us omit now the first term in the left-hand side of Eq. (1.1.1), which is from the first glance is small as containing small coefficient δ:

$$\varepsilon\dot{x} + x = 0 \tag{1.1.59}$$

and

$$x = e^{-(1/\varepsilon)t}. \tag{1.1.60}$$

Solutions (1.1.58) and (1.1.60) have the same asymptotic $x \to 0$ if $t \to \infty$. But function x corresponding to Eq. (1.1.58) is decreasing very fast in comparison with function

(1.1.60). It means that coefficient in front of senior derivative leads to formation of "boundary layer" with very fast function's change.

Then terms with small coefficients in front of senior derivatives cannot be omitted.

It remains only to notice that the MMS has many variants and the efficiency of MMS application is related with specificity of the problem.

1.2. Hierarchy of Bogolubov kinetic equations

Let us consider a closed physical system that consists of N particles. We investigate non-relativistic gas motion that adheres to the laws of classical mechanics. Introduce the N-particle distribution function (DF) f_N in a $6N$-dimensional space, in such a way

$$\begin{aligned} \mathrm{d}W &= f_N(t; \hat{\mathbf{r}}_1, \ldots, \hat{\mathbf{r}}_N; \hat{\mathbf{v}}_1, \ldots, \hat{\mathbf{v}}_N)\, \mathrm{d}\hat{\mathbf{r}}_1 \cdots \mathrm{d}\hat{\mathbf{r}}_N\, \mathrm{d}\hat{\mathbf{v}}_1 \cdots \mathrm{d}\hat{\mathbf{v}}_N \\ &= f_N\, \mathrm{d}\Omega_1 \cdots \mathrm{d}\Omega_N = f_N\, \mathrm{d}\Omega, \end{aligned} \tag{1.2.1}$$

where

$$\mathrm{d}\Omega_i = \mathrm{d}\hat{\mathbf{r}}_i\, \mathrm{d}\hat{\mathbf{v}}_i, \quad \mathrm{d}\hat{\mathbf{r}}_i \equiv \mathrm{d}r_{i1}\, \mathrm{d}r_{i2}\, \mathrm{d}r_{i3}, \quad \mathrm{d}\hat{\mathbf{v}}_i \equiv \mathrm{d}v_{i1}\, \mathrm{d}v_{i2}\, \mathrm{d}v_{i3}, \tag{1.2.2}$$

$$\mathrm{d}\Omega = \mathrm{d}\Omega_1 \cdots \mathrm{d}\Omega_N, \tag{1.2.3}$$

as to $\mathrm{d}W$ there is a probability to find the mentioned system at the time moment t in the element $\mathrm{d}\Omega$ of phase space. The Liouville equation is valid for the N-particle distribution function,

$$\frac{\partial f_N}{\partial t} + \sum_{i=1}^{N} \mathbf{v}_i \cdot \frac{\partial f_N}{\partial \mathbf{r}_i} + \sum_{i=1}^{N} \mathbf{F}_i \cdot \frac{\partial f_N}{\partial \mathbf{p}_i} = 0, \tag{1.2.4}$$

where $\mathbf{v}_i$, $\mathbf{p}_i$ are the velocity and momentum of a particle numbered i, $\mathbf{F}_i$ is the force acting on the ith particle whose mass is equal to m_i.

Consider the physical sense of Eq. (1.2.4). Let at a time moment t there be a unit volume in the considered phase space defined by positions of the vectors $\mathbf{r}_1, \ldots, \mathbf{r}_N; \mathbf{v}_1, \ldots, \mathbf{v}_N$. As it has written, the probability to find the system of this volume at the time moment t is f_N. Phase points – at the time moment t consisting on the surface that encloses the mentioned volume – change their positions in the time interval Δt; as a result, the position and form of the volume is also changing. At the time moment $t + \Delta t$, the volume will be found in the position in the vicinity of the $6N$-component vector with the coordinates

$$(\mathbf{r}_1 + \Delta\mathbf{r}_1, \ldots, \mathbf{r}_N + \Delta\mathbf{r}_N; \mathbf{v}_1 + \Delta\mathbf{v}_1, \ldots, \mathbf{v}_N + \Delta\mathbf{v}_N),$$

and the probability to find the system in this state is the same as at the time moment t, because the distribution function (DF) f_N does not change along the phase trajectory.

Then

$$f_N(t;\mathbf{r}_1,\ldots,\mathbf{r}_N;\mathbf{v}_1,\ldots,\mathbf{v}_N)$$
$$=f_N(t+\Delta t;\mathbf{r}_1+\Delta\mathbf{r}_1,\ldots,\mathbf{r}_N+\Delta\mathbf{r}_N;\mathbf{v}_1+\Delta\mathbf{v}_1,\ldots,\mathbf{v}_N+\Delta\mathbf{v}_N), \tag{1.2.5}$$

where the evolution of $\mathbf{r}_i$ and $\mathbf{v}_i$ $(i=1,\ldots,N)$ is defined by the equations of mechanical motion. Obviously, Eq. (1.2.4) is also valid for multi-component mixture of gases when the system contains the particles of different masses. But the mixture should be a non-reacting one. Really, if chemical reactions take place, the number of particles can be changed in the system. This fact can lead to the appearance of a source term in Eq. (1.2.4). Another treatment is also valid when molecules are considered as bounded states of atoms whose total number is the same during the system evolution. This approach is not simpler for the description of evolution on the kinetic level because, strictly speaking, the laws of classical mechanics are not sufficient for the calculation of reacting gases. As a result, on the lowest level characterized by the one-particle DF, the problem arises in explicit form for the integral of inelastic collisions (Alekseev, 1982). We begin with the set of kinetic equations for non-reacting gases.

Introduce an s-particle distribution function in accordance with the definition:

$$f_s=\int f_N\,\mathrm{d}\Omega_{s+1}\cdots\mathrm{d}\Omega_N$$
$$=\int f_N(t,\Omega_1,\ldots,\Omega_N)\,\mathrm{d}\Omega_{s+1}\cdots\mathrm{d}\Omega_N. \tag{1.2.6}$$

Integrate Eq. (1.2.4) with respect to $\Omega_{s+1},\ldots,\Omega_N$ with the aim of obtaining the equation for the s-particle distribution function (see Bogolyubov, 1946; Born and Green, 1946; Green, 1952; Kirkwood, 1947; Yvon, 1935)

$$\int\left(\frac{\partial f_N}{\partial t}+\sum_{i=1}^{N}\mathbf{v}_i\cdot\frac{\partial f_N}{\partial\vec{r}_i}+\sum_{i=1}^{N}\mathbf{F}_i\cdot\frac{\partial f_N}{\partial\mathbf{p}_i}\right)\mathrm{d}\Omega_{s+1}\cdots\mathrm{d}\Omega_N=0. \tag{1.2.7}$$

Generally speaking, on the right-hand side of Eq. (1.2.7) there could be a function $\psi(t,\Omega_1,\ldots,\Omega_s)$, but obviously $\psi(t,\Omega_1,\ldots,\Omega_s)=0$. Suppose that gas is kept in a vessel whose volume is known. This restriction is not of principal significance and can be removed at the final stage of consideration.

Consider

$$\int\frac{\partial f_N}{\partial t}\mathrm{d}\Omega_{s+1}\cdots\mathrm{d}\Omega_N. \tag{1.2.8}$$

We can change the order of integrating and differentiating because the limits of integration do not depend on time:

$$\int\frac{\partial f_N}{\partial t}\mathrm{d}\Omega_{s+1}\cdots\mathrm{d}\Omega_N=\frac{\partial}{\partial t}\int f_N\,\mathrm{d}\Omega_{s+1}\cdots\mathrm{d}\Omega_N=\frac{\partial f_s}{\partial t}. \tag{1.2.9}$$

Precisely in the same manner we change the order of operations in the second term for $i \leqslant s$:

$$\int \sum_{i=1}^{s} \mathbf{v}_i \cdot \frac{\partial f_N}{\partial \mathbf{r}_i} \, \mathrm{d}\Omega_{s+1} \cdots \mathrm{d}\Omega_N = \sum_{i=1}^{s} \mathbf{v}_i \cdot \frac{\partial}{\partial \mathbf{r}_i} \int f_N \, \mathrm{d}\Omega_{s+1} \cdots \mathrm{d}\Omega_N, \tag{1.2.10}$$

and for $i \leqslant s$ one obtains

$$\int \sum_{i=1}^{s} \mathbf{v}_i \cdot \frac{\partial f_N}{\partial \mathbf{r}_i} \, \mathrm{d}\Omega_{s+1} \cdots \mathrm{d}\Omega_N = \sum_{i=1}^{s} \mathbf{v}_i \cdot \frac{\partial f_s}{\partial \mathbf{r}_i}. \tag{1.2.11}$$

For $i \geqslant s+1$ we have

$$\begin{aligned} &\int \mathbf{v}_i \cdot \frac{\partial f_N}{\partial \mathbf{r}_i} \, \mathrm{d}\Omega_{s+1} \cdots \mathrm{d}\Omega_N \\ &= \int \mathbf{v}_i \cdot \frac{\partial f_{s+1}(t, \Omega_1, \ldots, \Omega_s, \Omega_i)}{\partial \mathbf{r}_i} \, \mathrm{d}\Omega_i \\ &= \int \left[\int \left(\frac{\partial f_{s+1} v_{i1}}{\partial r_{i1}} + \frac{\partial f_{s+1} v_{i2}}{\partial r_{i2}} + \frac{\partial f_{s+1} v_{i3}}{\partial r_{i3}} \right) \mathrm{d}r_{i1} \, \mathrm{d}r_{i2} \, \mathrm{d}r_{i3} \right] \mathrm{d}\mathbf{v}_i \\ &= \int \left[\int f_{s+1} \mathbf{v}_i \cdot \mathbf{n} \, \mathrm{d}s \right] \mathrm{d}\mathbf{v}_i. \end{aligned} \tag{1.2.12}$$

In the last equality the Gauss–Ostrogradskii theorem is used; the external – to the gas – direction is accepted as positive. The physical sense of the integral on the right side of Eq. (1.2.12) consists in the probability definition of the ith particle penetration through the wall of the vessel when s particles are in the states $\Omega_1, \ldots, \Omega_s$, respectively. However, the walls of the vessel are accepted as impermeable for the gas molecules and this probability and, as a consequence, this integral are equal to zero. This integral is also equal to zero in the case where the gas is found in a restricted area of space, but the integration is realized over an infinitely distant surface.

Let us consider now the last term in Eq. (1.2.7). Let $i \leqslant s+1$, then

$$\begin{aligned} &\int \mathbf{F}_i \cdot \frac{\partial f_N}{\partial \mathbf{p}_i} \, \mathrm{d}\Omega_{s+1} \cdots \mathrm{d}\Omega_N \\ &= \mathbf{F}_i \cdot \frac{\partial}{\partial \mathbf{p}_i} \int f_N \, \mathrm{d}\Omega_{s+1} \cdots \mathrm{d}\Omega_N = \mathbf{F}_i \cdot \frac{\partial f_s}{\partial \mathbf{p}_i}. \end{aligned} \tag{1.2.13}$$

Suppose further that force $\mathbf{F}_i$ does not depend on the velocity of the ith particle or depends in accordance with the Lorentz law. Then, for $i \geqslant s+1$, the following transformation is true

$$\int \mathbf{F}_i \cdot \frac{\partial f_N}{\partial \mathbf{p}_i}\,\mathrm{d}\Omega_{s+1}\cdots \mathrm{d}\Omega_N = \int \frac{\partial}{\partial \mathbf{p}_i}\cdot (\mathbf{F}_i f_N)\,\mathrm{d}\Omega_{s+1}\cdots \mathrm{d}\Omega_N$$
$$= \int \left[\int \frac{\partial}{\partial \mathbf{p}_i}\cdot (\mathbf{F}_i f_{s+1})\,\mathrm{d}\mathbf{v}_i\right] \mathrm{d}\mathbf{r}_i. \tag{1.2.14}$$

The last integral is equal to zero. Really, the internal integral can be transformed from the volume form to the surface one in the velocity space. Taking into account that the gas energy is finite, we conclude that the probability density f_N should tend to zero, if $v_i \to \infty$. Then the surface – integral over an infinite distant surface in the velocity space – turns into zero.

Denote the external force acting on the i-particle as $\mathbf{F}_i^b$. We have

$$\mathbf{F}_i = \mathbf{F}_i^b + \sum_{j=1}^{N} \mathbf{F}_{ij} \quad (\mathbf{F}_{ii} = 0), \tag{1.2.15}$$

where $\mathbf{F}_{ij}(\mathbf{r}_i, \mathbf{r}_j)$ is the force acting on the ith particle from the particle j. Taking into account (1.2.13)–(1.2.15), the third term of the integrand in relation (1.2.7) is written as follows:

$$\int \sum_{i=1}^{N} \mathbf{F}_i \cdot \frac{\partial f_N}{\partial \mathbf{p}_i}\,\mathrm{d}\Omega_{s+1}\cdots \mathrm{d}\Omega_N$$
$$= \int \sum_{i=1}^{s} \mathbf{F}_i \cdot \frac{\partial f_N}{\partial \mathbf{p}_i}\,\mathrm{d}\Omega_{s+1}\cdots \mathrm{d}\Omega_N$$
$$= \sum_{i=1}^{s} \mathbf{F}_i^b \cdot \frac{\partial f_s}{\partial \mathbf{p}_i} + \sum_{i=1}^{s} \int \sum_{j=1}^{s} \mathbf{F}_{ij} \cdot \frac{\partial f_N}{\partial \mathbf{p}_i}\,\mathrm{d}\Omega_{s+1}\cdots \mathrm{d}\Omega_N$$
$$+ \sum_{i=1}^{s} \int \sum_{j=s+1}^{N} \mathbf{F}_{ij} \cdot \frac{\partial f_N}{\partial \mathbf{p}_i}\,\mathrm{d}\Omega_{s+1}\cdots \mathrm{d}\Omega_N. \tag{1.2.16}$$

Consider the terms in (1.2.16)

$$\sum_{i=1}^{s} \int \sum_{j=1}^{s} \mathbf{F}_{ij} \cdot \frac{\partial f_N}{\partial \mathbf{p}_i}\,\mathrm{d}\Omega_{s+1}\cdots \mathrm{d}\Omega_N = \sum_{i,j=1}^{s} \mathbf{F}_{ij} \cdot \frac{\partial f_s}{\partial \mathbf{p}_i}. \tag{1.2.17}$$

We transform the term

$$\sum_{i=1}^{s} \int \sum_{j=s+1}^{N} \mathbf{F}_{ij} \cdot \frac{\partial f_N}{\partial \mathbf{p}_i}\,\mathrm{d}\Omega_{s+1}\cdots \mathrm{d}\Omega_N, \tag{1.2.18}$$

at first for the particular case of one-species gas.

With this aim write down the integral

$$\int \mathbf{F}_{ij} f_{s+1}(t, \Omega_1, \ldots, \Omega_s, \Omega_j)\, \mathrm{d}\Omega_j, \quad j \geqslant s+1,\ i = 1, \ldots, s. \tag{1.2.19}$$

The physical sense of this integral relates to the averaged force acting on the i-particle from the particle j when s particles are in the states $\Omega_1, \ldots, \Omega_s$. For one-species gas in the force of the particle identity, the value of this integral for $i \geqslant s+1$ does not depend on the number j, and we obtain the relation

$$\int \mathbf{F}_{ij} f_{s+1}(t, \Omega_1, \ldots, \Omega_s, \Omega_j) \mathrm{d}\Omega_j = \int \mathbf{F}_{i,s+1} f_{s+1}(t, \Omega_1, \ldots, \Omega_{s+1})\, \mathrm{d}\Omega_{s+1}$$
$$(j \geqslant s+1,\ i = 1, \ldots, s). \tag{1.2.20}$$

Now transform (1.2.18) using the derived relation (1.2.20):

$$\begin{aligned}
&\sum_{i=1}^{s} \int \sum_{j=s+1}^{N} \mathbf{F}_{ij} \cdot \frac{\partial f_N}{\partial \mathbf{p}_i}\, \mathrm{d}\Omega_{s+1} \cdots \mathrm{d}\Omega_N \\
&\quad = \sum_{i=1}^{s} \frac{\partial}{\partial \mathbf{p}_i} \cdot \int \sum_{j=s+1}^{N} \mathbf{F}_{ij} f_{s+1}\, \mathrm{d}\Omega_j \\
&\quad = \sum_{i=1}^{s} (N-s) \frac{\partial}{\partial \mathbf{p}_i} \cdot \int \mathbf{F}_{i,s+1} f_{s+1}\, \mathrm{d}\Omega_{s+1}.
\end{aligned} \tag{1.2.21}$$

Pay attention now to multi-component gas. In accordance with the chemical kinds of molecules, η groups of particles stand out. Within of each group all the particles are considered as identical. The transformation, analogous to (1.2.20), can be realized for the condition $s \ll N$:

$$\begin{aligned}
&\sum_{j=s+1}^{N} \int \mathbf{F}_{ij} f_{s+1}(t, \Omega_1, \ldots, \Omega_s, \Omega_j)\, \mathrm{d}\Omega_j \\
&\quad = \sum_{\delta=1}^{\eta} N_\delta \int \mathbf{F}_{i,j\in N_\delta} f_{s+1}(t, \Omega_1, \ldots, \Omega_s, \Omega_{j\in N_\delta})\, \mathrm{d}\Omega_{j\in N_\delta} \\
&\quad (i = 1, \ldots, s).
\end{aligned} \tag{1.2.22}$$

Some comments on the relation (1.2.22). Let, in a mixture consisting of N particles, N_δ particles belong to the same species of particles so that $\delta = 1, \ldots, \eta$, and $N = \sum_{\delta=1}^{\eta} N_\delta$. In every group all the particles are identical, therefore when integrating over the entire physical space and the velocity space it does not matter what number of particles is, used in every group. The integration over all $\Omega_{j\in N_\delta}$ means that integration is realized over all the phase volume of one of the particles belonging to the group N_δ.

As a result, for one-species gas, the following hierarchy of Bogolyubov–Born–Green–Kirkwood–Yvon kinetic equations (BBGKY-equations) takes place:

$$\frac{\partial f_s}{\partial t} + \sum_{i=1}^{s} \mathbf{v}_i \cdot \frac{\partial f_s}{\partial \mathbf{r}_i} + \sum_{i=1}^{s} \mathbf{F}_i^b \cdot \frac{\partial f_s}{\partial \mathbf{p}_i} + \sum_{i,j=1}^{s} \mathbf{F}_{ij} \cdot \frac{\partial f_s}{\partial \mathbf{p}_i}$$
$$= -\sum_{i=1}^{s} (N-s) \frac{\partial}{\partial \mathbf{p}_i} \cdot \int f_{s+1}(t, \Omega_1, \ldots, \Omega_{s+1}) \mathbf{F}_{i,s+1} \, \mathrm{d}\Omega_{s+1}. \qquad (1.2.23)$$

The internal forces $\mathbf{F}_{ij}$ that do not depend on the velocity can be factored out of the derivative by momentum $\mathbf{p}_i$.

The BBGKY-1 equation ($s = 1$)

$$\frac{\partial f_1}{\partial t} + \frac{\mathbf{p}_1}{m} \cdot \frac{\partial f_1}{\partial \mathbf{r}_1} + \mathbf{F}_1^b \cdot \frac{\partial f_1}{\partial \mathbf{p}_1} = (1-N) \frac{\partial}{\partial \mathbf{p}_1} \cdot \int \mathbf{F}_{12} f_2 \, \mathrm{d}\Omega_2. \qquad (1.2.24)$$

The BBGKY-2 equation ($s = 2$)

$$\frac{\partial f_2}{\partial t} + \left(\frac{\mathbf{p}_1}{m} \cdot \frac{\partial f_1}{\partial \mathbf{r}_1} + \frac{\mathbf{p}_2}{m} \cdot \frac{\partial f_2}{\partial \mathbf{r}_2} \right) + \mathbf{F}_1^b \cdot \frac{\partial f_2}{\partial \mathbf{p}_1} + \mathbf{F}_2^b \cdot \frac{\partial f_2}{\partial \mathbf{p}_2}$$
$$+ \mathbf{F}_{12}^b \cdot \frac{\partial f_2}{\partial \mathbf{p}_1} + \mathbf{F}_{21}^b \cdot \frac{\partial f_2}{\partial \mathbf{p}_2}$$
$$= (2-N) \left(\frac{\partial}{\partial \mathbf{p}_1} \cdot \int \mathbf{F}_{13} f_3 \, \mathrm{d}\Omega_3 + \frac{\partial}{\partial \mathbf{p}_2} \cdot \int \mathbf{F}_{23} f_3 \, \mathrm{d}\Omega_3 \right). \qquad (1.2.25)$$

For multi-species gas, in the absence of chemical reactions, we have

$$\frac{\partial f_s}{\partial t} + \sum_{i=1}^{s} \mathbf{v}_i \cdot \frac{\partial f_s}{\partial \mathbf{r}_i} + \sum_{i=1}^{s} \mathbf{F}_i^b \cdot \frac{\partial f_s}{\partial \mathbf{p}_i} + \sum_{i,j=1}^{s} \mathbf{F}_{ij} \cdot \frac{\partial f_s}{\partial \mathbf{p}_i}$$
$$= -\sum_{i=1}^{s} \sum_{j=s+1}^{N} \mathbf{F}_{ij} \cdot \frac{\partial}{\partial \mathbf{p}_i} f_{s+1}(t, \Omega_1, \ldots, \Omega_s, \Omega_j) \, \mathrm{d}\Omega_j$$
$$(s = 1, \ldots, N). \qquad (1.2.26)$$

If $s \ll N$ it is reasonable to write down Eq. (1.2.26) with an explicit extraction of species of particles in a multi-species mixture of gases:

$$\frac{\partial f_s}{\partial t} + \sum_{i=1}^{s} \mathbf{v}_i \cdot \frac{\partial f_s}{\partial \mathbf{r}_i} + \sum_{i=1}^{s} \mathbf{F}_i^b \cdot \frac{\partial f_s}{\partial \mathbf{p}_i} + \sum_{i,j=1}^{s} \mathbf{F}_{ij} \cdot \frac{\partial f_s}{\partial \mathbf{p}_i}$$
$$= -\sum_{i=1}^{s} \sum_{\delta=1}^{\eta} N_\delta \int \mathbf{F}_{i,j\in N_\delta} \cdot \frac{\partial}{\partial \mathbf{p}_i} f_{s+1}(t, \Omega_1, \ldots, \Omega_s, \Omega_{j\in N_\delta}) \, \mathrm{d}\Omega_{j\in N_\delta}. \qquad (1.2.27)$$

For reacting gases, Eqs. (1.2.26) should be corrected taking into account an additional source integral term related to chemical reactions. Possible approximations for this term are discussed in Alekseev (1982), but, generally speaking, the form of this term should be based on the quantum theory of inelastic collisions.

The used assumption that, for the convenience of consideration, the gas is taken as bounded in a vessel of volume V, is not significant. It is possible to use the limit $V \to \infty$, $N \to \infty$ under the condition $n = N/V = const$. In this case, hierarchy of equations has no changes.

The set of integro-differential equations turns out to be a linked one, so that in the lowest-order approximation the distribution function f_1 depends on f_2. This means formally that, strictly speaking, the solution procedure for such a set should be as follows. First find the distribution function f_N and then solve the set of BBGKY equations subsequently for decreasingly lower-order distributions. But if we know the function f_N, there is no need at all to solve the equations for f_s and it actually suffices to employ the definition (1.2.6) of the function f_s.

We thus conclude that the rigorous solution to the set of BBGKY equations is again equivalent to solving Liouville equations. On the other hand, the seemingly illogical solution procedure involving a search for the distribution function f_1 is of great significance in the kinetic theory and in non-equilibrium statistical mechanics. This approach involves breaking the BBGKY chain by introducing certain additional assumptions (which have a clear physical meaning, though). These assumptions are discussed in detail below.

1.3. Derivation of the generalized Boltzmann equation

We now proceed with the derivation of the generalized Boltzmann equation (GBE) (Alexeev, 1988, 1994, 1995c) by applying Bogolyubov's procedure and writing down once more Eq. (1.2.26), introducing the forces acting on the unit of mass of the particles. We conserve the previous notation for the kind of forces that cannot lead to misunderstandings.

$$\frac{\partial f_s}{\partial t} + \sum_{i=1}^{s} \mathbf{v}_i \cdot \frac{\partial f_s}{\partial \mathbf{r}_i} + \sum_{i=1}^{s} \mathbf{F}_i^b \cdot \frac{\partial f_s}{\partial \mathbf{v}_i} + \sum_{i,j=1}^{s} \mathbf{F}_{ij} \cdot \frac{\partial f_s}{\partial \mathbf{v}_i}$$

$$= -\sum_{i=1}^{s} \sum_{j=s+1}^{N} \int \mathbf{F}_{ij} \cdot \frac{\partial}{\partial \mathbf{v}_i} f_{s+1} \, \mathrm{d}\Omega_j, \tag{1.3.1}$$

where $\mathrm{d}\Omega_j = \mathrm{d}\mathbf{r}_j \, \mathrm{d}\mathbf{v}_j$, f_N is normalized on a unit. It means that the probability to find all N particles of a physical system at some points of physical space with some velocities is equal to a unit,

$$\int f_N \, \mathrm{d}\Omega_1 \, \mathrm{d}\Omega_2 \cdots \mathrm{d}\Omega_N = 1. \tag{1.3.2}$$

We now write down a dimensionless equation for the one-particle distribution function. In doing so, we follow the multi-scale method and introduce three groups of scales:

- at the r_b level – the particle interaction radius r_b, $r_b = (V_b)^{1/3}$, where V_b is the volume of particle interaction, which could be non-spherical in the general case; the characteristic collision velocity v_{0b}, and the characteristic collision time r_b/v_{0b};
- at the λ level – the mean free path λ, the mean free-flight velocity $v_{0\lambda}$, and the characteristic time scale $\lambda/v_{0\lambda}$, and
- at the L level – the characteristic hydrodynamic dimension L, the hydrodynamic velocity, v_{0L} and the hydrodynamic time L/v_{0L}.

Other notation: V is the character volume of a physical system, F_0 is the scale of internal forces of molecular interaction, $F_{0\lambda}$ is the scale of external forces.

A hat "^" over a symbol means that the quantity labelled like this is made dimensionless. From the normalizing condition (1.3.2) it follows that the dimensionless s-particle distribution function $\hat{f}_s$ can be written as

$$\hat{f}_s = f_s v_{0b}^{3s} V^s. \tag{1.3.3}$$

For other values, we have

$$\widehat{F}_{ij} = \frac{F_{ij}}{F_0}, \quad \hat{t}_b = \frac{t}{r_b v_{0b}^{-1}}, \quad \hat{\mathbf{r}}_{ib} = \frac{\mathbf{r}_i}{r_b}, \quad \widehat{\mathbf{F}}_i = \frac{\mathbf{F}_i}{F_{0\lambda}}.$$

Eq. (1.3.1) can be rewritten in the form:

$$\begin{aligned}
&\frac{v_{0b}}{r_b}\frac{\partial \hat{f}_s}{\partial \hat{t}_b} + \frac{v_{0b}}{r_b}\sum_{i=1}^{s}\hat{\mathbf{v}}_{ib}\cdot\frac{\partial \hat{f}_s}{\partial \hat{\mathbf{r}}_b} + \frac{F_{0\lambda}}{v_{0b}}\sum_{i=1}^{s}\widehat{\mathbf{F}}_i\cdot\frac{\partial \hat{f}_s}{\partial \hat{\mathbf{v}}_{ib}} + \frac{F_0}{v_{0b}}\sum_{i,j=1}^{s}\widehat{\mathbf{F}}_{ij}\cdot\frac{\partial \hat{f}_s}{\partial \hat{\mathbf{v}}_{ib}} \\
&\quad = -\frac{F_0}{v_{0b}}\left(v_{0b}^{-3}V^{-1}\right)r_b^3 v_{0b}^3\sum_{i=1}^{s}\sum_{j=s+1}^{N}\int \widehat{\mathbf{F}}_{ij}\cdot\frac{\partial}{\partial \hat{\mathbf{v}}_{ib}}\hat{f}_{s+1}\,\mathrm{d}\widehat{\Omega}_j
\end{aligned} \tag{1.3.4}$$

and after dividing of both sides of equations by v_{0b}/r_b, we find

$$\begin{aligned}
&\frac{\partial \hat{f}_s}{\partial \hat{t}_b} + \sum_{i=1}^{s}\hat{\mathbf{v}}_{ib}\cdot\frac{\partial \hat{f}_s}{\partial \hat{\mathbf{r}}_b} + \frac{F_{0\lambda}}{v_{0b}^2/r_b}\sum_{i=1}^{s}\widehat{\mathbf{F}}_i\cdot\frac{\partial \hat{f}_s}{\partial \hat{\mathbf{v}}_{ib}} + \frac{F_0}{v_{0b}^2/r_b}\sum_{i,j=1}^{s}\widehat{\mathbf{F}}_{ij}\cdot\frac{\partial \hat{f}_s}{\partial \hat{\mathbf{v}}_{ib}} \\
&\quad = -\frac{r_b^3}{V}\frac{F_0}{v_{0b}^2/r_b}\sum_{i=1}^{s}\sum_{j=s+1}^{N}\int \widehat{\mathbf{F}}_{ij}\cdot\frac{\partial}{\partial \hat{\mathbf{v}}_{ib}}\hat{f}_{s+1}\,\mathrm{d}\widehat{\Omega}_j.
\end{aligned} \tag{1.3.5}$$

As a scale for the force of molecular interaction – recall that this force is related to the unit of mass – choose the value v_{0b}^2/r_b, connected with the scale of collision velocity;

$F_0 = v_{0b}^2/r_b$ and then

$$\frac{\partial \hat{f}_s}{\partial \hat{t}_b} + \sum_{i=1}^{s} \hat{\mathbf{v}}_{ib} \cdot \frac{\partial \hat{f}_s}{\partial \hat{\mathbf{r}}_{ib}} + \alpha \sum_{i=1}^{s} \widehat{\mathbf{F}}_i \cdot \frac{\partial \hat{f}_s}{\partial \hat{\mathbf{v}}_{ib}} + \sum_{i,j=1}^{s} \widehat{\mathbf{F}}_{ij} \cdot \frac{\partial \hat{f}_s}{\partial \hat{\mathbf{v}}_{ib}}$$
$$= -\beta \sum_{i=1}^{s} \sum_{j=s+1}^{N} \int \widehat{\mathbf{F}}_{ij} \cdot \frac{\partial}{\partial \hat{\mathbf{v}}_{ib}} \hat{f}_{s+1} \, \mathrm{d}\widehat{\Omega}_j, \tag{1.3.6}$$

where the dimensionless parameters are introduced

$$\alpha = \frac{F_{0\lambda}}{F_0}, \qquad \beta = \frac{r_b^3}{V} = \frac{V_b}{V}. \tag{1.3.7}$$

For a one-particle DF, the $\hat{f}_1$ equation is valid

$$\frac{\partial \hat{f}_1}{\partial \hat{t}_b} + \hat{\mathbf{v}}_{1b} \cdot \frac{\partial \hat{f}_1}{\partial \hat{\mathbf{r}}_{1b}} + \alpha \widehat{\mathbf{F}}_1 \cdot \frac{\partial \hat{f}_s}{\partial \hat{\mathbf{v}}_{1b}}$$
$$+ \beta \sum_{j=2}^{N} \int \hat{\mathbf{F}}_{1j} \cdot \frac{\partial}{\partial \hat{\mathbf{v}}_1} \hat{f}_2(\Omega_1, \Omega_j, t) \, \mathrm{d}\widehat{\Omega}_j = 0. \tag{1.3.8}$$

If effects of the particle correlations could be completely omitted, the simplest approximation for a two-particle DF is valid

$$\hat{f}_2(\Omega_1, \Omega_j, t) = \hat{f}_1(\Omega_1, t) \hat{f}_{1j}(\Omega_j, t), \tag{1.3.9}$$

where j is the number of particle, $j = 2, \ldots, N$. From (1.3.8) it follows that

$$\frac{\partial \hat{f}_1}{\partial \hat{t}_b} + \hat{\mathbf{v}}_{1b} \cdot \frac{\partial \hat{f}_1}{\partial \hat{\mathbf{r}}_{1b}} + \alpha \widehat{\mathbf{F}}_1 \cdot \frac{\partial \hat{f}_1}{\partial \hat{\mathbf{v}}_{1b}}$$
$$+ \beta \frac{\partial \hat{f}_1}{\partial \hat{\mathbf{v}}_{1b}} \cdot \sum_{j=2}^{N} \int \widehat{\mathbf{F}}_{1j} \hat{f}_{1j}(\Omega_j, t) \, \mathrm{d}\widehat{\Omega}_j = 0. \tag{1.3.10}$$

The integral term in Eq. (1.3.10) is connected with the averaged force $\widehat{\mathbf{F}}_1^a$, acting on the first particle from all the other particles of the system:

$$\widehat{\mathbf{F}}_1^a = \sum_{j=2}^{N} \int \hat{\mathbf{F}}_{1j} \hat{f}_{1,j}(\Omega_j, t) \, \mathrm{d}\widehat{\Omega}_j. \tag{1.3.11}$$

The relation for the averaged force can be rewritten for a multi-component system consisting of η species, as follows

$$\widehat{\mathbf{F}}_1^a = \sum_{\delta=1}^{\eta} N_\delta \int \widehat{\mathbf{F}}_{1,\, j\in N_\delta} \hat{f}_{1,\, j\in N_\delta}(\Omega_j, t)\, \mathrm{d}\widehat{\Omega}_{j\in N_\delta}. \tag{1.3.12}$$

Using (1.3.11), (1.3.12), one obtains from (1.3.10)

$$\frac{\partial \hat{f}_1}{\partial \hat{t}_b} + \hat{\mathbf{v}}_{1b} \cdot \frac{\partial \hat{f}_1}{\partial \hat{\mathbf{r}}_{ib}} + \widehat{\mathbf{F}}_1^{\mathrm{sc}} \cdot \frac{\partial \hat{f}_1}{\partial \hat{\mathbf{v}}_{ib}} = 0, \tag{1.3.13}$$

where the self-consistent force $\widehat{\mathbf{F}}_1^{\mathrm{sc}}$, acting on the particle numbered one, is introduced

$$\widehat{\mathbf{F}}_1^{\mathrm{sc}} = \alpha \widehat{\mathbf{F}}_1 + \beta \widehat{\mathbf{F}}_1^a. \tag{1.3.14}$$

Kinetic equation (1.3.13) is a Vlasov equation widely used in plasma physics together with electrodynamics equations, which close the system of equations. In molecular dynamics of neutral gases the self-consistent forces are significant only on the r_b-scale – usually the smallest ones from all possible kinetic scales – and Eq. (1.3.13) is the same as the free molecular limit of the Boltzmann equation.

In what follows, we intend to construct the generalized Boltzmann physical kinetics outlining the statistical features of so called, “rarefied” gases. Under the term “rarefied gases” we understand the physical systems for the description of which one-particle distribution function f_1 is sufficient.

For multi-component mixture, Eq. (1.3.8) can be written as

$$\begin{aligned} &\frac{\partial \hat{f}_1}{\partial \hat{t}_b} + \hat{\mathbf{v}}_{1b} \cdot \frac{\partial \hat{f}_1}{\partial \hat{\mathbf{r}}_{1b}} + \alpha \widehat{\mathbf{F}}_1 \cdot \frac{\partial \hat{f}_1}{\partial \hat{\mathbf{v}}_{1b}} \\ &\quad = -\frac{r_b^3 N}{V} \sum_{\delta=1}^{\eta} \frac{N_\delta}{N} \int \widehat{\mathbf{F}}_{ij\in N_\delta} \cdot \frac{\partial}{\partial \hat{\mathbf{v}}_b} \hat{f}_{2,\, j\in N_\delta}(\Omega_j, t)\, \mathrm{d}\widehat{\Omega}_{j\in N_\delta}. \end{aligned} \tag{1.3.15}$$

For rarefied gas, the value

$$\varepsilon = \frac{r_b^3 N}{V} = n r_b^3 = n V_b, \tag{1.3.16}$$

defining the particle number in the volume of interaction, is a small parameter and the ratio N_δ / N is the number density of species in the physical system is given.

In the kinetic theory, other conditions of normalization of the DF f_s apart from (1.3.2), first of all the DF, are normalized to number density. Consider this question in detail.

Let

$$\int \tilde{f}_N \, \mathrm{d}\Omega_1 \cdots \mathrm{d}\Omega_N = N^N, \tag{1.3.2$'$}$$

where N is the total number of particles in the system which occupies the volume V, $\mathrm{d}\Omega_i = \mathrm{d}\mathbf{v}_i\, \mathrm{d}\mathbf{r}_i$.

For the s-particles distribution function

$$\int \tilde{f}_s\, \mathrm{d}\Omega_1 \cdots \mathrm{d}\Omega_s = N^s, \tag{1.3.17}$$

and finally for the one-particle DF, we have

$$\int \tilde{f}_1\, \mathrm{d}\Omega_1 = N. \tag{1.3.18}$$

Since

$$\int n\, \mathrm{d}\mathbf{r} = N, \tag{1.3.19}$$

where n is the number density, we get

$$\int \left(\int \tilde{f}_1\, \mathrm{d}\mathbf{v}_1 \right) \mathrm{d}\mathbf{r} = N \tag{1.3.20}$$

and

$$\int \tilde{f}_1\, \mathrm{d}\mathbf{v}_1 = n. \tag{1.3.21}$$

With the help of the DF $\tilde{f}_s$ Eq. (1.3.1) is written as

$$\begin{aligned} &\frac{\partial \tilde{f}_s}{\partial t} + \sum_{i=1}^{s} \mathbf{v}_i \cdot \frac{\partial \tilde{f}_s}{\partial \mathbf{r}_i} + \sum_{i=1}^{s} \mathbf{F}_i \cdot \frac{\partial \tilde{f}_s}{\partial \mathbf{v}_i} + \sum_{i,j=1}^{s} \mathbf{F}_{ij} \cdot \frac{\partial \tilde{f}_s}{\partial \mathbf{v}_i} \\ &\quad = -\frac{1}{N} \sum_{i=1}^{s} \sum_{j=s+1}^{N} \int \mathbf{F}_{ij} \cdot \frac{\partial}{\partial \mathbf{v}_i} \tilde{f}_{s+1}\, \mathrm{d}\Omega_{s+1}, \end{aligned} \tag{1.3.1$'$}$$

because $\tilde{f}_s = N^s f_s$.

Write down Eq. (1.3.1′) in a dimensionless form; introduce the same scale, however recall that the DF $\tilde{f}_s$ was normalized is another way (compare with (1.3.3)):

$$\hat{\tilde{f}}_s = \tilde{f}_s \frac{v_0^{3s}}{n^s}. \tag{1.3.3$'$}$$

Write down the analogue of Eq. (1.3.4)

$$\frac{n^s}{v_0^{3s}} \frac{v_0}{r_b} \left\{ \frac{\partial \hat{\tilde{f}}_s}{\partial \hat{t}_b} + \sum_{i=1}^{s} \hat{\mathbf{v}}_i \cdot \frac{\partial \hat{\tilde{f}}_s}{\partial \hat{\mathbf{r}}_{ib}} + \alpha \sum_{i=1}^{s} \widehat{\mathbf{F}}_i \cdot \frac{\partial \hat{\tilde{f}}_s}{\partial \hat{\mathbf{v}}_i} + \sum_{i,j=1}^{s} \widehat{\mathbf{F}}_{ij} \cdot \frac{\partial \hat{\tilde{f}}_s}{\partial \hat{\mathbf{v}}_i} \right\}$$

$$= -\frac{F_0}{v_0}\frac{n^{s+1}}{v_0^{3s+3}}\frac{1}{N}\sum_{i=1}^{s}\sum_{j=s+1}^{N}\int \widehat{\mathbf{F}}_{ij}\cdot\frac{\partial}{\partial\hat{\mathbf{v}}_{ib}}\hat{\tilde{f}}_{s+1}\,\mathrm{d}\widehat{\Omega}_{s+1}v_0^3r_b^3 \tag{1.3.4$'$}$$

or, after cancelling the factor before the curly brackets, we find

$$\frac{\partial\hat{\tilde{f}}_s}{\partial\hat{t}_b}+\sum_{i=1}^{s}\hat{\mathbf{v}}_{ib}\cdot\frac{\partial\hat{\tilde{f}}_s}{\partial\hat{\mathbf{r}}_{ib}}+\alpha\sum_{i=1}^{s}\widehat{\mathbf{F}}_i\cdot\frac{\partial\hat{\tilde{f}}_s}{\partial\hat{\mathbf{v}}_i}+\sum_{i,j=1}^{s}\widehat{\mathbf{F}}_{ij}\cdot\frac{\partial\hat{\tilde{f}}_s}{\partial\hat{\mathbf{v}}_i}$$

$$= -nr_b^3\frac{F_0}{v_0^2/r_b}\frac{1}{N}\sum_{i=1}^{s}\sum_{j=s+1}^{N}\int \widehat{\mathbf{F}}_{ij}\cdot\frac{\partial}{\partial\hat{\mathbf{v}}_{ib}}\hat{\tilde{f}}_{s+1}\,\mathrm{d}\widehat{\Omega}_{s+1}. \tag{1.3.22}$$

As before, the value v_0^2/r_b is chosen as the scale for the force F_0 and for multicomponent mixture we find the analogue of Eq. (1.3.15)

$$\frac{\partial\hat{\tilde{f}}_s}{\partial\hat{t}_b}+\hat{\mathbf{v}}_{1b}\cdot\frac{\partial\hat{\tilde{f}}_1}{\partial\hat{\mathbf{r}}_{1b}}+\alpha\widehat{\mathbf{F}}_1\cdot\frac{\partial\hat{\tilde{f}}_1}{\partial\hat{\mathbf{v}}_{1b}}$$

$$= -\varepsilon\sum_{\delta=1}^{\eta}\frac{N_\delta}{N}\int \widehat{\mathbf{F}}_{i,\,j\in N_\delta}\cdot\frac{\partial}{\partial\hat{\mathbf{v}}_{ib}}\hat{\tilde{f}}_{2,\,j\in N_\delta}(\Omega_1,\Omega_{j\in N_\delta},t)\,\mathrm{d}\widehat{\Omega}_{j\in N_\delta}. \tag{1.3.23}$$

In the following transformation of Eq. (1.3.23), for simplicity we omit the sign $\sim$ from DF returning to the details of normalization in case of necessity.

Of the numerous scales involved in the gas kinetics problems, three major groups of scales pertaining to length, time, and velocity deserve a special consideration. In this case, the particle interaction scale r_b presents only one of the scales (and the shortest one) in the scale hierarchy in molecular systems, where the λ scale, related to the particle mean free path, and the hydrodynamic L scale – for example, the length or the diameter of the flow channel, the characteristic size of the streamlined body, etc. – always exist.

In gas dynamics, the conditions

$$r_b \ll \lambda \ll L \tag{1.3.24}$$

are usually satisfied. If desired, inequalities (1.3.24) can be rewritten in terms of parameters such as the characteristic collision time, mean free time, and hydrodynamic flow time. Since the Boltzmann equation is valid only on the λ and L scales, a fundamental problem arises here how to adequately describe kinetic processes at all the three scales of a system's evolution.

In Section 1.1, we were able to ascertain that standard perturbation methods cannot present satisfactory results, if the system's evolution were investigated on very different time scales. In this case, it is natural to apply the method of many scales.

We assume that the arguments of the s-particle function $\hat{f}_s$ are the above three groups of scaled variables and the mentioned small parameter $\varepsilon = nr_b^3$ (compare with (1.1.21)):

$$\hat{f}_s = \hat{f}_s\big(\hat{t}_b,\hat{\mathbf{r}}_{ib},\hat{\mathbf{v}}_{ib};\hat{t}_\lambda,\hat{\mathbf{r}}_{i\lambda},\hat{\mathbf{v}}_{i\lambda};\hat{t}_L,\hat{\mathbf{r}}_{iL},\hat{\mathbf{v}}_{iL};\varepsilon\big). \tag{1.3.25}$$

As it has been indicated, a hat over a symbol means that the quantity labelled like this is made dimensionless. We now write down an asymptotic series for the function $\hat{f}_s$:

$$\hat{f}_s = \sum_{\nu=0}^{\infty} \hat{f}_s^{\nu}\left(\hat{t}_b, \hat{\mathbf{r}}_{ib}, \hat{\mathbf{v}}_{ib}; \hat{t}_\lambda, \hat{\mathbf{r}}_{i\lambda}, \hat{\mathbf{v}}_{i\lambda}; \hat{t}_L, \hat{\mathbf{r}}_{iL}, \hat{\mathbf{v}}_{iL}\right)\varepsilon^{\nu}, \tag{1.3.26}$$

which should be used for solving the equation

$$\begin{aligned} &\frac{\partial \hat{f}_s}{\partial \hat{t}_b} + \sum_{i=1}^{s} \hat{\mathbf{v}}_{ib} \cdot \frac{\partial \hat{f}_s}{\partial \hat{\mathbf{r}}_{ib}} + \sum_{i,j=1}^{s} \widehat{\mathbf{F}}_{ij} \cdot \frac{\partial \hat{f}_s}{\partial \hat{\mathbf{v}}_{ib}} + \alpha \sum_{i=1}^{s} \widehat{\mathbf{F}}_i \cdot \frac{\partial \hat{f}_s}{\partial \hat{\mathbf{v}}_{ib}} \\ &\quad = -\varepsilon \frac{1}{N} \sum_{i=1}^{s} \sum_{j=s+1}^{N} \int \widehat{\mathbf{F}}_{ij} \cdot \frac{\partial}{\partial \hat{\mathbf{v}}_{ib}} \hat{f}_{s+1}\left(\hat{t}, \widehat{\Omega}_1, \ldots, \widehat{\Omega}_s, \widehat{\Omega}_j\right) \mathrm{d}\widehat{\Omega}_j. \end{aligned} \tag{1.3.27}$$

Let us take the derivatives on the left-hand side of the sth BBGKY equation according to the rules intended for taking the derivatives of composite functions (compare with (1.1.22)):

$$\frac{\mathrm{d}\hat{f}_s}{\mathrm{d}\hat{t}_b} = \frac{\partial \hat{f}_s}{\partial \hat{t}_b} + \frac{\partial \hat{f}_s}{\partial \hat{t}_\lambda}\frac{\partial \hat{t}_\lambda}{\partial \hat{t}_b} + \frac{\partial \hat{f}_s}{\partial \hat{t}_L}\frac{\partial \hat{t}_L}{\partial \hat{t}_\lambda}\frac{\partial \hat{t}_\lambda}{\partial \hat{t}_b}, \tag{1.3.28}$$

$$\frac{\mathrm{d}\hat{f}_s}{\mathrm{d}\hat{\mathbf{r}}_{ib}} = \frac{\partial \hat{f}_s}{\partial \hat{\mathbf{r}}_{ib}} + \sum_{k=1}^{3} \frac{\partial \hat{f}_s}{\partial \hat{r}_{i\lambda,k}}\frac{\partial \hat{r}_{i\lambda,k}}{\partial \hat{\mathbf{r}}_{ib}} + \sum_{k,l=1}^{3} \frac{\partial \hat{f}_s}{\partial \hat{r}_{iL,k}}\frac{\partial \hat{r}_{iL,k}}{\partial \hat{r}_{i\lambda,l}}\frac{\partial \hat{r}_{i\lambda,l}}{\partial \hat{\mathbf{r}}_{ib}}, \tag{1.3.29}$$

$$\frac{\mathrm{d}\hat{f}_s}{\mathrm{d}\hat{\mathbf{v}}_{ib}} = \frac{\partial \hat{f}_s}{\partial \hat{\mathbf{v}}_{ib}} + \sum_{k=1}^{3} \frac{\partial \hat{f}_s}{\partial \hat{v}_{i\lambda,k}}\frac{\partial \hat{v}_{i\lambda,k}}{\partial \hat{\mathbf{v}}_{ib}} + \sum_{k,l=1}^{3} \frac{\partial \hat{f}_s}{\partial \hat{v}_{iL,k}}\frac{\partial \hat{v}_{iL,k}}{\partial \hat{v}_{i\lambda,l}}\frac{\partial \hat{v}_{i\lambda,l}}{\partial \hat{\mathbf{v}}_{ib}}. \tag{1.3.30}$$

Introduce the following parameters as a ratio of the scale factors. No limitations are introduced here for these parameters.

$$\varepsilon_1 = \frac{\lambda}{L} \quad \text{(Knudsen number)}, \tag{1.3.31}$$

$$\varepsilon_2 = \frac{v_{0\lambda}}{v_{0b}}, \qquad \varepsilon_3 = \frac{v_{0L}}{v_{0\lambda}}. \tag{1.3.32}$$

Then the approximated derivatives can be rewritten as follows:

$$\frac{\mathrm{d}\hat{f}_s}{\mathrm{d}\hat{t}_b} = \frac{\partial \hat{f}_s}{\partial \hat{t}_b} + \frac{\partial \hat{f}_s}{\partial \hat{t}_\lambda}\varepsilon\varepsilon_2 + \frac{\partial \hat{f}_s}{\partial \hat{t}_L}\varepsilon\varepsilon_1\varepsilon_2\varepsilon_3, \tag{1.3.33}$$

$$\frac{\mathrm{d}\hat{f}_s}{\mathrm{d}\hat{\mathbf{r}}_{ib}} = \frac{\partial \hat{f}_s}{\partial \hat{\mathbf{r}}_{ib}} + \frac{\partial \hat{f}_s}{\partial \hat{\mathbf{r}}_{i\lambda}}\varepsilon + \frac{\partial \hat{f}_s}{\partial \hat{\mathbf{r}}_{iL}}\varepsilon\varepsilon_1, \tag{1.3.34}$$

$$\frac{\mathrm{d}\hat{f}_s}{\mathrm{d}\hat{\mathbf{v}}_{ib}} = \frac{\partial \hat{f}_s}{\partial \hat{\mathbf{v}}_{ib}} + \frac{\partial \hat{f}_s}{\partial \hat{\mathbf{v}}_{i\lambda}}\varepsilon\varepsilon_2\frac{F_0}{F_{0\lambda}} + \frac{\partial \hat{f}_s}{\partial \hat{\mathbf{v}}_{iL}}\varepsilon\varepsilon_2\frac{F_0}{F_{0\lambda}}\varepsilon_3^{-1}, \tag{1.3.35}$$

because, for example,

$$
\begin{aligned}
&\frac{\partial \hat{t}_\lambda}{\partial \hat{t}_b} = \frac{M_{t_b}}{M_{t_\lambda}} = \frac{r_b}{v_{0b}}\frac{v_{0\lambda}}{\lambda} = \varepsilon_2 \frac{r_b}{\lambda} = \varepsilon_2 \frac{r_b}{(n r_b^2)^{-1}} = \varepsilon\varepsilon_2, \\
&\frac{\partial \hat{t}_L}{\partial \hat{t}_\lambda} = \frac{M_{t_\lambda}}{M_{t_L}} = \frac{v_{oL}\lambda}{v_{0\lambda} L} = \varepsilon_1\varepsilon_3, \\
&\frac{\partial \hat{v}_{i\lambda,k}}{\partial v_{ib,k}} = \frac{M_{v_{0b}}}{M_{v_{0\lambda}}} = \varepsilon \frac{M_{t_\lambda}}{M_{t_b}} = \varepsilon \frac{v_{0\lambda}}{F_{0\lambda}}\frac{F_0}{v_{0b}} = \varepsilon\varepsilon_2 \frac{F_0}{F_{0\lambda}},
\end{aligned}
\tag{1.3.36}
$$

where M denotes the scales of values written as the index. The scales of the free mean path and radius of interaction are associated with the known relation

$$
\lambda = \left(n r_b^2\right)^{-1}; \tag{1.3.37}
$$

the following relations are written for scales on internal forces F_0 related to the mass unit and the scale of velocity v_{0b}

$$
F_0 = \frac{v_{0b}^2}{r_b} \tag{1.3.38}
$$

and, analogously,

$$
F_{0\lambda} = \frac{v_{0\lambda}}{t_\lambda}. \tag{1.3.39}
$$

Substitute now the series (1.3.26) in the approximated derivations (1.3.33)–(1.3.35) and the obtained relations in Eq. (1.3.27). Furthermore, we use the expression for $\hat{f}_{s+1}$ on the right side of Eq. (1.3.27) as the mentioned series. Equating the coefficients of ε^0 and ε^1, ε^2 now yields

– at ε^0:

$$
\frac{\partial \hat{f}_s^0}{\partial \hat{t}_b} + \sum_{i=1}^{s} \hat{\mathbf{v}}_{ib} \cdot \frac{\partial \hat{f}_s^0}{\partial \hat{\mathbf{r}}_{ib}} + \sum_{i,j=1}^{s} \widehat{\mathbf{F}}_{ij} \cdot \frac{\partial \hat{f}_s^0}{\partial \hat{\mathbf{v}}_{ib}} + \alpha \sum_{i=1}^{s} \widehat{\mathbf{F}}_i \cdot \frac{\partial \hat{f}_s^0}{\partial \hat{\mathbf{v}}_{ib}} = 0, \tag{1.3.40}
$$

– at ε^1:

$$
\begin{aligned}
&\frac{\partial \hat{f}_s^1}{\partial \hat{t}_b} + \sum_{i=1}^{s} \hat{\mathbf{v}}_{ib} \cdot \frac{\partial \hat{f}_s^1}{\partial \hat{\mathbf{r}}_{ib}} + \sum_{i,j=1}^{s} \widehat{\mathbf{F}}_{ij} \cdot \frac{\partial \hat{f}_s^1}{\partial \hat{\mathbf{v}}_{ib}} + \alpha \sum_{i=1}^{s} \widehat{\mathbf{F}}_i \cdot \frac{\partial \hat{f}_s^1}{\partial \hat{\mathbf{v}}_{ib}} \\
&\quad + \varepsilon_2 \frac{\partial \hat{f}_s^0}{\partial \hat{t}_\lambda} + \sum_{i=1}^{s} \hat{\mathbf{v}}_{ib} \cdot \frac{\partial \hat{f}_s^0}{\partial \hat{\mathbf{r}}_{i\lambda}} + \varepsilon_2 \frac{F_0}{F_{0\lambda}} \sum_{i,j=1}^{s} \widehat{\mathbf{F}}_{ij} \cdot \frac{\partial \hat{f}_s^0}{\partial \hat{\mathbf{v}}_{i\lambda}}
\end{aligned}
$$

$$+\varepsilon_2\sum_{i=1}^{s}\widehat{\mathbf{F}}_i\cdot\frac{\partial\hat{f}_s^0}{\partial\hat{\mathbf{v}}_{i\lambda}}+\varepsilon_1\varepsilon_2\varepsilon_3\frac{\partial\hat{f}_s^0}{\partial\hat{t}_L}+\varepsilon_1\sum_{i=1}^{s}\hat{\mathbf{v}}_{ib}\cdot\frac{\partial\hat{f}_s^0}{\partial\hat{\mathbf{r}}_{iL}}$$

$$+\frac{\varepsilon_2}{\varepsilon_3}\frac{F_0}{F_{0\lambda}}\sum_{i,j=1}^{s}\widehat{\mathbf{F}}_{ij}\cdot\frac{\partial\hat{f}_s^0}{\partial\hat{\mathbf{v}}_{iL}}+\frac{\varepsilon_2}{\varepsilon_3}\sum_{i=1}^{s}\widehat{\mathbf{F}}_i\cdot\frac{\partial\hat{f}_s^0}{\partial\hat{\mathbf{v}}_{iL}}$$

$$=-\frac{1}{N}\sum_{i=1}^{s}\sum_{j=s+1}^{N}\int\widehat{\mathbf{F}}_{ij}\cdot\frac{\partial}{\partial\hat{\mathbf{v}}_{ib}}\hat{f}_{s+1}^0\,\mathrm{d}\widehat{\Omega}_j,\tag{1.3.41}$$

– at ε^2:

$$\frac{\partial\hat{f}_s^2}{\partial\hat{t}_b}+\sum_{i=1}^{s}\hat{\mathbf{v}}_{ib}\cdot\frac{\partial\hat{f}_s^2}{\partial\hat{\mathbf{r}}_{ib}}+\sum_{i,j=1}^{s}\widehat{\mathbf{F}}_{ij}\cdot\frac{\partial\hat{f}_s^2}{\partial\hat{\mathbf{v}}_{ib}}+\alpha\sum_{i=1}^{s}\widehat{\mathbf{F}}_i\cdot\frac{\partial\hat{f}_s^2}{\partial\hat{\mathbf{v}}_{ib}}$$

$$+\varepsilon_2\frac{\partial\hat{f}_s^1}{\partial\hat{t}_\lambda}+\sum_{i=1}^{s}\hat{\mathbf{v}}_{ib}\cdot\frac{\partial\hat{f}_s^1}{\partial\hat{\mathbf{r}}_{i\lambda}}+\varepsilon_2\frac{F_0}{F_{0\lambda}}\sum_{i,j=1}^{s}\widehat{\mathbf{F}}_{ij}\cdot\frac{\partial\hat{f}_s^1}{\partial\hat{\mathbf{v}}_{i\lambda}}$$

$$+\varepsilon_2\sum_{i=1}^{s}\widehat{\mathbf{F}}_i\cdot\frac{\partial\hat{f}_s^1}{\partial\hat{\mathbf{v}}_{i\lambda}}+\varepsilon_1\varepsilon_2\varepsilon_3\frac{\partial\hat{f}_s^1}{\partial\hat{t}_L}+\varepsilon_1\sum_{i=1}^{s}\hat{\mathbf{v}}_{ib}\cdot\frac{\partial\hat{f}_s^1}{\partial\hat{\mathbf{r}}_{iL}}$$

$$+\frac{\varepsilon_2}{\varepsilon_3}\frac{F_0}{F_{0\lambda}}\sum_{i,j=1}^{s}\widehat{\mathbf{F}}_{ij}\cdot\frac{\partial\hat{f}_s^1}{\partial\hat{\mathbf{v}}_{iL}}+\frac{\varepsilon_2}{\varepsilon_3}\sum_{i=1}^{s}\widehat{\mathbf{F}}_i\cdot\frac{\partial\hat{f}_s^1}{\partial\hat{\mathbf{v}}_{iL}}$$

$$=-\frac{1}{N}\sum_{i=1}^{s}\sum_{j=s+1}^{N}\int\widehat{\mathbf{F}}_{ij}\cdot\left\{\frac{\partial\hat{f}_{s+1}^1}{\partial\hat{\mathbf{v}}_{ib}}+\frac{F_0}{F_{0\lambda}}\varepsilon_2\frac{\partial\hat{f}_{s+1}^0}{\partial\hat{\mathbf{v}}_{i\lambda}}+\frac{\varepsilon_2}{\varepsilon_3}\frac{F_0}{F_{0\lambda}}\frac{\partial\hat{f}_{s+1}^0}{\partial\hat{\mathbf{v}}_{iL}}\right\}\mathrm{d}\widehat{\Omega}_j.\tag{1.3.42}$$

The next approximations are organized in an analogous way.

At $s=1$ one obtains ($F_{11}\equiv 0$)

$$\frac{\partial\hat{f}_1^0}{\partial\hat{t}_b}+\hat{\mathbf{v}}_{ib}\cdot\frac{\partial\hat{f}_1^0}{\partial\hat{\mathbf{r}}_{ib}}+\alpha\widehat{\mathbf{F}}_1\cdot\frac{\partial\hat{f}_1^0}{\partial\hat{\mathbf{v}}_{ib}}=0,\tag{1.3.43}$$

$$\frac{\partial\hat{f}_1^1}{\partial\hat{t}_b}+\hat{\mathbf{v}}_{1b}\cdot\frac{\partial\hat{f}_1^1}{\partial\hat{\mathbf{r}}_{1b}}+\alpha\widehat{\mathbf{F}}_1\cdot\frac{\partial\hat{f}_1^1}{\partial\hat{\mathbf{v}}_{1b}}+\varepsilon_2\frac{\partial\hat{f}_1^0}{\partial\hat{t}_\lambda}+\hat{\mathbf{v}}_{1b}\cdot\frac{\partial\hat{f}_1^0}{\partial\hat{\mathbf{r}}_{1\lambda}}+\varepsilon_2\widehat{\mathbf{F}}_1\cdot\frac{\partial\hat{f}_1^0}{\partial\hat{\mathbf{v}}_{1\lambda}}$$

$$+\varepsilon_1\varepsilon_2\varepsilon_3\frac{\partial\hat{f}_1^0}{\partial\hat{t}_L}+\varepsilon_1\hat{\mathbf{v}}_{1b}\cdot\frac{\partial\hat{f}_1^0}{\partial\hat{\mathbf{r}}_{1L}}+\frac{\varepsilon_2}{\varepsilon_3}\widehat{\mathbf{F}}_1\cdot\frac{\partial\hat{f}_1^0}{\partial\hat{\mathbf{v}}_{1L}}$$

$$=-\sum_{\delta=1}^{\eta}\frac{N_\delta}{N}\int\widehat{\mathbf{F}}_{1,\,j\in N_\delta}\cdot\frac{\partial}{\partial\hat{\mathbf{v}}_{1b}}\hat{f}_{2,\,j\in N_\delta}^0\,\mathrm{d}\widehat{\Omega}_{j\in N_\delta},\tag{1.3.44}$$

$$\frac{\partial \hat{f}_1^2}{\partial \hat{t}_b} + \hat{\mathbf{v}}_1 \cdot \frac{\partial \hat{f}_1^2}{\partial \hat{\mathbf{r}}_{1b}} + \alpha \widehat{\mathbf{F}}_1 \cdot \frac{\partial \hat{f}_1^2}{\partial \hat{\mathbf{v}}_{1b}} + \varepsilon_2 \frac{\partial \hat{f}_1^1}{\partial \hat{t}_\lambda} + \hat{\mathbf{v}}_{1b} \cdot \frac{\partial \hat{f}_1^1}{\partial \hat{\mathbf{r}}_{1\lambda}} + \varepsilon_2 \widehat{\mathbf{F}}_1 \cdot \frac{\partial \hat{f}_1^1}{\partial \hat{\mathbf{v}}_{1\lambda}}$$
$$+ \varepsilon_1 \varepsilon_2 \varepsilon_3 \frac{\partial \hat{f}_1^1}{\partial \hat{t}_L} + \varepsilon_1 \hat{\mathbf{v}}_{1b} \cdot \frac{\partial \hat{f}_1^1}{\partial \hat{\mathbf{r}}_{1L}} + \frac{\varepsilon_2}{\varepsilon_3} \widehat{\mathbf{F}}_1 \cdot \frac{\partial \hat{f}_1^1}{\partial \hat{\mathbf{v}}_{1L}}$$
$$= -\sum_{\delta=1}^{\eta} \frac{N_\delta}{N} \int \widehat{\mathbf{F}}_{1,\, j\in N_\delta} \cdot \left\{ \frac{\partial \hat{f}_2^1}{\partial \hat{\mathbf{v}}_{1b}} + \frac{F_0}{F_{0\lambda}} \varepsilon_2 \frac{\partial \hat{f}_2^0}{\partial \hat{\mathbf{v}}_{1\lambda}} + \frac{\varepsilon_2}{\varepsilon_3} \frac{F_0}{F_{0\lambda}} \frac{\partial \hat{f}_2^0}{\partial \hat{\mathbf{v}}_{1L}} \right\} \mathrm{d}\widehat{\Omega}_{j\in N_\delta}. \tag{1.3.45}$$

Some conclusions can be drawn at this step of investigation.

It follows from Eq. (1.3.40) that on the r_b-scale the function $\hat{f}_s^0$ has no change along the phase trajectory or, otherwise, after integrating on the r_b-scale

$$\hat{f}_s^0 = \hat{f}_s^0\big(\hat{t}_\lambda, \hat{\mathbf{v}}_{i\lambda}, \hat{\mathbf{r}}_{i\lambda}; \hat{t}_L, \hat{\mathbf{v}}_{iL}, \hat{\mathbf{r}}_{iL}\big). \tag{1.3.46}$$

If function (1.3.46) is known, the function $\hat{f}_s^1$ has to be found from Eq. (1.3.41). This is possible if certain additional assumptions are posed on the function $\hat{f}_{s+1}^0$ entering the integral right-hand side of expression (1.3.41). Eq. (1.3.42) can be used for calculating $\hat{f}_s^2$ if not only the functions $\hat{f}_s^0$, $\hat{f}_s^1$ are known from the preceding equations, but also $\hat{f}_{s+1}^1$. Thus, we see that the system of equations contains linked terms and if, in Eq. (1.3.41), we need to introduce assumptions concerning the function $\hat{f}_{s+1}^0$, then in Eq. (1.3.42) – concerning $\hat{f}_{s+1}^1$.

In real life, the dependence (1.3.46) is unknown beforehand. Then Eq. (1.3.41) can serve to determine $\hat{f}_s^0$ on λ- and L-scales, but, in this case, it becomes doubly linked, with respect to both the lower index $s+1$ and the upper index 1. As a result, the problem of breaking the linked terms arises.

In what follows, we intend to deal with systems admitting the one-particle description by means of Eqs. (1.3.43)–(1.3.45). To this ends we transform the integral term in (1.3.44).

Let us write Eq. (1.3.40) at $s = 2$ for two particles "1" and "j".

$$\frac{\partial \hat{f}_2^0}{\partial \hat{t}_b} + \hat{\mathbf{v}}_{1b} \cdot \frac{\partial \hat{f}_2^0}{\partial \hat{\mathbf{r}}_{1b}} + \hat{\mathbf{v}}_{j\in N_\delta, b} \cdot \frac{\partial \hat{f}_2^0}{\partial \hat{\mathbf{r}}_{j\in N_\delta, b}} + \widehat{\mathbf{F}}_{1,\, j\in N_\delta} \cdot \frac{\partial \hat{f}_2^0}{\partial \hat{\mathbf{v}}_{1b}}$$
$$+ \widehat{\mathbf{F}}_{j\in N_\delta, 1} \cdot \frac{\partial \hat{f}_2^0}{\partial \hat{\mathbf{v}}_{j\in N_\delta}} + \alpha \widehat{\mathbf{F}}_1 \cdot \frac{\partial \hat{f}_2^0}{\partial \mathbf{v}_{1b}} + \alpha \widehat{\mathbf{F}}_{j\in N_\delta} \cdot \frac{\partial \hat{f}_2^0}{\partial \hat{\mathbf{v}}_{j\in N_\delta, b}} = 0. \tag{1.3.47}$$

Introducing a new variable $\hat{\mathbf{x}}_{1,\, j\in N_\delta} = \hat{\mathbf{r}}_{1b} - \hat{\mathbf{r}}_{j\in N_\delta, b}$, we find from Eq. (1.3.47) that

$$-\widehat{\mathbf{F}}_{1,\, j\in N_\delta} \cdot \frac{\partial}{\partial \hat{\mathbf{v}}_{1b}} \hat{f}_{2,\, j\in N_\delta}^0$$
$$= \frac{\partial \hat{f}_2^0}{\partial \hat{t}_b} + \hat{\mathbf{v}}_{1b} \cdot \frac{\partial \hat{f}_2^0}{\partial \hat{\mathbf{r}}_{1b}} + (\hat{\mathbf{v}}_{1b} - \hat{\mathbf{v}}_{j\in N_\delta, b}) \cdot \frac{\partial \hat{f}_2^0}{\partial \hat{\mathbf{x}}_{1,\, j\in N_\delta}}$$

$$+\widehat{\mathbf{F}}_{j\in N_\delta,1}\cdot\frac{\partial \hat{f}_2^0}{\partial \hat{\mathbf{v}}_{j\in N_\delta}}+\alpha\widehat{\mathbf{F}}_1\cdot\frac{\partial \hat{f}_2^0}{\partial \mathbf{v}_{1b}}+\alpha\widehat{\mathbf{F}}_{j\in N_\delta}\cdot\frac{\partial \hat{f}_2^0}{\partial \hat{\mathbf{v}}_{j\in N_\delta,b}}. \tag{1.3.48}$$

Using the last equation, we obtain the following representation for the integral in Eq. (1.3.44)

$$\begin{aligned}
-\int \widehat{\mathbf{F}}_{1,j\in N_\delta}\cdot\frac{\partial}{\partial \hat{\mathbf{v}}_{1b}}\hat{f}^0_{2,j\in N_\delta}\,\mathrm{d}\widehat{\Omega}_{j\in N_\delta} &= \int \left(\hat{\mathbf{v}}_{1b}-\hat{\mathbf{v}}_{j\in N_\delta,b}\right)\cdot\frac{\partial \hat{f}_2^0}{\partial \hat{\mathbf{x}}_{1,j\in N_\delta}}\,\mathrm{d}\widehat{\Omega}_{j\in N_\delta}\\
&+\int\left(\frac{\partial \hat{f}_2^0}{\partial \hat{t}_b}+\hat{\mathbf{v}}_{1b}\cdot\frac{\partial \hat{f}_2^0}{\partial \hat{\bar{r}}_{1b}}+\alpha\widehat{\mathbf{F}}_1\cdot\frac{\partial \hat{f}_2^0}{\partial \hat{\mathbf{v}}_{1b}}+\alpha\widehat{\mathbf{F}}_{j\in N_\delta}\cdot\frac{\partial \hat{f}_2^0}{\partial \hat{\mathbf{v}}_{j\in N_\delta,b}}\right)\mathrm{d}\widehat{\Omega}_{j\in N_\delta}\\
&+\int \widehat{\mathbf{F}}_{j\in N_\delta,1}\cdot\frac{\partial \hat{f}_2^0}{\partial \hat{\mathbf{v}}_{j\in N_\delta}}\,\mathrm{d}\widehat{\Omega}_{j\in N_\delta}.
\end{aligned} \tag{1.3.49}$$

The last integral on the right-hand side of Eq. (1.3.49) can be written in the form

$$\begin{aligned}
&\int \widehat{\mathbf{F}}_{j\in N_\delta,1}\cdot\frac{\partial \hat{f}_2^0}{\partial \hat{\mathbf{v}}_{j\in N_\delta}}\,\mathrm{d}\widehat{\Omega}_{j\in N_\delta}\\
&=\int\left[\int\frac{\partial}{\partial \hat{\mathbf{v}}_{j\in N_\delta}}\cdot\left(\widehat{\mathbf{F}}_{j\in N_\delta,1}\hat{f}_2^0\right)\mathrm{d}\hat{\mathbf{v}}_{j\in N_\delta}\right]\mathrm{d}\hat{\mathbf{r}}_{j\in N_\delta}.
\end{aligned} \tag{1.3.50}$$

But the inner integral can be transformed by the Gauss theorem into an integral over an infinitely distant surface in the velocity space, which vanishes because $\hat{f}_2^0\to 0$ as $\hat{v}_j\to\infty$.

Let us introduce an assumption for rarefied gas of neutral particles that, on the λ- and L-scales, the positions and velocities of particles 1 and j are not correlated, i.e.,

$$\hat{f}_2^0\left(\hat{t},\widehat{\Omega}_1,\widehat{\Omega}_j\right)=\hat{f}_1^0\left(\hat{t},\widehat{\Omega}_1\right)\hat{f}^0_{j\in N_\delta}\left(\hat{t},\widehat{\Omega}_{j\in N_\delta}\right). \tag{1.3.51}$$

This assumption is discussed in detail in Chapter 8.

$$\begin{aligned}
&\int\left(\frac{\partial \hat{f}_2^0}{\partial \hat{t}_b}+\hat{\mathbf{v}}_{1b}\cdot\frac{\partial \hat{f}_2^0}{\partial \hat{\mathbf{r}}_{1b}}+\alpha\widehat{\mathbf{F}}_1\cdot\frac{\partial \hat{f}_2^0}{\partial \mathbf{v}_{1b}}+\alpha\widehat{\mathbf{F}}_{j\in N_\delta}\cdot\frac{\partial \hat{f}_2^0}{\partial \hat{\mathbf{v}}_{j\in N_\delta,b}}\right)\mathrm{d}\widehat{\Omega}_{j\in N_\delta}\\
&=\int\left[\hat{f}^0_{j\in N_\delta}\left(\frac{\partial \hat{f}_1^0}{\partial \hat{t}_b}+\hat{\mathbf{v}}_{1b}\cdot\frac{\partial \hat{f}_1^0}{\partial \hat{\mathbf{r}}_{1b}}+\alpha\widehat{\mathbf{F}}_1\cdot\frac{\partial \hat{f}_1^0}{\partial \mathbf{v}_{1b}}\right)\right]\mathrm{d}\widehat{\Omega}_{j\in N_\delta}\\
&\quad+\int \hat{f}_1^0\frac{\partial \hat{f}^0_{j\in N_\delta}}{\partial \hat{t}_b}\,\mathrm{d}\widehat{\Omega}_j+\alpha\int\frac{\partial}{\partial \mathbf{v}_{j\in N_\delta}}\cdot\left(\widehat{\mathbf{F}}_{j\in N_\delta}\hat{f}_1^0\hat{f}^0_{j\in N_\delta}\right)\mathrm{d}\widehat{\Omega}_{j\in N_\delta}.
\end{aligned} \tag{1.3.52}$$

In expression (1.3.52), the first integral on the right is zero, because relation (1.3.43) and the third integral is zero for the same reasons as in Eq. (1.3.50). The second integral is equal to zero only if the influence of a self-consistent field can be ignored in comparison with the external forces acting in the physical system. This problem is investigated in Chapter 8, devoted to applications of the generalized Boltzmann physical kinetics in plasma physics.

The integral term in (1.3.49) is written as

$$\begin{aligned}
-\int \widehat{\mathbf{F}}_{1,\, j\in N_\delta} \cdot \frac{\partial}{\partial \hat{\mathbf{v}}_{1b}} \hat{f}^{0}_{2,\, j\in N_\delta}\, \mathrm{d}\widehat{\Omega}_{j\in N_\delta}
&= \int (\hat{\mathbf{v}}_{1b} - \hat{\mathbf{v}}_{j\in N_\delta,\, b}) \cdot \frac{\partial \hat{f}^{0}_{2}}{\partial \hat{\mathbf{x}}_{1,\, j\in N_\delta}}\, \mathrm{d}\widehat{\Omega}_{j\in N_\delta} \\
&= \int (\hat{\mathbf{v}}_{1b} - \hat{\mathbf{v}}_{j\in N_\delta,\, b}) \cdot \frac{\partial \hat{f}^{0}_{2}}{\partial \hat{\mathbf{x}}_{1,\, j\in N_\delta}}\, \mathrm{d}\hat{\mathbf{r}}_{j\in N_\delta,\, b}\, \mathrm{d}\hat{\mathbf{v}}_{j\in N_\delta,\, b}.
\end{aligned} \tag{1.3.53}$$

Using $\hat{\mathbf{x}}_{1,\, j\in N_\delta}$ as an integrating variable, we arrive at the expression for the collision term $\hat{J}^{\mathrm{st},0}$

$$\begin{aligned}
\hat{J}^{\mathrm{st},0} &= -\sum_{\delta=1}^{\eta} \frac{N_\delta}{N} \int \widehat{\mathbf{F}}_{1,\, j\in N_\delta} \cdot \frac{\partial}{\partial \hat{\mathbf{v}}_{1b}} \hat{f}^{0}_{2,\, j\in N_\delta}\, \mathrm{d}\widehat{\Omega}_{j\in N_\delta} \\
&= \sum_{\delta=1}^{\eta} \frac{N_\delta}{N} \int \hat{\mathbf{g}}_{1,\, j\in N_\delta} \cdot \frac{\partial \hat{f}^{0}_{2}}{\partial \hat{\mathbf{x}}_{1,\, j\in N_\delta}}\, \mathrm{d}\hat{\mathbf{x}}_{1,\, j\in N_\delta,\, b}\, \mathrm{d}\hat{\mathbf{v}}_{j\in N_\delta,\, b},
\end{aligned} \tag{1.3.54}$$

where the relative velocity of particles 1 and j is introduced:

$$\hat{\mathbf{g}}_{j\in N_\delta,\, 1} = \hat{\mathbf{v}}_{j\in N_\delta,\, b} - \hat{\mathbf{v}}_{1b}. \tag{1.3.55}$$

Introduce the cylindrical coordinate system $\hat{l}$, $\hat{b}$, φ with the origin of coordinates at the point $\hat{\mathbf{r}}_{1b}$ and the axis $\hat{l}$, parallel to the relative velocity of encountered particles i and j; corresponding the dimensionless impact parameter $\hat{b}$ and the azimuth angle φ,

$$\begin{aligned}
\hat{J}^{\mathrm{st},0} &= \sum_{\delta=1}^{\eta} \frac{N_\delta}{N} \int \hat{g}_{j\in N_\delta,\, 1} \left[\int_{-\infty}^{+\infty} \frac{\partial \hat{f}^{0}_{2}}{\partial \hat{l}}\, \mathrm{d}\hat{l} \right] \hat{b}\, \mathrm{d}\hat{b}\, \mathrm{d}\dot{\varphi}\, \mathrm{d}\hat{\mathbf{v}}_{j\in N_\delta,\, b} \\
&= \sum_{\delta=1}^{\eta} \frac{N_\delta}{N} \int \left[\hat{f}^{0}_{2}(+\infty) - \hat{f}^{0}_{2}(-\infty) \right] \hat{g}_{j\in N_\delta,\, 1} \hat{b}\, \mathrm{d}\hat{b}\, \mathrm{d}\dot{\varphi}\, \mathrm{d}\hat{\mathbf{v}}_{j\in N_\delta,\, b},
\end{aligned} \tag{1.3.56}$$

where integration is realized on the r_b-scale, i.e., the functions $\hat{f}^{0}_{2}(+\infty)$ and $\hat{f}^{0}_{2}(-\infty)$ are calculated for velocities $\hat{\mathbf{v}}'_{j\in N_\delta}$, $\hat{\mathbf{v}}'_1$ and $\hat{\mathbf{v}}_{j\in N_\delta}$, $\hat{\mathbf{v}}_1$, when particles are placed out of the interaction zone. As usual the particle velocities after collision are indicated by prime.

Using the assumption that the condition of molecular chaos is valid on the λ-scale for encountering particles – and, as a consequence, relation (1.3.51) – we obtain

$$\hat{J}^{\mathrm{st},0} = \sum_{\delta=1}^{\eta} \frac{N_\delta}{N} \int \left[\hat{f}_1^{0\prime} \hat{f}_{j\in N_\delta}^{0\prime} - \hat{f}_1^{0} \hat{f}_{j\in N_\delta}^{0} \right] \hat{g}_{j\in N_\delta,\,1} \hat{b}\, \mathrm{d}\hat{b}\, \mathrm{d}\varphi\, \mathrm{d}\hat{\mathbf{v}}_{j\in N_\delta} \tag{1.3.57}$$

the Boltzmann collision integral is written for the multi-component gas.

By manipulating Eq. (1.3.44), we obtain

$$\frac{D_1 \hat{f}_1^1}{D\hat{t}_b} + \frac{\mathrm{d}_1 \hat{f}_1^0}{\mathrm{d}\hat{t}_{\lambda,L}} = \hat{J}^{\mathrm{st},0}, \tag{1.3.58}$$

where we have introduced the notation

$$\frac{D_1 \hat{f}_1^1}{D\hat{t}_b} = \frac{\partial \hat{f}_1^1}{\partial \hat{t}_b} + \hat{\mathbf{v}}_{1b} \cdot \frac{\partial \hat{f}_1^1}{\partial \hat{\mathbf{r}}_{1b}} + \alpha \widehat{\mathbf{F}}_1 \cdot \frac{\partial \hat{f}_1^1}{\partial \hat{\mathbf{v}}_{1b}}, \tag{1.3.59}$$

$$\frac{\mathrm{d}_1 \hat{f}_1^0}{\mathrm{d}\hat{t}_{\lambda,L}} = \varepsilon_2 \frac{\partial \hat{f}_1^0}{\partial \hat{t}_\lambda} + \hat{\mathbf{v}}_{1b} \cdot \frac{\partial \hat{f}_1^0}{\partial \hat{\mathbf{r}}_{1\lambda}} + \varepsilon_2 \widehat{\mathbf{F}}_1 \cdot \frac{\partial \hat{f}_1^0}{\partial \hat{\mathbf{v}}_{1\lambda}} + \varepsilon_1 \varepsilon_2 \varepsilon_3 \frac{\partial \hat{f}_1^0}{\partial \hat{t}_L}$$
$$+ \varepsilon_1 \hat{\mathbf{v}}_{1b} \cdot \frac{\partial \hat{f}_1^0}{\partial \hat{\mathbf{r}}_{1L}} + \frac{\varepsilon_2}{\varepsilon_3} \widehat{\mathbf{F}}_1 \cdot \frac{\partial \hat{f}_1^0}{\partial \hat{\mathbf{v}}_{1L}}. \tag{1.3.60}$$

The following remarks are of fundamental importance in connection with the theory being developed.

(1) Until now no restrictions are imposed on the values of ε_1, ε_2, ε_3, including the Knudsen number ε_1.
(2) Eq. (1.3.58) contains linking not only with respect to the lower, but also to the upper index, implying that in order to employ the kinetic equation, additional assumptions should be made to reduce the equation to one dependent variable.
(3) The collision integral $\hat{J}^{\mathrm{st},0}$ transforms to the Boltzmann collision integral if the pair correlation functions in the zero-order ε-expansion vanish and if one can ignore, at the r_b-scale, the explicit effect, on a given trial particle, of the self-consistent force of internal origin. We shall address this point in more detail below, when discussing the relationship between the generalized Boltzmann equations and alternative derivations of kinetic equations. The zero-order two-particle distribution function entering the Boltzmann collision integral is calculated at the λ-scale and is presented, as usual, as a product of zero-order one-particle functions; this means that interacting particles are not correlated prior to a collision.
(4) The use of this representation makes it possible to express the collision integral $\hat{J}^{\mathrm{st},0}$ in the Boltzmannian form. The presence of superscript "0" in $\hat{J}^{\mathrm{st},0}$ is physically meant that even though the variation of the distribution function on the r_b-scale is taken into account [the first term on the right-hand side of Eq. (1.3.58)], the form of the Boltzmann collision integral containing the function f_1^0 remains unchanged.

(5) It is crucial that the term $D_1 \hat{f}_1^1 / D\hat{t}_B$ in Eq. (1.3.58), accounting for the variation of the distribution function on the r_b-scale, is of the same order of magnitude as the λ- and L-scale terms. This has nothing to do with whatever approximations for $D_1 \hat{f}_1^1 / D\hat{t}_B$ may later be made to break the Bogolyubov chain. The (unjustified) formal neglect of the term $D_1 \hat{f}_1^1 / D\hat{t}_B$ reduces Eq. (1.3.58) to the Boltzmann equation. This means, in turn, that the r_b-scale distribution function is left out of consideration in the Boltzmann kinetic theory; particles featuring in the Boltzmann kinetic theory are point-like and structureless. The system can be described in terms of the independent variables **r**, **p**, t, and the change in the distribution function due to collisions is instantaneous and is accounted for by the source term $\hat{J}^{\mathrm{st},0}$.

We intend to employ Eq. (1.3.58) for describing the evolution of the distribution function $\hat{f}_1^0$ on λ- and L-scales. But kinetic equation (1.3.58) contains the linking term $D_1 \hat{f}_1^1 / D\hat{t}_b$ with respect to the upper index, implying that in order to employ the kinetic equations, a problem arises concerning the approximation of this term, in a definite sense, analogous to the problem, which led us to approximations (1.3.9) or (1.3.51).

We now proceed to break the Bogolyubov chain on the r_b-scale with respect to the superscript in $D_1 \hat{f}_1^1 / D\hat{t}_b$. This term allows the exact representation, using the series (1.3.26).

$$\frac{D_1 \hat{f}_1^1}{D\hat{t}_b} = \frac{D_1}{D\hat{t}_b}\left[\frac{\partial \hat{f}_1}{\partial \varepsilon}\right]_{\varepsilon=0}. \tag{1.3.61}$$

Note, however, that in the "field" description, the distribution function f_1 in the interaction on the r_b-scale depends on ε through the dynamical variables **r**, **v**, t, interrelated by the laws of classical mechanics. We can therefore use the approximation

$$\begin{aligned}
\frac{D_1}{D\hat{t}_b}\left[\left(\frac{\partial \hat{f}_1}{\partial \varepsilon}\right)_{\varepsilon=0}\right] &\cong \frac{D_1}{D(-\hat{t}_b)}\left[\frac{\partial \hat{f}_1}{\partial(-\hat{t}_b)}\left(\frac{\partial(-\hat{t}_b)}{\partial \varepsilon}\right)_{\varepsilon=0}\right. \\
&\quad \left. + \frac{\partial \hat{f}_1}{\partial \hat{\mathbf{r}}_b}\cdot\frac{\partial \hat{\mathbf{r}}_b}{\partial(-\hat{t}_b)}\left(\frac{\partial(-\hat{t}_b)}{\partial \varepsilon}\right)_{\varepsilon=0} + \frac{\partial \hat{f}_1}{\partial \hat{\mathbf{v}}_b}\cdot\frac{\partial \hat{\mathbf{v}}_b}{\partial(-\hat{t}_b)}\left(\frac{\partial(-\hat{t}_b)}{\partial \varepsilon}\right)_{\varepsilon=0}\right] \\
&= -\frac{D_1}{D\hat{t}_b}\left[\left(\frac{\partial \hat{t}_b}{\partial \varepsilon}\right)_{\varepsilon=0}\frac{D_1 \hat{f}_1}{D\hat{t}_b}\right] \cong -\frac{D_1}{D\hat{t}_b}\left[\left(\frac{\partial \hat{t}_b}{\partial \varepsilon}\right)_{\varepsilon=0}\frac{D_1 \hat{f}_1^0}{D\hat{t}_b}\right].
\end{aligned} \tag{1.3.62}$$

The approximation introduced here proceeds against the course of time and corresponds to the condition that there be no correlations as to $t_0 \to -\infty$, where there is some instant of time t_0 on the r_b-scale at which the particles start to interact with each other. In the Boltzmann kinetic theory, the condition of correlation weakening has the form (Bogolyubov, 1946)

$$\lim_{t_0 \to -\infty} W_2\big[\mathbf{r}_1 - \mathbf{v}_1(t - t_0), \mathbf{v}_1; \mathbf{r}_2 - \mathbf{v}_2(t - t_0), \mathbf{v}_2; t_0\big] = 0, \tag{1.3.63}$$

where W_2 is the pair correlation function. As $t_0 \to -\infty$ (but not for $t_0 \to +\infty$!), the condition (1.3.63) of correlation weakening, together with the approximation (1.3.62), single out a time direction and lead to the time irreversibility in real physical processes (Alexeev, 1995c). The next section of this paper discusses this point in detail in connection with the proof of the generalized H-theorem. In the sequel, we will turn back to the chain of relations (1.3.62) discussing this approximation from disparate positions.

Return to the dimensional form of Eq. (1.3.44) taking into account the new normalization condition for the DF f_1, i.e.,

$$\tilde{\tilde{f}}_\delta = \tilde{f}_1 \frac{N_\delta}{N}. \tag{1.3.64}$$

Then

$$\int \tilde{\tilde{f}}_\delta \, \mathbf{dv}_\delta \, \mathbf{dr} = \int \tilde{f}_1 \, \mathbf{dv} \, \mathbf{dr} \frac{N_\delta}{N} = N_\delta, \tag{1.3.65}$$

or

$$\int \tilde{\tilde{f}}_\delta \, \mathbf{dv}_\delta = n_\delta, \tag{1.3.66}$$

and

$$\int n_\delta \, \mathbf{dr} = N_\delta. \tag{1.3.67}$$

The DF normalization introduced for multi-component mixture allows us to treat the function $\tilde{\tilde{f}}_\delta$ as the mathematical expectation of a number of particles (species δ) for which centers of mass at the time moment t are localed in the unit volume in the vicinity of $\mathbf{r}$, and velocities belong to the unit interval in the vicinity of $\mathbf{v}_\delta$. This condition will be used in what follows as a rule, and as a consequence of this fact, we omit the upper sign $\approx$. But the lower index "1" in the DF f_1 corresponds not only to the notation of the one-particle DF, but also to the number of particle among all the numbered N particles of the physical multi-component system considered. The lower index will be also omitted. As a result, for the η-species mixture, we have

$$\int f_\alpha \, d\mathbf{v}_\alpha = n_\alpha \quad (\alpha = 1, \ldots, \eta). \tag{1.3.68}$$

Let a particle, indicated as number one in the system, belong to species α in the mixture. It should be reflected by the investigation of the physical sense of the parameter $(\partial \hat{t}_b / \partial \varepsilon)_{\varepsilon=0}$ in approximation (1.3.62). Introduce the corresponding dimensional parameter

$$\tau_{1 \in N_\alpha} = \frac{\varepsilon}{[\partial \varepsilon / \partial t]_{\varepsilon=0}}, \tag{1.3.69}$$

where ε is the number of particles of all kinds that find themselves within the interaction volume of a particle numbered as 1 by the instant of time t; introducing ε^{eq} (the "equilibrium" particle density in the interaction volume), Eq. (1.3.69) can be written in a typical relaxation form

$$\frac{\partial \varepsilon}{\partial t} = -\frac{\varepsilon(t) - \varepsilon^{\mathrm{eq}}}{\tau_\alpha}. \tag{1.3.70}$$

The denominator in Eq. (1.3.69) is interpreted as the number of particles that find themselves within the interaction volume of a certain particle belonging to the αth component per unit time; the derivative is calculated under the additional condition $\varepsilon = 0$.

Clearly, this number is equal to the number of collisions occurring in the interaction volume per unit time. Hence, the parameter $\tau_{1 \in N_\alpha} = \varepsilon / [\partial \varepsilon / \partial t]_{\varepsilon=0}$ is the mean time *between* collisions of a particle of the αth sort with particles of all other sorts, and will be specified in the following as τ_α.

Reverting now to the dimensional form of Eq. (1.3.44) which becomes

$$\frac{Df_\alpha}{Dt} - \frac{D}{Dt}\left(\tau_\alpha \frac{Df_\alpha}{Dt}\right) = \sum_{\beta=1}^{\eta} \int \left[f'_\alpha f'_\beta - f_\alpha f_\beta\right] g_{\beta\alpha} b \, \mathrm{d}b \, \mathrm{d}\varphi \, \mathrm{d}\mathbf{v}_\beta$$
$$(\alpha = 1, \ldots, \eta), \tag{1.3.71}$$

where

$$\frac{D}{Dt} = \frac{\partial}{\partial t} + \mathbf{v}_\alpha \cdot \frac{\partial}{\partial \mathbf{r}} + \mathbf{F}_\alpha \cdot \frac{\partial}{\partial \mathbf{v}_\alpha}. \tag{1.3.72}$$

Here $g_{\beta\alpha}$ is the relative velocity of the colliding particles (α and β), b is the impact parameter, and φ is the azimuth angle.

It is extremely important that the parameter τ_α is the mean time *between* the successive collisions of the α particle with particles of all kinds, defined by

$$\tau_\alpha = \frac{n_\alpha}{\sum_{\beta=1}^{\eta} N_{\alpha\beta}}. \tag{1.3.73}$$

The number $N_{\alpha\beta}$ of collisions between particles of α and β sorts per unit volume in a unit time is calculated using the functions f_α and f_β. For the Maxwellian distribution functions (Chapman and Cowling, 1952; Hirschfelder, Curtiss and Bird, 1954)

$$N_{\alpha\beta} = 2 n_\alpha n_\beta \sigma_{\alpha\beta}^2 \left(\frac{2\pi k_B T}{m_{\alpha\beta}}\right)^{1/2}, \tag{1.3.74}$$

where σ_α is the diameter of the particle α and $m_{\alpha\beta}$ is the reduced mass $m_{\alpha\beta} = m_\alpha m_\beta / (m_\alpha + m_\beta)$. For the model of rigid spheres,

$$\sigma_{\alpha\beta} = \frac{1}{2}(\sigma_\alpha + \sigma_\beta). \tag{1.3.75}$$

For an arbitrary DF, the total number of collisions per units of volume and time for molecules of α and β species can be found as follows

$$N_{\alpha\beta} = \int f_\alpha f_\beta g_{\alpha\beta} b \, \mathrm{d}b \, \mathrm{d}\varphi \, \mathrm{d}\mathbf{v}_\alpha \, \mathrm{d}\mathbf{v}_\beta. \tag{1.3.76}$$

The *generalized Boltzmann kinetic equation* (GBE) (1.3.71) involves the additional integral parameter defined by the same DF.

In the hydrodynamic limit when the Knudsen numbers $Kn_\alpha = l_\alpha / L$ (l_α is the mean free path between collisions for the αth particles) are small, the mean time of the free path τ_α can be expressed as a function of dynamical viscosity μ_α of species α. Within the hard-sphere model, the first (Maxwellian) approximation yields (Chapman and Cowling, 1952)

$$\tau^{(0)} p = \Pi \mu, \tag{1.3.77}$$

where p is the static pressure, and μ denotes the dynamic viscosity. Successive approximations connected with Sonine's polynomials lead to a small correction of the coefficient, $\Pi_\tau = 0.786$.

In multi-species gas the approximate relations can be used

$$N_{\alpha\beta} = k_B T \frac{n_\alpha n_\beta}{\mu_{\alpha\beta}}, \tag{1.3.78}$$

$$\tau_\alpha^{-1} = k_B T \sum_\beta n_\beta \mu_{\alpha\beta}, \tag{1.3.79}$$

$$\mu = \sum_\beta n_\beta \tau_\beta = \sum_\beta \frac{n_\beta}{n_1 \mu_{\beta 1}^{-1} + n_2 \mu_{\beta 2}^{-1} + \cdots}. \tag{1.3.80}$$

The considered theory is related to multi-species non-reacting gases. An additional problem arises for an adequate description of the inelastic particle collisions. First of all, the conception of "species" should be scrutinized. In chemical reactions redistribution of atoms is realized. Then if, in the frame of the Liouville description, the motion of every such particle is traced, so the total number of particles in an ensemble of particles is not changing and the Liouville equation can be written in the standard form. This approach is widely applied in statistical physics (see, for example, Semenov, 1984). But, in the kinetic theory, this method leads to difficulties connected with expressions for collision integrals. As a result, another approach is used based on the classical concept of a chemical component. In this case, the inclusion of inelastic collision is realized with the aid of approximate collision integrals satisfying the law of mass conservation in non-relativistic chemical reactions. In fact we do not need any more, because the exactness of those cross-sections of inelastic processes usually is not high.

The elastic collision integral $\hat{J}^{\mathrm{st},0}$ in GBE (1.3.71) contains only a DF of the zeroth order in series by the density parameter ε. It means that, in the generalized Boltzmann kinetic theory (GBKT), the elastic collision integral can be used in the same form as in the classical BKT. This is also true with regard to the forms of inelastic collision integrals. It is significant to note that these affirmations are connected with local collision integrals. Below the effects of non-locality and time-delay will be discussed separately.

Several important remarks, which will also be discussed in detail in the following sections from different points of view:

(1) The generalized Boltzmann equation contains not only second derivatives with respect to time but also mixed (time-velocity and time-coordinate) partial derivatives. Introducing the "averaged" distribution function

$$f^{\alpha} = f - \tau \frac{Df}{Dt} \tag{1.3.81}$$

it assumes the form

$$\frac{Df^{\alpha}}{Dt} = J^{\mathrm{st}}(f) \tag{1.3.82}$$

similar to the Boltzmann equation (I.1). Now it becomes clear that the Boltzmann equation, which does not contain fluctuation terms, is not a closed one, and there is no rigorous solution (to put it mildly) to the closure problem for the system of moment equations in the theory of turbulence, based on hydrodynamical equations derived from the Boltzmann equations.

(2) The parameter τ in the generalized Boltzmann equation can be assigned a clear physical meaning and, unlike the so-called kinetically consistent difference schemes (Chetverushkin, 1999) to be discussed later, does not lead to secular terms.

(3) The generalized Boltzmann equation (GBE) in the dimensionless form is written as

$$\frac{D\hat{f}_{\alpha}}{D\hat{t}} - \frac{D}{D\hat{t}}\left(Kn\hat{\tau}_{\alpha}\frac{D\hat{f}_{\alpha}}{D\hat{t}}\right) = \frac{1}{Kn}\hat{J}_{\alpha}^{\mathrm{st}}. \tag{1.3.83}$$

From this it follows that the second term is of the order of the Knudsen number (Kn) and turns out to dominate the left-hand side of this equation as the Knudsen number increases. Needless to say, this is not going beyond the free-molecular limit of the equation because

$$\frac{D}{Dt}\frac{D\hat{f}_{\alpha}}{D\hat{t}} = 0, \quad \text{for } Kn \to \infty. \tag{1.3.84}$$

The solution of Eq. (1.3.84) is the equation of Knudsen flow

$$\frac{D\hat{f}_{\alpha}}{Dt} = 0, \tag{1.3.85}$$

i.e., the analogue of the Liouville equation for a one-particle distribution function.

(4) Note, however, that the second term in Eq. (1.3.83) cannot be ignored even for small Knudsen numbers because, in that case, Kn acts as a small coefficient of higher derivatives, with an unavoidable consequence that the effect of this term will be strong in some regions. The neglect of formally small terms is equivalent, in particular, to dropping the (small-scale) Kolmogorov turbulence out of consideration.

Consider now possibilities for simplification of GBE.

If cross-sections $\sigma_{\alpha\beta}$ and reduced masses $m_{\alpha\beta}$ are not too different, it is reasonable to suppose that the mean free times τ_α do not depend on the number of species α and to use the mixture viscosity for calculation of τ. For the model of hard spheres and Maxwellian DF it suffices to fulfill the condition (see relations (1.3.73)–(1.3.75)) for arbitrary α

$$\Gamma_\alpha = \sum_{\beta=1}^{\eta} \frac{\sigma_{\alpha\beta}^2}{\sqrt{m_{\alpha\beta}}} = const \quad (\alpha = 1, \ldots, \eta).$$

For example, for mixture $N_2(N1)$ and $O_2(N2)$ we have: $\sigma_1 = 3.7$ Å, $\sigma_2 = 3.5$ Å, (Hirschfelder, Curtiss and Bird, 1954), $\Gamma_1/\Gamma_2 = 1.09$; then the approach $\tau_\alpha = \tau$ is acceptable.

In this case, one obtains from the equation

$$\frac{Df_\alpha}{Dt} - \frac{D}{Dt}\left(\tau_\alpha \frac{Df_\alpha}{Dt}\right) = J_\alpha^{\mathrm{st,el}} + J_\alpha^{\mathrm{st,inel}} \tag{1.3.71$'$}$$

a simplified form

$$\frac{Df_\alpha}{Dt} - \frac{D}{Dt}\left(\tau \frac{Df_\alpha}{Dt}\right) = J_\alpha^{\mathrm{st,el}} + J_\alpha^{\mathrm{st,inel}}. \tag{1.3.86}$$

The right-hand side of (1.3.86) contains integrals of elastic and inelastic collisions written in a symbolic form.

This equation can be simplified after omitting the terms, which are proportional to the logarithm of hydrodynamic quantities. Really

$$\begin{aligned}\frac{Df_\alpha}{Dt} - \frac{D}{Dt}\left(\tau \frac{Df_\alpha}{Dt}\right) &= \frac{Df_\alpha}{Dt} - \tau \frac{D}{Dt}\frac{Df_\alpha}{Dt} - \frac{D\tau}{Dt}\frac{Df_\alpha}{Dt} \\ &= \frac{Df_\alpha}{Dt} - \tau\left(\frac{D}{Dt}\frac{Df_\alpha}{Dt} + \frac{D\ln\tau}{Dt}\frac{Df_\alpha}{Dt}\right) \\ &= \frac{Df_\alpha}{Dt} - \tau\left(\frac{D}{Dt}\frac{Df_\alpha}{Dt} + \frac{D\ln(\mu/p)}{Dt}\frac{Df_\alpha}{Dt}\right).\end{aligned} \tag{1.3.87}$$

Usually, the derivative of the logarithm of hydrodynamic quantities is a close-to-zero quantity, and one can ignore the second term in brackets in (1.3.87) as compared to the first term and write the GBE in the form

$$\frac{Df_\alpha}{Dt} - \tau \frac{D}{Dt}\frac{Df_\alpha}{Dt} = J_\alpha^{\text{st,el}} + J_\alpha^{\text{st,inel}}, \tag{1.3.88}$$

especially if the mean collision times for different components do not differ too much from one another. These are the possibilities for simplifying GBE that stand out and may be used in applications.

There exists a possibility of analytical GBE integrating if the model collision integral can be used. For this aim, we choose the BGK-approximation for the collision integral:

$$J^{\text{st,el}} = \frac{f^{(0)} - f}{k\tau}, \tag{1.3.89}$$

where $f^{(0)}$ is the local equilibrium DF, and $k\tau$ is a time of relaxation ($k \geqslant 1$). Approach (1.3.89) has been widely used in physics of ionized gases for a number of years through the work of Bhatnagar, Gross and Krook (1954).

Let us consider a one-dimensional case in the absence of external forces, $\tau = const$. GBE leads to the equation

$$\frac{\partial f}{\partial t} + v\frac{\partial f}{\partial x} - \tau\left(\frac{\partial^2 f}{\partial t^2} + 2v\frac{\partial^2 f}{\partial x \partial t} + v^2\frac{\partial^2 f}{\partial x^2}\right) = \frac{f^{(0)} - f}{k\tau}. \tag{1.3.90}$$

Write down (1.3.90) in the form

$$a_{11}\frac{\partial^2 f}{\partial x^2} + 2a_{12}\frac{\partial^2 f}{\partial x \partial t} + a_{22}\frac{\partial^2 f}{\partial t^2} + b_1\frac{\partial f}{\partial x} + b_2\frac{\partial f}{\partial t} = \frac{f - f^{(0)}}{k\tau}, \tag{1.3.91}$$

where

$$a_{11} = \tau v^2, \quad a_{12} = \tau v, \quad a_{22} = \tau, \quad b_1 = -v, \quad b_2 = -1.$$

Eq. (1.3.91) can be brought to the canonical form. Since $\Delta = a_{12}^2 - a_{11}a_{22} = 0$, we obtain an equation of parabolic type and the following characteristic equation

$$a_{11}(\mathrm{d}t)^2 - 2a_{12}\,\mathrm{d}t\,\mathrm{d}x + a_{22}(\mathrm{d}x)^2 = 0. \tag{1.3.92}$$

This equation splits into two equations

$$\begin{aligned} \left(\frac{\mathrm{d}t}{\mathrm{d}x}\right)_1 &= \frac{a_{12} + \sqrt{a_{12}^2 - a_{11}a_{12}}}{a_{11}}, \\ \left(\frac{\mathrm{d}t}{\mathrm{d}x}\right)_2 &= \frac{a_{12} - \sqrt{a_{12}^2 - a_{11}a_{12}}}{a_{11}}. \end{aligned} \tag{1.3.93}$$

Because of $\varDelta = 0$, one obtains only one characteristic equation for the parabolic equation

$$\frac{\mathrm{d}t}{\mathrm{d}x} = \frac{a_{12}}{a_{11}} = \frac{1}{v}. \tag{1.3.94}$$

As a result, the following independent variables can be taken for the mentioned transformation:

$$\xi = x - vt, \qquad \zeta = t. \tag{1.3.95}$$

Using (1.3.95) we find:

$$\frac{\partial f}{\partial x} = \frac{\partial f}{\partial \xi}\frac{\partial \xi}{\partial x} + \frac{\partial f}{\partial \zeta}\frac{\partial \zeta}{\partial x} = \frac{\partial f}{\partial \xi} \tag{1.3.96}$$

and analogously the second derivatives

$$\begin{aligned}
&\frac{\partial^2 f}{\partial x^2} = \frac{\partial^2 f}{\partial \xi^2}, \qquad \frac{\partial f}{\partial t} = -v\frac{\partial f}{\partial \xi} + \frac{\partial f}{\partial \zeta}, \\
&\frac{\partial^2 f}{\partial t^2} = v^2\frac{\partial^2 f}{\partial \xi^2} - 2v\frac{\partial^2 f}{\partial \xi \partial \zeta} + \frac{\partial^2 f}{\partial \zeta^2}, \\
&\frac{\partial^2 f}{\partial x \partial t} = -v\frac{\partial^2 f}{\partial \xi^2} + \frac{\partial^2 f}{\partial \xi \partial \zeta}.
\end{aligned} \tag{1.3.97}$$

After substituting these derivatives in (1.3.91), one obtains

$$\tau\frac{\partial^2 f}{\partial \zeta^2} - \frac{\partial f}{\partial \zeta} = \frac{f - f^{(0)}}{k\tau}. \tag{1.3.98}$$

Introducing a deviation $\hat{f} = f - f^{(0)}$, we find

$$\tau\hat{f}'' - \hat{f}' - \frac{1}{k\tau}\hat{f} = 0. \tag{1.3.99}$$

Eq. (1.3.99) has the characteristic equation

$$\tau m^2 - m - \frac{1}{k\tau} = 0, \tag{1.3.100}$$

with the roots

$$m_{1,2} = \frac{1 \pm \sqrt{1 + 4k^{-1}}}{2\tau}.$$

From the physical sense of solution, the root $m = (1 - \sqrt{1 + 4k^{-1}})/(2\tau)$ will be chosen, and we have

$$f - f_0 = C(\xi)\,\mathrm{e}^{-((\sqrt{1+4k^{-1}}-1)/(2\tau))t}. \tag{1.3.101}$$

Obviously, the value $C(\xi)$ is defined for the time moment $t = 0$, and for $k = 1$

$$f = f_0 + C(\xi)\,\mathrm{e}^{-0.618t/\tau}. \tag{1.3.102}$$

Here the exponential growth of one of possible solutions is eliminated by choosing an arbitrary factor – which is equal to zero – in a respective solution. This situation will be discussed from the position of the generalized H-theorem in the next section.

Relations (1.3.101), (1.3.102) rule the process of gas relaxation in the form of a traveling decreasing wave to the local equilibrium, defined by DF $f^{(0)}$.

For the space homogeneous relaxation corresponding to the equation

$$\frac{\partial f}{\partial t} - \tau\frac{\partial^2 f}{\partial t^2} = \frac{f^{(0)} - f}{k\tau}, \tag{1.3.103}$$

one obtains

$$f = f_0 + C(v)\,\mathrm{e}^{-0.618t/\tau} \quad (k = 1), \tag{1.3.104}$$

$$f = f_0 + C(v)\,\mathrm{e}^{-((\sqrt{1+4k^{-1}}-1)/(2\tau))t}. \tag{1.3.105}$$

An analogous equation in the Boltzmann physical kinetics is

$$\frac{\partial f}{\partial t} = \frac{f^{(0)} - f}{k\tau}, \tag{1.3.106}$$

which has the solution indicated by the upper index BE ($k = 1$),

$$f^{\mathrm{BE}} = f_0 + C(v)\,\mathrm{e}^{-t/\tau} \tag{1.3.107}$$

or for $k \neq 1$

$$f^{\mathrm{BE}} = f_0 + C(v)\,\mathrm{e}^{-t/(k\tau)}. \tag{1.3.108}$$

The next solutions are presented for Eqs. (1.3.103) and (1.3.106) as $k = 5$ (the upper index GBE is used now and further for solution of the generalized Boltzmann equation):

$$f^{\mathrm{GBE}} = f_0 + C(v)\,\mathrm{e}^{-0.17t/\tau},$$

$$f^{\mathrm{BE}} = f_0 + C(v)\,\mathrm{e}^{-0.2t/\tau}.$$

In this case, both GBE and BE solutions are very close to each other.

1.4. Generalized Boltzmann H-theorem and the problem of irreversibility of time

The Boltzmann H-theorem is of principal importance for the kinetic theory and it provides, in fact, the kinetic substantiation of the theory. The generalized H-theorem was proven in 1992 (see, for example, Alexeev, 1995c); we present below the main fragments of the derivation, discuss it from the standpoint of the problem of irreversibility of time, and generalize to the case of multi-component reacting gas.

Consider first a simple gas consisting of spherical molecules. The state of the gas is assumed to be constant, with external forces absent. Then, the GB-equation is reduced to the form

$$\frac{\partial f}{\partial t} - \frac{\partial}{\partial t}\left(\tau \frac{\partial f}{\partial t}\right) = J^{\mathrm{st}}. \tag{1.4.1}$$

We will introduce the Boltzmann H-function,

$$H = \int f \ln f \, \mathrm{d}\mathbf{v} \tag{1.4.2}$$

multiply both parts of Eq. (1.4.1) by $\ln f$, and transform the equation to

$$\frac{\partial f}{\partial t} \ln f = \frac{\partial}{\partial t}(f \ln f) - \frac{\partial f}{\partial t}, \tag{1.4.3}$$

$$\frac{\partial^2 f}{\partial t^2} \ln f = \frac{\partial^2}{\partial t^2}(f \ln f) - \frac{1}{f}\left(\frac{\partial f}{\partial t}\right)^2 - \frac{\partial^2 f}{\partial t^2}. \tag{1.4.4}$$

From GBE (1.4.1) and Eqs. (1.4.3), (1.4.4) it follows:

$$\begin{aligned}&\frac{\partial}{\partial t}(f \ln f) - \frac{\partial f}{\partial t} - \tau \frac{\partial^2}{\partial t^2}(f \ln f) + \tau \frac{1}{f}\left(\frac{\partial f}{\partial t}\right)^2 + \tau \frac{\partial^2 f}{\partial t^2} \\ &\quad - \ln f \frac{\partial \tau}{\partial t}\frac{\partial f}{\partial t} = J^{\mathrm{st}} \ln f\end{aligned} \tag{1.4.5}$$

or

$$\begin{aligned}&\frac{\partial}{\partial t}(f \ln f) - \tau \frac{\partial^2}{\partial t^2}(f \ln f) + \tau \frac{1}{f}\left(\frac{\partial f}{\partial t}\right)^2 - \frac{\partial \tau}{\partial t}\frac{\partial}{\partial t}(f \ln f) \\ &\quad = J^{\mathrm{st}} \ln f + \frac{\partial f}{\partial t} - \tau \frac{\partial^2 f}{\partial t^2} - \frac{\partial \tau}{\partial t}\frac{\partial f}{\partial t}.\end{aligned} \tag{1.4.6}$$

Using once more (1.4.1), we find

$$\begin{aligned}&\frac{\partial}{\partial t}(f \ln f) - \tau \frac{\partial^2}{\partial t^2}(f \ln f) + \tau \frac{1}{f}\left(\frac{\partial f}{\partial t}\right)^2 - \frac{\partial \tau}{\partial t}\frac{\partial}{\partial t}(f \ln f) \\ &\quad = (1 + \ln f) J^{\mathrm{st}}.\end{aligned} \tag{1.4.7}$$

We now integrate Eq. (1.4.7) term-by-term with respect to all the values of velocities and use the definition of the H-function,

$$\frac{\mathrm{d}H}{\mathrm{d}t} - \tau\frac{\mathrm{d}^2H}{\mathrm{d}t^2} - \frac{\mathrm{d}\tau}{\mathrm{d}t}\frac{\mathrm{d}H}{\mathrm{d}t} = -\tau\int\frac{1}{f}\left(\frac{\partial f}{\partial t}\right)^2\mathrm{d}\mathbf{v} + \int(1+\ln f)J^{\mathrm{st}}\,\mathrm{d}\mathbf{v}. \tag{1.4.8}$$

But the following inequality is valid

$$-\tau\int\frac{1}{f}\left(\frac{\partial f}{\partial t}\right)^2\mathrm{d}\mathbf{v} + \int(1+\ln f)J^{\mathrm{st}}\,\mathrm{d}\mathbf{v} \leqslant 0. \tag{1.4.9}$$

Really, the first integral in (1.4.9) is obviously not positive, for the second integral Boltzmann's transformation can be applied

$$\begin{aligned}\int(1+\ln f)J^{\mathrm{st}}\,\mathrm{d}\mathbf{v} &= \int(1+\ln f)\big(f'f_1' - ff_1\big)gb\,\mathrm{d}b\,\mathrm{d}\varphi\,\mathrm{d}\mathbf{v}\,\mathrm{d}\mathbf{v}_1\\ &= \frac{1}{4}\int\big(1+\ln f + 1 + \ln f_1 - 1 - \ln f' - 1 - \ln f_1'\big)\\ &\qquad\times\big(f'f_1' - ff_1\big)gb\,\mathrm{d}b\,\mathrm{d}\varphi\,\mathrm{d}\mathbf{v}\,d\mathbf{v}_1\\ &= \frac{1}{4}\int\ln\frac{ff_1}{f'f_1'}\big(f'f_1' - ff_1\big)gb\,\mathrm{d}b\,\mathrm{d}\varphi\,\mathrm{d}\mathbf{v}\,\mathrm{d}\mathbf{v}_1.\end{aligned} \tag{1.4.10}$$

The second relation in the chain of Eq. (1.4.10) is obtained by formal re-notation of forward and backward collisions by using the principle of microscopic reversibility which can be written, for this case, in the form

$$\mathrm{d}\mathbf{v}\,\mathrm{d}\mathbf{v}_1 = \mathrm{d}\mathbf{v}'\,\mathrm{d}\mathbf{v}_1'. \tag{1.4.11}$$

Following Boltzmann, we notice that the value $\ln(ff_1/(f'f_1'))$ is positive or negative depending on whether ff_1 are larger than $f'f_1'$ or smaller. In either case, the sign of $\ln(ff_1/(f'f_1'))$ is opposite to the sign of the difference $f'f_1' - ff_1$.

Then, we obtain

$$\frac{\mathrm{d}}{\mathrm{d}t}\left(H - \tau\frac{\mathrm{d}H}{\mathrm{d}t}\right) \leqslant 0. \tag{1.4.12}$$

We introduce the H^a-function in accordance with the definition

$$H^a = H - \tau\frac{\mathrm{d}H}{\mathrm{d}t}. \tag{1.4.13}$$

Then the inequality is valid that yields the conclusion of the generalized H-theorem,

$$\frac{\mathrm{d}H^a}{\mathrm{d}t} \leqslant 0. \tag{1.4.14}$$

If we suppose that τ is constant, not depending on time, then inequality (1.4.14) can be considered as a combination of two principles: Boltzmann's principle

$$\frac{\mathrm{d}H}{\mathrm{d}t} \leqslant 0 \tag{1.4.15}$$

and Prigogine's principle (Prigogine, 1962; Nikolis and Prigogine, 1977)

$$\frac{\mathrm{d}^2 H}{\mathrm{d}t^2} \geqslant 0. \tag{1.4.16}$$

For closed physical systems, the H-function is a limited function, in particular, this function will be restricted from below. In other words, the integral $\int f \ln f \,\mathrm{d}\mathbf{v}$ does not tend to $-\infty$ as $v \to \infty$, i.e., integral converges.

For this aim, consider the integral

$$\int f \frac{1}{2} m v^2 \,\mathrm{d}\mathbf{v} = \frac{1}{2}\rho \overline{v^2}, \tag{1.4.17}$$

where ρ is the density and upper line connected with averaging of the corresponding value.

Integral (1.4.17) is the value of kinetic energy for a unit volume and therefore it is finite. Suppose that the function f decreases as $v \to \infty$ faster than $\mathrm{e}^{-mv^2/(2k_B T)}$, i.e.,

$$f < \mathrm{e}^{-mv^2/(2k_B T)}, \qquad \ln f < -\frac{mv^2}{2k_B T}.$$

Then $\int f \ln f \,\mathrm{d}\mathbf{v} > -\infty$.

If f decreases as $v \to \infty$ slower than $\mathrm{e}^{-mv^2/(2k_B T)}$, i.e.,

$$-\ln f < \frac{mv^2}{2k_B T},$$

the convergence of the integral $\int f \ln f \,\mathrm{d}\mathbf{v} > -\infty$ is defined by means of (1.4.17). Really, in this case, $-f \ln f < (1/(2k_B T)) m v^2 f$ and the integral of this value should be limited by virtue of (1.4.17), because the kinetic energy of a physical system is limited. In all cases, the integral in (1.4.17) – connected with the kinetic energy of a closed system – is a limited function.

During time evolution the decreasing H-function – and therefore the H^a-function – is limited as $v \to \infty$.

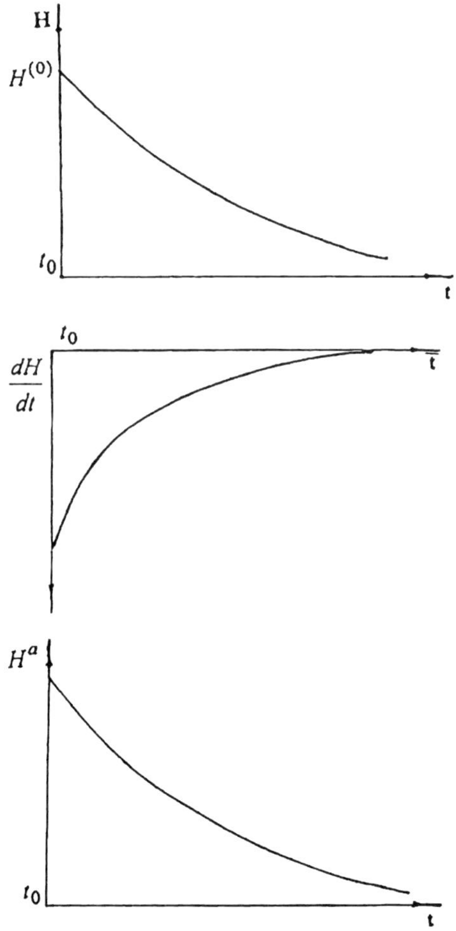

Fig. 1.2. Hypothetical time evolution of the H- and H^a-functions.

At this stage of investigation we can state that the generalized H-function (H^a) is not an increasing function for all hypothetical manners of the H-function behavior.

Let us consider the possible evolution of the H-function from this point of view.

If the H-function is a decreasing function, this function is restricted from below. Consider the time evolution of H- and H^a-functions. The H-function is a monotonically decreasing function, as it is shown in Figure 1.2. Figure 1.2 also shows the dependence $\mathrm{d}H/\mathrm{d}t$ on time, as a result of graphical differentiating, and the time evolution of H^a, as a consequence of these two mentioned graphs and definition (1.4.13). As we see, H^a is also a monotonically decreasing function.

Maybe another situation shown in Figure 1.3 could be realized? This dependence corresponds to the local growth of the H-function. It leads, as a result of graphical constructions, to the growth of H^a, but it is forbidden by inequality (1.4.14).

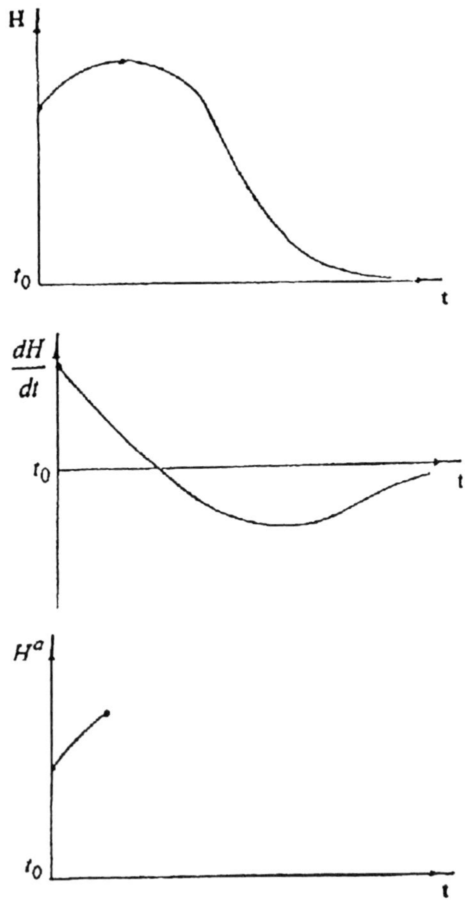

Fig. 1.3. Hypothetical time evolution of the H- and H^a-functions.

One proves (Alexeev, 1995c) that, if

(I) $\mathrm{d}H/\mathrm{d}t \leqslant 0$, then $\mathrm{d}H^a/\mathrm{d}t \leqslant 0$ as well.

If in some portion of the evolution curve

(II) $\mathrm{d}H/\mathrm{d}t > 0$, then $\mathrm{d}H^a/\mathrm{d}t > 0$, as well, which is forbidden by inequality (1.4.14).

The system of inequalities (I), (II) does not forbid the evolution shown in Figure 1.4. In this case, $\mathrm{d}^2H/\mathrm{d}t^2 \geqslant 0$ only in a "linear" region and entropy production ($\sim \mathrm{d}H/\mathrm{d}t$) evolve in a non-monotonic manner.

The possibility of appearance of the exponentially increasing in time H-function in the equilibrium state (if $H^a = 0$) – and infinite growth of energy of a closed system – should be excluded by turning into zero the constant of integration. In this case, the model of the space homogeneous physical system is not correct.

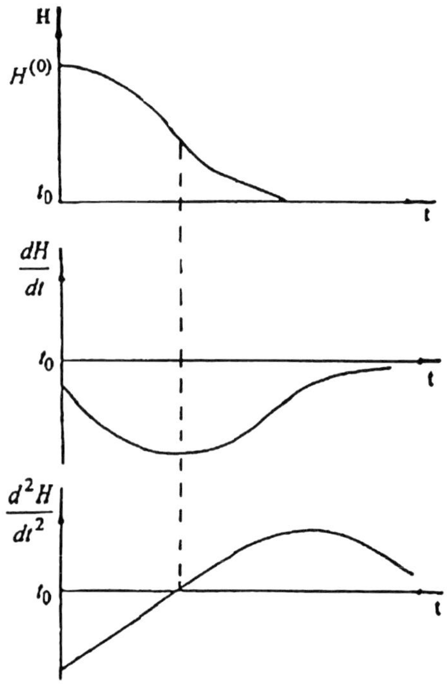

Fig. 1.4. Hypothetical time evolution of the H-function.

Now what happens to the fluctuations that develop in the system? To see this, consider the generalized equation of continuity, which was "guessed" in Introduction and which – as it will be shown in the next chapter – is a direct consequence of GBE in the hydrodynamic limit

$$\frac{\partial}{\partial t}\left\{\rho-\tau\left[\frac{\partial\rho}{\partial t}+\frac{\partial}{\partial\mathbf{r}}\cdot(\rho\mathbf{v}_0)\right]\right\}$$
$$+\frac{\partial}{\partial\mathbf{r}}\cdot\left\{\rho\mathbf{v}_0-\tau\left[\frac{\partial}{\partial t}(\rho\mathbf{v}_0)+\frac{\partial}{\partial\mathbf{r}}\cdot\rho\mathbf{v}_0\mathbf{v}_0+\overleftrightarrow{I}\cdot\frac{\partial p}{\partial\mathbf{r}}-\rho\mathbf{a}\right]\right\}=0. \quad (1.4.18)$$

We write here this equation in the generalized Eulerian formulation under the assumption of no external forces and for the one-dimensional unsteady case:

$$\frac{\partial}{\partial t}\left\{\rho-\tau^{(0)}\left[\frac{\partial\rho}{\partial t}+\frac{\partial}{\partial x}(\rho v_0)\right]\right\}+\frac{\partial}{\partial x}(\rho v_0)$$
$$=\frac{\partial}{\partial x}\left\{\tau^{(0)}\left[\frac{\partial}{\partial t}(\rho v_0)+\frac{\partial}{\partial x}(p+\rho v_0^2)\right]\right\}, \quad (1.4.19)$$

where $\tau^{(0)}$ is the mean time between collisions calculated in the locally Maxwellian approximation: $\tau^{(0)}p=\Pi\mu$ (as indicated, the factor Π being of order unity; for the hard-sphere model, $\Pi = 0.8$ to the first-order approximation in Sonine polynomials (Chapman and Cowling, 1952)).

We assume that except for shock-wave-type regions (to be discussed below within the framework of the generalized Boltzmann equation), hydrodynamical quantities vary not too rapidly on the scale of the order of the mean time between collisions:

$$\frac{\rho}{\tau^{(0)}} \gg \frac{\partial \rho}{\partial t}, \qquad \frac{\rho}{\tau^{(0)}} \gg \frac{\partial}{\partial x}(\rho v_0),$$

the temperature variations are small, the convective transfer is negligible, and the chaotic motion is highly energetic as compared to the kinetic energy of the flow, i.e., $\overline{V^2}/v_0^2 \gg 1$ (for example, for hydrogen at normal pressures and temperatures, we have $v_0 = 10\ \mathrm{cm \cdot s^{-1}}$, giving 3.4×10^8 for this ratio). Consequently, Eq. (1.4.19) becomes

$$\frac{\partial \rho}{\partial t} = \frac{\partial}{\partial x}\left(\frac{\tau^{(0)} p}{\rho}\frac{\partial \rho}{\partial x}\right) \tag{1.4.20}$$

or

$$\frac{\partial \rho}{\partial t} = \frac{\partial}{\partial x}\left(D\frac{\partial \rho}{\partial x}\right), \tag{1.4.21}$$

where $D = \Pi\mu/\rho$ is the self-diffusion coefficient (Hirschfelder, Curtiss and Bird, 1954). Eq. (1.4.21) is the diffusion equation, with the implication that (a) a locally increasing density fluctuation immediately activates the diffusion mechanism which smoothes it out, and (b) the generalized Boltzmann H-theorem proved above ensures that the smoothed fluctuations come to equilibrium.

Relation (1.4.8) can be rewritten as:

$$\frac{\mathrm{d}H^a}{\mathrm{d}t} = -\tau \int \frac{1}{f}\left(\frac{\partial f}{\partial t}\right)^2 \mathrm{d}\mathbf{v} + \frac{1}{4}\int \ln \frac{f f_1}{f' f_1'}\left(f' f_1' - f f_1\right) g b \,\mathrm{d}b \,\mathrm{d}\varepsilon \,\mathrm{d}\mathbf{v} \,\mathrm{d}\mathbf{v}_1. \tag{1.4.22}$$

The second integral on the right-hand side of Eq. (1.4.15) is smaller or equal to zero because of the obvious inequality

$$(b-a)\ln\frac{a}{b} \leqslant 0,$$

where $b = f' f_1'$, $a = f f_1$. Then, in a stationary state the equality

$$f_1' f' = f f_1, \tag{1.4.23}$$

is valid which leads to the Maxwellian distribution function.

If the Boltzmann H-function defined by the distribution function f is obtained from GBE, in the hydrodynamic limit the value $\tau \mathrm{d}H/\mathrm{d}t$ should be considered as fluctuation of the H-function on the Kolmogorov level of the turbulence description; then H^a in the relation

$$H^a = H - \tau \frac{\mathrm{d}H}{\mathrm{d}t} \tag{1.4.24}$$

is an averaged value of the H-function.

For multi-component gas, the analog of Eq. (1.4.1) has the form

$$\frac{\partial f_\alpha}{\partial t} - \frac{\partial}{\partial t}\left(\tau_\alpha \frac{\partial f_\alpha}{\partial t}\right) = J_\alpha^{\mathrm{st,el}}. \tag{1.4.25}$$

As a result, the H-function for the component α is written as

$$H_\alpha = \int f_\alpha \ln f_\alpha \,\mathrm{d}\mathbf{v}_\alpha. \tag{1.4.26}$$

Subsequent mathematics is analogous, and therefore, the inequality assumes the following form:

$$\frac{\mathrm{d}}{\mathrm{d}t}\left(H_\alpha - \tau_\alpha \frac{\mathrm{d}H_\alpha}{\mathrm{d}t}\right) \leqslant 0 \tag{1.4.27}$$

or

$$\frac{\mathrm{d}H_\alpha^a}{\mathrm{d}t} \leqslant 0. \tag{1.4.14$'$}$$

The summation over all the components leads to the H-function for the mixture,

$$\begin{aligned} H &= \sum_\alpha \int f_\alpha \ln f_\alpha \,\mathrm{d}\mathbf{v}_\alpha, \\ H^a &= \sum_\alpha \left(H_\alpha - \tau_\alpha \frac{\mathrm{d}H_\alpha}{\mathrm{d}t}\right) = H - \sum_{\alpha=1}^{\eta} \tau_\alpha \frac{\mathrm{d}H_\alpha}{\mathrm{d}t} \end{aligned} \tag{1.4.28}$$

and to the inequality

$$\frac{\mathrm{d}H^a}{\mathrm{d}t} \leqslant 0,$$

which results from (1.4.14$'$).

Let chemical reactions proceed in a multi-component gas mixture, for which the integral of bimolecular collisions has the form (Alekseev, 1982; Alekseev and Grushin, 1994) $(\beta, \gamma, \delta = 1, \ldots, \eta)$

$$J_\alpha^{\text{st,nel}} = \frac{1}{2}\sum_r \sum_{\beta,\gamma,\delta} \int \left[\frac{s_\alpha s_\beta}{s_\gamma s_\delta} \left(\frac{m_\alpha m_\beta}{m_\gamma m_\delta} \right)^3 f'_\gamma f'_\delta - f_\alpha f_\beta \right] g_{\alpha\beta}\, \mathrm{d}\sigma_{\alpha\beta}^{\gamma\delta_r}\, \mathrm{d}\mathbf{v}_\beta, \tag{1.4.29}$$

where $\mathrm{d}\sigma_{\alpha\beta}^{\gamma\delta_r}$ is the differential cross-section of inelastic collision in the rth reaction

$$A_\alpha + A_\beta \to A_\gamma + A_\delta,$$

and s_α, s_β, s_γ, s_δ are statistical weights of the energy state of particles A_α, A_β, A_γ, A_δ (Fowler and Guggenheim, 1939). Here GBE has the form:

$$\begin{aligned} \frac{\partial f_\alpha}{\partial t} - \frac{\partial}{\partial t}\left(\tau_\alpha \frac{\partial f_\alpha}{\partial t} \right) &= \sum_j \int [f'_\alpha f'_j - f_\alpha f_j] g_{\alpha j}\, \mathrm{d}\sigma_{\alpha j}^{\alpha j}\, \mathrm{d}\mathbf{v}_j \\ &+ \frac{1}{2}\sum_r \sum_{\beta,\gamma,\delta} \int [\xi_{\alpha\beta}^{\gamma\delta} f'_\gamma f'_\delta - f_\alpha f_\beta] g_{\alpha\beta}\, \mathrm{d}\sigma_{\alpha\beta}^{\gamma\delta_r}\, \mathrm{d}\mathbf{v}_\beta. \end{aligned} \tag{1.4.30}$$

Now, the analog of (1.4.8), after summation over α, is written as

$$\begin{aligned} &\sum_\alpha \left[\frac{\mathrm{d}H_\alpha}{\mathrm{d}t} - \tau_\alpha \frac{\mathrm{d}^2 H_\alpha}{\mathrm{d}t^2} - \frac{\mathrm{d}\tau_\alpha}{\mathrm{d}t}\frac{\mathrm{d}H_\alpha}{\mathrm{d}t} \right] \\ &= -\sum_\alpha \tau_\alpha \int \frac{1}{f_\alpha}\left(\frac{\partial f_\alpha}{\partial t} \right)^2 \mathrm{d}\mathbf{v}_\alpha + \sum_\alpha \int (1 + \ln f_\alpha) J_\alpha^{\text{st,el}}\, \mathrm{d}\mathbf{v}_\alpha \\ &\quad + \frac{1}{2} \sum_{r,\beta,\gamma,\delta,\alpha} \int (1 + \ln f_\alpha)\big(\xi_{\alpha\beta}^{\gamma\delta} f'_\gamma f'_\delta - f_\alpha f_\beta\big) g_{\alpha\beta}\, \mathrm{d}\sigma_{\alpha\beta}^{\gamma\delta_r}\, \mathrm{d}\mathbf{v}_\alpha\, \mathrm{d}\mathbf{v}_\beta. \end{aligned} \tag{1.4.31}$$

Obviously, in the integral sum on the right-hand side of (1.4.31) indices α, β, γ, δ are dummy and the transformation is valid:

$$\begin{aligned} &\sum_r \sum_{\alpha,\beta,\gamma,\delta} \int (1 + \ln f_\alpha)\big(\xi_{\alpha\beta}^{\gamma\delta} f'_\gamma f'_\delta - f_\alpha f_\beta\big) g_{\alpha\beta}\, \mathrm{d}\sigma_{\alpha\beta}^{\gamma\delta_r}\, \mathrm{d}\mathbf{v}_\alpha\, \mathrm{d}\mathbf{v}_\beta \\ &= \frac{1}{4}\sum_r \sum_{\alpha,\beta,\gamma,\delta} \int \ln \frac{f_\alpha f_\beta}{f_\gamma f_\delta} \big(\xi_{\alpha\beta}^{\gamma\delta} f'_\gamma f'_\delta - f_\alpha f_\beta\big) g_{\alpha\beta}\, \mathrm{d}\sigma_{\alpha\beta}^{\gamma\delta_r}\, \mathrm{d}\mathbf{v}_\alpha\, \mathrm{d}\mathbf{v}_\beta. \end{aligned} \tag{1.4.32}$$

We use the principle of microscopic reversibility

$$\xi_{\alpha\beta}^{\gamma\delta} g_{\alpha\beta}\, \mathrm{d}\sigma_{\alpha\beta}^{\gamma\delta_r}\, \mathrm{d}\mathbf{v}_\alpha\, \mathrm{d}\mathbf{v}_\beta = g'_{\gamma\delta}\, \mathrm{d}\sigma_{\gamma\delta}^{\alpha\beta_r}\, \mathrm{d}\mathbf{v}'_\gamma\, \mathrm{d}\mathbf{v}'_\delta \tag{1.4.33}$$

and once again arrive at the formulation of the H-theorem

$$\frac{dH^a}{dt} \leqslant 0, \qquad H^a = H - \sum_{\alpha=1}^{\eta} \tau_\alpha \frac{dH_\alpha}{dt}, \tag{1.4.34}$$

although, as we can see, inequality (1.4.14′) may prove invalid in the presence of chemical reactions.

In thermodynamics for equilibrium systems, entropy is introduced by the relation

$$S = -k_B \int w(\Omega_1, \ldots, \Omega_N) \ln w(\Omega_1, \ldots, \Omega_N)\, d\Omega_1 \cdots d\Omega_N, \tag{1.4.35}$$

where $w(\Omega_1, \ldots, \Omega_N)$ is the probability density for canonical distribution. For ideal gas using the assumption about the statistical independence of molecules, the simplification of (1.4.34) can be realized with the aid of the probability density

$$W_s(\Omega_s) = \int w(\Omega_1, \ldots, \Omega_N)\, d\Omega_1 \cdots d\Omega_{s-1}\, d\Omega_{s+1} \cdots d\Omega_N, \tag{1.4.36}$$

defining the appearance of the sth particle in the state Ω_s, when other particles occupy arbitrary states admitted by the physical system.

Then

$$w(\Omega_1, \ldots, \Omega_N) = \prod_{s=1}^{N} W_s(\Omega_s) \tag{1.4.37}$$

and, after substituting in (1.4.35), we obtain

$$\begin{aligned} S &= -k_B \sum_{s=1}^{N} \int \prod_{k=1}^{N} W_k(\Omega_k) \ln W_s(\Omega_s)\, d\Omega_1 \cdots d\Omega_N \\ &= -k_B \sum_{s=1}^{N} \int W_s(\Omega_s) \ln W_s(\Omega_s)\, d\Omega_s. \end{aligned} \tag{1.4.38}$$

Relation (1.4.37) is written for unit normalization of W_k. If we pass over to the one-particle distribution functions $f(\mathbf{r}, \mathbf{v}, t)$ used by us, then, with an accuracy within the non-principal constant S_0 that is associated only with the level of entropy count, we have the classical relation

$$S = -k_B H + S_0, \tag{1.4.39}$$

which leads to the thermodynamic inequality

$$\frac{dS}{dt} \geqslant 0. \tag{1.4.40}$$

We will now investigate the thermodynamic inequality (1.4.40) from the standpoint of the existing causal relations and direction of time. To this end, one needs to answer the question of how it happened that inequality (1.4.34), which is the one leading to the increase in entropy (1.4.40) and to the existence of irreversible processes, appeared in our generalized Boltzmann physical kinetics. This effect is a direct result of approximation (1.3.62), into which the motion in the direction opposite to the "time arrow" was introduced, so that the state of the system at the given moment of time is defined in a determinate manner by collisions that occurred in the past.

We will introduce the physical principle of causality as some operator which "cuts out", from all the events possible at the present moment of time, only a certain event whose causes exist in the past, and which transfers the certain event under consideration in the present into the class of causal relations for some possible event in the future. Thereby, the irreversibility of time is introduced as well. In other words, one cannot speak of the principle of causality without using the concept of irreversibility of time.

What may be the result of formal rejection of the principle of causality in this particular case? If one abandons the additional statement that the cause precedes the effect, τ in relation (1.4.8) may be replaced by $(-\tau)$,

$$\frac{\mathrm{d}H^{a\prime}}{\mathrm{d}t} - \tau\int\frac{1}{f}\left(\frac{\partial f}{\partial t}\right)^2 \mathrm{d}\mathbf{v} = \frac{1}{4}\int \ln\frac{ff_1}{f'f_1'}(f'f_1' - ff_1)gb\,\mathrm{d}b\,\mathrm{d}\varphi\,\mathrm{d}\mathbf{v}\,\mathrm{d}\mathbf{v}_1, \tag{1.4.41}$$

where

$$H^{a\prime} = H + \tau\frac{\mathrm{d}H}{\mathrm{d}t} \tag{1.4.42}$$

or, to put it differently,

$$\frac{\mathrm{d}H^{a\prime}}{\mathrm{d}t} - \tau\int\frac{1}{f}\left(\frac{\partial f}{\partial t}\right)^2 \mathrm{d}\mathbf{v} \leqslant 0. \tag{1.4.43}$$

Nothing can now be said about the sign of the derivative $\mathrm{d}H^{a\prime}/\mathrm{d}t$ in (1.4.43). Inequality (1.4.43) may also hold in case $\mathrm{d}H^{a\prime}/\mathrm{d}t > 0$, because, from this value, the nonnegative integral $\tau\int(1/f)(\partial f/\partial t)^2\,\mathrm{d}\mathbf{v}$ is subtracted, which does not vary when t is replaced by $(-t)$.

Therefore, the principle of entropy increase follows directly from the principle of irreversibility of time. After introducing the approximation of the two-particle distribution function via the product of one-particle functions and using the fact of correlation of the dynamic variables on the r_b-scale, the reversibility remains or is eliminated from treatment, depending on the approximation of the DF with respect to the hypothetical future or determinate past and, on the formal side, depending on the choice of sign before τ. The probability of reversible processes in closed dissipative physical systems is vanishingly small, and one must use GBE in the form of (1.3.71) for investigating the transport processes in applied problems. However, as regards gigantic (in comparison with human life) time intervals, it is of interest to investigate dissipative and, nevertheless, reversible systems.

In order to understand what the evolution of such a system may look like, consider the particular case of an alternative generalized Boltzmann equation,

$$\frac{Df}{Dt}+\frac{D}{Dt}\left(\tau\frac{Df}{Dt}\right)=J^{\mathrm{st}}(f), \tag{1.4.44}$$

which corresponds to the evolution of a one-dimensional non-stationary system with the collision integral in the BGK form. The appropriate equation has the form

$$\frac{\partial f}{\partial t}+v\frac{\partial f}{\partial x}+\tau\left(\frac{\partial^2 f}{\partial t^2}+2v\frac{\partial^2 f}{\partial x\partial t}+v^2\frac{\partial^2 f}{\partial x^2}\right)=\frac{f^{(0)}-f}{k\tau}. \tag{1.4.45}$$

Eq. (1.4.45), like analogous equation (1.3.90), is parabolic and, with the aid of transformation $\xi=x-vt$, $\zeta=t$, reduces to the equation

$$\tau\frac{\partial^2\hat f}{\partial\zeta^2}+\frac{\partial\hat f}{\partial\zeta}+\frac{\hat f}{k\tau}=0, \tag{1.4.46}$$

where $\hat f=f-f^{(0)}$.

Its characteristic equation

$$\tau m^2+m+\frac{1}{k\tau}=0, \tag{1.4.47}$$

with the roots

$$m_{1,2}=\frac{-1\pm\sqrt{1-4k^{-1}}}{2\tau}. \tag{1.4.48}$$

It follows from (1.4.48) that, with $k<4$, the characteristic equation permits non-monotonic solutions. We will study one of them at $k=1$,

$$f_{(+\tau)}^{\mathrm{GBE}}=f_0+C(\xi)\,\mathrm{e}^{-0.5t/\tau}\cos\left(0.866\frac{t}{\tau}\right). \tag{1.4.49}$$

Obviously, the value $C(\xi)$ is defined for the time moment $t=0$, and for $k=1$ the corresponding solutions of the BE and (1.3.99) have the form

$$f=f_0+C(\xi)\,\mathrm{e}^{-t/\tau}, \tag{1.4.50}$$

$$f=f_0+C(\xi)\,\mathrm{e}^{-0.618t/\tau}. \tag{1.4.51}$$

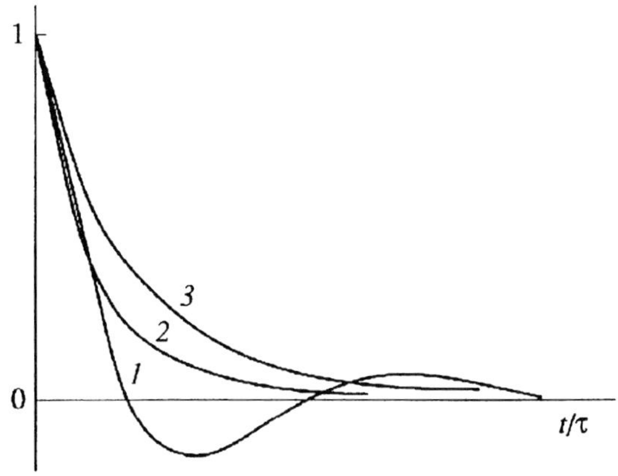

Fig. 1.5. Evolution of the time parts of distribution function (1.4.49)–(1.4.51): (1) $e^{-0.5t/\tau}\cos(0.866t/\tau)$, (2) $e^{-t/\tau}$, (3) $e^{-0.618t/\tau}$.

Figure 1.5 illustrates the evolution of time parts of distribution functions (1.4.51)–(1.4.53), namely, $e^{-0.5t/\tau}\cos(0.866t/\tau)$, $e^{-t/\tau}$, and $e^{-0.618t/\tau}$. It follows from Figure 1.5 that, given a global approximation and the state of thermodynamic equilibrium, the DF may repeatedly assume the same values and, consequently, the system may repeatedly pass through the same states. Moreover, in the process of evolution the system may find itself in the state of thermodynamic equilibrium and, nevertheless, leave the latter state. The respective "apocalyptic" points $A_1, A_2, \ldots$ are indicated in Figure 1.6. What happens at these points? The distribution function reaches its value corresponding to the thermodynamic equilibrium,

$$H = H^{(0)} = \int f^{(0)} \ln f^{(0)} \, \mathrm{d}\mathbf{v}, \tag{1.4.52}$$

however, the derivative

$$\left(\frac{\partial H}{\partial t}\right)_{A_i} = \int \left(\frac{\partial f}{\partial t}\right)_{A_i} \left(\ln f^{(0)} + 1\right) \mathrm{d}\mathbf{v} \tag{1.4.53}$$

is other than zero, and the sign of the integral depends on the sign of the derivative $(\partial f/\partial t)_{A_i}$ $(i = 1, 2, \ldots)$.

Consequently, the Boltzmann entropy may experience vibrations, shown in Figure 1.6, when approaching the state of equilibrium. Figure 1.6 gives, as a result of qualitative graphic differentiation, the evolution of the derivative $\partial H/\partial t$ and that of generalized entropy H^a, which also vibrationally approaches the global equilibrium. We use (1.4.49) to readily find the values of H^a at the apocalyptic points,

$$H^a_{A_i} = H^{(0)} - 0.866\,e^{-0.5t_i/\tau}(-1)^i \int C(v)\left(\ln f^{(0)} + 1\right) \mathrm{d}\mathbf{v}$$
$$(i = 1, 2, \ldots) \tag{1.4.54}$$

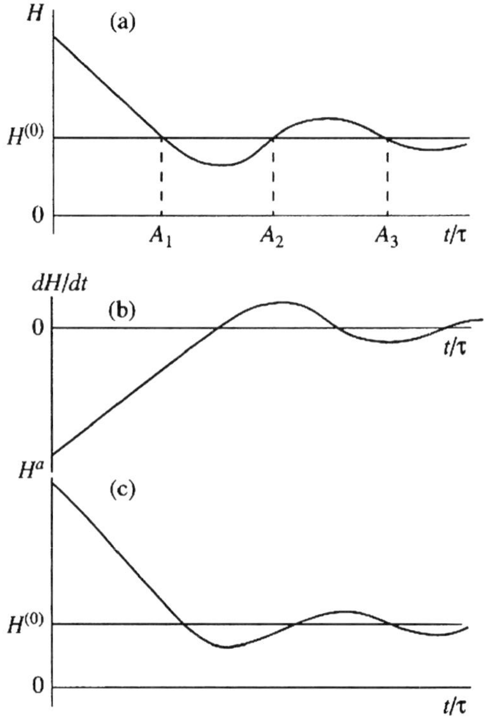

Fig. 1.6. Vibrations of the (a) Boltzmann and (c) generalized Boltzmann H-functions when approaching the state of equilibrium, as well as (b) evolution of derivative $\mathrm{d}H/\mathrm{d}t$ for the kinetic equation of the type of (1.4.44).

because

$$\left(\frac{\partial f}{\partial t}\right)_{A_i} = C\frac{0.866}{\tau}\,\mathrm{e}^{-0.5t_i/\tau}(-1)^i, \tag{1.4.55}$$

$$t_i = \frac{\pi\tau}{0.866}\left(i - \frac{1}{2}\right). \tag{1.4.56}$$

Condition (1.4.56) follows from vanishing of cos $0.866t/\tau$ at the apocalyptic points.

One can see now that the reversibility of processes shows up, on the level of one-particle description as well, as the reflection of reversible processes in a physical system, i.e., on the level of Liouville's equation written relative to the distribution function f_N.

By analogy with the notation of difference schemas with weight, one can replace Eqs. (1.3.71) and (1.4.44) by a generalized kinetic equation in the form

$$\frac{Df}{Dt} - \gamma_1\frac{D}{Dt}\left(\tau\frac{Df}{Dt}\right) + \gamma_2\frac{D}{Dt}\left(\tau\frac{Df}{Dt}\right) = J^{\mathrm{st}}(f), \tag{1.4.57}$$

where γ_1, γ_2 are some function of $\Omega_1, \ldots, \Omega_2$ proportional and correspondingly inversely proportional to T_{ret}, the time of return of the system to the initial state. For real dissipative processes, the time T_{ret} is so great, and the function γ_2 so small, as compared to a unity, that it suffices to use the GBE in the form of (1.3.71).

It is interesting to treat this problem also from the standpoint of the so-called "physical" derivation of the Boltzmann equation. Of course, subsequent reasoning will no longer be rigorous, but it will be nevertheless useful for understanding the situation. For this purpose, we will treat the variation of the number of particles of the sort α, which were initially, at the moment of time t, found in the volume $\mathrm{d}\mathbf{r}^t\, \mathrm{d}\mathbf{v}_\alpha^t$ of the phase space. After the interval of time $\mathrm{d}t$, the particles in the absence of collisions will be found in the volume $\mathrm{d}\mathbf{r}^{t+\mathrm{d}t}\, \mathrm{d}\mathbf{v}_\alpha^{t+\mathrm{d}t}$ and in so doing, the difference

$$f_\alpha\big[\mathbf{r}(t+\mathrm{d}t), \mathbf{v}_\alpha(t+\mathrm{d}t), t+\mathrm{d}t\big]\,\mathrm{d}\mathbf{r}^{t+\mathrm{d}t}\,\mathrm{d}\mathbf{v}_\alpha^{t+\mathrm{d}t} - f_\alpha(\mathbf{r}, \mathbf{v}_\alpha, t)\,\mathrm{d}\mathbf{r}^t\,\mathrm{d}\mathbf{v}_\alpha^t$$

will go to zero. In the presence of external forces $\mathbf{F}_\alpha$ such as Lorentz forces, there is no reason, generally speaking, to assume that the elements of phase volume hold out in the course of time. It can be shown that within the accuracy $\mathrm{O}[(\mathrm{d}t)^2]$ the elements of phase volumes can be transformed as

$$\mathrm{d}\mathbf{r}^{t+\mathrm{d}t} = \left[1 + \frac{\partial}{\partial \mathbf{r}} \cdot \mathbf{F}_\alpha (\mathrm{d}t)^2\right] \mathrm{d}\mathbf{r}^t. \tag{1.4.58}$$

But, in the general case, a six-dimensional Jacobian should be introduced

$$\frac{\mathrm{d}[\mathbf{r}^{t+\mathrm{d}t}, \mathbf{v}_\alpha^{t+\mathrm{d}t}]}{\mathrm{d}[\mathbf{r}^t, \mathbf{v}_\alpha^t]}.$$

Write the balance equation:

$$\begin{aligned} &f_\alpha\left[\mathbf{r} + \mathbf{v}_\alpha\,\mathrm{d}t + \frac{1}{2}\mathbf{F}_\alpha(\mathrm{d}t)^2, \mathbf{v}_\alpha + \mathbf{F}_\alpha\,\mathrm{d}t + \frac{1}{2}\frac{\partial \mathbf{F}_\alpha}{\partial t}(\mathrm{d}t)^2, t+\mathrm{d}t\right] \cdot \frac{\mathrm{d}[\mathbf{r}^{t+\mathrm{d}t}, \mathbf{v}_\alpha^{t+\mathrm{d}t}]}{\mathrm{d}[\mathbf{r}^t, \mathbf{v}_\alpha^t]} \\ &\quad - f_\alpha(\mathbf{r}, \mathbf{v}_\alpha, t) = J_\alpha^{\mathrm{st}}\,\mathrm{d}t, \end{aligned} \tag{1.4.59}$$

containing the terms of $\mathrm{O}[(\mathrm{d}t)^2]$ on the left-hand side of (1.4.59).

Calculation of the indicated Jacobian leads to the result

$$\frac{\mathrm{d}[\mathbf{r}^{t+\mathrm{d}t}, \mathbf{v}_\alpha^{t+\mathrm{d}t}]}{\mathrm{d}[\mathbf{r}^t, \mathbf{v}_\alpha^t]} = 1 + \left[\left(\frac{q_\alpha}{m_\alpha}\right)^2 B^2 - \frac{1}{2}\frac{\partial}{\partial \mathbf{r}} \cdot \mathbf{F}_\alpha\right](\mathrm{d}t)^2, \tag{1.4.60}$$

where q_α is charge of the αth particle, B is magnetic induction, $\mathbf{F}_\alpha$ is the external force, acting on the particle of species α.

Expanding the distribution function as a power series and conserving the terms of order $\mathrm{O}[(\mathrm{d}t)^2]$, the integro-differential difference equation is obtained (Alekseev, 1987, 1993):

$$\begin{aligned}
&\frac{\partial f_\alpha}{\partial t} + \mathbf{v}_\alpha \cdot \frac{\partial f_\alpha}{\partial \mathbf{r}} + \mathbf{F}_\alpha \cdot \frac{\partial f_\alpha}{\partial \mathbf{v}_\alpha} + \tau \left\{ \mathbf{F}_\alpha \cdot \frac{\partial f_\alpha}{\partial \mathbf{r}} + \frac{\partial \mathbf{F}_\alpha}{\partial t} \cdot \frac{\partial f_\alpha}{\partial \mathbf{v}_\alpha} \right.\\
&\quad + \frac{\partial f_\alpha}{\partial \mathbf{v}_\alpha} \mathbf{F}_\alpha : \frac{\partial}{\partial \mathbf{v}_\alpha} \mathbf{F}_\alpha + \frac{\partial f_\alpha}{\partial \mathbf{v}_\alpha} \mathbf{v}_\alpha : \frac{\partial}{\partial \mathbf{r}} \mathbf{F}_\alpha + \frac{\partial^2 f_\alpha}{\partial t^2} + 2\left(\frac{q_\alpha}{m_\alpha}\right) B^2 f_\alpha \\
&\quad - f_\alpha \frac{\partial}{\partial \mathbf{r}} \cdot \mathbf{F}_\alpha + \frac{\partial^2 f_\alpha}{\partial \mathbf{r} \partial \mathbf{r}} : \mathbf{v}_\alpha \mathbf{v}_\alpha + \frac{\partial^2 f_\alpha}{\partial \mathbf{v}_\alpha \partial \mathbf{v}_\alpha} : \mathbf{F}_\alpha \mathbf{F}_\alpha \\
&\quad \left. + 2 \frac{\partial^2 f_\alpha}{\partial \mathbf{v}_\alpha \partial \mathbf{r}} : \mathbf{v}_\alpha \mathbf{F}_\alpha + 2 \frac{\partial^2 f_\alpha}{\partial \mathbf{v}_\alpha \partial t} \mathbf{F}_\alpha + 2 \frac{\partial^2 f_\alpha}{\partial \mathbf{r} \partial t} \cdot \mathbf{v}_\alpha \right\} = J_\alpha^{\mathrm{st}},
\end{aligned} \tag{1.4.61}$$

where $\tau = \Delta t/2$ is half a time step in the finite difference approximation.

Eq. (1.4.61) must be treated as the source of difference-differential approximations for the left-hand side of the BE. Of course, the difference-differential operator in (1.4.61) is not the same as the differential operator of GBE (1.3.71); τ is simply the time step of the difference schema, and with formal growth, secular terms appear on the left-hand side of (1.4.61).

Now, let $F_\alpha \equiv 0$, then Eq. (1.4.47) takes the form

$$\frac{Df_\alpha}{Dt} + \tau \frac{D}{Dt}\frac{Df_\alpha}{Dt} = J_\alpha^{\mathrm{st}}, \tag{1.4.62}$$

where $D/Dt = \partial/\partial t + \mathbf{v}_\alpha \cdot \partial/\partial \mathbf{r}$ and the question of approximation of high accuracy for J_α^{st} formally remains open. In Eq. (1.4.62), τ is a constant with the opposite sign as compared to the analogous term in (1.3.86). Is it possible to derive a difference-differential approximation with the negative sign at the second substantial derivative? No doubt, this is possible.

We will now treat the approximation "backward" in time in the form

$$\begin{aligned}
&f_\alpha(t, \mathbf{r}, \mathbf{v}_\alpha) - f_\alpha(t - \Delta t, \mathbf{r} - \Delta \mathbf{r}, \mathbf{v}_\alpha - \Delta \mathbf{v}_\alpha) \\
&\quad = \left[\frac{Df_\alpha}{Dt}\right]_t \Delta t - \frac{(\Delta t)^2}{2} \left[\frac{D}{Dt}\frac{Df_\alpha}{Dt}\right]_t + \cdots
\end{aligned} \tag{1.4.63}$$

which leads to the balance equation:

$$\frac{Df_\alpha}{Dt} - \tau \frac{D}{Dt}\frac{Df_\alpha}{Dt} = J_\alpha^{\mathrm{st}}, \quad \tau = \frac{\Delta t}{2}. \tag{1.4.64}$$

From the mathematical standpoint, both difference-differential approximations are completely equivalent, and one or the other approximation may be preferred only from the standpoint of stability of the difference schema employed. We will now use the assumptions, which have brought us to the approximate notation of the GBE in the form of (1.3.86). In this case, the difference-differential approximation is the same in these assumptions as in the GBE provided, naturally, that the time step $\Delta t = const$ of the difference schema from the generalized Boltzmann kinetic theory is formally used as a doubled mean time between collisions.

Therefore, two mathematically equivalent difference approximations (1.4.59) and (1.4.63) have a different physical meaning (see (1.3.62)). One of them corresponds to the approximation to "predicted future", and the other one, to "determinate past" which is consistent with the principle of irreversibility of time.

It is significant that this fact is of no importance as regards the "physical derivation" of the Boltzmann equation, because no consideration at all is given to the DF evolution on the r_b-scale, and as a consequence, both approximations lead to one and the same result.

So, Boltzmann's result (treated as the effect of phenomenological derivation of BE) $\mathrm{d}H/\mathrm{d}t \leqslant 0$ may be derived without explicitly using the assumption on the irreversibility of time, while the analogous result in the theory of GBE calls for explicit use of the assumption on the irreversibility of time or, which is the same, of the principle of causality. Thereby, one of the main paradoxes, providing a subject for discussions in the Boltzmann physical kinetics, is resolved.

We will now treat the theory of kinetically consistent difference schemes (KCDS) (see, for example, Elizarova and Chetverushkin, 1986, 1989; Elizarova, 1992). The ideas of this theory date back to the studies by Reitz (in particular, Reitz, 1981), who used the splitting method to solve problems of the theory of transfer to the kinetic and hydrodynamic stages. As distinct from Reitz (1981), the KCDS theory uses the DF expansion in Taylor series with respect to the parameter $v\tau$, where τ is some arbitrary parameter defined by the ratio h of the network step in space to the characteristic hydrodynamic velocity V_{hyd} with the accuracy within third-order terms

$$f^{j+1}(\mathbf{r}, \mathbf{v}, t^{j+1}) = f_0^j - \tau \sum_{\alpha=1}^{3} \frac{\partial f_0^j}{\partial r_\alpha} v_\alpha + \frac{\tau^2}{2} \sum_{\alpha,\beta=1}^{3} \frac{\partial^2 f_0^j}{\partial r_\alpha \partial r_\beta} v_\alpha v_\beta + \cdots. \tag{1.4.65}$$

Here, f_0 is Maxwell's function, and j defines the number of step in time. In recent studies dealing with the KCDS (see, for example, Elizarova, 1992), three parameters have been introduced, $\tau^x = h^x/(2V_{\mathrm{hyd}})$, $\tau^y = h^y/(2V_{\mathrm{hyd}})$, $\tau^z = h^z/(2V_{\mathrm{hyd}})$, which are defined by space steps on the coordinates x, y, and z.

In order to derive the values of gas-dynamic parameters on the new time layer $t = t^{j+1}$, the performed expansion is multiplied by summator invariants and integrated with respect to the velocities of molecules of one-component gas. As a result, we derive a system of integro-differential equations with additional terms (on the right-hand sides, as distinct from the classical hydrodynamic equations), which present, by virtue of selected approximation, a combination of second space derivatives multiplied by the time step.

This approach does not lead to a new hydrodynamic description. Moreover, as it follows from derivation (1.4.61) (see also Alekseev, 1987, 1993), it does not even provide, in the general case, a second-order approximation for the Boltzmann equation or for generalized hydrodynamic equations, which in reality result directly from the GBE. Attempts at substantiating the KCDS on the basis of modified BE with an additional relaxation term are inadequate, because the BE "works" at times of the order of the relaxation time. In particular, with this approximation, as compared to the generalized hydrodynamic equations and GBE,

(a) all cross derivatives with respect to space and time are absent as well as second time derivatives, and, as a result, the KCDS cannot be used correctly to simulate turbulent flows;
(b) the KCDS cannot be used to construct the generalized Navier–Stokes approximation;
(c) external forces cannot be introduced in the KCDS;
(d) it is impossible to clarify the physical meaning of the parameter τ, this leading to the appearance of secular terms in the equations;
(e) it is impossible to estimate the contribution to hydrodynamic equations due to inevitable modification of the collision term;
(f) as noted in Klimontovich (1992), "the common drawback of Elizarova and Chetverushkin (1986, 1989) and Elizarova (1992) consists, in particular, in that the introduced additional terms disturb the invariance of the kinetic equation relative to Galilean transformations. In this case as well, the additional terms were introduced without adequate substantiation."

The following analogy may be drawn: let the finite-difference approximation of Newton's second law $\ddot{x} = F/m$ be written; depending on the accuracy of the scheme used, finite-difference increments of the third and higher orders may appear in the finite-difference approximation. This does not mean, naturally, that the return to differential notation gives a new law of nature of the type of $\ddot{x} + \tau\dddot{x} = F/m$. The reason for this situation is quite obvious: it is impossible to obtain a qualitatively new physical description by using a formally higher difference approximation for the classical equation.

In the approach developed by Klimontovich (1995, 1997), the Liouville equation is replaced by another kinetic equation with a source term (or "priming" term, using the terminology of Klimontovich (1995, p. 319)), that differs from the Liouville equation by the source term (Klimontovich, 1995) written in the τ form as

$$\frac{f_N(r, v, t) - \tilde{f}_N(r', v', t)}{\tau_{\rm ph}}.$$

According to Klimontovich, this term describes the "adjustment" of microscopic distribution of particles to the appropriate smoothed distribution. Henceforward, the value of $\tau_{\rm ph}$ is selected to be equal to that of τ during transition to the one-particle description. The resultant equation proves to be a combination of the Boltzmann and Fokker–Planck description (the differential part of the BE remains unchanged) with an additional "collision integral" (Klimontovich, 1995, p. 251),

$$I_{(R)}(R, v, t) = \frac{\partial}{\partial R}\left[D\frac{\partial f}{\partial R} - bF(R)f\right] \tag{1.4.66}$$

with due regard to smoothing over the "point" dimensions, where D is one of the three kinetic coefficients (kinematic viscosity υ_k, thermal diffusivity χ, and self-diffusion D), and b is the mobility.

It is assumed that all the three coefficients are identical, and the difference between them may be taken into account by using some other, more complex, smoothing function. One can perceive an analogy between the continuity equations of Slezkin (I.20) and Klimontovich (1995). We see little point in discussing the remaining analogous HE.

In fact, the source term in the Liouville equation may only appear in the case of incomplete statistical description of the reacting system, or in the presence of special non-holonomic links and radiation, while the size of the "point" (using Klimontovich terminology) is defined by the r_b-scale in the BE that has been previously not accounted for.

In closing this section we want to emphasize the fundamental point that the introduction of the third scale, which describes the distribution function variations on a time scale of the order of the collision time, leads to the single-order terms in the Boltzmann equation prior to Bogolyubov-chain-decoupling approximations, and to terms proportional to the mean time between collisions after these approximations. It follows that the Boltzmann equation requires a radical modification, which, in our opinion, is exactly what the generalized Boltzmann equation provides.

1.5. Generalized Boltzmann equation and iterative construction of higher-order equations in the Boltzmann kinetic theory

Let us consider the relation between the generalized Boltzmann equation and the iterative construction of higher-order equations in the Boltzmann kinetic theory. Neglecting external forces, the Boltzmann equation for a spatially homogeneous case, with the right-hand side taken in the Bhatnagar–Gross–Krook (BGK) form, is given by

$$\frac{\partial f}{\partial t} = -\frac{f - f_0}{\tau_{\text{rel}}}, \tag{1.5.1}$$

where τ_{rel} is the relaxation time, and f_0 is the equilibrium distribution function. From Eq. (1.5.1) it follows that

$$f = f_0 - \tau_{\text{rel}} \frac{\partial f}{\partial t} \cong f_0 - \tau_{\text{rel}} \frac{\partial f_0}{\partial t}. \tag{1.5.2}$$

The second iteration is constructed in a similar fashion giving

$$f = f_0 - \tau_{\text{rel}} \frac{\partial}{\partial t}\left(f_0 - \tau_{\text{rel}} \frac{\partial f}{\partial t} \right) \cong f_0 - \tau_{\text{rel}} \frac{\partial f_0}{\partial t} + \tau_{\text{rel}}^2 \frac{\partial^2 f_0}{\partial t^2}. \tag{1.5.3}$$

Thus, we obtain for the distribution function the series representation

$$f=\sum_{i=0}^{\infty}(-1)^i\frac{\partial^i f_0}{\partial t^i}\tau_{\text{rel}}^i, \tag{1.5.4}$$

where the zero-order derivative operator corresponds to the distribution function f_0.

From Eq. (1.5.3) there follows an analogue of the kinetic equation (1.5.1) for the second approximation:

$$-\tau_{\text{rel}}\frac{\partial^2 f_0}{\partial t^2}+\frac{\partial f_0}{\partial t}=-\frac{f-f_0}{\tau_{\text{rel}}}. \tag{1.5.5}$$

It is important to note that the second time derivative of the distribution function in Eq. (1.5.5) occurs with a minus sign. In the general case, we have the expansion

$$\sum_{i=1}^{\infty}\tau_{\text{rel}}^{i-1}(-1)^{i-1}\frac{\partial^i f_0}{\partial t^i}=-\frac{f-f_0}{\tau_{\text{rel}}}. \tag{1.5.6}$$

We now proceed to show that the generalized Boltzmann equation permits an iterative procedure similar to that just described. To this end, we can write the second approximation in the form (using the notation of approximation (1.3.62))

$$\begin{aligned}\frac{D_1\hat{f}_1^1}{D\hat{t}_b}&=-\frac{D_1}{D\hat{t}_b}\left(\hat{\tau}\frac{D_1\hat{f}_1}{D\hat{t}_b}\right)\cong-\frac{D_1}{D\hat{t}_b}\left[\hat{\tau}\frac{D_1}{D\hat{t}_b}\left(\hat{f}_1^0+\hat{f}_1^1\right)\right]\\&=-\frac{D_1}{D\hat{t}_b}\left(\hat{\tau}\frac{D_1\hat{f}_1^0}{D\hat{t}_b}+\hat{\tau}\frac{D_1\hat{f}_1^1}{D\hat{t}_b}\right)\\&=-\frac{D_1}{D\hat{t}_b}\left(\hat{\tau}\frac{D_1\hat{f}_1^0}{D\hat{t}_b}\right)+\frac{D_1}{D\hat{t}_b}\left[\hat{\tau}\frac{D_1}{D\hat{t}_b}\left(\hat{\tau}\frac{D_1\hat{f}_1^0}{D\hat{t}_b}\right)\right].\end{aligned} \tag{1.5.7}$$

Higher approximations follow the same pattern. Thus, with the notation of relation (1.3.62) we obtain

$$\begin{aligned}\frac{D_1\hat{f}_1^1}{D\hat{t}_b}&=-\frac{D_1}{D\hat{t}_b}\left(\hat{\tau}\frac{D_1\hat{f}_1^0}{D\hat{t}_b}\right)+\frac{D_1}{D\hat{t}_b}\left[\hat{\tau}\frac{D_1}{D\hat{t}_b}\left(\hat{\tau}\frac{D_1\hat{f}_1^0}{D\hat{t}_b}\right)\right]\\&\quad-\frac{D_1}{D\hat{t}_b}\left\{\hat{\tau}\frac{D_1}{D\hat{t}_b}\left[\hat{\tau}\frac{D_1}{D\hat{t}_b}\left(\hat{\tau}\frac{D_1\hat{f}_1^0}{D\hat{t}_b}\right)\right]\right\}+\cdots.\end{aligned} \tag{1.5.8}$$

It is of interest to estimate the accuracy of substitution of the zeroth-order term of the series f_1^0 instead of f_1 in (1.3.62)

$$\frac{D_1}{D\hat{t}_b}\left[\hat{\tau}\frac{D_1\hat{f}_1}{D\hat{t}_b}\right]\cong\frac{D_1}{D\hat{t}_b}\left[\hat{\tau}\frac{D_1\hat{f}_1^0}{D\hat{t}_b}\right]. \tag{1.5.9}$$

For this aim we use the model BGK collision integral:

$$\frac{Df_1}{Dt} = -\frac{f_1 - f_1^0}{\tau_{\rm rel}}. \tag{1.5.10}$$

The exact solution of Eq. (1.5.10) has the form of a non-local integral with the time-delay:

$$f_1(\mathbf{v}, \mathbf{r}, t) = \int_0^\infty f_1^0(\mathbf{v} - \mathbf{F}\delta t, \mathbf{r} - \mathbf{v}\delta t, t - \delta t)\frac{1}{\tau_{\rm rel}} \exp\left(-\frac{\delta t}{\tau_{\rm tel}}\right) \mathrm{d}\delta t, \tag{1.5.11}$$

where δt corresponds to the time-delay.

This result can be easily proved by using the integration in (1.5.11) by parts. Relation (1.5.11) contains a convergent infinite integral as containing exponential function $\exp(-\delta t/\tau_{\rm rel})$, which tends to zero as $\delta t \to \infty$.

In a linear approximation, expanding $f_1^0(\mathbf{v} - \mathbf{F\delta t}, \mathbf{r} - \mathbf{v}\delta t, t - \delta t)$ in powers of δt and neglecting $(\delta t)^2$ and higher powers, we obtain

$$f_1(\mathbf{v}, \mathbf{r}, t) = f_1^0(\mathbf{v}, \mathbf{r}, t) - (\delta t)\frac{Df_1^0(\mathbf{v}, \mathbf{r}, t)}{Dt}. \tag{1.5.12}$$

Comparing (1.5.2) with (1.5.12), we see that in the linear approximation $\delta t \sim \tau_{\rm rel}$. In the next section, these results will be discussed in the general theory of correlation functions.

Return now to a spatially homogeneous system free from forces. We obtain from (1.5.8) with $\tau = const$ that

$$\frac{\partial f_1^1}{\partial t} = \sum_{i=2}^{\infty} \tau^{i-1}(-1)^{i-1}\frac{\partial^i f_1^0}{\partial t^i}. \tag{1.5.13}$$

It follows that, in this particular case, the generalized Boltzmann equation takes the form

$$\sum_{i=2}^{\infty} \tau^{i-1}(-1)^{i-1}\frac{\partial^i f_1^0}{\partial t^i} + \frac{\partial f_1^0}{\partial t} = J^{\rm st,0} \tag{1.5.14}$$

or, collecting the terms on the left, one finds

$$\sum_{i=1}^{\infty} \tau^{i-1}(-1)^{i-1}\frac{\partial^i f_1^0}{\partial t^i} = J^{\rm st,0}. \tag{1.5.15}$$

The analogy between Eqs. (1.5.6) and (1.5.15) is clearly seen.

In solid-state problems, concerning, for example, charge and energy transfer in non-degenerate semiconductors, one solves the Boltzmann equation by iterations for a spatially homogeneous system in the presence of an external electromagnetic field. For the BGK-approximated collision integral, the Boltzmann equation becomes

$$F\frac{\partial f}{\partial v_z} = -\frac{f - f_0}{\tau_{\mathrm{rel}}} \tag{1.5.16}$$

(for a z-directed external force F), and the distribution function is written as

$$f = f_0 - \tau_{\mathrm{rel}} F \frac{\partial f}{\partial v_z}. \tag{1.5.17}$$

In the first approximation, we obtain

$$f = f_0 - \tau_{\mathrm{rel}} F \frac{\partial f_0}{\partial v_z}. \tag{1.5.18}$$

The substitution of Eq. (1.5.18) into the left-hand side of Eq. (1.5.17) yields the second-order approximation

$$f = f_0 - \tau_{\mathrm{rel}} F \frac{\partial f_0}{\partial v_z} + \tau_{\mathrm{rel}}^2 F^2 \frac{\partial^2 f_0}{\partial v_z^2} + \cdots, \tag{1.5.19}$$

provided the external force F, acting on the particle, is velocity-independent. The dots in this equation indicate that the procedure of constructing the series may be continued by this algorithm. From Eq. (1.5.19), the second-order accurate equation is

$$F\frac{\partial f_0}{\partial v_z} - \tau_{\mathrm{rel}} F^2 \frac{\partial^2 f_0}{\partial v_z^2} = -\frac{f - f_0}{\tau_{\mathrm{rel}}}. \tag{1.5.20}$$

This equation turns out to be a particular case of the generalized Boltzmann equation if the system under study is stationary, spatially homogeneous, and if the applied field is sufficiently weak, giving hope for the fast convergence of the mentioned series, in which the corresponding derivatives are taken of the equilibrium distribution function. The representation of the distribution function in a series form, Eqs. (1.5.4) or (1.5.19), is only possible when one uses the BGK model for the Boltzmann collision integral.

Thus, the generalized Boltzmann equation automatically captures the second iteration in the Boltzmann theory for $\tau = \tau_{\mathrm{rel}}$, but it does not, of course, presuppose the fulfillment of all the conditions listed. Note also that the appearance of the minus sign on the right-hand sides of Eqs. (1.5.1) and (1.5.16) in the BGK approximation has a deep physical meaning: this sign makes it possible to prove the H-theorem for the BGK-approximated Boltzmann equation and is related directly to the approximation proceeded against the course of time.

1.6. Generalized Boltzmann equation and the theory of non-local kinetic equations with time delay

It is of interest to examine the relation between the Boltzmann equation and the theory of kinetic equations accounting for time delay effects. We resort to the Bogolyubov equation (1.3.1′) for determining the evolution of the s-particle distribution function in a one-component gas:

$$\frac{\partial f_s}{\partial t} + \sum_{i=1}^{s} \mathbf{v}_i \cdot \frac{\partial f_s}{\partial \mathbf{r}_i} + \sum_{i=1}^{s} \mathbf{F}_i \cdot \frac{\partial f_s}{\partial \mathbf{v}_i} + \sum_{i,j=1}^{s} \mathbf{F}_{ij} \cdot \frac{\partial f_s}{\partial \mathbf{v}_i}$$
$$= -\frac{1}{N} \sum_{i=1}^{s} \sum_{j=s+1}^{N} \mathbf{F}_{ij} \cdot \frac{\partial f_s}{\partial \mathbf{v}_i}. \tag{1.6.1}$$

In writing Eq. (1.6.1), the normalization condition

$$\int f_s \, \mathrm{d}\Omega_1 \cdots \mathrm{d}\Omega_s = N^s \tag{1.6.2}$$

is used and it has also been assumed that the dynamic state of the system is fully described by the phase variables Ω_i.

Introducing the correlation functions W, the two-particle distribution function may be written as

$$f_2(\Omega_1, \Omega_2, t) = f_1(\Omega_1, t) f_1(\Omega_2, t) + W_2(\Omega_1, \Omega_2, t). \tag{1.6.3}$$

On the r_b-scale, the variables Ω_1 and Ω_2 turn out to be correlated, but because of definition (1.6.3) this effect is accounted for by the function W_2. Consequently, in this approach, it is the integral term containing W_2 which must lead to the Boltzmann (or a more general) collision integral. The BBGKY-1 equation has the form

$$\frac{\partial f_1}{\partial t} + \mathbf{v}_1 \cdot \frac{\partial f_1}{\partial \mathbf{r}_1} + \mathbf{F}_1 \cdot \frac{\partial f_1}{\partial \mathbf{v}_1} + \frac{1}{N} \sum_{j=2}^{N} \frac{\partial f_1}{\partial \mathbf{v}_1} \cdot \int \mathbf{F}_{1j} f_1(2) \, \mathrm{d}\Omega_2$$
$$= -\frac{1}{N} \sum_{j=2}^{N} \int \mathbf{F}_{1j} \cdot \frac{\partial W_2}{\partial \mathbf{v}_1} \, \mathrm{d}\Omega_2. \tag{1.6.4}$$

The internal force $\mathbf{F}_1^{(\mathrm{in})}$ exerted on a given particle 1 from the side of particle 2 at its arbitrary location in the phase space may be written as

$$\frac{1}{N} \sum_{j=2}^{N} \int \mathbf{F}_{1j} f_1(2) \, \mathrm{d}\Omega_2 = \mathbf{F}_1^{(\mathrm{in})}. \tag{1.6.5}$$

Here, as usual, symbol "2", the argument of the one-particle distribution function $f_1(2)$, denotes the phase variables of particle 2. For identical particles, one finds

$$\mathbf{F}_1^{(\text{in})} = \frac{N-1}{N}\int \mathbf{F}_{12} f_1(2)\,\mathrm{d}\Omega_2 \cong \int \mathbf{F}_{12} f_1(2)\,\mathrm{d}\Omega_2$$
$$= \int \mathbf{F}_{1j} f_1(j)\,\mathrm{d}\Omega_j, \quad j = 2, 3, \ldots. \tag{1.6.6}$$

If the self-consistent force $\mathbf{F}_1^{(\text{sc},1)}$, acting on a probe particle in the one-particle picture, is introduced as the sum

$$\mathbf{F}_1^{(\text{sc},1)} = \mathbf{F}_1 + \mathbf{F}_1^{(\text{in})}, \tag{1.6.7}$$

of the external force $\mathbf{F}_1$ and the internal force $\mathbf{F}_1^{(\text{in})}$, defined by Eq. (1.6.5), then we arrive at the equation

$$\frac{\partial f_1}{\partial t} + \mathbf{v}_1\cdot\frac{\partial f_1}{\partial \mathbf{r}_1} + \mathbf{F}_1^{(\text{sc},1)}\cdot\frac{\partial f_1}{\partial \mathbf{v}_1} = -\frac{1}{N}\sum_{j=2}^{N}\mathbf{F}_{1j}\cdot\frac{\partial W_2}{\partial \mathbf{v}_1}\,\mathrm{d}\Omega_2. \tag{1.6.8}$$

The BBGKY-2 equation has the form

$$\frac{\partial f_2}{\partial t} + \mathbf{v}_1\cdot\frac{\partial f_2}{\partial \mathbf{r}_1} + \mathbf{v}_2\cdot\frac{\partial f_2}{\partial \mathbf{r}_2} + \mathbf{F}_1\cdot\frac{\partial f_2}{\partial \mathbf{v}_1} + \mathbf{F}_2\cdot\frac{\partial f_2}{\partial \mathbf{v}_2} + \mathbf{F}_{12}\cdot\frac{\partial f_2}{\partial \mathbf{v}_1} + \mathbf{F}_{21}\cdot\frac{\partial f_2}{\partial \mathbf{v}_2}$$
$$= -\frac{1}{N}\sum_{j=3}^{N}\int\left[\mathbf{F}_{1j}\cdot\frac{\partial f_3}{\partial \mathbf{v}_1} + \mathbf{F}_{2j}\cdot\frac{\partial f_3}{\partial \mathbf{v}_2}\right]\mathrm{d}\Omega_3. \tag{1.6.9}$$

We next express the distribution function f_3 in terms of the correlation functions as

$$f_3(\Omega_1,\Omega_2,\Omega_3,t) = f_1(\Omega_1,t)f_1(\Omega_2,t)f_1(\Omega_3,t) + f_1(\Omega_1,t)W_2(\Omega_2,\Omega_3,t)$$
$$+ f_1(\Omega_2,t)W_2(\Omega_1,\Omega_3,t) + f_1(\Omega_3,t)W_2(\Omega_1,\Omega_2,t)$$
$$+ W_3(\Omega_1,\Omega_2,\Omega_3,t) \tag{1.6.10}$$

and apply the theory of correlation functions to obtain the approximation for collision integrals.

ASSUMPTION 1. *The correlation function W_3 may be neglected.*

Using Eq. (1.6.2), Eq. (1.6.10) can be put into the form

$$f_3(1,2,3) = f_1(3)f_2(1,2) + f_1(2)W_2(1,3) + f_1(1)W_2(2,3). \tag{1.6.11}$$

Substituting Eq. (1.6.11) into Eq. (1.6.9) and introducing self-consistent forces in the framework of a two-particle description ($j = 3, 4, 5, \ldots$), viz.

$$\mathbf{F}_1^{(\mathrm{sc},2)} = \mathbf{F}_1 + \mathbf{F}_{12} + \int \mathbf{F}_{1j} f_1(j)\, \mathrm{d}\Omega_j, \quad j = 3, 4, 5, \ldots, \tag{1.6.12}$$

and analogously

$$\mathbf{F}_2^{(\mathrm{sc},2)} = \mathbf{F}_2 + \mathbf{F}_{21} + \int \mathbf{F}_{2j} f_1(j)\, \mathrm{d}\Omega_j, \quad j = 3, 4, 5, \ldots, \tag{1.6.13}$$

we arrive at the equation for $f_2(1,2)$:

$$\begin{aligned}
&\frac{\partial f_2}{\partial t} + \mathbf{v}_1 \cdot \frac{\partial f_2}{\partial \mathbf{r}_1} + \mathbf{v}_2 \cdot \frac{\partial f_2}{\partial \mathbf{r}_2} + \mathbf{F}_1^{(\mathrm{sc},2)} \cdot \frac{\partial f_2}{\partial \mathbf{v}_1} + \mathbf{F}_2^{(\mathrm{sc},2)} \cdot \frac{\partial f_2}{\partial \mathbf{v}_2} \\
&= f_1(2)\left[\frac{\partial f_1(1)}{\partial t} + \mathbf{v}_1 \cdot \frac{\partial f_1(1)}{\partial \mathbf{r}_1} + \mathbf{F}_1^{(\mathrm{sc},1)} \cdot \frac{\partial f_1(1)}{\partial \mathbf{v}_1}\right] \\
&\quad + f_1(1)\left[\frac{\partial f_1(2)}{\partial t} + \mathbf{v}_2 \cdot \frac{\partial f_1(2)}{\partial \mathbf{r}_2} + \mathbf{F}_2^{(\mathrm{sc},1)} \cdot \frac{\partial f_1(2)}{\partial \mathbf{v}_2}\right]
\end{aligned} \tag{1.6.14}$$

making use of the results

$$\frac{\partial f_1(1)}{\partial t} + \mathbf{v}_1 \cdot \frac{\partial f_1(1)}{\partial \mathbf{r}_1} + \mathbf{F}_1^{(\mathrm{sc},1)} \cdot \frac{\partial f_1(1)}{\partial \mathbf{v}_1} = -\int \mathbf{F}_{13} \cdot \frac{\partial W_2(1,3)}{\partial \mathbf{v}_1}\, \mathrm{d}\Omega_3, \tag{1.6.15}$$

$$\frac{\partial f_1(2)}{\partial t} + \mathbf{v}_2 \cdot \frac{\partial f_1(2)}{\partial \mathbf{r}_2} + \mathbf{F}_2^{(\mathrm{sc},1)} \cdot \frac{\partial f_1(2)}{\partial \mathbf{v}_2} = -\int \mathbf{F}_{23} \cdot \frac{\partial W_2(2,3)}{\partial \mathbf{v}_2}\, \mathrm{d}\Omega_3. \tag{1.6.16}$$

In writing Eq. (1.6.14), we have used the following assumption.

ASSUMPTION 2. *The polarization effects leading to the integrals*

$$-\frac{1}{N}\sum_{j=3}^{N}\int\left[\mathbf{F}_{1j} \cdot \frac{\partial}{\partial \mathbf{v}_1}\big(f_1(1) W_2(2,3)\big)\right] \mathrm{d}\Omega_3,$$

$$-\frac{1}{N}\sum_{j=3}^{N}\int\left[\mathbf{F}_{2j} \cdot \frac{\partial}{\partial \mathbf{v}_2}\big(f_1(2) W_2(1,3)\big)\right] \mathrm{d}\Omega_3$$

may be ignored.

We next introduce the substantial derivatives

$$\frac{Df_2(1,2)}{Dt} = \frac{\partial f_2}{\partial t} + \mathbf{v}_1 \cdot \frac{\partial f_2}{\partial \mathbf{r}_1} + \mathbf{v}_2 \cdot \frac{\partial f_2}{\partial \mathbf{r}_2} + \mathbf{F}_1^{(\mathrm{sc},2)} \cdot \frac{\partial f_2}{\partial \mathbf{v}_1}$$

$$+ \mathbf{F}_2^{(\mathrm{sc},2)} \cdot \frac{\partial f_2}{\partial \mathbf{v}_2}, \tag{1.6.17}$$

$$\frac{D_1 f_1(1)}{Dt} = \frac{\partial f_1}{\partial t} + \mathbf{v}_1 \cdot \frac{\partial f_1}{\partial \mathbf{r}_1} + \mathbf{F}_1^{(\mathrm{sc},1)} \cdot \frac{\partial f_1}{\partial \mathbf{v}_1}, \tag{1.6.18}$$

$$\frac{D_2 f_1(2)}{Dt} = \frac{\partial f_1}{\partial t} + \mathbf{v}_2 \cdot \frac{\partial f_1}{\partial \mathbf{r}_2} + \mathbf{F}_2^{(\mathrm{sc},1)} \cdot \frac{\partial f_1}{\partial \mathbf{v}_2} \tag{1.6.19}$$

which, when substituted into Eq. (1.6.14), yield

$$\frac{Df_2(1,2)}{Dt} = f_1(2)\frac{D_1 f_1(1)}{Dt} + f_1(1)\frac{D_2 f_1(2)}{Dt}. \tag{1.6.20}$$

Let us now integrate with respect to time along the phase trajectory in a six-dimensional space:

$$f_2(1,2) = f_{2,0}(1,2) + \int_{t_0}^{t_0+\tau} f_1(2)\frac{D_1 f_1(1)}{Dt}\,\mathrm{d}\tau + \int_{t_0}^{t_0+\tau} f_1(1)\frac{D_2 f_1(2)}{Dt}\,\mathrm{d}t, \tag{1.6.21}$$

where $f_{2,0}(1,2)$ denotes the initial value of the two-particle distribution function.

ASSUMPTION 3. *We resort to the Bogolyubov condition of the weakening of initial correlations corresponding to a certain initial instant of time* t_0 *(see Eq.* (1.3.62)):

$$\lim_{t_0 \to -\infty} W_2\big[\mathbf{r}_1(t_0 - t), \mathbf{v}_1(t_0 - t); \mathbf{r}_2(t_0 - t), \mathbf{v}_2(t_0 - t); t_0 - t\big] = 0. \tag{1.6.22}$$

This condition implies that

(a) *we are dealing with infinite motion in a two-body problem,*
(b) *we may speak of the condition of molecular chaos being fulfilled prior to the collision of particles* 1 *and* 2, *which corresponds to the approximation in Eq.* (1.3.62) *proceeded against the course of time, and*
(c) *Eq.* (1.6.22) *is written on the* r_b*-scale even though no scale is introduced explicitly.*

Because of Assumption 3, Eq. (1.6.21) may be represented in the form

$$f_2(1,2) = f_1\big[\mathbf{r}_1(t_0), \mathbf{v}_1(t_0), t_0\big] f_1\big[\mathbf{r}_2(t_0), \mathbf{v}_2(t_0), t_0\big] + \int_{t_0}^{t_0+\tau} f_1(2)\frac{D_1 f_1(1)}{Dt}\,\mathrm{d}t + \int_{t_0}^{t_0+\tau} f_1(1)\frac{D_2 f_1(2)}{Dt}\,\mathrm{d}t. \tag{1.6.23}$$

ASSUMPTION 4. *The collision of the probe particles,* 1 *and* 2, *is dominated by the forces of their internal interaction, so that (see Eqs.* (1.6.7) *and* (1.6.12)) *one obtains*

$$\mathbf{F}_1^{(\mathrm{sc},2)} = \mathbf{F}_1^{(\mathrm{sc},1)}, \qquad \mathbf{F}_2^{(\mathrm{sc},2)} = \mathbf{F}_2^{(\mathrm{sc},1)}. \tag{1.6.24}$$

Eq. (1.6.23) then becomes

$$f_2(1,2) = f_1\big[\mathbf{r}_1(t_0), \mathbf{v}_1(t_0), t_0\big] f_2\big[\mathbf{r}_2(t_0), \mathbf{v}_2(t_0), t_0\big] + \int_{t_0}^{t_0+\tau} \frac{D_{12}}{Dt}[f_1 f_2]\,\mathrm{d}t. \tag{1.6.25}$$

Integrating by parts, we find

$$\begin{aligned} f_2(1,2) &= f_1\big[\mathbf{r}_1(t_0), \mathbf{v}_1(t_0), t_0\big] f_2\big[\mathbf{r}_2(t_0), \mathbf{v}_2(t_0), t_0\big] \\ &\quad + \tau\left[\frac{D_{12}}{Dt}\big[f_1(\mathbf{r}_1(t_0), \mathbf{v}_1(t_0), t_0) f_1(\mathbf{r}_2(t_0), \mathbf{v}_2(t_0), t_0)\big]\right]_{t=t_0} \\ &\quad - \int_{t_0}^{t_0+\tau} t\frac{D_{12}}{Dt}\frac{D_{12}}{Dt}\big[f_1(1) f_1(2)\big]\,\mathrm{d}t. \end{aligned} \tag{1.6.26}$$

ASSUMPTION 5. *Delay is sufficiently small, so that the linearization in the delay time can be used.*

The sum of the first two terms in Eq. (1.6.26) determines the product $f_1(1)f_2(2)$ at the instant of time t in the linear approximation in τ, the velocities of particles 1 and 2 corresponding to their initial values at time t_0 (taken to be $t_0 = -\infty$ on the r_b-scale).

If we now substitute $f_2(1,2)$ from Eq. (1.6.26) into the BBGKY-1 equation, we obtain

$$\begin{aligned} &\frac{\partial f_1(1)}{\partial t} + \mathbf{v}_1 \cdot \frac{\partial f_1(1)}{\partial \mathbf{r}_1} + \mathbf{F}_1 \cdot \frac{\partial f_1(1)}{\partial \mathbf{v}_1} \\ &\quad = -\int \mathbf{F}_{12} \cdot \frac{\partial}{\partial \mathbf{v}_1}\big[f_1(\mathbf{r}_1, \mathbf{v}_1(-\infty), t) f_1(\mathbf{r}_2, \mathbf{v}_2(-\infty), t)\big]\,\mathrm{d}\Omega_2 \\ &\qquad + \int \mathbf{F}_{12} \cdot \frac{\partial}{\partial \mathbf{v}_1}\left\{\int_{t_0}^{t_0+\tau} t\frac{D_{12}}{Dt}\frac{D_{12}}{Dt}\big[f_1(1) f_1(2)\big]\,\mathrm{d}t\right\}\mathrm{d}\Omega_2. \end{aligned} \tag{1.6.27}$$

The first integral on the right corresponds to the classical form of the Bogolyubov collision integral and can be transformed in the usual manner to the Boltzmann collision integral (Bogolyubov, 1946). The second collision integral accounts for the time delay effect and is amenable to a differential approximation analogous to Eq. (1.3.62). To obtain this approximation, the following assumption is made.

ASSUMPTION 6. *For an arbitrary location of particle* 2 *in the phase space of interacting particles* 1 *and* 2, *the dependence on the integrand inside the braces in the time delay integral*

$$J_2^{\mathrm{st}} = \int \mathbf{F}_{12} \cdot \frac{\partial}{\partial \mathbf{v}_1}\left\{\int_{t_0}^{t_0+\tau} t\frac{D_{12}}{Dt}\frac{D_{12}}{Dt}\big[f_1(1) f_1(2)\big]\,\mathrm{d}t\right\}\mathrm{d}\Omega_2 \tag{1.6.28}$$

is determined by the acting internal force $\mathbf{F}_{12}$ *via the change in the particle velocities.*

This assumption was used by Bogolyubov (see Silin, 1971, p. 203).

From Eq. (1.6.28) we have

$$\begin{aligned}
J_2^{\mathrm{st}} &= \int \mathbf{F}_{12}\cdot\frac{\partial}{\partial\mathbf{v}_1}\left\{\int_{t_0}^{t} t'\frac{D_{12}}{Dt'}\frac{D_{12}}{Dt'}\left[f_1(1)f_1(2)\right]\mathrm{d}t'\right\}\mathrm{d}\Omega_2 \\
&= \int\left(\mathbf{F}_{12}\cdot\frac{\partial}{\partial\mathbf{v}_1}+\mathbf{F}_{21}\cdot\frac{\partial}{\partial\mathbf{v}_2}\right)\left\{\int_{t_0}^{t} t'\frac{D_{12}}{Dt'}\frac{D_{12}}{Dt'}\left[f_1(1)f_1(2)\right]\mathrm{d}t'\right\}\mathrm{d}\Omega_2 \\
&\cong \tau_d\int\frac{D_{12}}{Dt}\frac{D_{12}}{Dt}\left[f_1(1)f_1(2)\right]\mathrm{d}\Omega_2 = \tau_d\frac{D_1}{Dt}\frac{D_1 f_1(1)}{Dt},
\end{aligned}\tag{1.6.29}$$

where Assumption 5 has been used again and an effective delay time τ_d introduced.

Generally speaking, integration with respect to time in Eq. (1.6.29) is "eliminated" by the substantial derivative, which also contains spatial differentiation. However, to the linear approximation in the delay time this contribution is negligible. It can be seen from the relations

$$\begin{aligned}
J_2^{\mathrm{st}} &= \int\left\{\left[\mathbf{F}_{12}\cdot\frac{\partial}{\partial\mathbf{v}_1}+\mathbf{F}_{21}\cdot\frac{\partial}{\partial\mathbf{v}_2}+(\mathbf{v}_2-\mathbf{v}_1)\cdot\frac{\partial}{\partial\mathbf{x}_{21}}\right]\right. \\
&\quad\left.\times\int_{t_0}^{t} t'\frac{D_{12}}{Dt'}\frac{D_{12}}{Dt'}\left[f_1(1)f_1(2)\right]\mathrm{d}t'\right\}\mathrm{d}\Omega_2 \\
&\quad-\int\left\{(\mathbf{v}_2-\mathbf{v}_1)\cdot\frac{\partial}{\partial\mathbf{x}_{21}}\int_{t_0}^{t} t'\frac{D_{12}}{Dt'}\frac{D_{12}}{Dt'}\left[f_1(1)f_1(2)\right]\mathrm{d}t'\right\}\mathrm{d}\Omega_2 \\
&= \tau_d\frac{D_1}{Dt}\frac{D_1 f_1(1)}{Dt} \\
&\quad-\int\left\{(\mathbf{v}_2-\mathbf{v}_1)\cdot\frac{\partial}{\partial\mathbf{x}_{21}}\int_{t_0}^{t} t'\frac{D_{12}}{Dt'}\frac{D_{12}}{Dt'}\left[f_1(1)f_1(2)\right]\mathrm{d}t'\right\}\mathrm{d}\Omega_2 \\
&= \tau_d\frac{D_1}{Dt}\frac{D_1 f_1(1)}{Dt}+\int(\mathbf{v}_2-\mathbf{v}_1)\cdot\frac{\partial}{\partial\mathbf{x}_{21}}\left[\left[f_1(1)f_1(2)\right]-\left[f_1(1)f_1(2)\right]_{t_0}\right. \\
&\quad\left.-\tau_d\frac{D_1}{Dt}\left[f_1(1)f_1(2)\right]_{t_0}\right]\mathrm{d}\Omega_2 \\
&\cong \tau_d\frac{D_1}{Dt}\frac{D_1 f_1(1)}{Dt},
\end{aligned}\tag{1.6.30}$$

where $\mathbf{x}_{21}=\mathbf{r}_2-\mathbf{r}_1$.

Thus, the appearance of the second substantial derivative with respect to time in the generalized Boltzmann equation may be considered as a differential approximation to the time delay integral that emerges in the theory of correlation functions for kinetic equations.

It may appear that the above theory does not require at all that we apply the method of many scales and expand the distribution function in a power series of a small parameter

$\varepsilon = nr_b^3$. However, this is not the case. As we have seen above, the integration on the r_b-scale must be employed anyway, and giving up the ε-expansion of the distribution function, on the other hand, makes it impossible to estimate the value of τ_d. Each of the approaches outlined above actually complements one another and is interrelated with one another. The generalized Boltzmann equation can be treated both from the point of view of a higher-order Boltzmann theory and as a result of differential approximations to the collision integral accounting for time delay effects.

There is another point to be made. From Eq. (1.3.40), the equation for the distribution function f_2^0 accurate to the zeroth order in ε is

$$\frac{\partial \hat{f}_2^0}{\partial \hat{t}_b} + \hat{\mathbf{v}}_{1b} \cdot \frac{\partial \hat{f}_2^0}{\partial \hat{\mathbf{r}}_{1b}} + \hat{\mathbf{v}}_{j\in N_\delta} \cdot \frac{\partial \hat{f}_2^0}{\partial \hat{\mathbf{r}}_{j\in N_\delta, b}} + \widehat{\mathbf{F}}_{1,\, j\in N_\delta} \cdot \frac{\partial \hat{f}_2^0}{\partial \hat{\mathbf{v}}_{1b}}$$
$$+ \widehat{\mathbf{F}}_{j\in N_\delta, 1} \cdot \frac{\partial \hat{f}_2^0}{\partial \hat{\mathbf{v}}_{j\in N_\delta, b}} + \alpha \widehat{\mathbf{F}}_1 \cdot \frac{\partial \hat{f}_2^0}{\partial \hat{\mathbf{v}}_{1b}} + \alpha \widehat{\mathbf{F}}_{j\in N_\delta} \cdot \frac{\partial \hat{f}_2^0}{\partial \hat{\mathbf{v}}_{j\in N_\delta, b}} = 0. \tag{1.6.31}$$

Comparing this with Eqs. (1.6.9) and (1.6.10) shows that the correlation functions, accurate to the zeroth order in ε, are zero and that forces exerted on the colliding particles 1 and 2 from the side of other particles are not considered in the zero-order approximation on the r_b-scale. This result is used for transforming the collision integral $\hat{J}^{\mathrm{st},0}$ to the Boltzmann form in the multiscale method.

We may summarize then by saying that, the derivation of the kinetic equation in the context of the theory of correlation functions for one-particle distribution functions, leads to a kinetic equation of the form

$$\frac{Df}{Dt} = J^B + J^{\mathrm{td}}, \tag{1.6.32}$$

where J^B and J^{td} are the Boltzmann collision integral and the collision integral accounting for time delay effects, respectively.

The popularity of the BGK approximation to the Boltzmann collision integral:

$$J^B = \frac{f^{(0)} - f}{\tau} \tag{1.6.33}$$

is due to the drastic simplifications it affords. Essentially, the generalized Boltzmann physical kinetics offers a local approximation for the second collision integral

$$J^{\mathrm{td}} = \frac{D}{Dt}\left(\tau \frac{Df}{Dt}\right). \tag{1.6.34}$$

Thus, Eq. (1.6.32) in its "simplest" version takes the form

$$\frac{Df}{Dt} = \frac{f^{(0)} - f}{\tau} + \frac{D}{Dt}\left(\tau \frac{Df}{Dt}\right). \tag{1.6.35}$$

Since the ratio of the second to the first term on the right of this equation is $J^{\mathrm{td}}/J^{B} \approx \mathrm{O}(Kn^2)$, Kn being the Knudsen number, it would seem that the second term could be neglected for the small Knudsen numbers in hydrodynamic regime. However, in the transition to the hydrodynamic limit (after multiplying the kinetic equation by the collision invariants and subsequently integrating over velocities), the Boltzmann integral term vanishes, while the second term on the right-hand side of Eq. (1.6.35) gives a single-order contribution in the generalized Navier–Stokes description (let alone the effect of the small parameter of the higher derivative).

A well-known example of a local approximation to a non-local integral term in the kinetic theory is the Enskog theory of transport processes in a dense gas composed of hard spheres. To obtain a local version of the theory, Enskog used the expansion in terms of the small parameter δ/λ, where δ is the molecular diameter, and λ is the mean free path (see Chapman and Cowling, 1952). For example, for hydrogen at normal pressures and temperatures $\delta \approx 3 \times 10^{-8}$ cm and $\lambda \approx 1.1 \times 10^{-5}$ cm, and the resulting $\delta/\lambda = 2.7 \times 10^{-3}$ corresponds to the typical hydrodynamic-valid range of the Knudsen number variation. In the case of the expansion in terms of Kn, δ/λ would imply $L \approx 0.4 \times 10^{-2}$ cm as the characteristic hydrodynamic length whereat the smoothing is proceeding.

Finally we can state that introduction of control volume by the reduced description for ensemble of particles of finite diameters leads to fluctuations of velocity moments in the volume. This fact can be considered in a definite sense as a classical analog of Heisenberg indeterminacy principle of quantum mechanics. Successive application of this consideration leads not only to non-local time-delay effects (connected with molecules which centers of mass are inside of control volume) but also to "ghosts" – particles, which are (at a time moment) partly inside in control volume without presence of their center of mass in this volume. These "pass ahead" effects are squared in τ in Eq. (1.5.8). In rarefied media both effects lead to Johnson's flicker noise observed in 1925 for the first time by J.B. Johnson by the measurement of current fluctuations of thermo-electron emission.

CHAPTER 2

Theory of Generalized Hydrodynamic Equations

2.1. Transport of molecular characteristics

Consider a mixture of gases, which consists of η components. Note $\mathbf{v}_\alpha$ as molecule velocity in an immobile coordinate system. The mean velocity of molecules of α-species is defined by the relation

$$\bar{\mathbf{v}}_\alpha = \frac{1}{n_\alpha} \int \mathbf{v}_\alpha f_\alpha(\mathbf{r}, \mathbf{v}_\alpha, t)\, \mathrm{d}\mathbf{v}_\alpha. \tag{2.1.1}$$

Mean mass-velocity $\mathbf{v}_0$ of the gas mixture is

$$\mathbf{v}_0 = \frac{1}{\rho} \sum_\alpha m_\alpha n_\alpha \bar{\mathbf{v}}_\alpha. \tag{2.1.2}$$

Thermal velocity $\mathbf{V}_\alpha$ of particle is the velocity of this particle in a coordinate system moving with mean mass velocity

$$\mathbf{V}_\alpha = \mathbf{v}_\alpha - \mathbf{v}_0. \tag{2.1.3}$$

Diffusive velocity $\overline{\mathbf{V}}_\alpha$ is mean molecule velocity of α-component in a coordinate system moving with mean mass velocity

$$\overline{\mathbf{V}}_\alpha = \bar{\mathbf{v}}_\alpha - \mathbf{v}_0, \tag{2.1.4}$$

i.e., $\overline{\mathbf{V}}_\alpha$ is mean thermal velocity of α-molecules.

On the whole, if $\psi_\alpha(\mathbf{r}, \mathbf{v}_\alpha, t)$ is arbitrary scalar, vector or tensor function then the mean value of $\psi_\alpha(\mathbf{r}, \mathbf{v}_\alpha, t)$ is noted as $\bar{\psi}_\alpha$ and defined by relation

$$\bar{\psi}_\alpha(\mathbf{r}, t) = \frac{1}{n_\alpha} \int \psi_\alpha(\mathbf{r}, \mathbf{v}_\alpha, t) f_\alpha(\mathbf{r}, \mathbf{v}_\alpha, t)\, \mathrm{d}\mathbf{v}_\alpha. \tag{2.1.5}$$

Now we can introduce diffusive flux $\mathbf{J}_\alpha$ of α-species

$$\mathbf{J}_\alpha = m_\alpha n_\alpha \overline{\mathbf{V}}_\alpha, \quad \alpha = 1, \ldots, \eta. \tag{2.1.6}$$

After summation of the left and right-hand sides of Eq. (2.1.6) over all α ($\alpha = 1, \ldots, \eta$) and using (2.1.2), (2.1.3), one obtains

$$\sum_{\alpha=1}^{\eta} \mathbf{J}_\alpha = \sum_{\alpha=1}^{\eta} \rho_\alpha \overline{\mathbf{V}}_\alpha = \sum_{\alpha=1}^{\eta} \rho_\alpha (\bar{\mathbf{v}}_\alpha - \mathbf{v}_0) = 0. \tag{2.1.7}$$

Consider now in the gas an elementary surface ds, moving with mean mass-velocity $\mathbf{v}_0$ relatively to a chosen immovable coordinate system, and introduce in ds a positive normal direction $\mathbf{n}$. The transport of molecular characteristics across a surface can be found with the help of distribution function (DF) $f_\alpha(\mathbf{r}, \mathbf{v}_\alpha, t)$, which defines the probable position of mass center of molecules in phase space. Let us obtain the number of mass centers of α-molecules crossing this area ds in positive direction $\mathbf{n}$ in time dt. Let Θ denote the angle between the positive direction of $\mathbf{n}$ and vector $\mathbf{V}_\alpha$. Because the own velocity of α-molecules relative to ds is $\mathbf{V}_\alpha$, in time dt the area ds will be crossed by all molecules belonging to the volume $\mathrm{d}\mathbf{r} = V_\alpha \cos\Theta \,\mathrm{d}s\,\mathrm{d}t$ ($\mathrm{d}\mathbf{r} \equiv \mathrm{d}x\,\mathrm{d}y\,\mathrm{d}z$). The number of those particles the velocities of which belong to the interval $\mathbf{V}_\alpha, \mathbf{V}_\alpha + \mathrm{d}\mathbf{v}_\alpha$ is equal to $f_\alpha(\mathbf{r}, \mathbf{v}_\alpha, t) V_\alpha \cos\Theta \,\mathrm{d}\mathbf{V}_\alpha \,\mathrm{d}t\,\mathrm{d}s$. Transportation of mass, momentum and energy is realized by flux of α-molecules across ds. Arbitrary functions $\psi_\alpha(\mathbf{r}, \mathbf{v}_\alpha, t)$ of this kind can be called molecular markers. Flux $\mathrm{d}\Gamma_n^{(+)\psi_\alpha}$ of scalar molecular marker in $\mathbf{n}$-direction is given by

$$\mathrm{d}\Gamma_n^{(+)\psi_\alpha} = \psi_\alpha(\mathbf{r}, \mathbf{v}_\alpha, t) f_\alpha(\mathbf{r}, \mathbf{v}_\alpha, t) V_\alpha \cos\Theta \,\mathrm{d}\mathbf{V}_\alpha \,\mathrm{d}t\,\mathrm{d}s.$$

If $\psi_\alpha(\mathbf{r}, \mathbf{v}_\alpha, t)$ is vector function (momentum, for example), it is convenient to consider the fluxes of scalar components of the mentioned marker ψ_α. Flux $\Gamma_n^{(+)\psi_\alpha}$ is given by integration over all velocities groups for which $V_{\alpha n} > 0$ ($V_{\alpha n}$ is projection of $\mathbf{V}_\alpha$ on normal direction $\mathbf{n}$):

$$\Gamma_n^{(+)\psi_\alpha} = \mathrm{d}s\,\mathrm{d}t \int_{V_{\alpha n}>0} \psi_\alpha f_\alpha V_{\alpha n} \,\mathrm{d}\mathbf{V}_\alpha. \tag{2.1.8}$$

Similarly, the flux $\Gamma_n^{(-)\psi_\alpha}$ of markers from the positive side to the negative is

$$\Gamma_n^{(-)\psi_\alpha} = -\mathrm{d}s\,\mathrm{d}t \int_{V_{\alpha n}<0} \psi_\alpha f_\alpha V_{\alpha n} \,\mathrm{d}\mathbf{V}_\alpha. \tag{2.1.9}$$

The full flux $\Gamma_n^{\psi_\alpha}$ of marker ψ_α in positive direction of $\mathbf{n}$ is

$$\Gamma_n^{\psi_\alpha} = \Gamma_n^{(+)\psi_\alpha} - \Gamma_n^{(-)\psi_\alpha} = \mathrm{d}s\,\mathrm{d}t \int \psi_\alpha f_\alpha V_{\alpha n} \,\mathrm{d}\mathbf{V}_\alpha, \tag{2.1.10}$$

where integrating is realized over all $\mathbf{V}_\alpha$. The specific flux $\gamma_n^{\psi_\alpha}$ across the unit area in unit time is equal to

$$\gamma_n^{\psi_\alpha} = \int \psi_\alpha f_\alpha V_{\alpha n} \,\mathrm{d}\mathbf{V}_\alpha. \tag{2.1.11}$$

Consider transport of mass, momentum and energy in a gas.

(1) Let $\psi_\alpha^{(1)} = m_\alpha$, then from (2.1.11) follows the relation for diffusive flux $\mathbf{J}_\alpha$ projection on the normal direction $\mathbf{n}$

$$J_{\alpha n} = m_\alpha \int f_\alpha V_{\alpha n}\, \mathrm{d}\mathbf{V}_\alpha = \rho_\alpha \overline{V_{\alpha n}}. \tag{2.1.12}$$

(2) If $\psi_\alpha^{(2)} = m V_{\alpha i},\ i = 1, 2, 3$, then relation (2.1.11) defines momentum transport in direction of $\mathbf{n}$.

Let $\mathbf{n}$ be an alternate direction coinciding with positive directions of coordinate axes in physical space. Then obviously the momentum transport is defined by symmetric tensor of the second range

$$P_\alpha = \rho_\alpha \overline{\mathbf{V}_\alpha \mathbf{V}_\alpha}, \tag{2.1.13}$$

where $\overline{\mathbf{V}_\alpha \mathbf{V}_\alpha}$ is diada with nine components,

$$P_\alpha = \left\{ \begin{matrix} \rho_\alpha \overline{V_{\alpha 1} V_{\alpha 1}}, \rho_\alpha \overline{V_{\alpha 1} V_{\alpha 2}}, \rho_\alpha \overline{V_{\alpha 1} V_{\alpha 3}} \\ \rho_\alpha \overline{V_{\alpha 2} V_{\alpha 1}}, \rho_\alpha \overline{V_{\alpha 2} V_{\alpha 2}}, \rho_\alpha \overline{V_{\alpha 2} V_{\alpha 3}} \\ \rho_\alpha \overline{V_{\alpha 3} V_{\alpha 1}}, \rho_\alpha \overline{V_{\alpha 3} V_{\alpha 2}}, \rho_\alpha \overline{V_{\alpha 3} V_{\alpha 3}} \end{matrix} \right\}. \tag{2.1.14}$$

The sum of tensors of partial pressures gives pressure tensor for gas mixture

$$P = \sum_\alpha \rho_\alpha \overline{\mathbf{V}_\alpha \mathbf{V}_\alpha}. \tag{2.1.15}$$

Vector of pressure $\mathbf{p}$ for surface element cannot coincide with normal direction $\mathbf{n}$ to this surface. It is easy to check that normal component of pressure for arbitrary oriented surface element in a gas is essentially positive:

$$\mathbf{p} \cdot \mathbf{n} = \sum_\alpha \rho_\alpha \overline{\mathbf{V}_\alpha V_{\alpha n}} \cdot \mathbf{n} = \sum_\alpha \rho_\alpha \overline{V_{\alpha n}^2}. \tag{2.1.16}$$

By definition, the mean value of normal pressure acting on three arbitrary but reciprocally orthogonal planes, equals the static pressure in gas

$$p = \frac{1}{3} \sum_\alpha \rho_\alpha \overline{V_{\alpha n}^2}. \tag{2.1.17}$$

(3) Let $\psi_\alpha^{(3)} = m_\alpha V_\alpha^2/2 + \varepsilon_\alpha$, where ε_α is internal energy of particle of species α.

Then relation (2.1.11) defines projection of heat flux $\mathbf{q}$ on the normal direction $\mathbf{n}$ to an element of surface, moving with the mean mass-velocity $\mathbf{v}_0$

$$q_n = \sum_\alpha q_{\alpha n} = \frac{1}{2} \sum_\alpha \rho_\alpha \overline{V_\alpha^2 V_{\alpha n}} + \sum_\alpha \varepsilon_\alpha. \tag{2.1.18}$$

Consider the quantity of kinetic energy E_k in unit volume at a time moment t

$$E_k = \sum_\alpha \int \frac{1}{2} m_\alpha v_\alpha^2 f_\alpha \, \mathrm{d}\mathbf{v}_\alpha = \frac{1}{2} \sum_\alpha \rho_\alpha \overline{v_\alpha^2}. \tag{2.1.19}$$

Taking into account that

$$\mathbf{v}_\alpha = \mathbf{v}_0 + \mathbf{V}_\alpha \tag{2.1.20}$$

and using (2.1.7), one obtains

$$E_k = \sum_\alpha \frac{1}{2} \rho_\alpha \overline{V_\alpha^2} + \frac{1}{2} \rho v_0^2 = E_{\mathrm{micro}} + E_{\mathrm{macro}}. \tag{2.1.21}$$

Estimate the order of this value. It is known from experiments that nitrogen density (at temperature $T = 273.16$ K and mercury pressure $P = 760$ mm) $\rho = 1.25 \times 10^{-3}$ $\mathrm{g \cdot cm^{-3}}$. Then from (2.1.17) it follows that for hydrodynamic velocity $v_0 \approx 1\ \mathrm{m \cdot s^{-1}}$ the ratio $E_{\mathrm{micro}}/E_{\mathrm{macro}} \approx 2.4 \times 10^5$. It means that in this case practically all energy of molecules corresponds to energy of chaos movement, $\sqrt{\overline{V^2}} \approx 1.5 \times 10^2\ \mathrm{m \cdot s^{-1}}$.

But apart from energy from translation movement, molecules have, generally speaking, vibration and rotation energy. Given chemical reactions, the transition of potential energy of molecules into kinetic energy of particles should be taken into account. All these kinds of energy including energy of chaotic movement identify as internal energy of gas. Temperature associated with translation movement can be defined as

$$\frac{3}{2} k_B T n = \sum_\alpha \frac{\rho_\alpha \overline{V_\alpha^2}}{2}, \tag{2.1.22}$$

where k_B is Boltzmann constant, n the number of particles per unit volume. It is natural to define temperature T_α of α-species as

$$m_\alpha \overline{V_\alpha^2} = 3 k_B T_\alpha. \tag{2.1.23}$$

In mixtures of chemical reacting gases, temperatures of species can differ significantly (Alekseev, 1982).

2.2. Hydrodynamic Enskog equations

Recall derivation of the Enskog hydrodynamic equations. Multiplying Boltzmann equation by molecular marker ψ_α and integrating over all $\mathbf{v}_\alpha$ yields

$$\int \psi_\alpha \left(\frac{\partial f_\alpha}{\partial t} + \mathbf{v}_\alpha \cdot \frac{\partial f_\alpha}{\partial \mathbf{r}} + \mathbf{F}_\alpha \cdot \frac{\partial f_\alpha}{\partial \mathbf{v}_\alpha} \right) \mathrm{d}\mathbf{v}_\alpha$$

$$= \sum_{j=1}^{\eta} \int \psi_\alpha (f'_\alpha f'_j - f_\alpha f_j) P_{\alpha j}^{\alpha j} g_{\alpha j} b \, \mathrm{d}b \, \mathrm{d}\varepsilon \, \mathrm{d}\mathbf{v}_\alpha \, \mathrm{d}\mathbf{v}_j$$

$$+ \int \psi_\alpha J_\alpha^{\mathrm{st,inel}} \, \mathrm{d}\mathbf{v}_\alpha, \tag{2.2.1}$$

where $P_{\alpha j}^{\alpha j}$ is probability density of elastic collisions of particles α and j, $J_\alpha^{\mathrm{st,inel}}$ is an integral of inelastic collisions. The use of notation for mean values leads to relations

$$\int \psi_\alpha \frac{\partial f_\alpha}{\partial t} \mathrm{d}\mathbf{v}_\alpha = \frac{\partial}{\partial t} \int \psi_\alpha f_\alpha \, \mathrm{d}\mathbf{v}_\alpha - \int f_\alpha \frac{\partial \psi_\alpha}{\partial t} \mathrm{d}\mathbf{v}_\alpha$$

$$= \frac{\partial n_\alpha \bar{\psi}_\alpha}{\partial t} - n_\alpha \overline{\frac{\partial \psi_\alpha}{\partial t}}, \tag{2.2.2}$$

$$\int \psi \mathbf{v}_\alpha \cdot \frac{\partial f_\alpha}{\partial \mathbf{r}} = \frac{\partial}{\partial \mathbf{r}} \cdot \left(n_\alpha \overline{\psi_\alpha \mathbf{v}_\alpha} \right) - n_\alpha \overline{\mathbf{v}_\alpha \cdot \frac{\partial \psi_\alpha}{\partial \mathbf{r}}}. \tag{2.2.3}$$

If external forces do not depend on the particle's velocities or corresponding Lorentz forces, the following transformations are valid

$$\int \psi_\alpha F_{\alpha 1} \frac{\partial f_\alpha}{\partial v_{\alpha 1}} \mathrm{d}\mathbf{v}_\alpha = \int [F_{\alpha 1} \psi_\alpha f_\alpha]_{v_{\alpha 1}=-\infty}^{v_{\alpha 1}=+\infty} \mathrm{d}v_{\alpha 2} \, \mathrm{d}v_{\alpha 3} - \int F_{\alpha 1} f_\alpha \frac{\partial \psi_\alpha}{\partial v_{\alpha 1}} \mathrm{d}\mathbf{v}_\alpha$$

$$= -n_\alpha F_{\alpha 1} \overline{\frac{\partial \psi_\alpha}{\partial v_{\alpha 1}}}. \tag{2.2.4}$$

Hence Enskog's equations are written as

$$\frac{\partial n_\alpha \bar{\psi}_\alpha}{\partial t} + \frac{\partial}{\partial \mathbf{r}} \cdot \left(n_\alpha \overline{\psi_\alpha \mathbf{v}_\alpha} \right) - n_\alpha \left[\overline{\frac{\partial \psi_\alpha}{\partial t}} + \overline{\mathbf{v}_\alpha \cdot \frac{\partial \psi_\alpha}{\partial \mathbf{r}}} + \overline{\mathbf{F}_\alpha \cdot \frac{\partial \psi_\alpha}{\partial \mathbf{v}_\alpha}} \right]$$

$$= \sum_j \int \psi_\alpha (f'_\alpha f'_j - f_\alpha f_j) P_{\alpha j}^{\alpha j} g_{\alpha j} b \, \mathrm{d}b \, \mathrm{d}\varepsilon \, \mathrm{d}\mathbf{v}_\alpha \, \mathrm{d}\mathbf{v}_j$$

$$+ \int \psi_\alpha J_\alpha^{\mathrm{st,inel}} \, \mathrm{d}\mathbf{v}_\alpha. \tag{2.2.5}$$

Enskog's equations (2.2.5) are integro-differential equations and, generally speaking, they are not more simple than initial Boltzmann equation. But for so-called summational invariants ($\psi_\alpha^{(1)}, \psi_\alpha^{(2)}, \psi_\alpha^{(3)}$), integral terms on the right-hand side of Eq. (2.2.5) could be significantly simplified. One proves (Hirschfelder, Curtiss and Bird, 1954), that for $\psi_\alpha^{(1)} = m_\alpha$, the first-mentioned term turns into zero and the second one leads to the mass rate of formation of α-species as a result of chemical reactions. For other invariants $\psi_\alpha^{(2)}$, $\psi_\alpha^{(3)}$, in view of conservation laws the integral terms are equal to zero after summation over all α ($\alpha = 1, \ldots, \eta$).

In the generalized Boltzmann kinetic theory (GBKT), as it was shown, local collision integrals can be written in the same form like in classical Boltzmann theory. Therefore, GBKT does not create additional difficulties connected with collision integrals.

2.3. Transformations of the generalized Boltzmann equation

The generalized Boltzmann equation (GBE) inevitably leads to formulation of new hydrodynamic equations, which are called generalized hydrodynamic equations (GHE). Classical hydrodynamic equations of Enskog, Euler and Navier–Stokes are particular cases of these equations. For the purpose of derivation of GHE let us transform GBE to the form convenient for further applications. Write down the second term on the left of GBE

$$\frac{Df_\alpha}{Dt} - \frac{D}{Dt}\left(\tau_\alpha \frac{Df_\alpha}{Dt}\right) = J_\alpha^{\mathrm{st,el}} + J_\alpha^{\mathrm{st,inel}}, \tag{2.3.1}$$

in explicit form

$$\frac{D}{Dt}\left(\tau_\alpha \frac{Df_\alpha}{Dt}\right) = \frac{D\tau_\alpha}{Dt}\frac{Df_\alpha}{Dt} + \tau_\alpha \frac{D}{Dt}\frac{Df_\alpha}{Dt},$$

where

$$\frac{D}{Dt} = \frac{\partial}{\partial t} + \mathbf{v}_\alpha \cdot \frac{\partial}{\partial \mathbf{r}} + \mathbf{F}_\alpha \cdot \frac{\partial}{\partial \mathbf{v}_\alpha}, \tag{2.3.2}$$

$$\begin{aligned}\frac{D}{Dt}\frac{Df_\alpha}{Dt} &= \frac{\partial^2 f_\alpha}{\partial t^2} + 2\mathbf{v}_\alpha \cdot \frac{\partial^2 f_\alpha}{\partial \mathbf{r}\partial t} + 2\mathbf{F}_\alpha \cdot \frac{\partial^2 f_\alpha}{\partial \mathbf{v}_\alpha \partial t} + \mathbf{v}_\alpha \cdot \frac{\partial}{\partial \mathbf{r}}\left(\mathbf{v}_\alpha \cdot \frac{\partial f_\alpha}{\partial \mathbf{r}}\right)\\ &\quad + \mathbf{F}_\alpha \cdot \frac{\partial}{\partial \mathbf{v}_\alpha}\left(\mathbf{v}_\alpha \cdot \frac{\partial f_\alpha}{\partial \mathbf{r}}\right) + \frac{\partial \mathbf{F}_\alpha}{\partial t} \cdot \frac{\partial f_\alpha}{\partial \mathbf{v}_\alpha} + \mathbf{v}_\alpha \cdot \frac{\partial}{\partial \mathbf{r}}\left(\mathbf{F}_\alpha \cdot \frac{\partial f_\alpha}{\partial \mathbf{v}_\alpha}\right)\\ &\quad + \mathbf{F}_\alpha \cdot \frac{\partial}{\partial \mathbf{v}_\alpha}\left(\mathbf{F}_\alpha \cdot \frac{\partial f_\alpha}{\partial \mathbf{v}_\alpha}\right).\end{aligned} \tag{2.3.3}$$

In doing so we should keep in mind that if the mean time between collisions of α-particles does not depend on velocity, then

$$\frac{D\tau_\alpha}{Dt} = \frac{\partial \tau_\alpha}{\partial t} + \mathbf{v}_\alpha \cdot \frac{\partial \tau_\alpha}{\partial \mathbf{r}}. \tag{2.3.4}$$

Transform terms in (2.3.3):

$$\mathbf{v}_\alpha \cdot \frac{\partial}{\partial \mathbf{r}}\left(\mathbf{v}_\alpha \cdot \frac{\partial f_\alpha}{\partial \mathbf{r}}\right) = \mathbf{v}_\alpha \mathbf{v}_\alpha : \frac{\partial^2 f_\alpha}{\partial \mathbf{r}\partial \mathbf{r}}, \tag{2.3.5}$$

$$\mathbf{F}_\alpha \cdot \frac{\partial}{\partial \mathbf{v}_\alpha}\left(\mathbf{v}_\alpha \cdot \frac{\partial f_\alpha}{\partial \mathbf{r}}\right) = \mathbf{F}_\alpha \mathbf{v}_\alpha : \frac{\partial^2 f_\alpha}{\partial \mathbf{r} \partial \mathbf{v}_\alpha} + \mathbf{F}_\alpha \cdot \frac{\partial f_\alpha}{\partial \mathbf{r}}, \tag{2.3.6}$$

$$\mathbf{v}_\alpha \cdot \frac{\partial}{\partial \mathbf{r}}\left(\mathbf{F}_\alpha \cdot \frac{\partial f_\alpha}{\partial \mathbf{v}_\alpha}\right) = \mathbf{v}_\alpha \mathbf{F}_\alpha : \frac{\partial^2 f_\alpha}{\partial \mathbf{v}_\alpha \partial \mathbf{r}} + \frac{\partial f_\alpha}{\partial \mathbf{v}_\alpha} \mathbf{v}_\alpha : \frac{\partial}{\partial \mathbf{r}} \mathbf{F}_\alpha, \tag{2.3.7}$$

$$\mathbf{F}_\alpha \cdot \frac{\partial}{\partial \mathbf{v}_\alpha}\left(\mathbf{F}_\alpha \cdot \frac{\partial f_\alpha}{\partial \mathbf{v}_\alpha}\right) = \mathbf{F}_\alpha \mathbf{F}_\alpha : \frac{\partial^2 f_\alpha}{\partial \mathbf{v}_\alpha \partial \mathbf{v}_\alpha} + \frac{\partial f_\alpha}{\partial \mathbf{v}_\alpha} \mathbf{F}_\alpha : \frac{\partial}{\partial \mathbf{v}_\alpha} \mathbf{F}_\alpha. \tag{2.3.8}$$

The sign ":" in (2.3.5)–(2.3.8) denotes, as usually, the double tensor product. For example, in (2.3.6)

$$\mathbf{F}_\alpha \mathbf{v}_\alpha : \frac{\partial^2 f_\alpha}{\partial \mathbf{r} \partial \mathbf{v}_\alpha} = \sum_{i,j=1}^{3} \mathbf{F}_{\alpha i} \mathbf{v}_{\alpha j} \frac{\partial^2 f_\alpha}{\partial r_j \partial v_{\alpha i}}. \tag{2.3.9}$$

The derivative of external forces $\mathbf{F}_\alpha$ ($\alpha = 1, \ldots, \eta$) with respect to velocity appears on the right side of (2.3.8); force $\mathbf{F}_\alpha$ – acting on the particle of species α – is related to the unit of mass of this particle. If $\mathbf{F}_\alpha$ does not depend on velocity, this derivative naturally turns into zero. In the following notation, the force independent of velocity is $\mathbf{F}_\alpha^{(1)}$. If $\mathbf{F}_\alpha$ includes the Lorentz force, noted as $\mathbf{F}_\alpha^B$ ($\mathbf{B}$ – magnetic induction),

$$\mathbf{F}_\alpha = \mathbf{F}_\alpha^{(1)} + \mathbf{F}_\alpha^B \tag{2.3.10}$$

the subsequent transformations of (2.3.8) can be realized:

$$\begin{aligned} &\frac{\partial f_\alpha}{\partial \mathbf{v}_\alpha} \mathbf{F}_\alpha : \frac{\partial}{\partial \mathbf{v}_\alpha} \mathbf{F}_\alpha \\ &\quad = \frac{\partial f_\alpha}{\partial \mathbf{v}_\alpha} \mathbf{F}_\alpha^{(1)} : \frac{\partial}{\partial \mathbf{v}_\alpha} \mathbf{F}_\alpha^B + \frac{\partial f_\alpha}{\partial \mathbf{v}_\alpha} \mathbf{F}_\alpha^B : \frac{\partial}{\partial \mathbf{v}_\alpha} \mathbf{F}_\alpha^B \\ &\quad = \left(\frac{q_\alpha}{m_\alpha}\right)^2 \frac{\partial f_\alpha}{\partial \mathbf{v}_\alpha} \cdot \left\{\mathbf{B}(\mathbf{v}_\alpha \cdot \mathbf{B}) - B^2 \mathbf{v}_\alpha\right\} + \frac{\partial f_\alpha}{\partial \mathbf{v}_\alpha} \mathbf{F}_\alpha^{(1)} : \frac{\partial}{\partial \mathbf{v}_\alpha} \mathbf{F}_\alpha^B, \end{aligned} \tag{2.3.11}$$

because

$$\mathbf{F}_\alpha^B = \frac{q_\alpha}{m_\alpha} [\mathbf{v}_\alpha \times \mathbf{B}], \tag{2.3.12}$$

where q_α is the charge of the particle of species α.

The last term on the right side of relation (2.3.11) is written as

$$\frac{\partial f_\alpha}{\partial \mathbf{v}_\alpha} \mathbf{F}_\alpha^{(1)} : \frac{\partial}{\partial \mathbf{v}_\alpha} \mathbf{F}_\alpha^B = \frac{q_\alpha}{m_\alpha} \frac{\partial f_\alpha}{\partial \mathbf{v}_\alpha} \cdot \left(\mathbf{F}_\alpha^{(1)} \times \mathbf{B}\right). \tag{2.3.13}$$

Sign "×" corresponds to a vector product. GBE can contain Umov–Pointing vector $\mathbf{S} \sim [\mathbf{E}, \mathbf{H}]$ ($\mathbf{H}$ is magnetic intensity) in explicit form because the force of non-magnetic origin $\mathbf{F}_\alpha^{(1)}$ can be connected with electric intensity $\mathbf{E}$.

As a result,

$$\frac{\partial f_\alpha}{\partial \mathbf{v}_\alpha}\mathbf{F}_\alpha : \frac{\partial}{\partial \mathbf{v}_\alpha}\mathbf{F}_\alpha$$
$$= \left(\frac{q_\alpha}{m_\alpha}\right)^2 \frac{\partial f_\alpha}{\partial \mathbf{v}_\alpha} \cdot \left\{\mathbf{B}(\mathbf{v}_\alpha \cdot \mathbf{B}) - B^2 \mathbf{v}_\alpha\right\} + \frac{q_\alpha}{m_\alpha}\frac{\partial f_\alpha}{\partial \mathbf{v}_\alpha} \cdot \left(\mathbf{F}_\alpha^{(1)} \times \mathbf{B}\right). \tag{2.3.14}$$

We reach the generalized Boltzmann equation in the form

$$\left(\frac{\partial f_\alpha}{\partial t} + \mathbf{v}_\alpha \cdot \frac{\partial f_\alpha}{\partial \mathbf{r}} + \mathbf{F}_\alpha \cdot \frac{\partial f_\alpha}{\partial \mathbf{v}_\alpha}\right)\left[1 - \left(\frac{\partial \tau_\alpha}{\partial t} + \mathbf{v}_\alpha \cdot \frac{\partial \tau_\alpha}{\partial \vec{r}}\right)\right]$$
$$- \tau_\alpha \left[\frac{\partial^2 f_\alpha}{\partial t^2} + 2\frac{\partial^2 f_\alpha}{\partial \mathbf{r}\partial t} \cdot \mathbf{v}_\alpha + \frac{\partial^2 f_\alpha}{\partial \mathbf{r}\partial \mathbf{r}} : \mathbf{v}_\alpha \mathbf{v}_\alpha + 2\frac{\partial^2 f_\alpha}{\partial \mathbf{v}_\alpha \partial t} \cdot \mathbf{F}_\alpha\right.$$
$$+ \frac{\partial \mathbf{F}_\alpha}{\partial t} \cdot \frac{\partial f_\alpha}{\partial \mathbf{v}_\alpha} + \mathbf{F}_\alpha \cdot \frac{\partial f_\alpha}{\partial \mathbf{r}} + \frac{q_\alpha}{m_\alpha}\frac{\partial f_\alpha}{\partial \mathbf{v}_\alpha} \cdot \left(\mathbf{F}_\alpha^{(1)} \times \mathbf{B}\right)$$
$$+ \left(\frac{q_\alpha}{m_\alpha}\right)^2 \frac{\partial f_\alpha}{\partial \mathbf{v}_\alpha} \cdot \left[\mathbf{B}(\mathbf{v}_\alpha \cdot \mathbf{B}) - B^2 \mathbf{v}_\alpha\right] + \frac{\partial f_\alpha}{\partial \mathbf{v}_\alpha}\mathbf{v}_\alpha : \frac{\partial}{\partial \mathbf{r}}\mathbf{F}_\alpha$$
$$\left.+ \frac{\partial^2 f_\alpha}{\partial \mathbf{v}_\alpha \partial \mathbf{v}_\alpha} : \mathbf{F}_\alpha \mathbf{F}_\alpha + 2\frac{\partial^2 f_\alpha}{\partial \mathbf{v}_\alpha \partial \mathbf{r}} : \mathbf{v}_\alpha \mathbf{F}_\alpha\right] = J_\alpha^{\mathrm{st,el}} + J_\alpha^{\mathrm{st,inel}}. \tag{2.3.15}$$

As is seen, the explicit form of differential part of GBE is much more complicated in comparison to Boltzmann equation. As a result, the transition to generalized hydrodynamic equations (GHE) requires more effort.

Let us go to this work following the standard procedure of obtaining the hydrodynamic description: multiply GBE by the particle collision invariants $\psi_\alpha^{(i)}$ $(i = 1, 2, 3)$ and integrate over all $\mathbf{v}_\alpha$. The result of this procedure leads to the generalized Enskog equations of continuity, momentum and energy.

2.4. Generalized continuity equation

Multiply GBE (2.3.15) by $\psi_\alpha^{(1)} = m_\alpha$ and realize term-by-term integration of left and right sides of equation. Formally 22 terms should be transformed, here and in the following we of necessity demonstrate only character and complicated elements of these transformations and restrict our consideration to some comments in more simple cases.

Consider integrals the transformation of which is realized by integration in parts. In doing so, the integrated part turns into zero because distribution function is equal to zero if $v_\alpha = \pm\infty$.

$$\int m_\alpha \frac{\partial f_\alpha}{\partial \mathbf{v}_\alpha}\mathbf{v}_\alpha : \frac{\partial}{\partial \mathbf{r}}\mathbf{F}_\alpha \, \mathrm{d}\mathbf{v}_\alpha$$

$$
\begin{aligned}
&= -\rho_\alpha \frac{\partial}{\partial \mathbf{r}} \cdot \mathbf{F}_\alpha^{(1)} + \int m_\alpha \frac{\partial f_\alpha}{\partial \mathbf{v}_\alpha} \mathbf{v}_\alpha : \frac{\partial}{\partial \mathbf{r}} \mathbf{F}_\alpha^B \, \mathrm{d}\mathbf{v}_\alpha \\
&= -\rho_\alpha \frac{\partial}{\partial \mathbf{r}} \cdot \mathbf{F}_\alpha^{(1)} + q_\alpha \int \left\{ \frac{\partial f_\alpha}{\partial \mathbf{v}_\alpha} \cdot \left[\mathbf{v}_\alpha \times \left[\left(\mathbf{v}_\alpha \cdot \frac{\partial}{\partial \mathbf{r}} \right) \mathbf{B} \right] \right] \right\} \mathrm{d}\mathbf{v}_\alpha \\
&= -\rho_\alpha \frac{\partial}{\partial \mathbf{r}} \cdot \mathbf{F}_\alpha^{(1)} + \frac{q_\alpha}{m_\alpha} \rho_\alpha \mathbf{v}_\alpha \cdot \operatorname{rot} \mathbf{B},
\end{aligned}
\tag{2.4.1}
$$

$$
\int m_\alpha \frac{\partial f_\alpha}{\partial \mathbf{v}_\alpha} \cdot \left\{ \mathbf{B}(\mathbf{v}_\alpha \cdot \mathbf{B}) - B^2 \mathbf{v}_\alpha \right\} \mathrm{d}\mathbf{v}_\alpha = 2B^2 \rho_\alpha \left(\frac{q_\alpha}{m_\alpha} \right)^2 . \tag{2.4.2}
$$

Consider also:

$$
\begin{aligned}
&m_\alpha \int \left(\mathbf{F}_\alpha \cdot \frac{\partial f_\alpha}{\partial \mathbf{v}_\alpha} \right) \left(\mathbf{v}_\alpha \cdot \frac{\partial \tau_\alpha}{\partial \mathbf{r}} \right) \mathrm{d}\mathbf{v}_\alpha \\
&= -m_\alpha \frac{\partial \tau_\alpha}{\partial \mathbf{r}} \cdot \int f_\alpha \frac{\partial}{\partial \mathbf{v}_\alpha} \cdot (\mathbf{F}_\alpha \mathbf{v}_\alpha) \, \mathrm{d}\mathbf{v}_\alpha = -m_\alpha \frac{\partial \tau_\alpha}{\partial \mathbf{r}} \cdot \int f_\alpha \left(\mathbf{F}_\alpha \cdot \frac{\partial}{\partial \mathbf{v}_\alpha} \right) \mathbf{v}_\alpha \, \mathrm{d}\mathbf{v}_\alpha \\
&= -m_\alpha \frac{\partial \tau_\alpha}{\partial \mathbf{r}} \cdot \int \mathbf{F}_\alpha f_\alpha \, \mathrm{d}\mathbf{v}_\alpha \\
&= -\rho_\alpha \frac{\partial \tau_\alpha}{\partial \mathbf{r}} \cdot \mathbf{F}_\alpha^{(1)} - \frac{q_\alpha}{m_\alpha} \rho_\alpha \frac{\partial \tau_\alpha}{\partial \mathbf{r}} \cdot [\bar{\mathbf{v}}_\alpha \times \mathbf{B}].
\end{aligned}
\tag{2.4.3}
$$

Collecting all – transformed in this way – terms, one obtains the generalized continuity equation

$$
\begin{aligned}
&\frac{\partial}{\partial t} \left\{ \rho_\alpha - \tau_\alpha \left[\frac{\partial \rho_\alpha}{\partial t} + \frac{\partial}{\partial \mathbf{r}} \cdot (\rho_\alpha \bar{\mathbf{v}}_\alpha) \right] \right\} + \frac{\partial}{\partial \mathbf{r}} \cdot \left\{ \rho_\alpha \bar{\mathbf{v}}_\alpha - \tau_\alpha \left[\frac{\partial}{\partial t} (\rho_\alpha \bar{\mathbf{v}}_\alpha) \right. \right. \\
&\quad \left. \left. + \frac{\partial}{\partial \mathbf{r}} \cdot \left(\rho_\alpha \overline{\mathbf{v}_\alpha \mathbf{v}_\alpha} \right) - \rho_\alpha \mathbf{F}_\alpha^{(1)} - \frac{q_\alpha}{m_\alpha} \rho_\alpha \bar{\mathbf{v}}_\alpha \times \mathbf{B} \right] \right\} = R_\alpha,
\end{aligned}
\tag{2.4.4}
$$

where R_α is mass rate of α-particles formation in presence of inelastic (also chemical) processes

$$
R_\alpha = m_\alpha \int J_\alpha^{\mathrm{st,nel}} \, \mathrm{d}\mathbf{v}_\alpha . \tag{2.4.5}
$$

Generalized continuity equation differs in radical way from the classical continuity equation. The origin of this distinction – for more simple case of one-component non-reacting gas – was discussed from qualitative positions in section "Historical introduction". Nevertheless, we repeat the main points of this consideration because of their importance.

Recall the phenomenological derivation of continuity equation. Control volume is defined in an area occupied by gas. The characteristic length of this control volume is much bigger than mean free path between collisions but much smaller than hydrodynamic scale. Write down the mass balance for this volume with transparent boundary surface. In another way, variation of mass in control volume could be connected only with fluxes of particles directed inside or outside of reference surface. This procedure leads to the well-known equation

$$\frac{\partial \rho}{\partial t} + \frac{\partial}{\partial \mathbf{r}} \cdot \rho \mathbf{v}_0 = 0. \tag{2.4.6}$$

Obviously, in the derivation of Eq. (2.4.6), the implicit presumption was that particles can be placed inside or outside of control volume. It means that only point-like particles are taken into consideration. This fact – as it was in Chapter 1 in the course of GBE derivation from the Bogolyubov chain – is principal restriction of Boltzmann kinetic theory. In the generalized Boltzmann kinetic theory (GBKT), considered particles have finite sizes and then at some instant of time can be placed partly inside, partly outside of reference surface. This fact, along with the use of the DF form oriented for describing the point structureless particles, leads to appearance of fluctuation terms, in particular in (2.4.4). For example, the term $\tau_\alpha[\partial \rho_\alpha/\partial t + (\partial/\partial \mathbf{r}) \cdot (\rho_\alpha \bar{\mathbf{v}}_\alpha)]$ reflects the fluctuation of density ρ_α, and the term

$$\tau_\alpha \left[\frac{\partial}{\partial t}(\rho_\alpha \bar{\mathbf{v}}_\alpha) + \frac{\partial}{\partial \mathbf{r}} \left(\rho_\alpha \overline{\mathbf{v}_\alpha \mathbf{v}_\alpha} \right) - \rho_\alpha \mathbf{F}_\alpha^{(1)} - \frac{q_\alpha}{m_\alpha} \rho_\alpha \bar{\mathbf{v}}_\alpha \times \mathbf{B} \right]$$

corresponds to the fluctuation of momentum $\rho_\alpha \bar{\mathbf{v}}_\alpha$. For the clarity, let us write down the generalized continuity equation in dimensionless form for particular case of one-component gas:

$$\frac{\partial}{\partial t} \left\{ \rho - \tau \left[\frac{\partial \rho}{\partial t} + \frac{\partial}{\partial \mathbf{r}} \cdot (\rho \mathbf{v}_0) \right] \right\} + \frac{\partial}{\partial \mathbf{r}} \cdot \left\{ \rho \mathbf{v}_0 - \tau \left[\frac{\partial}{\partial t} (\rho \mathbf{v}_0) + \frac{\partial}{\partial \mathbf{r}} \cdot \left(\rho \overline{\mathbf{v}\mathbf{v}} \right) \right] \right\} = 0. \tag{2.4.7}$$

Introduce the density scale ρ_∞, hydrodynamic lengths scale L, mean thermal velocity as the scale for particle velocity, and $v_{0\infty}$ as the scale for hydrodynamic velocity $\mathbf{v}_0$. Hydrodynamic time scale is defined as $t_H = L v_{0\infty}^{-1}$, and scale for τ is l_λ / v_T. In dimensionless form (2.4.7) is written as

$$\frac{\partial}{\partial \hat{t}} \left\{ \hat{\rho} - A_v \hat{\tau} \frac{l_\lambda}{L} \left[\frac{\partial \hat{\rho}}{\partial \hat{t}} + \frac{\partial}{\partial \hat{\mathbf{r}}} \cdot (\hat{\rho} \hat{\mathbf{v}}_0) \right] \right\} + \frac{\partial}{\partial \hat{\mathbf{r}}} \cdot \left\{ \hat{\rho} \hat{\mathbf{v}}_0 - A_v \hat{\tau} \frac{l_\lambda}{L} \left[\frac{\partial}{\partial \hat{t}} (\hat{\rho} \hat{\mathbf{v}}_0) + A_v^{-2} \frac{\partial}{\partial \hat{\mathbf{r}}} \cdot \left(\hat{\rho} \overline{\hat{\mathbf{v}} \hat{\mathbf{v}}} \right) \right] \right\} = 0. \tag{2.4.8}$$

But Knudsen number is $Kn = l_\lambda / L$, where l_λ is mean free path. Therefore, indicated fluctuation terms are proportional to Knudsen number and are really small in the case of small Knudsen numbers in the regime of continuum media. Qualitative pictures of these regimes are shown for small Kn (Figures 2.1 and 2.2) and large Kn (Figures 2.3 and 2.4).

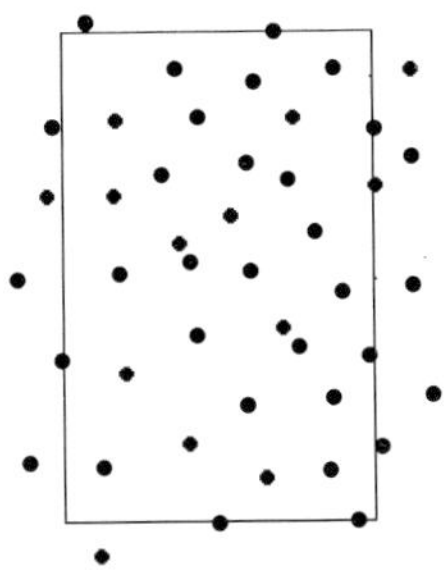

Fig. 2.1. Small Kn, particles of finite size.

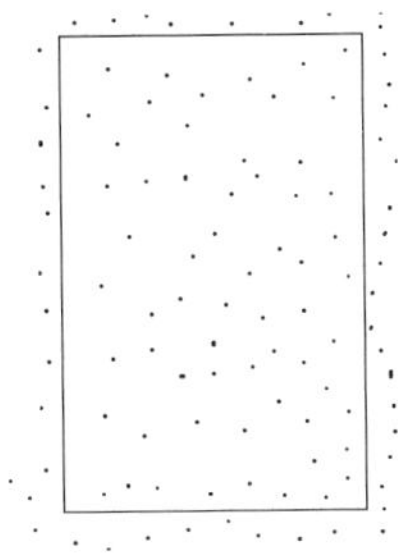

Fig. 2.2. Small Kn, point-like particles.

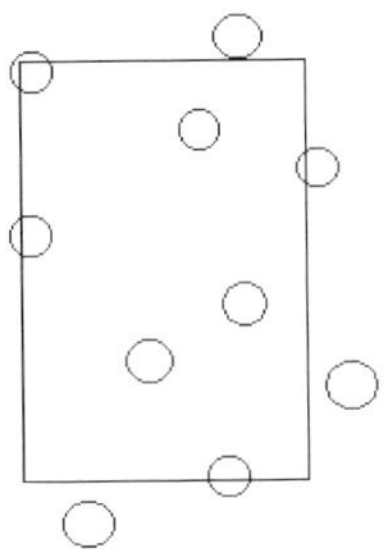

Fig. 2.3. Large Kn, particles of finite size.

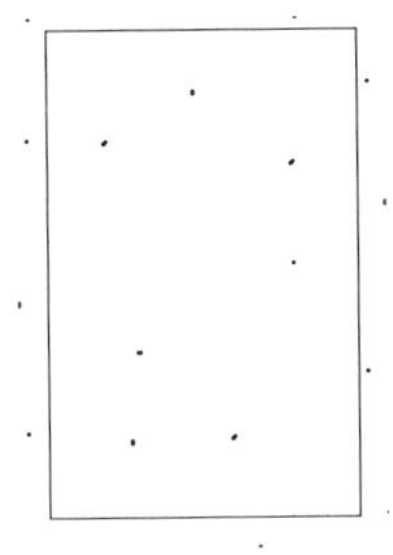

Fig. 2.4. Large Kn, point-like particles.

As is seen, for $Kn \sim 1$ fluctuations of mass in control volume are of the same order as basic terms and introduction of additional terms becomes more and more significant. But no reason to think that for $Kn \ll 1$ fluctuation terms could be omitted. As will be shown in the next chapter, the consideration of these terms is of principal significance for turbulence description on the Kolmogorov micro (or sub-grid) scale. Also should be noticed that for the case of $Kn \ll 1$ hydrodynamic equations (including the generalized continuity equation) belong to the class of differential equations with small parameter in view of senior derivatives. This fact could lead to "boundary layers" effects in solution of these equations.

2.5. Generalized momentum equation for component

Generalized momentum equation for α-component is obtained by multiplying GBE (2.3.15) by collision invariant $\psi_\alpha^{(2)} = m_\alpha \mathbf{v}_\alpha$ and the following integration over all $\mathbf{v}_\alpha$

$$\int m_\alpha \mathbf{v}_\alpha \left(\frac{\partial f_\alpha}{\partial t} + \mathbf{v}_\alpha \cdot \frac{\partial f_\alpha}{\partial \mathbf{r}} + \mathbf{F}_\alpha \cdot \frac{\partial f_\alpha}{\partial \mathbf{v}_\alpha} \right) \left[1 - \left(\frac{\partial \tau_\alpha}{\partial t} + \mathbf{v}_\alpha \cdot \frac{\partial \tau_\alpha}{\partial \mathbf{r}} \right) \right] \mathrm{d}\mathbf{v}_\alpha$$

$$- \tau_\alpha \int m_\alpha \mathbf{v}_\alpha \left[\frac{\partial^2 f_\alpha}{\partial t^2} + 2 \frac{\partial^2 f_\alpha}{\partial \mathbf{r} \partial t} \cdot \mathbf{v}_\alpha + \frac{\partial^2 f_\alpha}{\partial \mathbf{r} \partial \mathbf{r}} : \mathbf{v}_\alpha \mathbf{v}_\alpha + 2 \frac{\partial^2 f_\alpha}{\partial \mathbf{v}_\alpha \partial t} \cdot \mathbf{F}_\alpha \right.$$

$$+ \frac{\partial \mathbf{F}_\alpha}{\partial t} \cdot \frac{\partial f_\alpha}{\partial \mathbf{v}_\alpha} + \mathbf{F}_\alpha \cdot \frac{\partial f_\alpha}{\partial \mathbf{r}} + \frac{q_\alpha}{m_\alpha} \frac{\partial f_\alpha}{\partial \mathbf{v}_\alpha} \cdot \left(\mathbf{F}_\alpha^{(1)} \times \mathbf{B} \right)$$

$$+\left(\frac{q_\alpha}{m_\alpha}\right)^2\frac{\partial f_\alpha}{\partial \mathbf{v}_\alpha}\cdot\left[\mathbf{B}(\mathbf{v}_\alpha\cdot\mathbf{B})-B^2\mathbf{v}_\alpha\right]+\frac{\partial f_\alpha}{\partial \mathbf{v}_\alpha}\mathbf{v}_\alpha:\frac{\partial}{\partial \mathbf{r}}\mathbf{F}_\alpha$$

$$\left.+\frac{\partial^2 f_\alpha}{\partial \mathbf{v}_\alpha\partial \mathbf{v}_\alpha}:\mathbf{F}_\alpha\mathbf{F}_\alpha+2\frac{\partial^2 f_\alpha}{\partial \mathbf{v}_\alpha\partial \mathbf{r}}:\mathbf{v}_\alpha\mathbf{F}_\alpha\right]\mathrm{d}\mathbf{v}_\alpha$$

$$=\int m_\alpha\mathbf{v}_\alpha J_\alpha^{\mathrm{st,el}}\,\mathrm{d}\mathbf{v}_\alpha+\int m_\alpha\mathbf{v}_\alpha J_\alpha^{\mathrm{st,inel}}\,\mathrm{d}\mathbf{v}_\alpha. \tag{2.5.1}$$

Transformation of the first integral on the left side of (2.5.1) follows the main features of the standard transformations used in derivation of the Enskog momentum equation. For example ($i=1,2,3$),

$$\int m_\alpha v_{\alpha i}\left(\mathbf{F}_\alpha\cdot\frac{\partial f_\alpha}{\partial \mathbf{v}_\alpha}\right)\left(\mathbf{v}_\alpha\cdot\frac{\partial \tau_\alpha}{\partial \mathbf{r}}\right)\mathrm{d}\mathbf{v}_\alpha$$

$$=m_\alpha\sum_{k,l=1}^{3}\frac{\partial \tau_\alpha}{\partial r_k}\int v_{\alpha i}F_{\alpha l}v_{\alpha k}\frac{\partial f_\alpha}{\partial v_{\alpha l}}\,\mathrm{d}\mathbf{v}_\alpha$$

$$=m_\alpha\sum_{k,l}\frac{\partial \tau_\alpha}{\partial r_k}\int F_{\alpha l}f_\alpha\frac{\partial}{\partial v_{\alpha l}}(v_{\alpha i}v_{\alpha k})\,\mathrm{d}\mathbf{v}_\alpha$$

$$=-m_\alpha\sum_{k}\frac{\partial \tau_\alpha}{\partial r_k}\int F_{\alpha i}f_\alpha v_{\alpha k}\,\mathrm{d}\mathbf{v}_\alpha-m_\alpha\sum_{k}\frac{\partial \tau_\alpha}{\partial r_k}\int F_{\alpha k}f_\alpha v_{\alpha i}\,\mathrm{d}\mathbf{v}_\alpha. \tag{2.5.2}$$

In derivation of (2.5.2) it is taken into account that possible dependence of the external force on velocity corresponds to the Lorentz force; also, using (2.3.10), one obtains

$$\int m_\alpha\mathbf{v}_\alpha\left(\mathbf{F}_\alpha\cdot\frac{\partial f_\alpha}{\partial \mathbf{v}_\alpha}\right)\left(\mathbf{v}_\alpha\cdot\frac{\partial \tau_\alpha}{\partial \mathbf{r}}\right)\mathrm{d}\mathbf{v}_\alpha$$

$$=-\rho_\alpha\mathbf{F}_\alpha^{(1)}\frac{\partial \tau_\alpha}{\partial \mathbf{r}}\cdot\bar{\mathbf{v}}_\alpha-\rho_\alpha\bar{\mathbf{v}}_\alpha\mathbf{F}_\alpha^{(1)}\cdot\frac{\partial \tau_\alpha}{\partial \mathbf{r}}-\frac{q_\alpha}{m_\alpha}\overline{\left(\frac{\partial \tau_\alpha}{\partial \mathbf{r}}\cdot\rho_\alpha\mathbf{v}_\alpha\right)[\mathbf{v}_\alpha\times\mathbf{B}]}$$

$$-\frac{q_\alpha}{m_\alpha}\rho_\alpha\overline{\left\{[\mathbf{v}_\alpha\times\mathbf{B}]\cdot\frac{\partial \tau_\alpha}{\partial \mathbf{r}}\right\}\mathbf{v}_\alpha}. \tag{2.5.3}$$

As usual, top line in (2.5.3) denotes averaging over velocity with taking into account the rule (2.1.5).

Consider now transformations of the second integral in (2.5.1). Below are given examples of mentioned transformations.

$$\int m_\alpha\mathbf{v}_\alpha\frac{\partial}{\partial \mathbf{r}}\cdot\mathbf{F}_\alpha\,\mathrm{d}\mathbf{v}_\alpha=\rho_\alpha\bar{\mathbf{v}}_\alpha\frac{\partial}{\partial \mathbf{r}}\cdot\mathbf{F}_\alpha^{(1)}+q_\alpha\int f_\alpha\mathbf{v}_\alpha\frac{\partial}{\partial \mathbf{r}}\cdot(\mathbf{v}_\alpha\times\mathbf{B})\,\mathrm{d}\mathbf{v}_\alpha$$

$$=\rho_\alpha\bar{\mathbf{v}}_\alpha\frac{\partial}{\partial \mathbf{r}}\cdot\mathbf{F}_\alpha^{(1)}-\frac{q_\alpha}{m_\alpha}\rho_\alpha\overline{\mathbf{v}_\alpha\mathbf{v}_\alpha}\cdot\mathrm{rot}\,\mathbf{B}, \tag{2.5.4}$$

$$\int m_\alpha \mathbf{v}_\alpha \frac{\partial^2 f_\alpha}{\partial \mathbf{v}_\alpha \partial t} \cdot \mathbf{F}_\alpha \, \mathrm{d}\mathbf{v}_\alpha$$

$$= -m_\alpha \mathbf{F}_\alpha^{(1)} \frac{\partial n_\alpha}{\partial t} + q_\alpha \int \mathbf{v}_\alpha \frac{\partial}{\partial \mathbf{v}_\alpha} \cdot \left\{ [\mathbf{v}_\alpha \times \mathbf{B}] \frac{\partial f_\alpha}{\partial t} \right\} \mathrm{d}\mathbf{v}_\alpha$$

$$= -\mathbf{F}_\alpha^{(1)} \frac{\partial \rho_\alpha}{\partial t} + \frac{q_\alpha}{m_\alpha} \mathbf{B} \times \frac{\partial}{\partial t} (\rho_\alpha \bar{\mathbf{v}}_\alpha), \tag{2.5.5}$$

$$\int m_\alpha \mathbf{v}_\alpha \frac{\partial^2 f_\alpha}{\partial \mathbf{v}_\alpha \partial \mathbf{v}_\alpha} : \mathbf{F}_\alpha \mathbf{F}_\alpha \, \mathrm{d}\mathbf{v}_\alpha$$

$$= 2m_\alpha \sum_{ij} F_{\alpha i}^{(1)} \int \mathbf{v}_\alpha \frac{\partial^2 f_\alpha}{\partial v_{\alpha i} \partial v_{\alpha j}} F_{\alpha j}^B \, \mathrm{d}\mathbf{v}_\alpha + \int m_\alpha \mathbf{v}_\alpha \frac{\partial^2 f_\alpha}{\partial \mathbf{v}_\alpha \partial \mathbf{v}_\alpha} : \mathbf{F}_\alpha^B \mathbf{F}_\alpha^B \, \mathrm{d}\mathbf{v}_\alpha$$

$$= 2m_\alpha \sum_{ij} F_{\alpha i}^{(1)} \int \mathbf{v}_\alpha \frac{\partial^2 f_\alpha}{\partial v_{\alpha i} \partial v_{\alpha j}} F_{\alpha j}^B \, \mathrm{d}\mathbf{v}_\alpha$$

$$+ 2\left(\frac{q_\alpha}{m_\alpha}\right)^2 \rho_\alpha \big[\mathbf{B}(\mathbf{B} \cdot \bar{\mathbf{v}}_\alpha) - 2B^2 \bar{\mathbf{v}}_\alpha\big]. \tag{2.5.6}$$

Consider in detail the evaluation of integral on the right-hand side of (2.5.6) for velocity component $v_{\alpha 1}$; use the integration by parts

$$\sum_{i,j} m_\alpha F_{\alpha i}^{(1)} \int v_{\alpha 1} \frac{\partial^2 f_\alpha}{\partial v_{\alpha i} \partial v_{\alpha j}} F_{\alpha j}^B \, \mathrm{d}\mathbf{v}_\alpha$$

$$= \sum_{\substack{i,j \\ i \neq j,\ i \neq 1}} m_\alpha F_{\alpha i}^{(1)} \int v_{\alpha 1} \frac{\partial^2 f_\alpha}{\partial v_{\alpha i} \partial v_{\alpha j}} F_{\alpha j}^B \, \mathrm{d}\mathbf{v}_\alpha$$

$$= q_\alpha m_\alpha \big(F_{\alpha 2}^{(1)} B_3 - F_{\alpha 3}^{(1)} B_2\big). \tag{2.5.7}$$

Similarly transformed can be the terms related to other components of $\mathbf{v}_\alpha$, and (2.5.6) is written as

$$\int m_\alpha \mathbf{v}_\alpha \frac{\partial^2 f_\alpha}{\partial \mathbf{v}_\alpha \partial \mathbf{v}_\alpha} : \mathbf{F}_\alpha \mathbf{F}_\alpha \, \mathrm{d}\mathbf{v}_\alpha$$

$$= 2\rho_\alpha \left(\frac{q_\alpha}{m_\alpha}\right)^2 \big[\mathbf{B}(\mathbf{B} \cdot \bar{\mathbf{v}}_\alpha) - 2B^2 \bar{\mathbf{v}}_\alpha\big] + 2\frac{q_\alpha}{m_\alpha} \rho_\alpha \mathbf{F}_\alpha^{(1)} \times \mathbf{B}. \tag{2.5.8}$$

Consider now the integral

$$\int m_\alpha \mathbf{v}_\alpha \frac{\partial^2 f_\alpha}{\partial \mathbf{v}_\alpha \partial \mathbf{r}} : \mathbf{v}_\alpha \mathbf{F}_\alpha \, \mathrm{d}\mathbf{v}_\alpha$$

$$
\begin{aligned}
&= -m_\alpha \int \mathbf{F}_\alpha \left(\mathbf{v}_\alpha \cdot \frac{\partial f_\alpha}{\partial \mathbf{r}} \right) \mathrm{d}\mathbf{v}_\alpha - m_\alpha \int \mathbf{v}_\alpha \left(\mathbf{F}_\alpha \cdot \frac{\partial f_\alpha}{\partial \mathbf{r}} \right) \mathrm{d}\mathbf{v}_\alpha \\
&= -\mathbf{F}_\alpha^{(1)} \frac{\partial}{\partial \mathbf{r}} \cdot (\rho_\alpha \bar{\mathbf{v}}_\alpha) - \left(\mathbf{F}_\alpha^{(1)} \cdot \frac{\partial}{\partial \mathbf{r}} \right) (\rho_\alpha \bar{\mathbf{v}}_\alpha) \\
&\quad - m_\alpha \int \mathbf{F}_\alpha^B \left(\mathbf{v}_\alpha \cdot \frac{\partial f_\alpha}{\partial \mathbf{r}} \right) \mathrm{d}\mathbf{v}_\alpha - m_\alpha \int \mathbf{v}_\alpha \left(\mathbf{F}_\alpha^B \cdot \frac{\partial f_\alpha}{\partial \mathbf{r}} \right) \mathrm{d}\mathbf{v}_\alpha .
\end{aligned}
\tag{2.5.9}
$$

Integrals of (2.5.9) associated with magnetic field, can be transformed using formula for triple vector product

$$
(\mathbf{v}_\alpha \times \mathbf{B}) \cdot \frac{\partial f_\alpha}{\partial \mathbf{r}} = \left(\mathbf{B} \times \frac{\partial}{\partial \mathbf{r}} \right) \cdot f_\alpha \mathbf{v}_\alpha . \tag{2.5.10}
$$

We have:

$$
\begin{aligned}
m_\alpha \int \mathbf{F}_\alpha^B \left(\mathbf{v}_\alpha \cdot \frac{\partial f_\alpha}{\partial \mathbf{r}} \right) \mathrm{d}\mathbf{v}_\alpha &= q_\alpha \int [\mathbf{v}_\alpha \times \mathbf{B}] \left(\mathbf{v}_\alpha \cdot \frac{\partial f_\alpha}{\partial \mathbf{r}} \right) \mathrm{d}\mathbf{v}_\alpha \\
&= \frac{q_\alpha}{m_\alpha} \left[\frac{\partial}{\partial \mathbf{r}} \cdot \rho_\alpha \overline{\mathbf{v}_\alpha \mathbf{v}_\alpha} \right] \times \mathbf{B},
\end{aligned}
\tag{2.5.11}
$$

$$
\begin{aligned}
m_\alpha \int \mathbf{v}_\alpha \left(\mathbf{F}_\alpha^B \cdot \frac{\partial f_\alpha}{\partial \mathbf{r}} \right) \mathrm{d}\mathbf{v}_\alpha &= q_\alpha \int \mathbf{v}_\alpha \left[(\mathbf{v}_\alpha \times \mathbf{B}) \cdot \frac{\partial f_\alpha}{\partial \mathbf{r}} \right] \mathrm{d}\mathbf{v}_\alpha \\
&= \left[\mathbf{B} \times \frac{\partial}{\partial \mathbf{r}} \right] \cdot \frac{q_\alpha}{m_\alpha} \rho_\alpha \overline{\mathbf{v}_\alpha \mathbf{v}_\alpha} .
\end{aligned}
\tag{2.5.12}
$$

As a result, (2.5.9) will be brought into form:

$$
\begin{aligned}
\int m_\alpha \mathbf{v}_\alpha \frac{\partial^2 f_\alpha}{\partial \mathbf{v}_\alpha \partial \mathbf{r}} : \mathbf{v}_\alpha \mathbf{F}_\alpha \, \mathrm{d}\mathbf{v}_\alpha &= -\mathbf{F}_\alpha^{(1)} \frac{\partial}{\partial \mathbf{r}} \cdot (\rho_\alpha \bar{\mathbf{v}}_\alpha) - \left(\mathbf{F}_\alpha^{(1)} \cdot \frac{\partial}{\partial \mathbf{r}} \right) (\rho_\alpha \bar{\mathbf{v}}_\alpha) \\
&\quad - \left[\frac{\partial}{\partial \mathbf{r}} \cdot \frac{q_\alpha}{m_\alpha} \rho_\alpha \overline{\mathbf{v}_\alpha \mathbf{v}_\alpha} \right] \times \mathbf{B} - \left(\mathbf{B} \times \frac{\partial}{\partial \mathbf{r}} \right) \cdot \frac{q_\alpha}{m_\alpha} \rho_\alpha \overline{\mathbf{v}_\alpha \mathbf{v}_\alpha} .
\end{aligned}
\tag{2.5.13}
$$

Let us introduce the vector product of operator $\partial/\partial\mathbf{r}$ and diada $\overline{\mathbf{v}_\alpha \mathbf{v}_\alpha}$ as a tensor defined by vectors $(\partial/\partial\mathbf{r}) \times \overline{\mathbf{v}_\alpha v_{\alpha i}}$, in particular it means that

$$
\left(\mathbf{B} \times \frac{\partial}{\partial \mathbf{r}} \right) \cdot \rho_\alpha \overline{\mathbf{v}_\alpha \mathbf{v}_\alpha} = \mathbf{B} \cdot \frac{\partial}{\partial \mathbf{r}} \times \overline{\mathbf{v}_\alpha \mathbf{v}_\alpha} . \tag{2.5.14}
$$

We need also for the following transformations the relation:

$$
B^2 \bar{\mathbf{v}}_\alpha - \mathbf{B}(\mathbf{B} \cdot \bar{\mathbf{v}}_\alpha) = -(\bar{\mathbf{v}}_\alpha \times \mathbf{B}) \times \mathbf{B}. \tag{2.5.15}
$$

After substitution of all transformed integrals into (2.5.1), one obtains the form of the generalized momentum equation

$$\frac{\partial}{\partial t}\left\{\rho_\alpha\bar{\mathbf{v}}_\alpha-\tau_\alpha\left[\frac{\partial}{\partial t}(\rho_\alpha\bar{\mathbf{v}}_\alpha)+\frac{\partial}{\partial\mathbf{r}}\cdot\rho_\alpha\overline{\mathbf{v}_\alpha\mathbf{v}_\alpha}-\rho_\alpha\mathbf{F}_\alpha^{(1)}-\frac{q_\alpha}{m_\alpha}\rho_\alpha\bar{\mathbf{v}}_\alpha\times\mathbf{B}\right]\right\}$$
$$-\mathbf{F}_\alpha^{(1)}\left[\rho_\alpha-\tau_\alpha\left(\frac{\partial\rho_\alpha}{\partial t}+\frac{\partial}{\partial\mathbf{r}}\cdot(\rho_\alpha\bar{\mathbf{v}}_\alpha)\right)\right]-\frac{q_\alpha}{m_\alpha}\left\{\rho_\alpha\bar{\mathbf{v}}_\alpha-\tau_\alpha\left[\frac{\partial}{\partial t}(\rho_\alpha\bar{\mathbf{v}}_\alpha)\right.\right.$$
$$\left.\left.+\frac{\partial}{\partial\mathbf{r}}\cdot\rho_\alpha\overline{\mathbf{v}_\alpha\mathbf{v}_\alpha}-\rho_\alpha\mathbf{F}_\alpha^{(1)}-\frac{q_\alpha}{m_\alpha}\rho_\alpha\bar{\mathbf{v}}_\alpha\times\mathbf{B}\right]\right\}\times\mathbf{B}$$
$$+\frac{\partial}{\partial\mathbf{r}}\cdot\left\{\rho_\alpha\overline{\mathbf{v}_\alpha\mathbf{v}_\alpha}-\tau_\alpha\left[\frac{\partial}{\partial t}\rho_\alpha\overline{\mathbf{v}_\alpha\mathbf{v}_\alpha}+\frac{\partial}{\partial\mathbf{r}}\cdot\rho_\alpha\overline{(\mathbf{v}_\alpha\mathbf{v}_\alpha)\mathbf{v}_\alpha}-\mathbf{F}_\alpha^{(1)}\rho_\alpha\bar{\mathbf{v}}_\alpha\right.\right.$$
$$\left.\left.-\rho_\alpha\bar{\mathbf{v}}_\alpha\mathbf{F}_\alpha^{(1)}-\frac{q_\alpha}{m_\alpha}\rho_\alpha\overline{[\mathbf{v}_\alpha\times\mathbf{B}]\mathbf{v}_\alpha}-\frac{q_\alpha}{m_\alpha}\rho_\alpha\overline{\mathbf{v}_\alpha[\mathbf{v}_\alpha\times\mathbf{B}]}\right]\right\}$$
$$=\int m_\alpha\mathbf{v}_\alpha J_\alpha^{\mathrm{st,el}}\,\mathrm{d}\mathbf{v}_\alpha+\int m_\alpha\mathbf{v}_\alpha J_\alpha^{\mathrm{st,inel}}\,\mathrm{d}\mathbf{v}_\alpha. \tag{2.5.16}$$

By virtue of momentum conservation law, after summation the integral terms on the right side of (2.5.16) turn into zero

$$\sum_\alpha m_\alpha\int\mathbf{v}_\alpha J_\alpha^{\mathrm{st,el}}\,\mathrm{d}\mathbf{v}_\alpha=0, \tag{2.5.17}$$

$$\sum_\alpha m_\alpha\int\mathbf{v}_\alpha J_\alpha^{\mathrm{st,inel}}\,\mathrm{d}\mathbf{v}_\alpha=0. \tag{2.5.18}$$

Boltzmann collision integral satisfies these demands; moreover, it can be proved (Hirschfelder, Curtiss and Bird, 1954; Alekseev, 1982) that

$$\int\psi_\alpha^{(1)}J_\alpha^{\mathrm{st,el}}\,\mathrm{d}\mathbf{v}_\alpha=0,\qquad\sum_\alpha\int J_\alpha^{\mathrm{st,inel}}\psi_\alpha^{(1)}\,\mathrm{d}\mathbf{v}_\alpha=0, \tag{2.5.19}$$

$$\sum_\alpha\int\psi_\alpha^{(2)}J_\alpha^{\mathrm{st,el}}\,\mathrm{d}\mathbf{v}_\alpha=0,\qquad\sum_\alpha\int\psi_\alpha^{(2)}J_\alpha^{\mathrm{st,inel}}\,\mathrm{d}\mathbf{v}_\alpha=0, \tag{2.5.20}$$

$$\sum_\alpha\int\psi_\alpha^{(3)}J_\alpha^{\mathrm{st,el}}\,\mathrm{d}\mathbf{v}_\alpha=0,\qquad\sum_\alpha\int\psi_\alpha^{(3)}J_\alpha^{\mathrm{st,inel}}\,\mathrm{d}\mathbf{v}_\alpha=0, \tag{2.5.21}$$

where

$$\psi_\alpha^{(1)}=m_\alpha,\qquad\psi_\alpha^{(2)}=m_\alpha\mathbf{v}_\alpha,\qquad\psi_\alpha^{(3)}=\frac{m_\alpha v_\alpha^2}{2}+\varepsilon_\alpha.$$

The proof is based on inversion of forward and backward collisions and use of the mass, momentum and energy conservation laws for non-relativistic particle's collision.

But in plasma physics – in particular in strong electric fields – the use of hydrodynamic equations for gas mixtures is not sufficient for adequate description of physical system. In this case one uses transport equations for components introducing approximations for integrals of types

$$\int \psi_\alpha^{(2,3)} J_\alpha^{\mathrm{st,el}} \,\mathrm{d}\mathbf{v}_\alpha, \qquad \int \psi_\alpha^{(2,3)} J_\alpha^{\mathrm{st,inel}} \,\mathrm{d}\mathbf{v}_\alpha,$$

like the BGK approximation (Bhatnagar, Gross and Krook, 1954).

2.6. Generalized energy equation for component

Generalized energy equation for α-component ($\alpha = 1, \ldots, \eta$) is a result of a term-by-term multiplication of left and right sides of GBE (2.3.15) by collision invariant $\psi_\alpha^{(3)} = m_\alpha v_\alpha^2/2 + \varepsilon_\alpha$ and following integration over all velocities:

$$\begin{aligned}
&\int \left(\frac{m_\alpha v_\alpha^2}{2} + \varepsilon_\alpha\right)\left(\frac{\partial f_\alpha}{\partial t} + \mathbf{v}_\alpha \cdot \frac{\partial f_\alpha}{\partial \mathbf{r}} + \mathbf{F}_\alpha \cdot \frac{\partial f_\alpha}{\partial \mathbf{v}_\alpha}\right) \\
&\quad \times \left[1 - \left(\frac{\partial \tau_\alpha}{\partial t} + \mathbf{v}_\alpha \cdot \frac{\partial \tau_\alpha}{\partial \mathbf{r}}\right)\right]\mathrm{d}\mathbf{v}_\alpha \\
&\quad - \tau_\alpha \int \left(\frac{m_\alpha v_\alpha^2}{2} + \varepsilon_\alpha\right)\left[\frac{\partial^2 f_\alpha}{\partial t^2} + 2\frac{\partial^2 f_\alpha}{\partial \mathbf{r}\partial t} \cdot \mathbf{v}_\alpha + \frac{\partial^2 f_\alpha}{\partial \mathbf{r}\partial \mathbf{r}} : \mathbf{v}_\alpha \mathbf{v}_\alpha\right. \\
&\quad + 2\frac{\partial^2 f_\alpha}{\partial \mathbf{v}_\alpha \partial t} \cdot \mathbf{F}_\alpha + \frac{\partial \mathbf{F}_\alpha}{\partial t} \cdot \frac{\partial f_\alpha}{\partial \mathbf{v}_\alpha} + \mathbf{F}_\alpha \cdot \frac{\partial f_\alpha}{\partial \mathbf{r}} + \frac{q_\alpha}{m_\alpha}\frac{\partial f_\alpha}{\partial \mathbf{v}_\alpha} \cdot \left(\mathbf{F}_\alpha^{(1)} \times \mathbf{B}\right) \\
&\quad + \left(\frac{q_\alpha}{m_\alpha}\right)^2 \frac{\partial f_\alpha}{\partial \mathbf{v}_\alpha} \cdot \left[\mathbf{B}(\mathbf{v}_\alpha \cdot \mathbf{B}) - B^2 \mathbf{v}_\alpha\right] \\
&\quad \left. + \frac{\partial f_\alpha}{\partial \mathbf{v}_\alpha}\mathbf{v}_\alpha : \frac{\partial}{\partial \mathbf{r}}\mathbf{F}_\alpha + \frac{\partial^2 f_\alpha}{\partial \mathbf{v}_\alpha \partial \mathbf{v}_\alpha} : \mathbf{F}_\alpha \mathbf{F}_\alpha + 2\frac{\partial^2 f_\alpha}{\partial \mathbf{v}_\alpha \partial \mathbf{r}} : \mathbf{v}_\alpha \mathbf{F}_\alpha\right]\mathrm{d}\mathbf{v}_\alpha \\
&= \int \left(\frac{m_\alpha v_\alpha^2}{2} + \varepsilon_\alpha\right) J_\alpha^{\mathrm{st,el}} \,\mathrm{d}\mathbf{v}_\alpha + \int \left(\frac{m_\alpha v_\alpha^2}{2} + \varepsilon_\alpha\right) J_\alpha^{\mathrm{st,inel}} \,\mathrm{d}\mathbf{v}_\alpha. \qquad (2.6.1)
\end{aligned}$$

As reference material we present the result of transformation of all terms in the generalized energy equation (2.6.1):

$$\begin{aligned}
&\int \frac{\partial f_\alpha}{\partial t}\left(1 - \frac{\partial \tau_\alpha}{\partial t}\right)\left(\frac{m_\alpha v_\alpha^2}{2} + \varepsilon_\alpha\right)\mathrm{d}\mathbf{v}_\alpha \\
&= \left(1 - \frac{\partial \tau_\alpha}{\partial t}\right)\frac{\partial}{\partial t}\left(\frac{\rho_\alpha \overline{v_\alpha^2}}{2} + n_\alpha \varepsilon_\alpha\right), \qquad (2.6.2)
\end{aligned}$$

$$\int \frac{\partial f_\alpha}{\partial t}\left(\mathbf{v}_\alpha \cdot \frac{\partial \tau_\alpha}{\partial \mathbf{r}}\right)\left(\frac{m_\alpha v_\alpha^2}{2} + \varepsilon_\alpha\right)\mathrm{d}\mathbf{v}_\alpha$$

$$= \frac{\partial \tau_\alpha}{\partial \mathbf{r}} \cdot \frac{\partial}{\partial t}\left[\frac{1}{2}\rho_\alpha \overline{\mathbf{v}_\alpha v_\alpha^2} + \varepsilon_\alpha n_\alpha \bar{\mathbf{v}}_\alpha\right], \tag{2.6.3}$$

$$\int \left(\frac{m_\alpha v_\alpha^2}{2} + \varepsilon_\alpha\right)\left(\mathbf{v}_\alpha \cdot \frac{\partial f_\alpha}{\partial \mathbf{r}}\right)\mathrm{d}\mathbf{v}_\alpha = \frac{\partial}{\partial \mathbf{r}} \cdot \left\{\frac{1}{2}\rho_\alpha \overline{\mathbf{v}_\alpha v_\alpha^2} + \varepsilon_\alpha n_\alpha \bar{\mathbf{v}}_\alpha\right\}, \tag{2.6.4}$$

$$\int \left(\frac{m_\alpha v_\alpha^2}{2} + \varepsilon_\alpha\right)\left(\mathbf{v}_\alpha \cdot \frac{\partial f_\alpha}{\partial \mathbf{r}}\right)\left(\mathbf{v}_\alpha \cdot \frac{\partial \tau_\alpha}{\partial \mathbf{r}}\right)\mathrm{d}\mathbf{v}_\alpha$$

$$= \frac{1}{2}\frac{\partial \tau_\alpha}{\partial \mathbf{r}}\frac{\partial}{\partial \mathbf{r}} : \rho_\alpha \overline{v_\alpha^2 \mathbf{v}_\alpha \mathbf{v}_\alpha} + \frac{\partial \tau_\alpha}{\partial \mathbf{r}}\frac{\partial}{\partial \mathbf{r}} : \varepsilon_\alpha n_\alpha \overline{\mathbf{v}_\alpha \mathbf{v}_\alpha}, \tag{2.6.5}$$

$$\int \left(\frac{m_\alpha v_\alpha^2}{2} + \varepsilon_\alpha\right)\left(\mathbf{F}_\alpha \cdot \frac{\partial f_\alpha}{\partial \mathbf{v}_\alpha}\right)\mathrm{d}\mathbf{v}_\alpha = -\rho_\alpha \overline{\mathbf{F}_\alpha \cdot \mathbf{v}_\alpha}, \tag{2.6.6}$$

$$\int \left(\frac{m_\alpha v_\alpha^2}{2} + \varepsilon_\alpha\right)\left(\mathbf{F}_\alpha \cdot \frac{\partial f_\alpha}{\partial \mathbf{v}_\alpha}\right)\left(\mathbf{v}_\alpha \cdot \frac{\partial \tau_\alpha}{\partial \mathbf{r}}\right)\mathrm{d}\mathbf{v}_\alpha$$

$$= -\rho_\alpha \frac{\partial \tau_\alpha}{\partial \mathbf{r}} \cdot \left(\overline{\mathbf{F}_\alpha \cdot \mathbf{v}_\alpha \mathbf{v}_\alpha}\right) - \frac{1}{2}\frac{\partial \tau_\alpha}{\partial \mathbf{r}} \cdot \left(\rho_\alpha \overline{v_\alpha^2 \mathbf{F}_\alpha}\right) - \frac{\partial \tau_\alpha}{\partial \mathbf{r}} \cdot (\varepsilon_\alpha n_\alpha \mathbf{F}_\alpha), \tag{2.6.7}$$

$$\int \left(\frac{m_\alpha v_\alpha^2}{2} + \varepsilon_\alpha\right)\frac{\partial^2 f_\alpha}{\partial t^2}\,\mathrm{d}\mathbf{v}_\alpha = \frac{\partial^2}{\partial t^2}\left(\frac{\rho_\alpha \overline{v_\alpha^2}}{2} + \varepsilon_\alpha n_\alpha\right), \tag{2.6.8}$$

$$\int \left(\frac{m_\alpha v_\alpha^2}{2} + \varepsilon_\alpha\right)\frac{\partial^2 f_\alpha}{\partial \mathbf{r}\partial t} \cdot \mathbf{v}_\alpha\,\mathrm{d}\mathbf{v}_\alpha = \frac{\partial^2}{\partial \mathbf{r}\partial t} \cdot \left(\frac{1}{2}\rho_\alpha \overline{v_\alpha^2 \mathbf{v}_\alpha} + \varepsilon_\alpha n_\alpha \bar{\mathbf{v}}_\alpha\right), \tag{2.6.9}$$

$$\int \left(\frac{m_\alpha v_\alpha^2}{2} + \varepsilon_\alpha\right)\frac{\partial^2 f_\alpha}{\partial \mathbf{r}\partial \mathbf{r}} : \mathbf{v}_\alpha \mathbf{v}_\alpha\,\mathrm{d}\mathbf{v}_\alpha$$

$$= \frac{\partial^2}{\partial \mathbf{r}\partial \mathbf{r}} : \left[\frac{1}{2}\rho_\alpha \overline{\mathbf{v}_\alpha \mathbf{v}_\alpha v_\alpha^2} + \varepsilon_\alpha n_\alpha \overline{\mathbf{v}_\alpha \mathbf{v}_\alpha}\right], \tag{2.6.10}$$

$$\int \left(\frac{m_\alpha v_\alpha^2}{2} + \varepsilon_\alpha\right)\frac{\partial^2 f_\alpha}{\partial \mathbf{v}_\alpha \partial t} \cdot \mathbf{F}_\alpha\,\mathrm{d}\mathbf{v}_\alpha = -\mathbf{F}_\alpha^{(1)} \cdot \frac{\partial}{\partial t}(\rho_\alpha \bar{\mathbf{v}}_\alpha), \tag{2.6.11}$$

$$\int \left(\frac{m_\alpha v_\alpha^2}{2} + \varepsilon_\alpha\right)\frac{\partial \mathbf{F}_\alpha}{\partial t} \cdot \frac{\partial f_\alpha}{\partial \mathbf{v}_\alpha}\,\mathrm{d}\mathbf{v}_\alpha = -\frac{\partial \mathbf{F}_\alpha^{(1)}}{\partial t} \cdot \rho_\alpha \bar{\mathbf{v}}_\alpha, \tag{2.6.12}$$

$$\int \left(\frac{m_\alpha v_\alpha^2}{2} + \varepsilon_\alpha\right)\mathbf{F}_\alpha \cdot \frac{\partial f_\alpha}{\partial \mathbf{r}}\,\mathrm{d}\mathbf{v}_\alpha$$

$$= \frac{1}{2}\frac{\partial}{\partial \mathbf{r}} \cdot \left(\rho_\alpha \overline{\mathbf{F}_\alpha v_\alpha^2}\right) + \frac{\partial}{\partial \mathbf{r}} \cdot \left(\varepsilon_\alpha n_\alpha \overline{\mathbf{F}}_\alpha\right) - \frac{1}{2}\rho_\alpha \overline{v_\alpha^2 \frac{\partial}{\partial \mathbf{r}} \cdot \mathbf{F}_\alpha}, \tag{2.6.13}$$

$$\int \left(\frac{m_\alpha v_\alpha^2}{2} + \varepsilon_\alpha\right)\frac{\partial f_\alpha}{\partial \mathbf{v}_\alpha} \cdot \left(\mathbf{F}_\alpha^{(1)} \times \mathbf{B}\right)\mathrm{d}\mathbf{v}_\alpha = -\rho_\alpha \bar{\mathbf{v}}_\alpha \cdot \left[\mathbf{F}_\alpha^{(1)} \times \mathbf{B}\right], \tag{2.6.14}$$

$$\int\left(\frac{m_\alpha v_\alpha^2}{2}+\varepsilon_\alpha\right)\frac{\partial f_\alpha}{\partial \mathbf{v}_\alpha}\cdot\left[\mathbf{B}(\mathbf{v}_\alpha\cdot\mathbf{B})-B^2\mathbf{v}_\alpha\right]\mathrm{d}\mathbf{v}_\alpha$$
$$=\frac{m_\alpha}{2}\int\frac{\partial f_\alpha}{\partial \mathbf{v}_\alpha}v_\alpha^2\cdot\left[\mathbf{B}(\mathbf{v}_\alpha\cdot\mathbf{B})-B^2\mathbf{v}_\alpha\right]\mathrm{d}\mathbf{v}_\alpha+2\varepsilon_\alpha n_\alpha B^2$$
$$=-\rho_\alpha\overline{\mathbf{v}_\alpha\mathbf{v}_\alpha}:\mathbf{BB}-B^2\frac{\rho_\alpha\overline{v_\alpha^2}}{2}+2\varepsilon_\alpha n_\alpha B^2-\frac{m_\alpha}{2}B^2\int\frac{\partial f_\alpha}{\partial \mathbf{v}_\alpha}\cdot v_\alpha^2\mathbf{v}_\alpha\,\mathrm{d}\mathbf{v}_\alpha$$
$$=-\rho_\alpha\overline{\mathbf{v}_\alpha\mathbf{v}_\alpha}:\mathbf{BB}+2\varepsilon_\alpha n_\alpha B^2+4B^2\frac{\rho_\alpha\overline{v_\alpha^2}}{2},\tag{2.6.15}$$

$$\int\left(\frac{m_\alpha v_\alpha^2}{2}+\varepsilon_\alpha\right)\frac{\partial f_\alpha}{\partial \mathbf{v}_\alpha}\mathbf{v}_\alpha:\frac{\partial}{\partial\mathbf{r}}\mathbf{F}_\alpha\,\mathrm{d}\mathbf{v}_\alpha$$
$$=\int\left(\frac{m_\alpha v_\alpha^2}{2}+\varepsilon_\alpha\right)\frac{\partial f_\alpha}{\partial \mathbf{v}_\alpha}\mathbf{v}_\alpha:\frac{\partial}{\partial\mathbf{r}}\mathbf{F}_\alpha^{(1)}\,\mathrm{d}\mathbf{v}_\alpha$$
$$+\int\left(\frac{m_\alpha v_\alpha^2}{2}+\varepsilon_\alpha\right)\frac{\partial f_\alpha}{\partial \mathbf{v}_\alpha}\mathbf{v}_\alpha:\frac{\partial}{\partial\mathbf{r}}\mathbf{F}_\alpha^{B}\,\mathrm{d}\mathbf{v}_\alpha.\tag{2.6.16}$$

Transformation of integrals on the right side of (2.6.16):

$$\int\left(\frac{m_\alpha v_\alpha^2}{2}+\varepsilon_\alpha\right)\frac{\partial f_\alpha}{\partial \mathbf{v}_\alpha}\mathbf{v}_\alpha:\frac{\partial}{\partial\mathbf{r}}\mathbf{F}_\alpha^{(1)}\,\mathrm{d}\mathbf{v}_\alpha$$
$$=-\left(\frac{\rho_\alpha v_\alpha^2}{2}+n_\alpha\varepsilon_\alpha\right)\frac{\partial}{\partial\mathbf{r}}\cdot\mathbf{F}_\alpha^{(1)}-\rho_\alpha\overline{\mathbf{v}_\alpha\mathbf{v}_\alpha}:\frac{\partial}{\partial\mathbf{r}}\mathbf{F}_\alpha^{(1)},\tag{2.6.17}$$

$$\int\left(\frac{m_\alpha v_\alpha^2}{2}+\varepsilon_\alpha\right)\frac{\partial f_\alpha}{\partial \mathbf{v}_\alpha}\mathbf{v}_\alpha:\frac{\partial}{\partial\mathbf{r}}\mathbf{F}_\alpha^{B}\,\mathrm{d}\mathbf{v}_\alpha$$
$$=\int\frac{m_\alpha v_\alpha^2}{2}\frac{\partial f_\alpha}{\partial \mathbf{v}_\alpha}:\frac{\partial}{\partial\mathbf{r}}\mathbf{F}_\alpha^{B}\,\mathrm{d}\mathbf{v}_\alpha+\frac{q_\alpha}{m_\alpha}n_\alpha\varepsilon_\alpha\bar{\mathbf{v}}_\alpha\cdot\operatorname{rot}\mathbf{B}$$
$$=-\sum_{i=1}^{3}\frac{m_\alpha}{2}\int v_\alpha^2 f_\alpha\frac{\partial}{\partial r_i}F_{\alpha i}^{B}\,\mathrm{d}\mathbf{v}_\alpha$$
$$-\sum_{ij}m_\alpha\int v_{\alpha i}v_{\alpha j}f_\alpha\frac{\partial}{\partial r_j}F_{\alpha i}^{B}\,\mathrm{d}\mathbf{v}_\alpha+\frac{q_\alpha}{m_\alpha}n_\alpha\varepsilon_\alpha\bar{\mathbf{v}}_\alpha\cdot\operatorname{rot}\mathbf{B}$$
$$=\frac{1}{2}n_\alpha q_\alpha\overline{v_\alpha^2\mathbf{v}_\alpha}\cdot\operatorname{rot}\mathbf{B}+\frac{q_\alpha}{m_\alpha}n_\alpha\varepsilon_\alpha\bar{\mathbf{v}}_\alpha\cdot\operatorname{rot}\mathbf{B}.\tag{2.6.18}$$

Using (2.6.17), (2.6.18), we finish the transformation of (2.6.16):

$$\int\left(\frac{m_\alpha v_\alpha^2}{2}+\varepsilon_\alpha\right)\frac{\partial f_\alpha}{\partial \mathbf{v}_\alpha}\mathbf{v}_\alpha:\frac{\partial}{\partial\mathbf{r}}\mathbf{F}_\alpha\,\mathrm{d}\mathbf{v}_\alpha$$

$$
= -\rho_\alpha \overline{\mathbf{v}_\alpha \mathbf{v}_\alpha} : \frac{\partial}{\partial \mathbf{r}} \mathbf{F}_\alpha^{(1)} - n_\alpha \left(\frac{m_\alpha \overline{v_\alpha^2}}{2} + \varepsilon_\alpha \right) \frac{\partial}{\partial \mathbf{r}} \cdot \mathbf{F}_\alpha^{(1)}
$$

$$
+ \frac{1}{2} n_\alpha q_\alpha \overline{v_\alpha^2 \mathbf{v}_\alpha} \cdot \operatorname{rot} \mathbf{B} + n_\alpha q_\alpha \varepsilon_\alpha \bar{\mathbf{v}}_\alpha \cdot \operatorname{rot} \mathbf{B}, \tag{2.6.19}
$$

$$
\int \left(\frac{m_\alpha v_\alpha^2}{2} + \varepsilon_\alpha \right) \frac{\partial^2 f_\alpha}{\partial \mathbf{v}_\alpha \partial \mathbf{v}_\alpha} : \mathbf{F}_\alpha \mathbf{F}_\alpha \, d\mathbf{v}_\alpha
$$

$$
= \int \left(\frac{m_\alpha v_\alpha^2}{2} + \varepsilon_\alpha \right) \frac{\partial^2 f_\alpha}{\partial \mathbf{v}_\alpha \partial \mathbf{v}_\alpha} : \mathbf{F}_\alpha^{(1)} \mathbf{F}_\alpha^{(1)} \, d\mathbf{v}_\alpha
$$

$$
+ 2 \int \left(\frac{m_\alpha v_\alpha^2}{2} + \varepsilon_\alpha \right) \frac{\partial^2 f_\alpha}{\partial \mathbf{v}_\alpha \partial \mathbf{v}_\alpha} : \mathbf{F}_\alpha^{(1)} \mathbf{F}_\alpha^{B} \, d\mathbf{v}_\alpha
$$

$$
+ \int \left(\frac{m_\alpha v_\alpha^2}{2} + \varepsilon_\alpha \right) \frac{\partial^2 f_\alpha}{\partial \mathbf{v}_\alpha \partial \mathbf{v}_\alpha} : \mathbf{F}_\alpha^{B} \mathbf{F}_\alpha^{B} \, d\mathbf{v}_\alpha. \tag{2.6.20}
$$

Consider transformation of every of three integrals on the right side of (2.6.20):

$$
\int \left(\frac{m_\alpha v_\alpha^2}{2} + \varepsilon_\alpha \right) \frac{\partial^2 f_\alpha}{\partial \mathbf{v}_\alpha \partial \mathbf{v}_\alpha} : \mathbf{F}_\alpha^{(1)} \mathbf{F}_\alpha^{(1)} \, d\mathbf{v}_\alpha = \rho_\alpha \mathbf{F}_\alpha^{(1)2}, \tag{2.6.21}
$$

$$
\int \left(\frac{m_\alpha v_\alpha^2}{2} + \varepsilon_\alpha \right) \frac{\partial^2 f_\alpha}{\partial \mathbf{v}_\alpha \partial \mathbf{v}_\alpha} : \mathbf{F}_\alpha^{(1)} \mathbf{F}_\alpha^{B} \, d\mathbf{v}_\alpha
$$

$$
= m_\alpha \sum_{\substack{i,j=1 \\ i \neq j}}^{3} \int f_\alpha F_{\alpha i}^{(1)} \frac{\partial F_{\alpha j}^{B}}{\partial v_{\alpha i}} v_{\alpha j} \, d\mathbf{v}_\alpha + m_\alpha \sum_{i=1}^{3} \int f_\alpha F_{\alpha i}^{(1)} F_{\alpha i}^{B} \, d\mathbf{v}_\alpha. \tag{2.6.22}
$$

Using the explicit expression for the Lorentz force we find if

$$
\sum_{\substack{i,j \\ i \neq j}} \int f_\alpha F_{\alpha i}^{(1)} \frac{\partial F_{\alpha j}^{B}}{\partial v_{\alpha i}} v_{\alpha j} \, d\bar{\mathbf{v}}_\alpha = -\frac{q_\alpha}{m_\alpha} n_\alpha \mathbf{F}_\alpha^{(1)} \cdot [\bar{\mathbf{v}}_\alpha \times \mathbf{B}], \tag{2.6.23}
$$

then the right-hand side of relation (2.6.22) is equal to zero. The last of three mentioned integrals can be transformed as follows:

$$
\int \left(\frac{m_\alpha v_\alpha^2}{2} + \varepsilon_\alpha \right) \frac{\partial^2 f_\alpha}{\partial \mathbf{v}_\alpha \partial \mathbf{v}_\alpha} : \mathbf{F}_\alpha^{B} \mathbf{F}_\alpha^{B} \, d\mathbf{v}_\alpha
$$

$$
= \rho_\alpha \overline{\left(F_\alpha^{B} \right)^2} - 2 \left(\frac{\rho_\alpha \overline{v_\alpha^2}}{2} + \varepsilon_\alpha n_\alpha \right) \left(\frac{q_\alpha}{m_\alpha} \right)^2 B^2
$$

$$
+ 2 \left(\frac{q_\alpha}{m_\alpha} \right)^2 \rho_\alpha \left[\overline{(\mathbf{B} \cdot \mathbf{v}_\alpha)^2 - v_\alpha^2 B^2} \right]
$$

$$= \rho_\alpha \left(\frac{q_\alpha}{m_\alpha}\right)^2 \left[\overline{(\mathbf{B}\cdot\mathbf{v}_\alpha)^2 - 2v_\alpha^2 B^2}\right] - 2B^2\left(\frac{q_\alpha}{m_\alpha}\right)^2 \varepsilon_\alpha n_\alpha. \tag{2.6.24}$$

Using (2.6.21)–(2.6.24), one obtains

$$\int \left(\frac{m_\alpha v_\alpha^2}{2} + \varepsilon_\alpha\right) \frac{\partial^2 f_\alpha}{\partial \mathbf{v}_\alpha \partial \mathbf{v}_\alpha} : \mathbf{F}_\alpha \mathbf{F}_\alpha \, \mathrm{d}\mathbf{v}_\alpha$$
$$= \rho_\alpha F_\alpha^{(1)2} + \left(\frac{q_\alpha}{m_\alpha}\right)^2 \rho_\alpha \left\{\overline{(\mathbf{B}\cdot\mathbf{v}_\alpha)^2} - 2\overline{v_\alpha^2} B^2\right\}$$
$$- 2B^2\left(\frac{q_\alpha}{m_\alpha}\right)^2 \varepsilon_\alpha n_\alpha. \tag{2.6.25}$$

Calculate now the last integral of the left side of (2.6.1)

$$\int \left(\frac{m_\alpha v_\alpha^2}{2} + \varepsilon_\alpha\right) \frac{\partial^2 f_\alpha}{\partial \mathbf{v}_\alpha \partial \mathbf{r}} : \mathbf{v}_\alpha \mathbf{F}_\alpha \, \mathrm{d}\mathbf{v}_\alpha$$
$$= \int \left(\frac{m_\alpha v_\alpha^2}{2} + \varepsilon_\alpha\right) \frac{\partial^2 f_\alpha}{\partial \mathbf{v}_\alpha \partial \mathbf{r}} : \mathbf{v}_\alpha \mathbf{F}_\alpha^{(1)} \, \mathrm{d}\mathbf{v}_\alpha$$
$$+ \int \left(\frac{m_\alpha v_\alpha^2}{2} + \varepsilon_\alpha\right) \frac{\partial^2 f_\alpha}{\partial \mathbf{v}_\alpha \partial \mathbf{r}} : \mathbf{v}_\alpha \mathbf{F}_\alpha^{B} \, \mathrm{d}\mathbf{v}_\alpha. \tag{2.6.26}$$

The first integral can be written in the form

$$\int \left(\frac{m_\alpha v_\alpha^2}{2} + \varepsilon_\alpha\right) \frac{\partial^2 f_\alpha}{\partial \mathbf{v}_\alpha \partial \mathbf{r}} : \mathbf{v}_\alpha \mathbf{F}_\alpha^{(1)} \, \mathrm{d}\mathbf{v}_\alpha$$
$$= -\mathbf{F}_\alpha^{(1)} \cdot \left\{\frac{\partial}{\partial \mathbf{r}} \cdot \rho_\alpha \overline{\mathbf{v}_\alpha \mathbf{v}_\alpha} + \frac{\partial}{\partial \mathbf{r}}\left(\frac{\rho_\alpha \overline{v_\alpha^2}}{2} + \varepsilon_\alpha n_\alpha\right)\right\}. \tag{2.6.27}$$

The second integral is transformed using explicit form of the Lorentz force

$$\int \left(\frac{m_\alpha v_\alpha^2}{2} + \varepsilon_\alpha\right) \frac{\partial^2 f_\alpha}{\partial \mathbf{v}_\alpha \partial \mathbf{r}} : \mathbf{v}_\alpha \mathbf{F}_\alpha^{B} \, \mathrm{d}\mathbf{v}_\alpha$$
$$= \frac{q_\alpha}{m_\alpha}\left\{\int f_\alpha \left(\frac{m_\alpha v_\alpha^2}{2} + \varepsilon_\alpha\right) \frac{\partial}{\partial \mathbf{r}} \cdot [\mathbf{v}_\alpha \times \mathbf{B}] \, \mathrm{d}\mathbf{v}_\alpha\right.$$
$$\left. - \frac{\partial}{\partial \mathbf{r}} \cdot \int f_\alpha [\mathbf{v}_\alpha \times \mathbf{B}] \left(\frac{m_\alpha v_\alpha^2}{2} + \varepsilon_\alpha\right) \mathrm{d}\mathbf{v}_\alpha\right\}. \tag{2.6.28}$$

The identity holds

$$\frac{\partial}{\partial \mathbf{r}} \cdot [\mathbf{v}_\alpha \times \mathbf{B}] = -\mathbf{v}_\alpha \cdot \mathrm{rot}\, \mathbf{B}. \tag{2.6.29}$$

Then (2.6.28) is written as

$$\int\left(\frac{m_\alpha v_\alpha^2}{2}+\varepsilon_\alpha\right)\frac{\partial^2 f_\alpha}{\partial \mathbf{v}_\alpha \partial \mathbf{r}}:\mathbf{v}_\alpha \mathbf{F}_\alpha^B\,\mathrm{d}\mathbf{v}_\alpha$$
$$=-\frac{1}{2}\operatorname{rot}\mathbf{B}\cdot\frac{q_\alpha}{m_\alpha}\rho_\alpha\overline{v_\alpha^2\mathbf{v}_\alpha}-\operatorname{rot}\mathbf{B}\cdot\frac{q_\alpha}{m_\alpha}\varepsilon_\alpha n_\alpha\bar{\mathbf{v}}_\alpha$$
$$-\frac{1}{2}\frac{\partial}{\partial \mathbf{r}}\cdot\rho_\alpha\frac{q_\alpha}{m_\alpha}\overline{[\mathbf{v}_\alpha\times\mathbf{B}]v_\alpha^2}-\frac{\partial}{\partial \mathbf{r}}\cdot\frac{q_\alpha}{m_\alpha}n_\alpha\varepsilon_\alpha[\bar{\mathbf{v}}_\alpha\times\mathbf{B}]. \qquad (2.6.30)$$

For integral (2.6.26) we have

$$\int\left(\frac{m_\alpha v_\alpha^2}{2}+\varepsilon_\alpha\right)\frac{\partial^2 f_\alpha}{\partial \mathbf{v}_\alpha \partial \mathbf{r}}:\mathbf{v}_\alpha \mathbf{F}_\alpha\,\mathrm{d}\mathbf{v}_\alpha$$
$$=-\mathbf{F}_\alpha^{(1)}\cdot\left\{\frac{\partial}{\partial \mathbf{r}}\cdot\rho_\alpha\overline{\mathbf{v}_\alpha\mathbf{v}_\alpha}+\frac{\partial}{\partial \mathbf{r}}\left(\frac{\rho_\alpha\overline{v_\alpha^2}}{2}+\varepsilon_\alpha n_\alpha\right)\right\}$$
$$-\frac{1}{2}\operatorname{rot}\mathbf{B}\cdot\frac{q_\alpha}{m_\alpha}\rho_\alpha\overline{v_\alpha^2\mathbf{v}_\alpha}-\operatorname{rot}\mathbf{B}\cdot\frac{q_\alpha}{m_\alpha}\varepsilon_\alpha n_\alpha\bar{\mathbf{v}}_\alpha$$
$$-\frac{1}{2}\frac{\partial}{\partial \mathbf{r}}\cdot\rho_\alpha\frac{q_\alpha}{m_\alpha}\overline{[\mathbf{v}_\alpha\times\mathbf{B}]v_\alpha^2}-\frac{\partial}{\partial \mathbf{r}}\cdot\frac{q_\alpha}{m_\alpha}n_\alpha\varepsilon_\alpha[\bar{\mathbf{v}}_\alpha\times\mathbf{B}]. \qquad (2.6.31)$$

Combining all results of integral transformation in (2.6.1), we find the following form of the generalized energy equation:

$$\frac{\partial}{\partial t}\left\{\frac{\rho_\alpha\overline{v_\alpha^2}}{2}+\varepsilon_\alpha n_\alpha-\tau_\alpha\left[\frac{\partial}{\partial t}\left(\frac{\rho_\alpha\overline{v_\alpha^2}}{2}+\varepsilon_\alpha n_\alpha\right)\right.\right.$$
$$\left.\left.+\frac{\partial}{\partial \mathbf{r}}\cdot\left(\frac{1}{2}\rho_\alpha\overline{v_\alpha^2\mathbf{v}_\alpha}+\varepsilon_\alpha n_\alpha\bar{\mathbf{v}}_\alpha\right)-\mathbf{F}_\alpha^{(1)}\cdot\rho_\alpha\bar{\mathbf{v}}_\alpha\right]\right\}$$
$$+\frac{\partial}{\partial \mathbf{r}}\cdot\left\{\frac{1}{2}\rho_\alpha\overline{v_\alpha^2\mathbf{v}_\alpha}+\varepsilon_\alpha n_\alpha\bar{\mathbf{v}}_\alpha-\tau_\alpha\left[\frac{\partial}{\partial t}\left(\frac{1}{2}\rho_\alpha\overline{v_\alpha^2\mathbf{v}_\alpha}+\varepsilon_\alpha n_\alpha\bar{\mathbf{v}}_\alpha\right)\right.\right.$$
$$+\frac{\partial}{\partial \mathbf{r}}\cdot\left(\frac{1}{2}\rho_\alpha\overline{v_\alpha^2\mathbf{v}_\alpha\mathbf{v}_\alpha}+\varepsilon_\alpha n_\alpha\overline{\mathbf{v}_\alpha\mathbf{v}_\alpha}\right)-\rho_\alpha\mathbf{F}_\alpha^{(1)}\cdot\overline{\mathbf{v}_\alpha\mathbf{v}_\alpha}-\frac{1}{2}\rho_\alpha\overline{v_\alpha^2\mathbf{F}_\alpha}$$
$$\left.\left.-\varepsilon_\alpha n_\alpha\overline{\mathbf{F}}_\alpha\right]\right\}-\left\{\rho_\alpha\mathbf{F}_\alpha^{(1)}\cdot\bar{\mathbf{v}}_\alpha-\tau_\alpha\left[\mathbf{F}_\alpha^{(1)}\cdot\left(\frac{\partial}{\partial t}(\rho_\alpha\bar{\mathbf{v}}_\alpha)\right.\right.\right.$$
$$\left.\left.\left.+\frac{\partial}{\partial \mathbf{r}}\cdot\rho_\alpha\overline{\mathbf{v}_\alpha\mathbf{v}_\alpha}-\rho_\alpha\mathbf{F}_\alpha^{(1)}-q_\alpha n_\alpha\bar{\mathbf{v}}_\alpha\times\mathbf{B}\right)\right]\right\}$$
$$=\int\left(\frac{m_\alpha v_\alpha^2}{2}+\varepsilon_\alpha\right)J_\alpha^{\mathrm{st,el}}\,\mathrm{d}\mathbf{v}_\alpha+\int\left(\frac{m_\alpha v_\alpha^2}{2}+\varepsilon_\alpha\right)J_\alpha^{\mathrm{st,inel}}\,\mathrm{d}\mathbf{v}_\alpha. \qquad (2.6.32)$$

2.7. Generalized hydrodynamic Euler equations

Formulate now the summary of the generalized Enskog hydrodynamic equations for components.

Continuity equation

$$\frac{\partial}{\partial t}\left\{\rho_\alpha - \tau_\alpha\left[\frac{\partial \rho_\alpha}{\partial t} + \frac{\partial}{\partial \mathbf{r}}\cdot(\rho_\alpha \bar{\mathbf{v}}_\alpha)\right]\right\}
+ \frac{\partial}{\partial \mathbf{r}}\cdot\left\{\rho_\alpha \bar{\mathbf{v}}_\alpha - \tau_\alpha\left[\frac{\partial}{\partial t}(\rho_\alpha \bar{\mathbf{v}}_\alpha)
+ \frac{\partial}{\partial \mathbf{r}}\cdot\left(\rho_\alpha \overline{\mathbf{v}_\alpha \mathbf{v}_\alpha}\right) - \rho_\alpha \mathbf{F}_\alpha^{(1)} - \frac{q_\alpha}{m_\alpha}\rho_\alpha \bar{\mathbf{v}}_\alpha \times \mathbf{B}\right]\right\} = R_\alpha. \tag{2.7.1}$$

Momentum equation

$$\frac{\partial}{\partial t}\left\{\rho_\alpha \bar{\mathbf{v}}_\alpha - \tau_\alpha\left[\frac{\partial}{\partial t}(\rho_\alpha \bar{\mathbf{v}}_\alpha) + \frac{\partial}{\partial \mathbf{r}}\cdot \rho_\alpha \overline{\mathbf{v}_\alpha \mathbf{v}_\alpha} - \rho_\alpha \mathbf{F}_\alpha^{(1)} - \frac{q_\alpha}{m_\alpha}\rho_\alpha \bar{\mathbf{v}}_\alpha \times \mathbf{B}\right]\right\}
- \mathbf{F}_\alpha^{(1)}\left[\rho_\alpha - \tau_\alpha\left(\frac{\partial \rho_\alpha}{\partial t} + \frac{\partial}{\partial \mathbf{r}}\cdot \rho_\alpha \bar{\mathbf{v}}_\alpha\right)\right]
- \frac{q_\alpha}{m_\alpha}\left\{\rho_\alpha \bar{\mathbf{v}}_\alpha - \tau_\alpha\left[\frac{\partial}{\partial t}(\rho_\alpha \bar{\mathbf{v}}_\alpha) + \frac{\partial}{\partial \mathbf{r}}\cdot \rho_\alpha \overline{\mathbf{v}_\alpha \mathbf{v}_\alpha} - \rho_\alpha \mathbf{F}_\alpha^{(1)}
- \frac{q_\alpha}{m_\alpha}\rho_\alpha \bar{\mathbf{v}}_\alpha \times \mathbf{B}\right]\right\} \times \mathbf{B} + \frac{\partial}{\partial \mathbf{r}}\cdot\left\{\rho_\alpha \overline{\mathbf{v}_\alpha \mathbf{v}_\alpha} - \tau_\alpha\left[\frac{\partial}{\partial t}\left(\rho_\alpha \overline{\mathbf{v}_\alpha \mathbf{v}_\alpha}\right)
+ \frac{\partial}{\partial \mathbf{r}}\cdot \rho_\alpha \overline{(\mathbf{v}_\alpha \mathbf{v}_\alpha)\mathbf{v}_\alpha} - \mathbf{F}_\alpha^{(1)}\rho_\alpha \bar{\mathbf{v}}_\alpha - \rho_\alpha \bar{\mathbf{v}}_\alpha \mathbf{F}_\alpha^{(1)}
- \frac{q_\alpha}{m_\alpha}\rho_\alpha \overline{[\mathbf{v}_\alpha \times \mathbf{B}]\mathbf{v}_\alpha} - \frac{q_\alpha}{m_\alpha}\rho_\alpha \overline{\mathbf{v}_\alpha[\mathbf{v}_\alpha \times \mathbf{B}]}\right]\right\}
= \int m_\alpha \mathbf{v}_\alpha J_\alpha^{\mathrm{st,el}}\,\mathrm{d}\mathbf{v}_\alpha + \int m_\alpha \mathbf{v}_\alpha J_\alpha^{\mathrm{st,inel}}\,\mathrm{d}\mathbf{v}_\alpha. \tag{2.7.2}$$

Eenergy equation

$$\frac{\partial}{\partial t}\left\{\frac{\rho_\alpha \overline{v_\alpha^2}}{2} + \varepsilon_\alpha n_\alpha - \tau_\alpha\left[\frac{\partial}{\partial t}\left(\frac{\rho_\alpha \overline{v_\alpha^2}}{2} + \varepsilon_\alpha n_\alpha\right)
+ \frac{\partial}{\partial \mathbf{r}}\cdot\left(\frac{1}{2}\rho_\alpha \overline{v_\alpha^2 \mathbf{v}_\alpha} + \varepsilon_\alpha n_\alpha \bar{\mathbf{v}}_\alpha\right) - \mathbf{F}_\alpha^{(1)}\cdot \rho_\alpha \bar{\mathbf{v}}_\alpha\right]\right\}
+ \frac{\partial}{\partial \mathbf{r}}\cdot\left\{\frac{1}{2}\rho_\alpha \overline{\mathbf{v}_\alpha v_\alpha^2} + \varepsilon_\alpha n_\alpha \bar{\mathbf{v}}_\alpha - \tau_\alpha\left[\frac{\partial}{\partial t}\left(\frac{1}{2}\rho_\alpha \overline{v_\alpha^2 \mathbf{v}_\alpha} + \varepsilon_\alpha n_\alpha \bar{\mathbf{v}}_\alpha\right)\right.\right.$$

$$
\begin{aligned}
&+\frac{\partial}{\partial \mathbf{r}}\cdot\left(\frac{1}{2}\rho_\alpha\overline{v_\alpha^2\mathbf{v}_\alpha\mathbf{v}_\alpha}+\varepsilon_\alpha n_\alpha\overline{\mathbf{v}_\alpha\mathbf{v}_\alpha}\right)\\
&-\rho_\alpha\mathbf{F}_\alpha^{(1)}\cdot\overline{\mathbf{v}_\alpha\mathbf{v}_\alpha}-\frac{1}{2}\rho_\alpha\overline{v_\alpha^2\mathbf{F}_\alpha}-\varepsilon_\alpha n_\alpha\overline{\mathbf{F}}_\alpha\Bigg]\Bigg\}\\
&-\Bigg\{\rho_\alpha\mathbf{F}_\alpha^{(1)}\cdot\bar{\mathbf{v}}_\alpha-\tau_\alpha\Bigg[\mathbf{F}_\alpha^{(1)}\cdot\left(\frac{\partial}{\partial t}(\rho_\alpha\bar{\mathbf{v}}_\alpha)+\frac{\partial}{\partial \mathbf{r}}\cdot\rho_\alpha\overline{\mathbf{v}_\alpha\mathbf{v}_\alpha}\right.\\
&\left.-\rho_\alpha\mathbf{F}_\alpha^{(1)}-q_\alpha n_\alpha\bar{\mathbf{v}}_\alpha\times\mathbf{B}\right)\Bigg]\Bigg\}\\
=&\int\left(\frac{m_\alpha v_\alpha^2}{2}+\varepsilon_\alpha\right)J_\alpha^{\mathrm{st,el}}\,\mathrm{d}\mathbf{v}_\alpha+\int\left(\frac{m_\alpha v_\alpha^2}{2}+\varepsilon_\alpha\right)J_\alpha^{\mathrm{st,inel}}\,\mathrm{d}\mathbf{v}_\alpha. \qquad (2.7.3)
\end{aligned}
$$

GHE for mixture of gases are significantly more simple because after summation of the left and right sides of (2.7.1)–(2.7.3) over all components α ($\alpha = 1, \ldots, \eta$), the right integral parts of these equations reduce to zero on the strength of conservation laws. Then we have

$$
\begin{aligned}
&\frac{\partial}{\partial t}\left\{\rho-\sum_\alpha\tau_\alpha\left[\frac{\partial\rho_\alpha}{\partial t}+\frac{\partial}{\partial \mathbf{r}}\cdot(\rho_\alpha\bar{\mathbf{v}}_\alpha)\right]\right\}\\
&+\frac{\partial}{\partial \mathbf{r}}\cdot\Bigg\{\rho\mathbf{v}_0-\sum_\alpha\tau_\alpha\Bigg[\frac{\partial}{\partial t}(\rho_\alpha\bar{\mathbf{v}}_\alpha)+\frac{\partial}{\partial \vec{r}}\cdot\left(\rho_\alpha\overline{\mathbf{v}_\alpha\mathbf{v}_\alpha}\right)-\rho_\alpha\mathbf{F}_\alpha^{(1)}\\
&-\frac{q_\alpha}{m_\alpha}\rho_\alpha\bar{\mathbf{v}}_\alpha\times\mathbf{B}\Bigg]\Bigg\}=0, \qquad (2.7.4)
\end{aligned}
$$

$$
\begin{aligned}
&\frac{\partial}{\partial t}\left\{\rho\mathbf{v}_0-\sum_\alpha\tau_\alpha\left[\frac{\partial}{\partial t}(\rho_\alpha\bar{\mathbf{v}}_\alpha)+\frac{\partial}{\partial \mathbf{r}}\cdot\rho_\alpha\overline{\mathbf{v}_\alpha\mathbf{v}_\alpha}-\rho_\alpha\mathbf{F}_\alpha^{(1)}-\frac{q_\alpha}{m_\alpha}\rho_\alpha\bar{\mathbf{v}}_\alpha\times\mathbf{B}\right]\right\}\\
&-\sum_\alpha\mathbf{F}_\alpha^{(1)}\left[\rho_\alpha-\tau_\alpha\left(\frac{\partial\rho_\alpha}{\partial t}+\frac{\partial}{\partial \mathbf{r}}\cdot\rho_\alpha\bar{\mathbf{v}}_\alpha\right)\right]\\
&-\sum_\alpha\frac{q_\alpha}{m_\alpha}\Bigg\{\rho_\alpha\bar{\mathbf{v}}_\alpha-\tau_\alpha\Bigg[\frac{\partial}{\partial t}(\rho_\alpha\bar{\mathbf{v}}_\alpha)+\frac{\partial}{\partial \mathbf{r}}\cdot\rho_\alpha\overline{\mathbf{v}_\alpha\mathbf{v}_\alpha}-\rho_\alpha\mathbf{F}_\alpha^{(1)}\\
&-\frac{q_\alpha}{m_\alpha}\rho_\alpha\bar{\mathbf{v}}_\alpha\times\mathbf{B}\Bigg]\Bigg\}\times\mathbf{B}+\frac{\partial}{\partial \mathbf{r}}\cdot\Bigg\{\sum_\alpha\rho_\alpha\overline{\mathbf{v}_\alpha\mathbf{v}_\alpha}\\
&-\sum_\alpha\tau_\alpha\Bigg[\frac{\partial}{\partial t}\left(\rho_\alpha\overline{\mathbf{v}_\alpha\mathbf{v}_\alpha}\right)+\frac{\partial}{\partial \mathbf{r}}\cdot\rho_\alpha\overline{(\mathbf{v}_\alpha\mathbf{v}_\alpha)\mathbf{v}_\alpha}-\mathbf{F}_\alpha^{(1)}\rho_\alpha\bar{\mathbf{v}}_\alpha-\rho_\alpha\bar{\mathbf{v}}_\alpha\mathbf{F}_\alpha^{(1)}\\
&-\frac{q_\alpha}{m_\alpha}\rho_\alpha\overline{[\mathbf{v}_\alpha\times\mathbf{B}]\mathbf{v}_\alpha}-\frac{q_\alpha}{m_\alpha}\rho_\alpha\overline{\mathbf{v}_\alpha[\mathbf{v}_\alpha\times\mathbf{B}]}\Bigg]\Bigg\}=0, \qquad (2.7.5)
\end{aligned}
$$

$$\frac{\partial}{\partial t}\left\{\sum_\alpha\left(\frac{\rho_\alpha\overline{v_\alpha^2}}{2}+\varepsilon_\alpha n_\alpha\right)-\sum_\alpha\tau_\alpha\left[\frac{\partial}{\partial t}\left(\frac{\rho_\alpha\overline{v_\alpha^2}}{2}+\varepsilon_\alpha n_\alpha\right)\right.\right.$$

$$\left.\left.+\frac{\partial}{\partial \mathbf{r}}\cdot\left(\frac{1}{2}\rho_\alpha\overline{v_\alpha^2\mathbf{v}_\alpha}+\varepsilon_\alpha n_\alpha\bar{\mathbf{v}}_\alpha\right)-\mathbf{F}_\alpha^{(1)}\cdot\rho_\alpha\bar{\mathbf{v}}_\alpha\right]\right\}$$

$$+\frac{\partial}{\partial \mathbf{r}}\cdot\left\{\sum_\alpha\left(\frac{1}{2}\rho_\alpha\overline{\mathbf{v}_\alpha v_\alpha^2}+\varepsilon_\alpha n_\alpha\bar{\mathbf{v}}_\alpha\right)-\sum_\alpha\tau_\alpha\left[\frac{\partial}{\partial t}\left(\frac{1}{2}\rho_\alpha\overline{\mathbf{v}_\alpha v_\alpha^2}+\varepsilon_\alpha n_\alpha\bar{\mathbf{v}}_\alpha\right)\right.\right.$$

$$+\frac{\partial}{\partial \mathbf{r}}\cdot\left(\frac{1}{2}\rho_\alpha\overline{v_\alpha^2\mathbf{v}_\alpha\mathbf{v}_\alpha}+\varepsilon_\alpha n_\alpha\overline{\mathbf{v}_\alpha\mathbf{v}_\alpha}\right)-\rho_\alpha\mathbf{F}_\alpha^{(1)}\cdot\overline{\mathbf{v}_\alpha\mathbf{v}_\alpha}$$

$$\left.\left.-\frac{1}{2}\rho_\alpha\overline{v_\alpha^2\mathbf{F}_\alpha}-\varepsilon_\alpha n_\alpha\overline{\mathbf{F}}_\alpha\right]\right\}-\left\{\sum_\alpha\rho_\alpha\mathbf{F}_\alpha^{(1)}\cdot\bar{\mathbf{v}}_\alpha-\sum_\alpha\tau_\alpha\left[\mathbf{F}_\alpha^{(1)}\cdot\left(\frac{\partial}{\partial t}(\rho_\alpha\bar{\mathbf{v}}_\alpha)\right.\right.\right.$$

$$\left.\left.\left.+\frac{\partial}{\partial \mathbf{r}}\cdot\rho_\alpha\overline{\mathbf{v}_\alpha\mathbf{v}_\alpha}-\rho_\alpha\mathbf{F}_\alpha^{(1)}-q_\alpha n_\alpha\bar{\mathbf{v}}_\alpha\times\mathbf{B}\right)\right]\right\}=0. \tag{2.7.6}$$

Generalized hydrodynamic Enskog equations (GHEnE) are very complicated. In particular, GEnE are of higher order than classical Enskog equations. This brings up immediately two problems: (1) is it possible to simplify these equations? (2) what about the additional boundary and initial conditions for these equations?

Let us begin by looking at the first question. Extraordinary features of GHEnE consist in appearance of terms proportional to τ_α in all Eqs. (2.7.1)–(2.7.6). Mean time between collisions τ_α of particles belonging to α-species can be calculated if DF f_α is known. For the model of hard spheres in one-component gas

$$\tau_\alpha^{(0)}=\frac{1}{\pi\sqrt{2}\bar{v}_\alpha n_\alpha\sigma_\alpha^2}, \tag{2.7.7}$$

where $\bar{v}_\alpha$ is mean velocity of α-molecules, σ_α is their diameter, upper index (0) underlines that calculation of $\tau_\alpha^{(0)}$ is realized for local Maxwellian function

$$\bar{v}_\alpha=\sqrt{\frac{8k_BT}{\pi m_\alpha}}. \tag{2.7.8}$$

Then

$$\tau_\alpha^{(0)}=\frac{\sqrt{m_\alpha}}{4\sqrt{\pi}n_\alpha\sigma_\alpha^2\sqrt{k_BT}}. \tag{2.7.9}$$

Multiplying term-by-term (2.7.9) by $p_\alpha=n_\alpha kT$, one obtains

$$\tau_\alpha^{(0)}p_\alpha=\frac{\sqrt{m_\alpha k_BT}}{4\sqrt{\pi}\sigma_\alpha^2}. \tag{2.7.10}$$

But for simple gas in the first ("Navier–Stokes") approximation dynamical viscosity for the hard spheres model is written as (Hirschfelder, Curtiss and Bird, 1954)

$$[\mu]_1 = \frac{5}{16}\frac{\sqrt{m_\alpha k_B T}}{\sqrt{\pi}\sigma_\alpha^2}. \tag{2.7.11}$$

Of course, the subject under discussion is changing of the transport coefficients calculation in GBKT but this question we leave to Chapter 5. Now we state that mean time between collisions in local Maxwellian approximation $\tau_\alpha^{(0)}$ can be presented using viscosity $[\mu_\alpha]_1$, calculated in the next successive approximation. In other words,

$$\tau_\alpha^{(0)} p_\alpha = 0.8[\mu_\alpha]_1. \tag{2.7.12}$$

This fact leads to very interesting conclusions in the theory of turbulence considered in Chapter 5 from positions of GBKT.

The use in BKT instead of the first approximation, the convergence series in Sonine polynomials for hard spheres leads to slightly changing of mentioned coefficient. One obtains 0.786 instead of 0.8.

Consider a simple gas under no forces in BGK approximation for classical Boltzmann equation (BE)

$$\frac{\partial f}{\partial t} + \mathbf{v}\cdot\frac{\partial f}{\partial \mathbf{r}} = -\frac{f - f^{(0)}}{\tau}. \tag{2.7.13}$$

From Eq. (2.7.13) follows the expression for DF f

$$f = f^{(0)} - \tau\frac{\partial f}{\partial t} + \tau\mathbf{v}\cdot\frac{\partial f}{\partial \mathbf{r}}. \tag{2.7.14}$$

For a gas consisting of point-like particles in the case of slow changing of parameters of state, $f - f^{(0)}$ must be small and Eq. (2.7.15) becomes

$$f = f^{(0)} - \tau^{(0)}\frac{\partial f^{(0)}}{\partial t} + \tau^{(0)}\mathbf{v}\cdot\frac{\partial f^{(0)}}{\partial \mathbf{r}}. \tag{2.7.15}$$

In one-dimensional stationary approximation

$$f = f^{(0)} + \tau^{(0)}\mathbf{v}\cdot\frac{\partial f^{(0)}}{\partial \mathbf{r}}, \tag{2.7.16}$$

and for a simple gas of uniform density and temperature, streaming parallel to Ox, but with hydrodynamic velocity changing along the axis Oy it is reasonable to suppose that changing of local Maxwellian DF $f^{(0)}$ is governed by evolution of hydrodynamic velocity v_0. As a result,

$$f = f^{(0)} + \tau^{(0)} v_y \frac{\partial v_0}{\partial y}\frac{\partial f^{(0)}}{\partial v_0}. \tag{2.7.17}$$

The viscous stress in the y-direction across a plane $x = const$ is

$$p_{xy} = \int m(v_x - v_0)v_y f\,\mathrm{d}\mathbf{v}. \tag{2.7.18}$$

Substitute f from (2.7.17) into Eq. (2.7.18)

$$p_{xy} = \int m(v_x - v_0)v_y f^{(0)}\,\mathrm{d}\mathbf{v} + \tau^{(0)}\frac{\partial v_0}{\partial y}\int v_y \frac{\partial f^{(0)}}{\partial v_0} m(v_x - v_0)v_y\,\mathrm{d}\mathbf{v}. \tag{2.7.19}$$

Because the first integrand is an odd function of $v_x - v_0$, the first integral vanishes and

$$p_{xy} = \tau^{(0)}\int v_y \frac{\partial f^{(0)}}{\partial v_0} m(v_x - v_0)v_y\,\mathrm{d}\mathbf{v}\frac{\partial v_0}{\partial y}, \tag{2.7.20}$$

or

$$p_{xy} = \mu\frac{\partial v_0}{\partial y}, \tag{2.7.21}$$

where coefficient of viscosity μ is equal to

$$\mu = \tau^{(0)}\int v_y \frac{\partial f^{(0)}}{\partial v_0} m(v_x - v_0)v_y\,\mathrm{d}\mathbf{v}, \tag{2.7.22}$$

or

$$\mu = \tau^{(0)}\frac{\partial}{\partial v_0}\int m f^{(0)}(v_x - v_0)v_y^2\,\mathrm{d}\mathbf{v} + \tau^{(0)}\int m f^{(0)} v_y^2\,\mathrm{d}\mathbf{v}. \tag{2.7.23}$$

The first integral in (2.7.23) vanishes as containing an odd function in the integrand, but

$$p = \int m f^{(0)} v_y^2\,\mathrm{d}\mathbf{v}. \tag{2.7.24}$$

It means that in considering stiff restrictions the valid formula is

$$\tau^{(0)} p = \mu. \tag{2.7.25}$$

Relation (2.7.25) is also a consequence of the so-called elementary theory of gases.

Finally we can state that

$$\tau_\alpha^{(0)} p_\alpha = \Pi\mu_\alpha, \quad \alpha = 1, \ldots, \eta, \tag{2.7.26}$$

where numerical coefficient Π reflects the characteristic features of particle interactions.

In the theory of generalized hydrodynamic equations, relations (2.7.26) allow to close GHE leaving all calculations – in the definite sense – on the macroscopic level of physical system's description.

If a mixture of gases contains particles of which the masses are not too different, it is possible to set in Eqs. (2.7.1)–(2.7.6) τ_α as independent of number of species. Therefore, the problem of the τ_α calculation is not difficult and exactly the same as in classical BKT.

The difficulties of principal character leads to the problem calculation of averaged values in (2.7.1)–(2.7.6) for which we need explicit expression for DF. But for local Maxwellian DF all averaged values in (2.7.1)–(2.7.6) can be calculated in finite form and lead to generalized Euler equations (GEuE).

Calculate averaged values for the velocities moments of DF $f_\alpha^{(0)}$. Summary of results follows.

For continuity equation:

$$\begin{aligned}\bar{\mathbf{v}}_\alpha &= \left(\frac{m_\alpha}{2\pi k_B T}\right)^{3/2} \int \mathbf{v}_\alpha \, \mathrm{e}^{-m_\alpha V_\alpha^2/(2k_B T)} \, d\mathbf{v}_\alpha \\ &= \left(\frac{m_\alpha}{2\pi k_B T}\right)^{3/2} \int \mathbf{v}_0 \, \mathrm{e}^{-m_\alpha V_\alpha^2/(2k_B T)} \, d\mathbf{v}_\alpha = \mathbf{v}_0,\end{aligned} \tag{2.7.27}$$

$$\begin{aligned}\overline{\mathbf{v}_\alpha \mathbf{v}_\alpha} &= \left(\frac{m_\alpha}{2\pi k_B T}\right)^{3/2} \int \mathbf{v}_\alpha \mathbf{v}_\alpha \, \mathrm{e}^{-m_\alpha V_\alpha^2/(2k_B T)} \, d\mathbf{v}_\alpha \\ &= \mathbf{v}_0 \mathbf{v}_0 + \frac{1}{n_\alpha} \int f_\alpha^{(0)} \mathbf{V}_\alpha \mathbf{V}_\alpha \, d\mathbf{v}_\alpha = \mathbf{v}_0 \mathbf{v}_0 + \frac{p_\alpha}{\rho_\alpha} \overset{\leftrightarrow}{I},\end{aligned} \tag{2.7.28}$$

where p_α is static pressure of α-species, $\overset{\leftrightarrow}{I}$ is unit tensor. In calculations realized in (2.7.27), (2.7.28) take into account vanishing integrals containing odd functions as integrand.

We have the following generalized Euler continuity equation (compare with (2.7.1), (2.7.4)):

$$\begin{aligned}&\frac{\partial}{\partial t}\left\{\rho_\alpha - \tau_\alpha^{(0)}\left[\frac{\partial \rho_\alpha}{\partial t} + \frac{\partial}{\partial \mathbf{r}} \cdot (\rho_\alpha \mathbf{v}_0)\right]\right\} \\ &\quad + \frac{\partial}{\partial \mathbf{r}} \cdot \left\{\rho_\alpha \mathbf{v}_0 - \tau_\alpha^{(0)}\left[\frac{\partial}{\partial t}(\rho_\alpha \mathbf{v}_0) + \frac{\partial}{\partial \mathbf{r}} \cdot (\rho_\alpha \mathbf{v}_0 \mathbf{v}_0) + \overset{\leftrightarrow}{I} \cdot \frac{\partial p_\alpha}{\partial \mathbf{r}}\right.\right. \\ &\quad \left.\left. - \rho_\alpha \mathbf{F}_\alpha^{(1)} - \frac{q_\alpha}{m_\alpha} \rho_\alpha \mathbf{v}_0 \times \mathbf{B}\right]\right\} = R_\alpha,\end{aligned} \tag{2.7.29}$$

and for a mixture

$$\frac{\partial}{\partial t}\left\{\rho - \sum_\alpha \tau_\alpha^{(0)}\left[\frac{\partial \rho_\alpha}{\partial t} + \frac{\partial}{\partial \mathbf{r}} \cdot (\rho_\alpha \mathbf{v}_0)\right]\right\}$$

$$+\frac{\partial}{\partial \mathbf{r}}\cdot\left\{\rho\mathbf{v}_0-\sum_\alpha\tau_\alpha^{(0)}\left[\frac{\partial}{\partial t}(\rho_\alpha\mathbf{v}_0)+\frac{\partial}{\partial \mathbf{r}}\cdot(\rho_\alpha\mathbf{v}_0\mathbf{v}_0)+\overleftrightarrow{I}\cdot\frac{\partial p_\alpha}{\partial \mathbf{r}}\right.\right.$$

$$\left.\left.-\rho_\alpha\mathbf{F}_\alpha^{(1)}-\frac{q_\alpha}{m_\alpha}\rho_\alpha\mathbf{v}_0\times\mathbf{B}\right]\right\}=0. \tag{2.7.30}$$

Generalized Euler momentum equations for species can be obtained in a similar way:

$$\frac{\partial}{\partial t}\left\{\rho_\alpha\mathbf{v}_0-\tau_\alpha^{(0)}\left[\frac{\partial}{\partial t}(\rho_\alpha\mathbf{v}_0)+\frac{\partial}{\partial \mathbf{r}}\cdot\rho_\alpha\mathbf{v}_0\mathbf{v}_0+\frac{\partial p_\alpha}{\partial \mathbf{r}}-\rho_\alpha\mathbf{F}_\alpha^{(1)}\right.\right.$$

$$\left.\left.-\left(\frac{q_\alpha}{m_\alpha}\right)\rho_\alpha\mathbf{v}_0\times\mathbf{B}\right]\right\}-\mathbf{F}_\alpha^{(1)}\left[\rho_\alpha-\tau_\alpha^{(0)}\left(\frac{\partial\rho_\alpha}{\partial t}+\frac{\partial}{\partial \mathbf{r}}\cdot(\rho_\alpha\mathbf{v}_0)\right)\right]$$

$$-\frac{q_\alpha}{m_\alpha}\left\{\rho_\alpha\mathbf{v}_0-\tau_\alpha^{(0)}\left[\frac{\partial}{\partial t}(\rho_\alpha\mathbf{v}_0)+\frac{\partial}{\partial \mathbf{r}}\cdot\rho_\alpha\mathbf{v}_0\mathbf{v}_0+\frac{\partial p_\alpha}{\partial \mathbf{r}}-\rho_\alpha\mathbf{F}_\alpha^{(1)}\right.\right.$$

$$\left.\left.-\frac{q_\alpha}{m_\alpha}\rho_\alpha\mathbf{v}_0\times\mathbf{B}\right]\right\}\times\mathbf{B}+\frac{\partial}{\partial \mathbf{r}}\cdot\left\{\rho_\alpha\mathbf{v}_0\mathbf{v}_0+p_\alpha\overleftrightarrow{I}-\tau_\alpha^{(0)}\left[\frac{\partial}{\partial t}\Big(\rho_\alpha\mathbf{v}_0\mathbf{v}_0\right.\right.$$

$$+p_\alpha\overleftrightarrow{I}\Big)+\frac{\partial}{\partial \mathbf{r}}\cdot\Big(\rho_\alpha(\mathbf{v}_0\mathbf{v}_0)\mathbf{v}_0+\rho_\alpha\overline{(\mathbf{v}_0\mathbf{V}_\alpha)\mathbf{V}_\alpha}+\rho_\alpha\overline{(\mathbf{V}_\alpha\mathbf{v}_0)\mathbf{V}_\alpha}$$

$$+\rho_\alpha\overline{(\mathbf{V}_\alpha\mathbf{V}_\alpha)}\mathbf{v}_0\Big)-\mathbf{F}_\alpha^{(1)}\rho_\alpha\mathbf{v}_0-\rho_\alpha\mathbf{v}_0\mathbf{F}_\alpha^{(1)}-\frac{q_\alpha}{m_\alpha}\rho_\alpha[\mathbf{v}_0\times\mathbf{B}]\mathbf{v}_0$$

$$\left.\left.-\frac{q_\alpha}{m_\alpha}\rho_\alpha\overline{[\mathbf{V}_\alpha\times\mathbf{B}]\mathbf{V}_\alpha}-\frac{q_\alpha}{m_\alpha}\rho_\alpha\mathbf{v}_0[\mathbf{v}_0\times\mathbf{B}]-\frac{q_\alpha}{m_\alpha}\rho_\alpha\overline{\mathbf{V}_\alpha[\mathbf{V}_\alpha\times\mathbf{B}]}\right]\right\}$$

$$=\int m_\alpha\mathbf{v}_\alpha J_\alpha^{\mathrm{st,el}}\,\mathrm{d}\mathbf{v}_\alpha+\int m_\alpha\mathbf{v}_\alpha J_\alpha^{\mathrm{st,inel}}\,\mathrm{d}\mathbf{v}_\alpha. \tag{2.7.31}$$

The following averaged expressions should be calculated in local Maxwellian approximation:

$$\frac{\partial^2}{\partial\mathbf{r}\partial\mathbf{r}}:\rho_\alpha\overline{(\mathbf{V}_\alpha\mathbf{v}_0)\mathbf{V}_\alpha}=\frac{\partial}{\partial \mathbf{r}}\left[\frac{\partial}{\partial \mathbf{r}}\cdot(p_\alpha\mathbf{v}_0)\right], \tag{2.7.32}$$

$$\frac{\partial^2}{\partial\mathbf{r}\partial\mathbf{r}}:\rho_\alpha\overline{(\mathbf{V}_\alpha\mathbf{V}_\alpha)}\mathbf{v}_0=\frac{\partial^2}{\partial\mathbf{r}\partial\mathbf{r}}:\overleftrightarrow{P}_\alpha^{(0)}\mathbf{v}_0=\frac{\partial^2}{\partial\mathbf{r}\partial\mathbf{r}}:p_\alpha\overleftrightarrow{I}\mathbf{v}_0=\Delta(p_\alpha\mathbf{v}_0), \tag{2.7.33}$$

where Δ is the Laplacian,

$$\frac{\partial}{\partial \mathbf{r}}\cdot\rho_\alpha\overline{[\mathbf{V}_\alpha\times\mathbf{B}]\mathbf{V}_\alpha}=-\frac{\partial}{\partial \mathbf{r}}\times(p_\alpha\mathbf{B}), \tag{2.7.34}$$

$$\frac{\partial}{\partial \mathbf{r}}\cdot\rho_\alpha\overline{\mathbf{V}_\alpha[\mathbf{V}_\alpha\times\mathbf{B}]}=\frac{\partial}{\partial \mathbf{r}}\times(p_\alpha\mathbf{B}). \tag{2.7.35}$$

Then summation of two last expressions leads to canceling of each other.

Write down the generalized Euler momentum equation for α-species:

$$\frac{\partial}{\partial t}\left\{\rho_\alpha \mathbf{v}_0 - \tau_\alpha^{(0)}\left[\frac{\partial}{\partial t}(\rho_\alpha \mathbf{v}_0) + \frac{\partial}{\partial \mathbf{r}}\cdot \rho_\alpha \mathbf{v}_0\mathbf{v}_0 + \frac{\partial p_\alpha}{\partial \mathbf{r}} - \rho_\alpha \mathbf{F}_\alpha^{(1)}\right.\right.$$
$$\left.\left. - \frac{q_\alpha}{m_\alpha}\rho_\alpha \mathbf{v}_0 \times \mathbf{B}\right]\right\} - \mathbf{F}_\alpha^{(1)}\left[\rho_\alpha - \tau_\alpha^{(0)}\left(\frac{\partial \rho_\alpha}{\partial t} + \frac{\partial}{\partial \mathbf{r}}\cdot(\rho_\alpha \mathbf{v}_0)\right)\right]$$
$$- \frac{q_\alpha}{m_\alpha}\left\{\rho_\alpha \mathbf{v}_0 - \tau_\alpha^{(0)}\left[\frac{\partial}{\partial t}(\rho_\alpha \mathbf{v}_0) + \frac{\partial}{\partial \mathbf{r}}\cdot \rho_\alpha \mathbf{v}_0\mathbf{v}_0 + \frac{\partial p_\alpha}{\partial \mathbf{r}} - \rho_\alpha \mathbf{F}_\alpha^{(1)}\right.\right.$$
$$\left.\left. - \frac{q_\alpha}{m_\alpha}\rho_\alpha \mathbf{v}_0 \times \mathbf{B}\right]\right\} \times \mathbf{B} + \frac{\partial}{\partial \mathbf{r}}\cdot\left\{\rho_\alpha \mathbf{v}_0\mathbf{v}_0 + p_\alpha \overleftrightarrow{I} - \tau_\alpha^{(0)}\left[\frac{\partial}{\partial t}(\rho_\alpha \mathbf{v}_0\mathbf{v}_0\right.\right.$$
$$+ p_\alpha \overleftrightarrow{I}) + \frac{\partial}{\partial \mathbf{r}}\cdot \rho_\alpha(\mathbf{v}_0\mathbf{v}_0)\mathbf{v}_0 + 2\overleftrightarrow{I}\left(\frac{\partial}{\partial \mathbf{r}}\cdot(p_\alpha \mathbf{v}_0)\right) + \frac{\partial}{\partial \mathbf{r}}\cdot(\overleftrightarrow{I} p_\alpha \mathbf{v}_0)$$
$$\left.\left. - \mathbf{F}_\alpha^{(1)}\rho_\alpha \mathbf{v}_0 - \rho_\alpha \mathbf{v}_0\mathbf{F}_\alpha^{(1)} - \frac{q_\alpha}{m_\alpha}\rho_\alpha[\mathbf{v}_0 \times \mathbf{B}]\mathbf{v}_0 - \frac{q_\alpha}{m_\alpha}\rho_\alpha \mathbf{v}_0[\mathbf{v}_0 \times \mathbf{B}]\right]\right\}$$
$$= \int m_\alpha \mathbf{v}_\alpha J_\alpha^{\mathrm{st,el}}\,\mathrm{d}\mathbf{v}_\alpha + \int m_\alpha \mathbf{v}_\alpha J_\alpha^{\mathrm{st,inel}}\,\mathrm{d}\mathbf{v}_\alpha. \tag{2.7.36}$$

After summation of (2.7.36) over all species, one obtains the generalized Euler momentum equation for mixture of gases

$$\frac{\partial}{\partial t}\left\{\rho \mathbf{v}_0 - \sum_\alpha \tau_\alpha^{(0)}\left[\frac{\partial}{\partial t}(\rho_\alpha \mathbf{v}_0) + \frac{\partial}{\partial \mathbf{r}}\cdot \rho_\alpha \mathbf{v}_0\mathbf{v}_0 + \frac{\partial p_\alpha}{\partial \mathbf{r}} - \rho_\alpha \mathbf{F}_\alpha^{(1)}\right.\right.$$
$$\left.\left. - \frac{q_\alpha}{m_\alpha}\rho_\alpha \mathbf{v}_0 \times \mathbf{B}\right]\right\} - \sum_\alpha \mathbf{F}_\alpha^{(1)}\left[\rho_\alpha - \tau_\alpha^{(0)}\left(\frac{\partial \rho_\alpha}{\partial t} + \frac{\partial}{\partial \mathbf{r}}\cdot(\rho_\alpha \mathbf{v}_0)\right)\right]$$
$$- \sum_\alpha \frac{q_\alpha}{m_\alpha}\left\{\rho_\alpha \mathbf{v}_0 - \tau_\alpha^{(0)}\left[\frac{\partial}{\partial t}(\rho_\alpha \mathbf{v}_0) + \frac{\partial}{\partial \mathbf{r}}\cdot \rho_\alpha \mathbf{v}_0\mathbf{v}_0 + \frac{\partial p_\alpha}{\partial \mathbf{r}} - \rho_\alpha \mathbf{F}_\alpha^{(1)}\right.\right.$$
$$\left.\left. - \frac{q_\alpha}{m_\alpha}\rho_\alpha \mathbf{v}_0 \times \mathbf{B}\right]\right\} \times \mathbf{B} + \frac{\partial}{\partial \mathbf{r}}\cdot\left\{\rho \mathbf{v}_0\mathbf{v}_0 + p\overleftrightarrow{I} - \sum_\alpha \tau_\alpha^{(0)}\left[\frac{\partial}{\partial t}(\rho_\alpha \mathbf{v}_0\mathbf{v}_0\right.\right.$$
$$+ p_\alpha \overleftrightarrow{I}) + \frac{\partial}{\partial \mathbf{r}}\cdot \rho_\alpha(\mathbf{v}_0\mathbf{v}_0)\mathbf{v}_0 + 2\overleftrightarrow{I}\left(\frac{\partial}{\partial \mathbf{r}}\cdot(p_\alpha \mathbf{v}_0)\right) + \frac{\partial}{\partial \mathbf{r}}\cdot(\overleftrightarrow{I} p_\alpha \mathbf{v}_0)$$
$$- \mathbf{F}_\alpha^{(1)}\rho_\alpha \mathbf{v}_0 - \rho_\alpha \mathbf{v}_0\mathbf{F}_\alpha^{(1)} - \frac{q_\alpha}{m_\alpha}\rho_\alpha[\mathbf{v}_0 \times \mathbf{B}]\mathbf{v}_0$$
$$\left.\left. - \frac{q_\alpha}{m_\alpha}\rho_\alpha \mathbf{v}_0[\mathbf{v}_0 \times \mathbf{B}]\right]\right\} = 0. \tag{2.7.37}$$

Let us get to the derivation of the generalized Euler energy equation. Write down this equation omitting all integral terms, which are obviously equal to zero as containing odd integrands of the thermal velocities $\mathbf{V}_\alpha$.

$$\begin{aligned}
&\frac{\partial}{\partial t}\left\{\frac{\rho_\alpha v_0^2}{2}+\frac{\rho_\alpha \overline{V_\alpha^2}}{2}+\varepsilon_\alpha n_\alpha-\tau_\alpha^{(0)}\left[\frac{\partial}{\partial t}\left(\frac{\rho_\alpha v_0^2}{2}+\frac{\rho_\alpha \overline{V_\alpha^2}}{2}+\varepsilon_\alpha n_\alpha\right)\right.\right.\\
&\quad+\frac{\partial}{\partial \mathbf{r}}\cdot\left(\frac{1}{2}\rho_\alpha v_0^2\mathbf{v}_0+\frac{1}{2}\rho_\alpha\overline{V_\alpha^2}\mathbf{v}_0+\rho_\alpha\overline{(\mathbf{v}_0\cdot\mathbf{V}_\alpha)\mathbf{V}_\alpha}+\varepsilon_\alpha n_\alpha\mathbf{v}_0\right)\\
&\quad\left.\left.-\mathbf{F}_\alpha^{(1)}\cdot\rho_\alpha\mathbf{v}_0\right]\right\}\\
&\quad+\frac{\partial}{\partial \mathbf{r}}\cdot\left\{\frac{1}{2}\rho_\alpha v_0^2\mathbf{v}_0+\rho_\alpha\overline{(\mathbf{v}_0\cdot\mathbf{V}_\alpha)\mathbf{V}_\alpha}+\frac{1}{2}\rho_\alpha\overline{V_\alpha^2}\mathbf{v}_0+\varepsilon_\alpha n_\alpha\mathbf{v}_0\right.\\
&\quad-\tau_\alpha^{(0)}\left[\frac{\partial}{\partial t}\left(\frac{1}{2}\rho_\alpha v_0^2\mathbf{v}_0+\rho_\alpha\overline{(\mathbf{v}_0\cdot\mathbf{V}_\alpha)\mathbf{V}_\alpha}+\frac{1}{2}\rho_\alpha\overline{V_\alpha^2}\mathbf{v}_0+\varepsilon_\alpha n_\alpha\mathbf{v}_0\right)\right.\\
&\quad+\frac{\partial}{\partial \mathbf{r}}\cdot\left[\frac{1}{2}\rho_\alpha v_0^2\mathbf{v}_0\mathbf{v}_0+\frac{1}{2}\rho_\alpha\overline{V_\alpha^2}\mathbf{v}_0\mathbf{v}_0+\frac{1}{2}\rho_\alpha v_0^2\overline{\mathbf{V}_\alpha\mathbf{V}_\alpha}\right.\\
&\quad+\rho_\alpha\overline{(\mathbf{v}_0\cdot\mathbf{V}_\alpha)\mathbf{V}_\alpha}\mathbf{v}_0+\rho_\alpha\overline{(\mathbf{v}_0\cdot\mathbf{V}_\alpha)\mathbf{v}_0\mathbf{V}_\alpha}\\
&\quad\left.+\frac{1}{2}\rho_\alpha\overline{V_\alpha^2\mathbf{V}_\alpha\mathbf{V}_\alpha}+\varepsilon_\alpha n_\alpha\mathbf{v}_0\mathbf{v}_0+\varepsilon_\alpha n_\alpha\overline{\mathbf{V}_\alpha\mathbf{V}_\alpha}\right]\\
&\quad\left.\left.-\rho_\alpha\mathbf{F}_\alpha^{(1)}\cdot\mathbf{v}_0\mathbf{v}_0-\rho_\alpha\mathbf{F}_\alpha^{(1)}\cdot\overline{\mathbf{V}_\alpha\mathbf{V}_\alpha}-\frac{1}{2}\rho_\alpha\overline{v_\alpha^2}\mathbf{F}_\alpha-\varepsilon_\alpha n_\alpha\overline{\mathbf{F}}_\alpha\right]\right\}\\
&\quad-\left\{\rho_\alpha\mathbf{F}_\alpha^{(1)}\cdot\mathbf{v}_0-\tau_\alpha^{(0)}\left[\mathbf{F}_\alpha^{(1)}\cdot\left(\frac{\partial}{\partial t}(\rho_\alpha\mathbf{v}_0)+\frac{\partial}{\partial \mathbf{r}}\cdot\rho_\alpha\mathbf{v}_0\mathbf{v}_0\right.\right.\right.\\
&\quad\left.\left.\left.+\frac{\partial}{\partial \mathbf{r}}\cdot\rho_\alpha\overline{\mathbf{V}_\alpha\mathbf{V}_\alpha}-\rho_\alpha\mathbf{F}_\alpha^{(1)}-q_\alpha n_\alpha\mathbf{v}_0\times\mathbf{B}\right)\right]\right\}\\
&=\int\left(\frac{m_\alpha v_\alpha^2}{2}+\varepsilon_\alpha\right)J_\alpha^{\mathrm{st,el}}\,\mathrm{d}\mathbf{v}_\alpha+\int\left(\frac{m_\alpha v_\alpha^2}{2}+\varepsilon_\alpha\right)J_\alpha^{\mathrm{st,inel}}\,\mathrm{d}\mathbf{v}_\alpha.
\end{aligned}\tag{2.7.38}$$

Let us calculate all average values connected with local Maxwellian DF

$$f_\alpha^{(0)}=n_\alpha\left(\frac{m_\alpha}{2\pi k_B T}\right)^{3/2}\mathrm{e}^{-m_\alpha V_\alpha^2/(2k_B T)}.\tag{2.7.39}$$

We have

$$\frac{\rho_\alpha\overline{V_\alpha^2}}{2}=\frac{3}{2}n_\alpha k_B T,\tag{2.7.40}$$

$$\frac{\partial}{\partial \mathbf{r}}\cdot\rho_\alpha\overline{(\mathbf{v}_0\cdot\mathbf{V}_\alpha)\mathbf{V}_\alpha}=m_\alpha\frac{\partial}{\partial \mathbf{r}}\cdot\left\{\mathbf{v}_0\cdot\int\mathbf{V}_\alpha\mathbf{V}_\alpha\,\mathrm{e}^{-W_\alpha^2}\,\mathrm{d}\mathbf{V}_\alpha\,n_\alpha\left(\frac{m_\alpha}{2\pi k_B T}\right)^{3/2}\right\}$$

$$
\begin{aligned}
&= \frac{4\pi}{\pi^{3/2}} \frac{2k_B}{m_\alpha} m_\alpha \sum_{i,j=1}^{3} \frac{\partial}{\partial r_i} \left\{ n_\alpha v_{0j} T \int W_{\alpha j} W_i W_\alpha^2 \mathrm{e}^{-W_\alpha^2} \, \mathrm{d}W_\alpha \right\} \\
&= \frac{8k_B}{\sqrt{\pi}} \sum_i \frac{\partial}{\partial r_i} \left\{ n_\alpha v_{0i} T \int W_{\alpha i}^2 W_\alpha^2 \mathrm{e}^{-W_\alpha^2} \, \mathrm{d}W_\alpha \right\} \\
&= \frac{8k_B}{3\sqrt{\pi}} \sum_i \frac{\partial}{\partial r_i} \left\{ n_\alpha v_{0i} T \int_0^\infty W_\alpha^4 \mathrm{e}^{-W_\alpha^2} \, \mathrm{d}W_\alpha \right\} \\
&= \frac{8}{3\sqrt{\pi}} \frac{\sqrt{\pi}}{2} \frac{1}{2} \frac{3}{2} \frac{\partial}{\partial \mathbf{r}} \cdot (\mathbf{v}_0 p_\alpha) = \frac{\partial}{\partial \mathbf{r}} \cdot (p_\alpha \mathbf{v}_0).
\end{aligned} \tag{2.7.41}
$$

Other integrals in (2.7.38) can be evaluated in a similar way:

$$
\frac{\partial}{\partial \mathbf{r}} \cdot \left\{ \overline{\mathbf{V}_\alpha \mathbf{V}_\alpha} \rho_\alpha v_0^2 \right\} = \frac{\partial}{\partial \mathbf{r}} \cdot \left(p_\alpha v_0^2 \overleftrightarrow{I} \right), \tag{2.7.42}
$$

$$
\begin{aligned}
&\frac{\partial}{\partial \mathbf{r}} \cdot \left\{ \rho_\alpha \overline{\left(\mathbf{v}_0 \cdot \mathbf{V}_\alpha \right) \mathbf{V}_\alpha} \mathbf{v}_0 \right\} \\
&\quad = \frac{\partial}{\partial \mathbf{r}} \cdot \left\{ \rho_\alpha \overline{\left(\mathbf{v}_0 \cdot \mathbf{V}_\alpha \right) \mathbf{v}_0 \mathbf{V}_\alpha} \right\} = \frac{\partial}{\partial \mathbf{r}} \cdot p_\alpha \mathbf{v}_0 \mathbf{v}_0,
\end{aligned} \tag{2.7.43}
$$

$$
\frac{\partial}{\partial \mathbf{r}} \cdot \rho_\alpha \overline{V_\alpha^2 \mathbf{V}_\alpha \mathbf{V}_\alpha} = 5 \frac{\partial}{\partial \mathbf{r}} \cdot \left(\frac{p_\alpha^2}{\rho_\alpha} \overleftrightarrow{I} \right), \tag{2.7.44}
$$

$$
\varepsilon_\alpha n_\alpha \overline{\mathbf{F}}_\alpha = \varepsilon_\alpha \frac{q_\alpha}{m_\alpha} n_\alpha [\mathbf{v}_0 \times \mathbf{B}] + \varepsilon_\alpha n_\alpha \mathbf{F}_\alpha^{(1)}. \tag{2.7.45}
$$

Calculate also an Euler approximation

$$
\begin{aligned}
&\frac{\partial}{\partial \mathbf{r}} \cdot \rho_\alpha \overline{v_\alpha^2 \mathbf{F}_\alpha} \\
&= \frac{\partial}{\partial \mathbf{r}} \cdot \left\{ m_\alpha \int v_\alpha^2 \mathbf{F}_\alpha^{(1)} f_\alpha^{(0)} \, \mathrm{d}\mathbf{v}_\alpha + q_\alpha \int v_\alpha^2 [\mathbf{v}_\alpha \times \mathbf{B}] f_\alpha^{(0)} \, \mathrm{d}\mathbf{v}_\alpha \right\} \\
&= \frac{\partial}{\partial \mathbf{r}} \cdot \left\{ \rho_\alpha \mathbf{F}_\alpha^{(1)} v_0^2 + \mathbf{F}_\alpha^{(1)} m_\alpha \int V_\alpha^2 f_\alpha^{(0)} \, \mathrm{d}\mathbf{V}_\alpha + \frac{q_\alpha}{m_\alpha} \rho_\alpha [\mathbf{v}_0 \times \mathbf{B}] v_0^2 \right. \\
&\qquad \left. + 2 q_\alpha \mathbf{v}_0 \cdot \int \mathbf{V}_\alpha [\mathbf{V}_\alpha \times \mathbf{B}] f_\alpha^{(0)} \, \mathrm{d}\mathbf{V}_\alpha + q_\alpha \int V_\alpha^2 [\mathbf{v}_0 \times \mathbf{B}] f_\alpha^{(0)} \, \mathrm{d}\mathbf{V}_\alpha \right\} \\
&= \frac{\partial}{\partial \mathbf{r}} \cdot \left\{ \rho_\alpha \mathbf{F}_\alpha^{(1)} v_0^2 + 3 \mathbf{F}_\alpha^{(1)} p_\alpha + \rho_\alpha \mathbf{v}_0^2 \frac{q_\alpha}{m_\alpha} [\mathbf{v}_0 \times \mathbf{B}] \right. \\
&\qquad \left. + 5 p_\alpha \frac{q_\alpha}{m_\alpha} [\mathbf{v}_0 \times \mathbf{B}] \right\}.
\end{aligned} \tag{2.7.46}
$$

As a result, the following form of the generalized Euler energy equation for species can be stated:

$$\frac{\partial}{\partial t}\left\{\frac{\rho_\alpha v_0^2}{2}+\frac{3}{2}p_\alpha+\varepsilon_\alpha n_\alpha-\tau_\alpha^{(0)}\left[\frac{\partial}{\partial t}\left(\frac{\rho_\alpha v_0^2}{2}+\frac{3}{2}p_\alpha+\varepsilon_\alpha n_\alpha\right)\right.\right.$$
$$\left.\left.+\frac{\partial}{\partial \mathbf{r}}\cdot\left(\frac{1}{2}\rho_\alpha v_0^2\mathbf{v}_0+\frac{5}{2}p_\alpha\mathbf{v}_0+\varepsilon_\alpha n_\alpha\mathbf{v}_0\right)-\mathbf{F}_\alpha^{(1)}\cdot\rho_\alpha\mathbf{v}_0\right]\right\}$$
$$+\frac{\partial}{\partial \mathbf{r}}\cdot\left\{\frac{1}{2}\rho_\alpha v_0^2\mathbf{v}_0+\frac{5}{2}p_\alpha\mathbf{v}_0+\varepsilon_\alpha n_\alpha\mathbf{v}_0-\tau_\alpha^{(0)}\left[\frac{\partial}{\partial t}\left(\frac{1}{2}\rho_\alpha v_0^2\mathbf{v}_0\right.\right.\right.$$
$$\left.+\frac{5}{2}p_\alpha\mathbf{v}_0+\varepsilon_\alpha n_\alpha\mathbf{v}_0\right)+\frac{\partial}{\partial \mathbf{r}}\cdot\left(\frac{1}{2}\rho_\alpha v_0^2\mathbf{v}_0\mathbf{v}_0+\frac{7}{2}p_\alpha\mathbf{v}_0\mathbf{v}_0+\frac{1}{2}p_\alpha v_0^2\overleftrightarrow{I}\right.$$
$$\left.+\frac{5}{2}\frac{p_\alpha^2}{\rho_\alpha}\overleftrightarrow{I}+\varepsilon_\alpha n_\alpha\mathbf{v}_0\mathbf{v}_0+\varepsilon_\alpha\frac{p_\alpha}{m_\alpha}\overleftrightarrow{I}\right)-\rho_\alpha\mathbf{F}_\alpha^{(1)}\cdot\mathbf{v}_0\mathbf{v}_0-p_\alpha\mathbf{F}_\alpha^{(1)}\cdot\overleftrightarrow{I}$$
$$-\frac{1}{2}\rho_\alpha v_0^2\mathbf{F}_\alpha^{(1)}-\frac{3}{2}\mathbf{F}_\alpha^{(1)}p_\alpha-\frac{\rho_\alpha v_0^2}{2}\frac{q_\alpha}{m_\alpha}[\mathbf{v}_0\times\mathbf{B}]-\frac{5}{2}p_\alpha\frac{q_\alpha}{m_\alpha}[\mathbf{v}_0\times\mathbf{B}]$$
$$\left.\left.-\varepsilon_\alpha n_\alpha\frac{q_\alpha}{m_\alpha}[\mathbf{v}_0\times\mathbf{B}]-\varepsilon_\alpha n_\alpha\mathbf{F}_\alpha^{(1)}\right]\right\}-\left\{\rho_\alpha\mathbf{F}_\alpha^{(1)}\cdot\mathbf{v}_0\right.$$
$$-\tau_\alpha^{(0)}\left[\mathbf{F}_\alpha^{(1)}\left(\frac{\partial}{\partial t}(\rho_\alpha\mathbf{v}_0)+\frac{\partial}{\partial \mathbf{r}}\cdot\rho_\alpha\mathbf{v}_0\mathbf{v}_0+\frac{\partial}{\partial \mathbf{r}}\cdot p_\alpha\overleftrightarrow{I}-\rho_\alpha\mathbf{F}_\alpha^{(1)}\right.\right.$$
$$\left.\left.\left.-q_\alpha n_\alpha[\mathbf{v}_0\times\mathbf{B}]\right)\right]\right\}$$
$$=\int\left(\frac{m_\alpha v_\alpha^2}{2}+\varepsilon_\alpha\right)J_\alpha^{\mathrm{st,el}}\,\mathrm{d}\mathbf{v}_\alpha+\int\left(\frac{m_\alpha v_\alpha^2}{2}+\varepsilon_\alpha\right)J_\alpha^{\mathrm{st,inel}}\,\mathrm{d}\mathbf{v}_\alpha. \tag{2.7.47}$$

Finally, after summation of Eqs. (2.7.47) over all species we reach the generalized Euler energy equation for mixture of gases:

$$\frac{\partial}{\partial t}\left\{\frac{\rho v_0^2}{2}+\frac{3}{2}p+\sum_\alpha\varepsilon_\alpha n_\alpha-\sum_\alpha\tau_\alpha^{(0)}\left[\frac{\partial}{\partial t}\left(\frac{\rho_\alpha v_0^2}{2}+\frac{3}{2}p_\alpha+\varepsilon_\alpha n_\alpha\right)\right.\right.$$
$$\left.\left.+\frac{\partial}{\partial \mathbf{r}}\cdot\left(\frac{1}{2}\rho_\alpha v_0^2\mathbf{v}_0+\frac{5}{2}p_\alpha\mathbf{v}_0+\varepsilon_\alpha n_\alpha\mathbf{v}_0\right)-\mathbf{F}_\alpha^{(1)}\cdot\rho_\alpha\mathbf{v}_0\right]\right\}$$
$$+\frac{\partial}{\partial \mathbf{r}}\cdot\left\{\frac{1}{2}\rho v_0^2\mathbf{v}_0+\frac{5}{2}p\mathbf{v}_0+\mathbf{v}_0\sum_\alpha\varepsilon_\alpha n_\alpha-\sum_\alpha\tau_\alpha^{(0)}\left[\frac{\partial}{\partial t}\left(\frac{1}{2}\rho_\alpha v_0^2\mathbf{v}_0\right.\right.\right.$$
$$\left.+\frac{5}{2}p_\alpha\mathbf{v}_0+\varepsilon_\alpha n_\alpha\mathbf{v}_0\right)+\frac{\partial}{\partial \mathbf{r}}\cdot\left(\frac{1}{2}\rho_\alpha v_0^2\mathbf{v}_0\mathbf{v}_0+\frac{7}{2}p_\alpha\mathbf{v}_0\mathbf{v}_0+\frac{1}{2}p_\alpha v_0^2\overleftrightarrow{I}\right.$$
$$\left.+\frac{5}{2}\frac{p_\alpha^2}{\rho_\alpha}\overleftrightarrow{I}+\varepsilon_\alpha n_\alpha\mathbf{v}_0\mathbf{v}_0+\varepsilon_\alpha\frac{p_\alpha}{m_\alpha}\overleftrightarrow{I}\right)-\rho_\alpha\mathbf{F}_\alpha^{(1)}\cdot\mathbf{v}_0\mathbf{v}_0-p_\alpha\mathbf{F}_\alpha^{(1)}\cdot\overleftrightarrow{I}$$

$$
- \frac{1}{2}\rho_\alpha v_0^2 \mathbf{F}_\alpha^{(1)} - \frac{3}{2}\mathbf{F}_\alpha^{(1)} p_\alpha - \frac{\rho_\alpha v_0^2}{2}\frac{q_\alpha}{m_\alpha}[\mathbf{v}_0 \times \mathbf{B}] - \frac{5}{2}p_\alpha \frac{q_\alpha}{m_\alpha}[\mathbf{v}_0 \times \mathbf{B}]
$$

$$
- \varepsilon_\alpha n_\alpha \frac{q_\alpha}{m_\alpha}[\mathbf{v}_0 \times \mathbf{B}] - \varepsilon_\alpha n_\alpha \mathbf{F}_\alpha^{(1)}\bigg]\bigg]\bigg\} - \bigg\{\mathbf{v}_0 \cdot \sum_\alpha \rho_\alpha \mathbf{F}_\alpha^{(1)}
$$

$$
- \sum_\alpha \tau_\alpha^{(0)}\bigg[\mathbf{F}_\alpha^{(1)} \cdot \bigg(\frac{\partial}{\partial t}(\rho_\alpha \mathbf{v}_0) + \frac{\partial}{\partial \mathbf{r}} \cdot \rho_\alpha \mathbf{v}_0\mathbf{v}_0 + \frac{\partial}{\partial \mathbf{r}} \cdot p_\alpha \overleftrightarrow{I} - \rho_\alpha \mathbf{F}_\alpha^{(1)}
$$

$$
- q_\alpha n_\alpha [\mathbf{v}_0 \times \mathbf{B}]\bigg)\bigg]\bigg\} = 0. \tag{2.7.48}
$$

Write down the system of GHE in the Euler approach taking into account the force of gravitation:

– continuity equation

$$
\frac{\partial}{\partial t}\bigg\{\rho - \Pi\frac{\mu}{p}\bigg[\frac{\partial \rho}{\partial t} + \frac{\partial}{\partial \mathbf{r}} \cdot (\rho \mathbf{v}_0)\bigg]\bigg\} + \frac{\partial}{\partial \mathbf{r}} \cdot \bigg\{\rho \mathbf{v}_0 - \Pi\frac{\mu}{p}\bigg[\frac{\partial}{\partial t}(\rho \mathbf{v}_0)
$$

$$
+ \frac{\partial}{\partial \mathbf{r}} \cdot \rho \mathbf{v}_0\mathbf{v}_0 + \overleftrightarrow{I} \cdot \frac{\partial p}{\partial \mathbf{r}} - \rho \mathbf{g}\bigg]\bigg\} = 0, \tag{2.7.49}
$$

– momentum equation

$$
\frac{\partial}{\partial t}\bigg\{\rho \mathbf{v}_0 - \Pi\frac{\mu}{p}\bigg[\frac{\partial}{\partial t}(\rho \mathbf{v}_0) + \frac{\partial}{\partial \mathbf{r}} \cdot \rho \mathbf{v}_0\mathbf{v}_0 + \frac{\partial p}{\partial \mathbf{r}} - \rho \mathbf{g}\bigg]\bigg\}
$$

$$
- \mathbf{g}\bigg[\rho - \Pi\frac{\mu}{p}\bigg(\frac{\partial \rho}{\partial t} + \frac{\partial}{\partial \mathbf{r}} \cdot (\rho \mathbf{v}_0)\bigg)\bigg]
$$

$$
+ \frac{\partial}{\partial \mathbf{r}} \cdot \bigg\{\rho \mathbf{v}_0\mathbf{v}_0 + p\overleftrightarrow{I} - \Pi\frac{\mu}{p}\bigg[\frac{\partial}{\partial t}(\rho \mathbf{v}_0\mathbf{v}_0 + p\overleftrightarrow{I}) + \frac{\partial}{\partial \mathbf{r}} \cdot \rho(\mathbf{v}_0\mathbf{v}_0)\mathbf{v}_0
$$

$$
+ 2\overleftrightarrow{I}\bigg[\frac{\partial}{\partial \mathbf{r}} \cdot (p\mathbf{v}_0)\bigg] + \frac{\partial}{\partial \mathbf{r}} \cdot (\overleftrightarrow{I} p\mathbf{v}_0) - \mathbf{g}\rho \mathbf{v}_0 - \mathbf{v}_0\mathbf{g}\rho\bigg]\bigg\} = 0, \tag{2.7.50}
$$

– energy equation

$$
\frac{\partial}{\partial t}\bigg\{\frac{\rho v_0^2}{2} + \frac{3}{2}p - \Pi\frac{\mu}{p}\bigg[\frac{\partial}{\partial t}\bigg(\frac{\rho v_0^2}{2} + \frac{3}{2}p\bigg)
$$

$$
+ \frac{\partial}{\partial \mathbf{r}} \cdot \bigg(\frac{1}{2}\rho v_0^2 \mathbf{v}_0 + \frac{5}{2}p\mathbf{v}_0\bigg) - \mathbf{g} \cdot \rho \mathbf{v}_0\bigg]\bigg\}
$$

$$
+ \frac{\partial}{\partial \mathbf{r}} \cdot \bigg\{\frac{1}{2}\rho v_0^2 \mathbf{v}_0 + \frac{5}{2}p\mathbf{v}_0 - \Pi\frac{\mu}{p}\bigg[\frac{\partial}{\partial t}\bigg(\frac{1}{2}\rho v_0^2 \mathbf{v}_0 + \frac{5}{2}p\mathbf{v}_0\bigg)
$$

$$
+ \frac{\partial}{\partial \mathbf{r}} \cdot \bigg(\frac{1}{2}\rho v_0^2 \mathbf{v}_0\mathbf{v}_0 + \frac{7}{2}p\mathbf{v}_0\mathbf{v}_0 + \frac{1}{2}p v_0^2 \overleftrightarrow{I} + \frac{5}{2}\frac{p^2}{\rho}\overleftrightarrow{I}\bigg) - \rho \mathbf{g} \cdot \mathbf{v}_0\mathbf{v}_0
$$

$$- p\mathbf{g}\cdot\overleftrightarrow{I} - \frac{1}{2}\rho v_0^2\mathbf{g} - \frac{3}{2}\mathbf{g}p\Big]\Big\} - \Big\{\rho\mathbf{g}\cdot\mathbf{v}_0 - \Pi\frac{\mu}{p}\Big[\mathbf{g}\cdot\Big(\frac{\partial}{\partial t}(\rho\mathbf{v}_0)$$

$$+\frac{\partial}{\partial\mathbf{r}}\cdot\rho\mathbf{v}_0\mathbf{v}_0 + \frac{\partial}{\partial\mathbf{r}}\cdot p\overleftrightarrow{I} - \rho\mathbf{g}\Big)\Big]\Big\} = 0, \tag{2.7.51}$$

where $\mathbf{g}$ is acceleration in gravitational field, $\tau^{(0)} = \Pi\mu/p$ (for hard spheres model $\Pi = 0.8$), Δ is the Laplacian.

Let us summarize generalized Euler hydrodynamic equations the simplest case of one-dimensional motion in absence of external forces:

– continuity equation

$$\frac{\partial}{\partial t}\Big\{\rho - \tau^{(0)}\Big[\frac{\partial\rho}{\partial t} + \frac{\partial}{\partial x}(\rho v_0)\Big]\Big\}$$

$$+\frac{\partial}{\partial x}\Big\{\rho v_0 - \tau^{(0)}\Big[\frac{\partial}{\partial t}(\rho v_0) + \frac{\partial}{\partial x}(\rho v_0^2) + \frac{\partial p}{\partial x}\Big]\Big\} = 0, \tag{2.7.52}$$

– momentum equation

$$\frac{\partial}{\partial t}\Big\{\rho v_0 - \tau^{(0)}\Big[\frac{\partial}{\partial t}(\rho v_0) + \frac{\partial}{\partial x}(\rho v_0^2) + \frac{\partial p}{\partial x}\Big]\Big\}$$

$$+\frac{\partial}{\partial x}\Big\{\rho v_0^2 + p - \tau^{(0)}\Big[\frac{\partial}{\partial t}(\rho v_0^2 + p) + \frac{\partial}{\partial x}(\rho v_0^3 + 3pv_0)\Big]\Big\} = 0, \tag{2.7.53}$$

– energy equation

$$\frac{\partial}{\partial t}\Big\{\rho v_0^2 + 3p - \tau^{(0)}\Big[\frac{\partial}{\partial t}(\rho v_0^2 + 3p) + \frac{\partial}{\partial x}(\rho v_0^3 + 5pv_0)\Big]\Big\}$$

$$+\frac{\partial}{\partial x}\Big\{\rho v_0^3 + 5pv_0 - \tau^{(0)}\Big[\frac{\partial}{\partial t}(\rho v_0^3 + 5pv_0)$$

$$+\frac{\partial}{\partial x}\Big(\rho v_0^4 + 8pv_0^2 + 5\frac{p^2}{\rho}\Big)\Big]\Big\} = 0. \tag{2.7.54}$$

2.8. Boundary conditions in the theory of the generalized hydrodynamic equations

Let us consider now the problem of the additional boundary conditions in the theory of GHE. With this aim write down once more Eq. (2.4.8) and demonstrate – using this equation as an example – how this problem can be solved.

$$\frac{\partial}{\partial t}\Big\{\hat\rho - Kn\,A_v\hat\tau^{(0)}\Big[\frac{\partial\hat\rho}{\partial\hat t} + \frac{\partial}{\partial\hat{\mathbf{r}}}\cdot(\hat\rho\hat{\mathbf{v}}_0)\Big]\Big\}$$

$$+\frac{\partial}{\partial \mathbf{r}} \cdot \left\{ \hat{\rho}\hat{\mathbf{v}}_0 - Kn\, A_\upsilon \hat{\tau}^{(0)} \left[\frac{\partial}{\partial \hat{t}}(\hat{\rho}\hat{\mathbf{v}}_0) + A_\upsilon^{-2} \frac{\partial}{\partial \hat{\mathbf{r}}} \cdot (\hat{\rho}\, \overline{\hat{\mathbf{v}}\hat{\mathbf{v}}}) \right] \right\} = 0. \tag{2.8.1}$$

Knudsen number is defined as a ratio of the mean free path ℓ near the wall to the character hydrodynamic scale L. The explicit form of coefficient A_υ is not significant for us here, notice only that the thermal velocity v_T is used as scale for molecular velocity near the surface. The length L can be significantly different and should be introduced in correspondence with concrete solving problem. For the boundary condition's construction, we consider an area near the streamed wall taking into account that L is the distance to the wall (Figure 2.5). Tend now L to zero using an arbitrary law of this tendency. It means that in (2.8.1) differential terms – for which *Kn* is coefficient – in square brackets are far more than algebraic terms from which they are subtracted. Therewith *Kn* can be considered as a value independent of $\hat{t}$ and $\hat{\mathbf{r}}$.

After dividing Eq. (2.8.1) by *Kn*, it becomes obvious that this equation can be identically satisfied for arbitrary tendency $L \to 0$ (and therefore $Kn \to \infty$) if

$$\frac{\partial \rho}{\partial t} + \frac{\partial}{\partial \mathbf{r}} \cdot \rho \mathbf{v}_0 = 0, \tag{2.8.2}$$

$$\frac{\partial}{\partial t}(\rho \mathbf{v}_0) + \frac{\partial}{\partial \mathbf{r}} \cdot \rho \overline{\mathbf{v}\mathbf{v}} = 0. \tag{2.8.3}$$

But Eqs. (2.8.2), (2.8.3) are classical continuity and momentum equations on the wall. This very remarkable fact will be discussed in the next chapter from turbulence positions.

On the whole we state that GHE can be written in the symbolic form (index "i" related to the continuity, momentum and energy equations)

$$\Phi_{1i} + Kn\, \Phi_{2i} = 0 \quad (i = 1, 2, 3)$$

and following considered method for $Kn \to \infty$ we reach conditions

$$\Phi_{2i,W} = 0, \tag{2.8.4}$$

where index "W" corresponds to the wall.

Relations (2.8.4) deliver additional boundary conditions we need, and have transparent physical sense – fluctuations disappear on the wall.

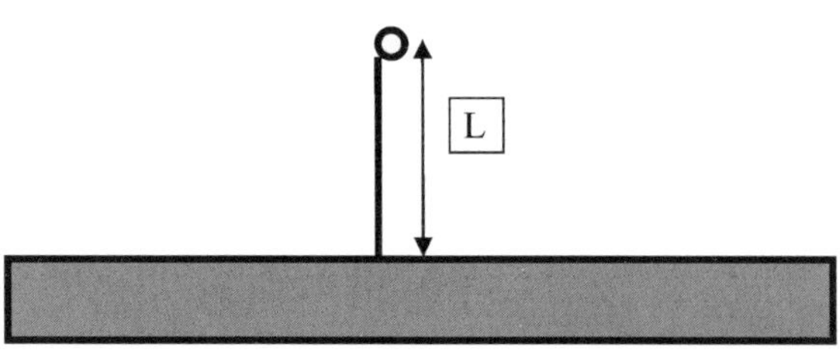

Fig. 2.5. Domain in the vicinity of the streamed wall.

In hydrodynamics of weakly rarefied gases the kinetic boundary conditions (in particular velocity slip) are used. These conditions are the direct consequence of the Boltzmann equation and allow to expand the area of the formal appliance of the Navier–Stokes description. Further we intend to show that GHE contain known kinetic boundary conditions as asymptotic solutions near the wall. It means on the one hand that GHE can be applied for description of the rarefied gas without additional kinetic conditions serving for adjusting kinetic and hydrodynamic solutions in boundary kinetic layer. On the other, we can state that "jump conditions" can be obtained as solution of *hydrodynamic* equations which can be used for adequate description of kinetic layer near the wall.

With the aim of analytical investigation we introduce reasonable assumptions often used in the theory of kinetic Knudsen layer. We suppose that the flow is one-dimensional (and therefore all gradients along streamed surface can be omitted), gravitation is not significant; static pressure in the layer is constant,

$$p = const. \tag{2.8.5}$$

The last presumption is not of principal significance, but allows to simplify the problem eliminating from consideration the energy equation and then the temperature jump.

One-dimensional Euler equations under these assumptions are

$$\frac{\partial}{\partial y}\left[\rho v - \Pi\frac{\mu}{p}\left(\frac{\partial}{\partial y}(\rho v^2) + \frac{\partial p}{\partial y}\right)\right] = 0, \tag{2.8.6}$$

$$\frac{\partial}{\partial y}\left\{\rho u v - \Pi\frac{\mu}{p}\left(\frac{\partial}{\partial y}(\rho u v^2) + \frac{\partial}{\partial y}(p u)\right)\right\} = 0, \tag{2.8.7}$$

$$\frac{\partial}{\partial y}\left\{p + \rho v^2 - \Pi\frac{\mu}{p}\left[\frac{\partial}{\partial y}(\rho v^3) + 3\frac{\partial}{\partial y}(p v)\right]\right\} = 0, \tag{2.8.8}$$

where y is in the direction normal to the streamed plane surface and u, v are components of hydrodynamic velocity. After integration over y one obtains

$$\rho v = \Pi\frac{\mu}{p}\frac{\mathrm{d}}{\mathrm{d}y}(\rho v^2) + C_1, \tag{2.8.9}$$

$$\rho u v = \Pi\frac{\mu}{p}\frac{\mathrm{d}}{\mathrm{d}y}(\rho u v^2) + \Pi\mu\frac{\mathrm{d}u}{\mathrm{d}y} + C_2, \tag{2.8.10}$$

$$\rho v^2 = \Pi\frac{\mu}{p}\frac{\mathrm{d}}{\mathrm{d}y}(\rho v^3) + 3\Pi\mu\frac{\mathrm{d}v}{\mathrm{d}y} + C_3 - p, \tag{2.8.11}$$

where C_1, C_2, C_3 are constants of integration.

By application of (2.8.9), Eq. (2.8.11) becomes

$$\Pi\frac{\mu}{p}(\rho v^2 + 3p)\frac{\mathrm{d}v}{\mathrm{d}y} = C_1 v - C_3 + p. \tag{2.8.12}$$

Write (2.8.9) in the form

$$1 = \Pi \frac{\mu}{p}\left[2\frac{\mathrm{d}v}{\mathrm{d}y} + v\frac{\mathrm{d}\ln\rho}{\mathrm{d}y}\right] + \frac{C_1}{\rho v}. \tag{2.8.13}$$

On the bottom of the Knudsen layer particles go into the stream after desorption from the surface with normal velocity component v_w, corresponding to the order of thermal velocity for wall temperature. Particles, directed to the surface, have on average hydrodynamic velocity, and a known relation

$$(\rho v)_+ = (\rho v)_- \tag{2.8.14}$$

defines impermeability of the surface. In kinetic layer the process of relaxation is being realized by adjusting the dynamic parameters of particles after desorption to the hydrodynamic parameters of flow. For us of interest is the process on the bottom of Knudsen layer and corresponding asymptotic GHE solutions. In this case we omit the second logarithmic in square brackets in (2.8.13) and eliminate density in Eq. (2.8.12):

$$6\Pi\mu\left(\frac{\mathrm{d}v}{\mathrm{d}y}\right)^2 = -\left[3C_1 v + 3p - 2C_3'\right]\frac{\mathrm{d}v}{\mathrm{d}y} + \left(C_1 v - C_3'\right)\frac{p}{\Pi\mu} = 0, \tag{2.8.15}$$

where $C_3' = C_3 - p$.

From (2.8.15) follows

$$\frac{\mathrm{d}v}{\mathrm{d}y} = \frac{3C_1 v + 3p - 2C_3' \pm \sqrt{(3C_1 v + 3p - 2C_3')^2 - 24p(C_1 v - C_3')}}{12\Pi\mu}. \tag{2.8.16}$$

Expression under square root in (2.8.16) can be transformed:

$$\begin{aligned}&\left(3C_1 v + 3p - 2C_3'\right)^2 - 24p\left(C_1 v - C_3'\right)\\ &\quad = \left(3C_1 v - 3p - 2C_3'\right)^2 + 12C_1 p v.\end{aligned} \tag{2.8.17}$$

For investigation of approximate solution one supposes that energy of directed hydrodynamic motion is much less than the energy of chaotic motion. As a result, we have

$$\begin{aligned}&\sqrt{(3C_1 v - 3p - 2C_3')^2 + 12C_1 p v}\\ &\quad = 3C_1 v - 3p - 2C_3' + \frac{6C_1 p v}{3C_1 v - 3p - 2C_3'}\end{aligned} \tag{2.8.18}$$

or

$$\frac{dv}{dy} = \frac{1}{2\tau}\left[1 - \frac{C_1 v}{3C_1 v - 3p - 2C_3'}\right]. \tag{2.8.19}$$

Further we use the simplest estimation – following from (2.8.19) – for normal component of velocity on the bottom of Knudsen layer,

$$v = \frac{y}{2\tau} + v_w, \tag{2.8.20}$$

where τ is mean time between collisions.

From Eq. (2.8.9) one obtains the equation defining the density change on the bottom of Knudsen layer

$$\left(\frac{y}{2\tau} + v_w\right)^2 \frac{d\rho}{dy} = -C_1 \frac{1}{\tau}, \tag{2.8.21}$$

solution of this equation:

$$\rho = \frac{2C_1}{y/(2\tau) + v_w} + D. \tag{2.8.22}$$

Constant of integration

$$D = \rho_w - \frac{2C_1}{v_w}, \tag{2.8.23}$$

then

$$\rho = \rho_w - \frac{y/(2\tau)}{y/(2\tau) + v_w} \frac{2C_1}{v_w}. \tag{2.8.24}$$

Let us calculate the transverse component of velocity in Knudsen layer. After substitution of (2.8.22), (2.8.24) in Eq. (2.8.10), one finds

$$\frac{1}{2}\frac{du}{dz}\left(Dz^2 + 2C_1 z + p\right) - uC_1 + C_2 = 0, \tag{2.8.25}$$

where

$$z = \frac{y}{2\tau} + v_w. \tag{2.8.26}$$

General solution of non-homogeneous differential equation (2.8.25) has the form

$$u = E\exp\left\{\int \frac{2C_1}{Dz^2 + 2C_1 z + p} dz\right\} + \frac{C_2}{C_1}, \tag{2.8.27}$$

where E is a constant of integration. Evaluate integral

$$I = \int \frac{\mathrm{d}z}{Dz^2 + 2C_1 z + p}. \tag{2.8.28}$$

Introduce

$$\delta = 4\left(pD - C_1^2\right). \tag{2.8.29}$$

In the theory of integrals (2.8.28) the sign of δ is important

$$\delta \cong p\left[\rho_w - \Delta\rho\left(1 + \frac{1}{2p}\frac{v_w^2 \Delta\rho}{2}\right)\right], \tag{2.8.30}$$

where $\Delta\rho$ is density change in Knudsen layer. From (2.8.30) follows

$$\delta \cong p\rho_\infty, \quad \delta > 0, \tag{2.8.31}$$

then integral can be evaluated in finite form

$$\begin{aligned} I &= \frac{1}{\sqrt{\delta}} \operatorname{arctg} \frac{2Dz + 2C_1}{\sqrt{\delta}} = \frac{1}{\sqrt{p\rho_\infty}} \operatorname{arctg}\left[\frac{2}{\sqrt{p\rho_\infty}}(Dz + C_1)\right] \\ &= \frac{1}{\sqrt{p\rho_\infty}} \operatorname{arctg}\left[\frac{2\rho_\infty z + v_w \Delta\rho}{\sqrt{p\rho_\infty}}\right], \end{aligned} \tag{2.8.32}$$

where

$$C_1 = \frac{\rho_w v_w - \rho_\infty v_w}{2} = \frac{v_w}{2}\Delta\rho, \qquad D = \rho_\infty. \tag{2.8.33}$$

Index ∞ relates to asymptotic hydrodynamic values outside of Knudsen layer. Returning to variable y, we find

$$I = \frac{1}{\sqrt{p\rho_\infty}} \operatorname{arctg} \frac{\rho_\infty y/\tau + v_w(\rho_w + \rho_\infty)}{\sqrt{p\rho_\infty}} \tag{2.8.34}$$

and corresponding general solution of Eq. (2.8.25)

$$u = E\,\mathrm{e}^{2IC_1} + \frac{C_2}{C_1}, \tag{2.8.35}$$

where

$$E = -\frac{C_2/C_1}{1 + (\Delta\rho/\rho_\infty)v_w(1/\sqrt{RT_\infty}) \operatorname{arctg}(v_w(1 + \rho_w/\rho_\infty)/\sqrt{RT_\infty})}. \tag{2.8.36}$$

Then in indicated suppositions GHE lead to the following profile of transversal velocity in Knudsen layer:

$$u = \frac{C_2}{C_1}\left\{1 - \frac{1}{1 + (\Delta\rho/\rho_\infty)v_w(1/\sqrt{RT_\infty})\varphi}\right. \\ \left. \times \left[1 + \frac{\Delta\rho}{\rho_\infty} v_w \frac{1}{\sqrt{RT_\infty}} \operatorname{arctg}\left[\frac{y/\tau + v_w(1 + \rho_w/\rho_\infty)}{\sqrt{RT_\infty}}\right]\right]\right\}, \tag{2.8.37}$$

where

$$\varphi = \operatorname{arctg} \frac{v_w(1 + \rho_w/\rho_\infty)}{\sqrt{RT_\infty}}. \tag{2.8.38}$$

Introduce notation $\overline{C}_2 = -C_2$, $\overline{C}_2 > 0$, and friction coefficient on the wall

$$f_w = \mu\left(\frac{\mathrm{d}u}{\mathrm{d}y}\right)_w, \tag{2.8.39}$$

then

$$\frac{C_1}{\overline{C}_2} = \frac{v_w\Delta\rho}{2\Pi f_w[1 + \rho_w v_w^2/p]} \tag{2.8.40}$$

or

$$\frac{C_1}{\overline{C}_2} \cong \frac{v_w\Delta\rho}{2\Pi f_w}. \tag{2.8.41}$$

Using (2.8.41), one obtains from (2.8.37)

$$u = f_w \frac{2\Pi}{\rho_\infty\sqrt{RT_\infty}}\left\{\operatorname{arctg}\left[\frac{y/\tau + v_w(1 + \rho_w/\rho_\infty)}{\sqrt{RT_\infty}}\right]\right. \\ \left. - \operatorname{arctg}\left[\frac{v_w(1 + \rho_w/\rho_\infty)}{\sqrt{RT_\infty}}\right]\right\} \\ \times \left\{1 + \frac{\Delta\rho}{\rho_\infty}\frac{v_w}{\sqrt{RT_\infty}} \operatorname{arctg}\left[\frac{v_w(1 + \rho_w/\rho_\infty)}{\sqrt{RT_\infty}}\right]\right\}^{-1}. \tag{2.8.42}$$

Profile of transversal velocity u – under formulated assumptions – is defined by trigonometric function arctg $\bar{y}$. To derive the expression (2.8.42), the assumption was introduced that transversal molecular velocity on the wall is equal to zero. In physical chemistry cases are known (Glasstone, Laidler and Eyring, 1941), when absorbed particles move with a velocity along the surface while being bound by mentioned surface. But evaluation of this "chemical" velocity slip is the problem of quantum chemistry and is not the subject of consideration here.

By small y profile is linear

$$\hat{u} = y\left\{\left[1 + \frac{\Delta\rho}{\rho_\infty}\frac{v_w}{\sqrt{RT_\infty}}\,\mathrm{arctg}\left(\frac{v_w(1+\rho_w/\rho_\infty)}{\sqrt{RT_\infty}}\right)\right]\right.$$
$$\left.\times\,\tau\sqrt{RT_\infty}\left[1 + \frac{v_w^2}{RT_\infty}\left(1 + \frac{\rho_w}{\rho_\infty}\right)^2\right]\right\}^{-1}, \tag{2.8.43}$$

where

$$\hat{u} = \frac{C_1}{\overline{C}_2}u. \tag{2.8.44}$$

Let us introduce now the kinetic velocity slip u_{sl} of flow as difference between transversal velocity on the external boundary of kinetic layer and transversal velocity on the wall. In classical hydrodynamics u_{sl} cannot be introduced without taking into account the Boltzmann equation. The origin of this fact is well known – classical hydrodynamics is not "working" in kinetic layer.

Formula (2.8.43), which is a consequence of GHE, can be rewritten in terminology of slip theory. With this aim introduce asymptotic velocity u_∞ as velocity on the external boundary of Knudsen layer ($u \to u_\infty$ if $y \to \infty$), and slip velocity

$$u_{\mathrm{sl}} = u_\infty - u_w. \tag{2.8.45}$$

As a result, formula (2.8.42) leads to the next slip velocity

$$u_{\mathrm{sl}} = f_w\frac{2\Pi}{\rho_\infty\sqrt{RT_\infty}}\left[\frac{\pi}{2} - \mathrm{arctg}\left(\frac{v_w(1+\rho_w/\rho_\infty)}{\sqrt{RT_\infty}}\right)\right]$$
$$\times\left[1 + \frac{\Delta\rho}{\rho_\infty}\frac{v_w}{\sqrt{RT_\infty}}\,\mathrm{arctg}\left(\frac{v_w(1+\rho_w/\rho_\infty)}{\sqrt{RT_\infty}}\right)\right]^{-1}. \tag{2.8.46}$$

Let us compare now the defined slip velocity u_{sl} from (2.8.46) with the known expressions of the Boltzmann kinetic theory, particularly with data being given by Cercignani (1975).

$$u_{\mathrm{sl}} = 1.1466l\left(\frac{\mathrm{d}u}{\mathrm{d}y}\right)_w = 1.1466\frac{\mu}{\rho_\infty}\sqrt{\frac{\pi}{2RT_\infty}}\left(\frac{\mathrm{d}u}{\mathrm{d}y}\right)_w$$
$$= 1.1466\frac{1}{\rho_\infty}\sqrt{\frac{\pi}{2RT_\infty}}f_w = 1.4370\frac{1}{\rho_\infty\sqrt{RT_\infty}}f_w. \tag{2.8.47}$$

If density change across Knudsen layer is small, one obtains from (2.8.46)

$$u_{\mathrm{sl}} = f_w\frac{2\Pi}{\rho_\infty\sqrt{RT_\infty}}\left[\frac{\pi}{2} - \mathrm{arctg}\,\frac{v_w(1+\rho_w/\rho_\infty)}{\sqrt{RT_\infty}}\right]$$
$$\times\left[1 - \frac{\Delta\rho}{\rho_\infty}\frac{v_w}{\sqrt{RT_\infty}}\,\mathrm{arctg}\,\frac{v_w(1+\rho_w/\rho_\infty)}{\sqrt{RT_\infty}}\right]. \tag{2.8.48}$$

Therefore relations for velocity slip (2.8.46), (2.8.48) contain also density jump. Let us neglect density jump in Knudsen layer. It means

$$\Delta\rho \ll \rho_\infty, \tag{2.8.49}$$

or approximately on the wall

$$\left(\frac{\partial \rho}{\partial y}\right)_w = 0. \tag{2.8.50}$$

We have from (2.8.48)

$$u_{\text{sl}} = 2\Pi\left[\frac{\pi}{2} - \operatorname{arctg}\frac{2v_w}{\sqrt{RT_\infty}}\right]\frac{1}{\rho_\infty\sqrt{RT_\infty}} f_w. \tag{2.8.51}$$

Obviously relations (2.8.47) and (2.8.51) have analogous structure but differ by numerical factors. Multiplier

$$A = 2\Pi\left[\frac{\pi}{2} - \operatorname{arctg}\frac{2v_w}{\sqrt{RT_\infty}}\right] \tag{2.8.52}$$

in (2.8.51) can vary within wide limits, its value is defined by model of particles interaction – which is used for parameter Π evaluation – and velocity of the particles desorption v_w from the surface. For the hard spheres model of particles interactions one obtains $0 \leqslant A \leqslant 2.51$. Notice that indicated by Cercignani four ciphers after point in numerical factors in (2.8.47) have sense only as an orientation by comparison of results of different analytical models.

As a conclusion we state that GHE incorporate the kinetic effect of slip as an element of adjusting of hydrodynamic and kinetic regimes of flow and allow – avoiding artificial fashions – to describe flows by intermediate flow numbers. This peculiar feature of GHE will be demonstrated next, particularly in calculations of sound propagation in rarefied gas and shock wave structure for arbitrary Mach numbers.

CHAPTER 3

Strict Theory of Turbulence and Some Applications of the Generalized Hydrodynamic Theory

The generalized Boltzmann equation necessarily leads to a new formulation of the hydrodynamical equations, yielding what we call the generalized hydrodynamical equations. The classical Enskog, Euler, and Navier–Stokes equations of fluid dynamics are special cases of these new equations. The derivation of the generalized hydrodynamical equations is given in the previous chapter. But the area of applications of GHE and corresponding principles used for their derivation is much vaster. We intend to discuss these new possibilities of generalized hydrodynamic equations in the theory of turbulent flows, electrodynamics of continuum media and quantum mechanics.

3.1. About principles of classical theory of turbulent flows

The turbulent fluid motion has been the subject of intense research for over a hundred years because it has numerous applications in aerodynamics, hydraulics, combustion and explosion processes, and hence is of direct relevance to processes occurring in turbines, engines, compressors, and other modern-day machines. The scientific literature on this subject is enormous, and a detailed analysis of all the existing models is beyond the scope of this book. Here the object is to discuss the currently available turbulence concept in the context of generalized equations of fluid dynamics. In what follows we will discuss "classical" turbulence, usually treated starting from the Navier–Stokes equations, moment methods, and similarity theory. We will also see how this picture corresponds to the generalized Boltzmann kinetics and will try to find out which of the known approaches may be used and which should be abandoned.

It is commonly held that a fully developed turbulence may be characterized by the irregular variation of velocity with time at each point in the flow, and that hydrodynamic quantities undergo fluctuations (turbulent ones or pulsations). Their scale varies over the wide range from the external (using the terminology of Landau and Lifshitz, 1988) scale comparable to the characteristic flow size, to a small scale on which the dynamic fluid viscosity begins to dominate.

Because of the major role of the Reynolds criterion in the theory of turbulence, the study of fluid motion on various typical scales crucially depends on the construction of the Reynolds number

$$Re = \frac{\upsilon_l l}{\upsilon_k}.$$

Here, l is the fluctuation scale, υ_l – the characteristic velocity, and υ_k – the kinematic viscosity. If $l \sim L$, with L being the typical hydrodynamic size, then the Reynolds number Re is large and the effect of molecular viscosity small – so that one may neglect it altogether and apply the similarity theory (the Kolmogorov–Obukhov law) to get some idea of the fluctuations.

From the large-scale fluctuations, the energy goes (practically undissipated) to the small-scale ones, where viscous dissipation takes place (Richardson model of 1922). And even though the dissipation of mechanical energy ε (falling at the unit mass per unit time) occurs on the least possible scale l_K (referred to below as the Kolmogorov turbulence scale), it is believed that the quantity ε also determines the properties of turbulent motion on larger scales. Between the Kolmogorov (or, using the terminology of Landau and Lifshitz (1988), internal) scale l_k and the external scale $l_L \sim L$ there is an inertial interval where the typical size l satisfies the inequality

$$l_K \ll l \leqslant l_L.$$

For want of a better model, it is assumed that turbulent motion is described by the same equations of fluid mechanics (Navier–Stokes equations) used for laminar flows, with a consequence that turbulence emerges as a flow instability or, in this particular case, as an instability in the Navier–Stokes flow model. This gives rise to many inconsistencies, however. It is known, for example, that "although no comprehensive theoretical study has thus far been made for flows through a circular pipe, there is a compelling evidence that this motion is stable with respect to infinitesimal perturbations (in an absolute as well as a convective sense) for any Reynolds numbers" (Landau and Lifshitz, 1988). This contradicts experimental data.

In 1924, W. Heisenberg published a study on the instability of laminar flows (Heisenberg, 1924). A year later, E. Noether "published another paper" – we are quoting Heisenberg (1969) – "in which she proved with all mathematical rigour that the problem admits of no unstable solutions at all and that a flow must be everywhere stable... What about the rigorous mathematics then? I think that even now nobody knows what is wrong with Noether's work". It would appear that rather than Noether's mistake, the drawbacks of the Navier–Stokes flow model are to blame.

The notions of averaged and fluctuating motions prompted Reynolds (1894) to explicitly isolate the fluctuation terms in the Navier–Stokes equations and to subsequently average them over a certain time interval. But neither this approach nor the later technique of averaging over the masses of liquid volumes (sometimes called Favre averaging (Favre, 1992)) provides close solutions, and indeed neither of them is adequate when it comes to physics because, as we will see below, the Navier–Stokes equations are not written for true physical quantities.

One further approach to the problem involves the evaluation of velocity correlation functions with the aim of establishing the relation between the velocities at two neighboring flow field points within the theory of local turbulence. For example, the simplest correlation function is the second-rank tensor

$$B_{rs} = \langle (v_{2r} - v_{1r})(v_{2s} - v_{1s}) \rangle, \tag{3.1.1}$$

where $\mathbf{v}_1$ and $\mathbf{v}_2$ are the fluid velocities at two neighboring points, and the angle brackets denote time averaging. A question remains, however, what exactly "neighboring points" means and how the time averaging procedure is to be carried out. The theory of correlation functions attracted a great deal of attention after Keller and Fridman first introduced them into the hydrodynamics of turbulent motions back in 1924.

In 1944, Landau gave a comprehensive assessment of this line of research. To quote him from Landau and Lifshitz (1988), "One would imagine that in principle it is possible to derive a universal formula, applicable to any turbulent motion, for determining B_{rr}, B_{tt} for all distances r small compared to l_L. In reality, however, such a formula cannot exist at all as the following argument shows. The instantaneous value of the quantity $(v_{2i} - v_{1i})(v_{2k} - v_{1k})$ could in principle be expressed in terms of the energy dissipation ε at the same instant of time t. However, the averaging of these expressions depends significantly on how ε varies in time throughout periods of large-scale (of order l_L) motions. But this variation is different for various specific cases of motion, so the result of such averaging cannot be universal".

One can but agree with this view. To put it another way, if the Kolmogorov scale admits an explicit universal formulation for turbulent fluctuations (as we will show later on), then large-scale fluctuations are determined by solving a specific boundary-value problem.

A. Kolmogorov advanced the hypothesis that the statistical regime of the small-scale components is universal and is determined by only two dimensional parameters, the average rate of energy dissipation ε and the kinematic viscosity v_k. From dimensional considerations it follows that the Kolmogorov fluctuation scale l_K is of the order of $v_k^{3/4}\varepsilon^{-1/4}$. The following estimations can be introduced:

$$\varepsilon \sim \frac{k_B T}{m\tau}, \qquad \left(\frac{p}{\rho}\right)^2 \sim \bar{v}^4, \tag{3.1.2}$$

where $\bar{v}$ is mean molecular velocity, then

$$l_K \sim \tau \bar{v} \tag{3.1.3}$$

and corresponds to the particle mean free path in a gas. But mean time between collision τ is proportional to viscosity, therefore we can wait that all fluctuations on the Kolmogorov micro-scale of turbulence will be proportional to viscosity.

3.2. Theory of turbulence and generalized Euler equations

We now apply the generalized hydrodynamical equations to the theory of turbulence and demonstrate that they enable one to write explicitly the fluctuations of all hydrodynamic quantities on the Kolmogorov turbulence scale l_K. Importantly, these turbulent fluctuations can be tabulated for any type of flow and in this sense can serve as "universal formulas" to use the terminology of monograph (Landau and Lifshitz, 1988). We start by writing down the generalized hydrodynamical equations and, for the sake of simplicity, employ the generalized Euler equations for the special case of a one-component gas flow in a gravitational field. To this end we multiply the generalized Boltzmann equations by the particles' elastic collision invariants $(m, mv, mv^2/2)$ and integrate the resulting equations term by term with respect to velocity.

The calculation of the moments using the Maxwellian distribution function yields the system of generalized Euler equations which includes:

– the continuity equation

$$\frac{\partial}{\partial t}\left\{\rho - \Pi\frac{\mu}{p}\left[\frac{\partial \rho}{\partial t} + \frac{\partial}{\partial \mathbf{r}}\cdot(\rho\mathbf{v}_0)\right]\right\} + \frac{\partial}{\partial \mathbf{r}}\cdot\left\{\rho\mathbf{v}_0 - \Pi\frac{\mu}{p}\left[\frac{\partial}{\partial t}(\rho\mathbf{v}_0) + \frac{\partial}{\partial \mathbf{r}}\cdot\rho\mathbf{v}_0\mathbf{v}_0 + \overleftrightarrow{I}\cdot\frac{\partial p}{\partial \mathbf{r}} - \rho\mathbf{g}\right]\right\} = 0, \quad (3.2.1)$$

– the equation of motion

$$\begin{aligned}
&\frac{\partial}{\partial t}\left\{\rho\mathbf{v}_0 - \Pi\frac{\mu}{p}\left[\frac{\partial}{\partial t}(\rho\mathbf{v}_0) + \frac{\partial}{\partial \mathbf{r}}\cdot\rho\mathbf{v}_0\mathbf{v}_0 + \frac{\partial p}{\partial \mathbf{r}} - \rho\mathbf{g}\right]\right\}\\
&\quad - \mathbf{g}\left[\rho - \Pi\frac{\mu}{p}\left(\frac{\partial \rho}{\partial t} + \frac{\partial}{\partial \mathbf{r}}\cdot(\rho\mathbf{v}_0)\right)\right]\\
&\quad + \frac{\partial}{\partial \mathbf{r}}\cdot\left\{\rho\mathbf{v}_0\mathbf{v}_0 + p\overleftrightarrow{I} - \Pi\frac{\mu}{p}\left[\frac{\partial}{\partial t}(\rho\mathbf{v}_0\mathbf{v}_0 + p\overleftrightarrow{I}) + \frac{\partial}{\partial \mathbf{r}}\cdot\rho(\mathbf{v}_0\mathbf{v}_0)\mathbf{v}_0\right.\right.\\
&\quad \left.\left. + 2\overleftrightarrow{I}\left[\frac{\partial}{\partial \mathbf{r}}\cdot(p\mathbf{v}_0)\right] + \frac{\partial}{\partial \mathbf{r}}\cdot(\overleftrightarrow{I}p\mathbf{v}_0) - \mathbf{g}\mathbf{v}_0\rho - \mathbf{v}_0\mathbf{g}\rho\right]\right\} = 0, \qquad (3.2.2)
\end{aligned}$$

– the equation of energy

$$\begin{aligned}
&\frac{\partial}{\partial t}\left\{\frac{\rho v_0^2}{2} + \frac{3}{2}p - \Pi\frac{\mu}{p}\left[\frac{\partial}{\partial t}\left(\frac{\rho v_0^2}{2} + \frac{3}{2}p\right)\right.\right.\\
&\quad \left.\left. + \frac{\partial}{\partial \mathbf{r}}\cdot\left(\frac{1}{2}\rho v_0^2\mathbf{v}_0 + \frac{5}{2}p\mathbf{v}_0\right) - \mathbf{g}\cdot\rho\mathbf{v}_0\right]\right\}\\
&\quad + \frac{\partial}{\partial \mathbf{r}}\cdot\left\{\frac{1}{2}\rho v_0^2\mathbf{v}_0 + \frac{5}{2}p\mathbf{v}_0 - \Pi\frac{\mu}{p}\left[\frac{\partial}{\partial t}\left(\frac{1}{2}\rho v_0^2\mathbf{v}_0 + \frac{5}{2}p\mathbf{v}_0\right)\right.\right.
\end{aligned}$$

$$+\frac{\partial}{\partial \mathbf{r}}\cdot\left(\frac{1}{2}\rho v_0^2\mathbf{v}_0\mathbf{v}_0+\frac{7}{2}p\mathbf{v}_0\mathbf{v}_0+\frac{1}{2}pv_0^2\overleftrightarrow{I}+\frac{5}{2}\frac{p^2}{\rho}\overleftrightarrow{I}\right)$$
$$-\rho\mathbf{g}\cdot\mathbf{v}_0\mathbf{v}_0-p\mathbf{g}\cdot\overleftrightarrow{I}-\frac{1}{2}\rho v_0^2\mathbf{g}-\frac{3}{2}\mathbf{g}p\Bigg]\Bigg\}-\Bigg\{\rho\mathbf{g}\cdot\mathbf{v}_0$$
$$-\Pi\frac{\mu}{p}\left[\mathbf{g}\cdot\left(\frac{\partial}{\partial t}(\rho\mathbf{v}_0)+\frac{\partial}{\partial \mathbf{r}}\cdot\rho\mathbf{v}_0\mathbf{v}_0+\frac{\partial}{\partial \mathbf{r}}\cdot p\overleftrightarrow{I}-\rho\mathbf{g}\right)\right]\Bigg\}=0, \tag{3.2.3}$$

where we have used the hydrodynamic approximation $\tau^{(0)}=\Pi\mu/p$ (for the hard-sphere model, $\Pi=0.8$) and where $\overleftrightarrow{I}$ is the unit tensor.

We next introduce ρ_∞, v_∞, p_∞, and μ_∞ as the density, velocity, pressure, and viscosity scales, respectively. We take the characteristic dimension to be L, and the time scale, L/v_∞. Then the dimensionless equation of continuity takes the form

$$\frac{\partial}{\partial \hat{t}}\left\{\hat{\rho}-\Pi\frac{\hat{\mu}}{\hat{p}}\frac{\mu_\infty v_\infty}{p_\infty L}\left[\frac{\partial\hat{\rho}}{\partial\hat{t}}+\frac{\partial}{\partial\hat{\mathbf{r}}}\cdot(\hat{\rho}\hat{\mathbf{v}}_0)\right]\right\}$$
$$+\frac{\partial}{\partial\hat{\mathbf{r}}}\cdot\Bigg\{\hat{\rho}\hat{\mathbf{v}}_0-\Pi\frac{\hat{\mu}}{\hat{p}}\frac{\mu_\infty}{p_\infty}\frac{v_\infty}{L}\Bigg[\frac{\partial}{\partial\hat{t}}(\hat{\rho}\hat{\mathbf{v}}_0)+\frac{\partial}{\partial\hat{\mathbf{r}}}\cdot\hat{\rho}\hat{\mathbf{v}}_0\hat{\mathbf{v}}_0$$
$$+\frac{p_\infty}{\rho_\infty v_\infty^2}\overleftrightarrow{I}\cdot\frac{\partial\hat{p}}{\partial\hat{\mathbf{r}}}-\frac{Lg}{v_\infty^2}\hat{\rho}\hat{g}\Bigg]\Bigg\}=0. \tag{3.2.4}$$

Dimensionless combinations of the scale quantities introduced above form the similarity criteria

$$\frac{\mu_\infty v_\infty}{p_\infty L}=\frac{\mu_\infty}{Lv_\infty\rho_\infty}\frac{\rho_\infty v_\infty^2}{p_\infty}=Re^{-1}Eu^{-1},\qquad \frac{v_\infty^2}{Lg}=Fr.$$

Thus, the continuity equation (3.2.4) contains the Reynolds, Euler, and Frud similarity criteria and may be rewritten as

$$\frac{\partial}{\partial \hat{t}}\left\{\hat{\rho}-\Pi\frac{\hat{\mu}}{\hat{p}}\frac{\mu_\infty v_\infty}{p_\infty L}\left[\frac{\partial\hat{\rho}}{\partial\hat{t}}+\frac{\partial}{\partial\hat{\mathbf{r}}}\cdot(\hat{\rho}\hat{\mathbf{v}}_0)\right]\right\}$$
$$+\frac{\partial}{\partial\hat{\mathbf{r}}}\cdot\Bigg\{\hat{\rho}\hat{\mathbf{v}}_0-\Pi\frac{\hat{\mu}}{\hat{p}}\frac{\mu_\infty}{p_\infty}\frac{v_\infty}{L}\Bigg[\frac{\partial}{\partial\hat{t}}(\hat{\rho}\hat{\mathbf{v}}_0)+\frac{\partial}{\partial\hat{\mathbf{r}}}\cdot\hat{\rho}\hat{\mathbf{v}}_0\hat{\mathbf{v}}_0$$
$$+\frac{p_\infty}{\rho_\infty v_\infty^2}\overleftrightarrow{I}\cdot\frac{\partial\hat{p}}{\partial\hat{\mathbf{r}}}-\frac{Lg}{v_\infty^2}\hat{\rho}\hat{g}\Bigg]\Bigg\}=0. \tag{3.2.5}$$

In a similar fashion we write the dimensionless equation of motion

$$\frac{\partial}{\partial\hat{t}}\left\{\hat{\rho}\hat{\mathbf{v}}_0-\Pi\frac{\hat{\mu}}{\hat{p}}\frac{Eu^{-1}}{Re}\left[\frac{\partial}{\partial\hat{t}}(\hat{\rho}\hat{\mathbf{v}}_0)+\frac{\partial}{\partial\hat{\mathbf{r}}}\cdot\hat{\rho}\hat{\mathbf{v}}_0\hat{\mathbf{v}}_0+Eu\frac{\partial\hat{p}}{\partial\hat{\mathbf{r}}}-\frac{1}{Fr}\hat{\rho}\hat{\mathbf{g}}\right]\right\}$$
$$-\frac{1}{Fr}\hat{\mathbf{g}}\left[\hat{\rho}-\Pi\frac{\hat{\mu}}{\hat{p}}\frac{Eu^{-1}}{Re}\left(\frac{\partial\hat{\rho}}{\partial\hat{t}}+\frac{\partial}{\partial\hat{\mathbf{r}}}\cdot(\hat{\rho}\hat{\mathbf{v}}_0)\right)\right]$$

$$
+\frac{\partial}{\partial \hat{\mathbf{r}}}\cdot\left\{\hat{\rho}\hat{\mathbf{v}}_0\hat{\mathbf{v}}_0+Eu\hat{p}\overleftrightarrow{I}-\Pi\frac{\hat{\mu}}{\hat{p}}\frac{Eu^{-1}}{Re}\left[\frac{\partial}{\partial \hat{t}}(\hat{\rho}\hat{\mathbf{v}}_0\hat{\mathbf{v}}_0Eu\hat{p}\overleftrightarrow{I})\right.\right.
$$

$$
+\frac{\partial}{\partial \hat{\mathbf{r}}}\cdot\hat{\rho}(\hat{\mathbf{v}}_0\hat{\mathbf{v}}_0)\hat{\mathbf{v}}_0+2\overleftrightarrow{I}\left[\frac{\partial}{\partial \hat{\mathbf{r}}}\cdot(\hat{p}\hat{\mathbf{v}}_0)\right]Eu+Eu\frac{\partial}{\partial \hat{\mathbf{r}}}\cdot(\overleftrightarrow{I}\hat{p}\hat{\mathbf{v}}_0)
$$

$$
\left.\left.-\frac{1}{Fr}\hat{\rho}\hat{\mathbf{g}}\hat{\mathbf{v}}_0-\frac{1}{Fr}\rho\hat{\mathbf{v}}_0\hat{\mathbf{g}}\right]\right\}=0, \tag{3.2.6}
$$

and the dimensionless equation of energy

$$
\frac{\partial}{\partial \hat{t}}\left\{\frac{\hat{\rho}\hat{v}_0^2}{2}+Eu\frac{3}{2}\hat{p}-\Pi\frac{\hat{\mu}}{\hat{p}}\frac{Eu^{-1}}{Re}\left[\frac{\partial}{\partial \hat{t}}\left(\frac{\hat{\rho}\hat{v}_0^2}{2}+Eu\frac{3}{2}\hat{p}\right)\right.\right.
$$

$$
\left.\left.+\frac{\partial}{\partial \hat{\mathbf{r}}}\cdot\left(\frac{1}{2}\hat{\rho}\hat{v}_0^2\hat{\mathbf{v}}_0+Eu\frac{5}{2}\hat{p}\hat{\mathbf{v}}_0\right)-\frac{1}{Fr}\hat{\bar{g}}\rho\hat{\mathbf{v}}_0\right]\right\}
$$

$$
+\frac{\partial}{\partial \hat{\mathbf{r}}}\cdot\left\{\frac{1}{2}\rho\hat{v}_0^2\hat{\mathbf{v}}_0+Eu\frac{5}{2}\hat{p}\hat{\mathbf{v}}_0-\Pi\frac{\hat{\mu}}{\hat{p}}\frac{Eu^{-1}}{Re}\left[\frac{\partial}{\partial t}\left(\frac{1}{2}\hat{\rho}\hat{v}_0^2\hat{\mathbf{v}}_0+Eu\frac{5}{2}\hat{p}\hat{\mathbf{v}}_0\right)\right.\right.
$$

$$
+\frac{\partial}{\partial \hat{\mathbf{r}}}\cdot\left(\frac{1}{2}\hat{\rho}\hat{v}_0^2\hat{\mathbf{v}}_0\hat{\mathbf{v}}_0+Eu\frac{7}{2}\hat{p}\hat{\mathbf{v}}_0\hat{\mathbf{v}}_0+\frac{1}{2}\hat{p}\hat{v}_0^2\overleftrightarrow{I}Eu+5Eu^2\frac{\hat{p}^2}{2\hat{\rho}}\overleftrightarrow{I}\right)
$$

$$
\left.\left.-\frac{1}{Fr}\hat{\rho}\hat{\mathbf{g}}\cdot\hat{\mathbf{v}}_0\hat{\mathbf{v}}_0-\frac{Eu}{Fr}\hat{p}\hat{\mathbf{g}}\cdot\overleftrightarrow{I}-\frac{1}{Fr}\frac{1}{2}\hat{\rho}\hat{v}_0^2\hat{\mathbf{g}}-\frac{Eu}{Fr}\frac{3}{2}\hat{\mathbf{g}}\hat{p}\right]\right\}
$$

$$
-\left\{\frac{1}{Fr}\hat{\rho}\hat{\mathbf{g}}\cdot\hat{\mathbf{v}}_0-\Pi\frac{\hat{\mu}}{\hat{p}}\frac{Eu^{-1}}{Re}\left[\frac{1}{Fr}\hat{\mathbf{g}}\cdot\left(\frac{\partial}{\partial \hat{t}}(\hat{\rho}\hat{\mathbf{v}}_0)+\frac{\partial}{\partial \hat{\mathbf{r}}}\cdot\hat{\rho}\hat{\mathbf{v}}_0\hat{\mathbf{v}}_0\right.\right.\right.
$$

$$
\left.\left.\left.+\frac{\partial}{\partial \hat{\mathbf{r}}}\cdot\hat{p}\overleftrightarrow{I}Eu-\hat{\rho}\hat{\mathbf{g}}\frac{1}{Fr}\right)\right]\right\}=0. \tag{3.2.7}
$$

Eqs. (3.2.5)–(3.2.7) are notable for their structure. All the generalized equations of Euler fluid dynamics contain the Reynolds, Euler, and Frud numbers (similarity criteria). Naturally, the inclusion of forces of electromagnetic origin would lead to additional similarity criteria. For each hydrodynamical quantity – density, energy, and momentum as well as their fluxes – there is a corresponding temporally and spatially fluctuating term which is proportional to Re^{-1} and, hence, to the viscosity.

For small-scale fluctuations (i.e., smaller characteristic dimension l in the Reynolds number), viscosity increases in importance and starts to determine ε, the dissipation of the mechanical energy.

Introduce the Kolmogorov scale length l_K as

$$
l_K=\left(\frac{\upsilon_k^3}{\varepsilon}\right)^{1/4} \tag{3.2.8}
$$

and find using (3.1.3) that the length l_K is of the mean free path order. The fluctuation terms thus determine turbulent Kolmogorov-scale fluctuations (small-scale fluctuations or, using the computational hydrodynamics term, sub-grid turbulence) which are of a universal nature and not problem specific.

To fully understand the situation, however, the following questions remain to be answered:

(1) Are there no contradictions in the system of fluctuations introduced in this way? In other words, is the set of fluctuations self-consistent?
(2) With a system of base (independent) fluctuations on hand, is it possible to derive explicit expressions for other hydrodynamical quantities and their moments?
(3) What do the generalized hydrodynamical equations for averaged quantities look like and how does the procedure for obtaining the averaged equations agree with the Reynolds procedure familiar from the theory of turbulence?

In answering the above questions, the generalized Euler equations for one-component gas will be employed for the sake of clarity. Implicit in the following analysis will be the fact, already noted above, that we are dealing with small-scale fluctuations.

The equations to be investigated are:

– continuity equation

$$\frac{\partial}{\partial t}\left\{\rho-\tau\left(\frac{\partial \rho}{\partial t}+\frac{\partial}{\partial \mathbf{r}}\cdot(\rho\mathbf{v}_0)\right)\right\}$$
$$+\frac{\partial}{\partial \mathbf{r}}\cdot\left\{\rho\mathbf{v}_0-\tau\left(\frac{\partial}{\partial t}(\rho\mathbf{v}_0)+\frac{\partial}{\partial \mathbf{r}}\cdot(\rho\mathbf{v}_0\mathbf{v}_0)+\frac{\partial p}{\partial \mathbf{r}}-\rho\mathbf{g}\right)\right\}=0, \tag{3.2.9}$$

– equation of motion

$$\frac{\partial}{\partial t}\left\{\rho v_{0\beta}-\tau\left[\frac{\partial}{\partial t}(\rho v_{0\beta})+\frac{\partial}{\partial r_\alpha}(p\delta_{\alpha\beta}+\rho v_{0\alpha})-\rho g_\beta\right]\right\}$$
$$-\left\{\rho-\tau\left[\frac{\partial \rho}{\partial t}+\frac{\partial}{\partial r_\alpha}(\rho v_{0\alpha})\right]\right\}g_\beta+\frac{\partial}{\partial r_\alpha}\left\{p\delta_{\alpha\beta}+\rho v_{0\alpha}v_{0\beta}\right.$$
$$-\tau\left[\frac{\partial}{\partial t}(p\delta_{\alpha\beta}+\rho v_{0\alpha}v_{0\beta})+\frac{\partial}{\partial r_\gamma}(p\delta_{\alpha\gamma}v_{0\beta}+pv_{0\alpha}\delta_{\beta\gamma}\right.$$
$$\left.\left.+pv_{0\gamma}\delta_{\alpha\beta}+\rho v_{0\alpha}v_{0\beta}v_{0\gamma})-g_\alpha\rho v_{0\beta}-g_\beta\rho v_{0\alpha}\right]\right\}=0, \tag{3.2.10}$$

where we employ the Einstein summation rule for recurrent subscripts $\alpha,\beta,\gamma=1,2,3$ referring to components of vectors in the Cartesian coordinate system,

– equation of energy

$$\frac{\partial}{\partial t}\left\{3p+\rho v_0^2-\tau\left[\frac{\partial}{\partial t}(3p+\rho v_0^2)+\frac{\partial}{\partial \mathbf{r}}\cdot\left(\mathbf{v}_0(\rho v_0^2+5p)\right)-2\mathbf{g}\cdot\rho\mathbf{v}_0\right]\right\}$$
$$+\frac{\partial}{\partial \mathbf{r}}\cdot\left\{\mathbf{v}_0(\rho v_0^2+5p)-\tau\left[\frac{\partial}{\partial t}\left(\mathbf{v}_0(\rho v_0^2+5p)\right)+\frac{\partial}{\partial \mathbf{r}}\cdot\left[\overleftrightarrow{I}pv_0^2+\rho v_0^2\mathbf{v}_0\mathbf{v}_0\right.\right.\right.$$

$$+7p\mathbf{v}_0\mathbf{v}_0+5\overset{\leftrightarrow}{I}\frac{p^2}{\rho}\Bigg]-2\rho\mathbf{v}_0\mathbf{v}_0\cdot\mathbf{g}-5p\overset{\leftrightarrow}{I}\cdot\mathbf{g}-\rho v_0^2\overset{\leftrightarrow}{I}\cdot\vec{g}\Bigg]\Bigg\}$$

$$-2\mathbf{g}\cdot\left\{\rho\mathbf{v}_0-\tau\left[\frac{\partial}{\partial t}(\rho\mathbf{v}_0)+\frac{\partial p}{\partial \mathbf{r}}+\frac{\partial}{\partial \mathbf{r}}\cdot(\rho\mathbf{v}_0\mathbf{v}_0)-\rho\mathbf{g}\right]\right\}=0. \quad (3.2.11)$$

To calculate hydrodynamic fluctuations, the Reynolds procedure will be employed. Thus, for example, the product of the true density ρ and the true velocity $\mathbf{v}_0$ can be used to obtain the fluctuation quantity $\mathbf{v}_0^f$. Indeed, we have

$$\rho\mathbf{v}_0=\left(\rho^a+\rho^f\right)\left(\mathbf{v}_0^a+\mathbf{v}_0^f\right), \quad (3.2.12)$$

where the superscript "a" denotes the average hydrodynamic quantities. Ignoring the fluctuation terms squared and keeping only first-order small quantities in relations of type (3.2.12) we find

$$(\rho\mathbf{v}_0)^f=\rho\mathbf{v}_0-\rho^a\mathbf{v}_0^a=\rho^a\mathbf{v}_0^f+\rho^f\mathbf{v}_0^a. \quad (3.2.13)$$

Thus one obtains

$$\mathbf{v}_0^f=\frac{(\rho\mathbf{v}_0)^f-\rho^f\mathbf{v}_0^a}{\rho^a}. \quad (3.2.14)$$

From the continuity equation (3.2.9) we have

$$\rho^f=\tau\left\{\frac{\partial\rho}{\partial t}+\frac{\partial}{\partial\mathbf{r}}\cdot(\rho\mathbf{v}_0)\right\}, \quad (3.2.15)$$

$$(\rho\mathbf{v}_0)^f=\tau\left\{\frac{\partial}{\partial t}(\rho\mathbf{v}_0)+\frac{\partial}{\partial\mathbf{r}}\cdot(\rho\mathbf{v}_0\mathbf{v}_0)+\frac{\partial p}{\partial\mathbf{r}}-\rho\mathbf{g}\right\} \quad (3.2.16)$$

and therefore from Eq. (3.2.14) it follows

$$\mathbf{v}_0^f=\tau\left\{\frac{\partial\mathbf{v}_0}{\partial t}+\left(\mathbf{v}_0\cdot\frac{\partial}{\partial\mathbf{r}}\right)\mathbf{v}_0+\frac{1}{\rho}\frac{\partial p}{\partial\mathbf{r}}-\mathbf{g}\right\}. \quad (3.2.17)$$

Using Eq. (3.2.10), we find that the fluctuation of the combined hydrodynamical quantity $p\delta_{\alpha\beta}+\rho v_{0\alpha}v_{0\beta}$ is given by

$$\begin{aligned}(p\delta_{\alpha\beta}&+\rho v_{0\alpha}v_{0\beta})^f\\&=\tau\left[\frac{\partial}{\partial t}(p\delta_{\alpha\beta}+\rho v_{0\alpha}v_{0\beta})+\frac{\partial}{\partial r_\gamma}(p\delta_{\alpha\gamma}v_{0\beta}+pv_{0\alpha}\delta_{\beta\gamma}\right.\\&\qquad\left.+pv_{0\gamma}\delta_{\alpha\beta}+\rho v_{0\alpha}v_{0\beta}v_{0\gamma})-g_\alpha\rho v_{0\beta}-g_\beta\rho v_{0\alpha}\right].\end{aligned} \quad (3.2.18)$$

With the help of Eq. (3.2.18) we obtain

$$\begin{aligned}&(3p+\rho v_0^2)^f\\&\quad=\tau\left[\frac{\partial}{\partial t}(3p+\rho v_0^2)+\frac{\partial}{\partial \mathbf{r}}\cdot\left(\mathbf{v}_0(\rho v_0^2+5p)\right)-2\mathbf{g}\cdot\rho\mathbf{v}_0\right].\end{aligned}\tag{3.2.19}$$

We now proceed to calculate $(\rho v_0^2)^f$:

$$\begin{aligned}(\rho v_0^2)^f&=\rho v_0^2-\rho^a v_0^{a\,2}=(\rho^a+\rho^f)(\mathbf{v}_0^a+\mathbf{v}_0^f)^2-\rho^a v_0^{a\,2}\\&\cong(\rho^a+\rho^f)(v_0^{a\,2}+2\mathbf{v}_0^a\cdot\mathbf{v}_0^f)-\rho^a v_0^{a\,2}\\&\cong\rho^f v_0^{a\,2}+2\rho^a\mathbf{v}_0^a\cdot\mathbf{v}_0^f\end{aligned}\tag{3.2.20}$$

which, when combined with Eqs. (3.2.15) and (3.2.17), yields

$$\begin{aligned}(\rho v_0^2)^f=\tau\Bigg\{&v_0^{a\,2}\left[\frac{\partial\rho}{\partial t}+\frac{\partial}{\partial\mathbf{r}}\cdot(\rho\mathbf{v}_0)\right]\\&+2\rho^a\mathbf{v}_0^a\cdot\left[\frac{\partial\mathbf{v}_0}{\partial t}+\left(\mathbf{v}_0\cdot\frac{\partial}{\partial\mathbf{r}}\right)\mathbf{v}_0+\frac{1}{\rho}\frac{\partial p}{\partial\mathbf{r}}-\mathbf{g}\right]\Bigg\}.\end{aligned}\tag{3.2.21}$$

From (3.2.19) fluctuation of static pressure is found

$$[3p+\rho v_0^2]^f=3p^f+(\rho v_0^2)^f,\tag{3.2.22}$$

$$\begin{aligned}3p^f=\tau\Bigg\{&\frac{\partial}{\partial t}(3p+\rho v_0^2)+\frac{\partial}{\partial\mathbf{r}}\cdot\left(\mathbf{v}_0(\rho v_0^2+5p)\right)-2\mathbf{g}\cdot\rho\mathbf{v}_0\\&-v_0^{a\,2}\left[\frac{\partial\rho}{\partial t}+\frac{\partial}{\partial\mathbf{r}}\cdot(\rho\mathbf{v}_0)\right]\\&-2\rho^a\mathbf{v}_0^a\cdot\left[\frac{\partial\mathbf{v}_0}{\partial t}+\left(\mathbf{v}_0\cdot\frac{\partial}{\partial\mathbf{r}}\right)\mathbf{v}_0+\frac{1}{\rho}\frac{\partial p}{\partial\mathbf{r}}-\mathbf{g}\right]\Bigg\}.\end{aligned}\tag{3.2.23}$$

Neglecting squared fluctuations – this tantamount to omitting terms proportional to τ^2 – we find:

$$\begin{aligned}3p^f&=\tau\Bigg\{\frac{\partial}{\partial t}(3p+\rho v_0^2)+\frac{\partial}{\partial\mathbf{r}}\cdot\left(\mathbf{v}_0(\rho v_0^2+5p)\right)\\&\qquad-v_0^2\left[\frac{\partial\rho}{\partial t}+\frac{\partial}{\partial\mathbf{r}}\cdot(\rho\mathbf{v}_0)\right]-2\rho\mathbf{v}_0\cdot\left[\frac{\partial\mathbf{v}_0}{\partial t}+\left(\mathbf{v}_0\cdot\frac{\partial}{\partial\mathbf{r}}\right)\mathbf{v}_0+\frac{1}{\rho}\frac{\partial p}{\partial\mathbf{r}}\right]\Bigg\}\\&=\tau\Bigg\{3\frac{\partial p}{\partial t}+\frac{\partial}{\partial\mathbf{r}}\cdot(\rho v_0^2\mathbf{v}_0)-v_0^2\frac{\partial}{\partial\mathbf{r}}\cdot(\rho\mathbf{v}_0)-2\rho\mathbf{v}_0\cdot\left(\mathbf{v}_0\cdot\frac{\partial}{\partial\mathbf{r}}\right)\mathbf{v}_0\\&\qquad+5\frac{\partial}{\partial\mathbf{r}}\cdot(p\mathbf{v}_0)-2\mathbf{v}_0\cdot\frac{\partial p}{\partial\mathbf{r}}\Bigg\}\end{aligned}$$

$$
\begin{aligned}
&= \tau\left\{3\frac{\partial p}{\partial t} + \left(\rho\mathbf{v}_0\cdot\frac{\partial}{\partial\mathbf{r}}\right)v_0^2 - 2\rho\mathbf{v}_0\cdot\left(\mathbf{v}_0\cdot\frac{\partial}{\partial\mathbf{r}}\right)\mathbf{v}_0 + 3\frac{\partial}{\partial\mathbf{r}}\cdot(p\mathbf{v}_0)\right.\\
&\qquad\left. + 2p\left(\frac{\partial}{\partial\mathbf{r}}\cdot\mathbf{v}_0\right)\right\}\\
&= \tau\left\{3\frac{\partial p}{\partial t} + 3\frac{\partial}{\partial\mathbf{r}}\cdot(p\mathbf{v}_0) + 2p\frac{\partial}{\partial\mathbf{r}}\cdot\mathbf{v}_0\right\},
\end{aligned}
$$

or

$$
p^f = \tau\left\{\frac{\partial p}{\partial t} + \frac{\partial}{\partial\mathbf{r}}\cdot(p\mathbf{v}_0) + \frac{2}{3}p\frac{\partial}{\partial\mathbf{r}}\cdot\mathbf{v}_0\right\}. \tag{3.2.24}
$$

Then formulated procedure allows to obtain fluctuations of the zero-order velocity's moment (3.2.15), fluctuations of moments of the first-order (3.2.17), and the second-order ones (3.2.20). Dependent fluctuations (for example, p^f) should be calculated using independent fluctuations of hydrodynamic values. Senior velocity's moment in the system (3.2.9)–(3.2.11) is $\mathbf{v}_0(\rho v_0^2 + 5p)$. Let us find the fluctuation of this moment.

$$
\left[\mathbf{v}_0\left(\rho v_0^2 + 5p\right)\right]^f = \mathbf{v}_0\left(\rho v_0^2 + 5p\right) - \mathbf{v}_0^a\left(\rho^a v_0^{a2} + 5p^a\right). \tag{3.2.25}
$$

The corresponding result is of principal significance and we deliver its derivation in detail.

$$
\begin{aligned}
\left[\mathbf{v}_0\left(\rho v_0^2 + 5p\right)\right]^f &\cong \left(\mathbf{v}_0^a + \mathbf{v}_0^f\right)\left(\rho^a + \rho^f\right)\left(v_0^{a2} + 2\mathbf{v}_0^a\cdot\mathbf{v}_0^f\right)\\
&\quad + 5\left(\mathbf{v}_0^a + \mathbf{v}_0^f\right)\left(p^a + p^f\right) - \mathbf{v}_0^a\left(\rho^a v_0^{a2} + 5p^a\right)\\
&\cong \left(\mathbf{v}_0^a\rho^f + \mathbf{v}_0^f\rho^a\right)v_0^{a2} + 5\mathbf{v}_0^f p^a + 5p^f\mathbf{v}_0^a + 2\mathbf{v}_0^a\cdot\mathbf{v}_0^f\mathbf{v}_0^a\rho^a.
\end{aligned} \tag{3.2.26}
$$

Use now (3.2.15), (3.2.17) and (3.2.25):

$$
\begin{aligned}
\left[\mathbf{v}_0\left(\rho v_0^2 + 5p\right)\right]^f &\cong \tau\left\{\mathbf{v}_0^a v_0^{a2}\left(\frac{\partial\rho}{\partial t} + \frac{\partial}{\partial\mathbf{r}}\cdot(\rho\mathbf{v}_0)\right)\right.\\
&\quad + \left(5p^a + \rho^a v_0^{a2}\right)\left[\frac{\partial\mathbf{v}_0}{\partial t} + \left(\mathbf{v}_0\cdot\frac{\partial}{\partial\mathbf{r}}\right)\mathbf{v}_0 + \frac{1}{\rho}\frac{\partial p}{\partial\mathbf{r}} - \mathbf{g}\right]\\
&\quad + 5\mathbf{v}_0^a\left(\frac{\partial p}{\partial t} + \frac{\partial}{\partial\mathbf{r}}\cdot(p\mathbf{v}_0) + \frac{2}{3}p\frac{\partial}{\partial\mathbf{r}}\cdot\mathbf{v}_0\right)\\
&\quad \left. + 2\mathbf{v}_0^a\rho^a\mathbf{v}_0^a\cdot\left[\frac{\partial\mathbf{v}_0}{\partial t} + \left(\mathbf{v}_0\cdot\frac{\partial}{\partial\mathbf{r}}\right)\mathbf{v}_0 + \frac{1}{\rho}\frac{\partial p}{\partial\mathbf{r}} - \mathbf{g}\right]\right\}.
\end{aligned} \tag{3.2.27}
$$

Consider first of all the temporal part of (3.2.27):

$$
\begin{aligned}
&\tau\left\{\mathbf{v}_0^a v_0^{a2}\frac{\partial\rho}{\partial t} + \left(5p^a + \rho^a v_0^{a2}\right)\frac{\partial\mathbf{v}_0}{\partial t} + 5\mathbf{v}_0^a\frac{\partial p}{\partial t} + 2\rho_a\mathbf{v}_0^a\mathbf{v}_0^a\cdot\frac{\partial\mathbf{v}_0}{\partial t}\right\}\\
&\quad\cong \tau\frac{\partial}{\partial t}\left\{\mathbf{v}_0\left(5p + \rho v_0^2\right)\right\}.
\end{aligned} \tag{3.2.28}
$$

The temporal part of fluctuation of the considered moment, calculated with help of fluctuations of the minor moments, coincides with the time derivative of this moment in right-hand side of Eq. (3.2.11).

Let us go now to the spatial part of fluctuation connected with the derivative $\partial/\partial\mathbf{r}$,

$$M^{<}_{v^3} = \tau\Bigg\{\mathbf{v}_0^a v_0^{a2}\frac{\partial}{\partial\mathbf{r}}\cdot(\rho\mathbf{v}_0) + \left(5p^a + \rho^a v_0^{a2}\right)\left[\left(\mathbf{v}_0\cdot\frac{\partial}{\partial\mathbf{r}}\right)\mathbf{v}_0 + \frac{1}{\rho}\frac{\partial p}{\partial\mathbf{r}} - \mathbf{g}\right]$$
$$+ 5\mathbf{v}_0^a\left[\frac{\partial}{\partial\mathbf{r}}\cdot(p\mathbf{v}_0) + \frac{2}{3}p\frac{\partial}{\partial\mathbf{r}}\cdot\mathbf{v}_0\right]$$
$$+ 2\mathbf{v}_0^a\rho^a\mathbf{v}_0^a\cdot\left[\frac{\partial\mathbf{v}_0}{\partial t} + \left(\mathbf{v}_0\cdot\frac{\partial}{\partial\mathbf{r}}\right)\mathbf{v}_0 + \frac{1}{\rho}\frac{\partial p}{\partial\mathbf{r}} - \mathbf{g}\right]\Bigg\}, \tag{3.2.29}$$

where $M^{<}_{v^3}$ denotes fluctuation of the third-order moment fond of help of minor moments. The corresponding term M_{v^3} in (3.2.11) has the form

$$M_{v^3} = \tau\Bigg\{\frac{\partial}{\partial\mathbf{r}}\cdot\left[\overleftrightarrow{I}pv_0^2 + \rho v_0^2\mathbf{v}_0\mathbf{v}_0 + 7p\mathbf{v}_0\mathbf{v}_0 + 5\overleftrightarrow{I}\frac{p^2}{\rho}\right]$$
$$- 2\rho\mathbf{v}_0\mathbf{v}_0\cdot\mathbf{g} - 5p\overleftrightarrow{I}\cdot\mathbf{g} - \rho v_0^2\overleftrightarrow{I}\cdot\mathbf{g}\Bigg\}. \tag{3.2.30}$$

Then the spatial fluctuation of the third-order velocity's moments contains the moments of the fourth-order.

Are they coincided, $M^{<}_{v^3}$ and M_{v^3}? It turns out that not. Investigate discrepancy, but before transform the derivative.

$$\frac{\partial}{\partial\mathbf{r}}\cdot\left(\rho v_0^2\mathbf{v}_0\mathbf{v}_0\right) = \rho v_0^2\frac{\partial}{\partial\mathbf{r}}\cdot(\mathbf{v}_0\mathbf{v}_0) + \left(\mathbf{v}_0\mathbf{v}_0\cdot\frac{\partial}{\partial\mathbf{r}}\right)\left(\rho v_0^2\right)$$
$$= \rho v_0^2\left(\mathbf{v}_0\cdot\frac{\partial}{\partial\mathbf{r}}\right)\mathbf{v}_0 + \rho v_0^2\mathbf{v}_0\left(\frac{\partial}{\partial\mathbf{r}}\cdot\mathbf{v}_0\right) + \left(\mathbf{v}_0\mathbf{v}_0\cdot\frac{\partial}{\partial\mathbf{r}}\right)\left(\rho v_0^2\right)$$
$$= \rho v_0^2\left(\mathbf{v}_0\cdot\frac{\partial}{\partial\mathbf{r}}\right)\mathbf{v}_0 + v_0^2\mathbf{v}_0\frac{\partial}{\partial\mathbf{r}}\cdot(\rho\mathbf{v}_0) + \left(\mathbf{v}_0\mathbf{v}_0\cdot\frac{\partial}{\partial\mathbf{r}}\right)\left(\rho v_0^2\right)$$
$$- v_0^2\mathbf{v}_0\left(\mathbf{v}_0\cdot\frac{\partial}{\partial\mathbf{r}}\right)\rho. \tag{3.2.31}$$

Let us consider the difference of two last terms in (3.2.31):

$$\left(\mathbf{v}_0\mathbf{v}_0\cdot\frac{\partial}{\partial\mathbf{r}}\right)\left(\rho v_0^2\right) - v_0^2\mathbf{v}_0\left(\mathbf{v}_0\cdot\frac{\partial}{\partial\mathbf{r}}\right)\rho$$
$$= \left(\mathbf{v}_0\mathbf{v}_0 v_0^2\cdot\frac{\partial}{\partial\mathbf{r}}\right)\rho + \left(\rho\mathbf{v}_0\mathbf{v}_0\cdot\frac{\partial}{\partial\mathbf{r}}\right)v_0^2 - v_0^2\mathbf{v}_0\left(\mathbf{v}_0\cdot\frac{\partial}{\partial\mathbf{r}}\right)\rho$$
$$= \left(\rho\mathbf{v}_0\mathbf{v}_0\cdot\frac{\partial}{\partial\mathbf{r}}\right)v_0^2 = 2\rho\mathbf{v}_0\mathbf{v}_0\cdot\left[\left(\mathbf{v}_0\cdot\frac{\partial}{\partial\mathbf{r}}\right)\mathbf{v}_0\right]. \tag{3.2.32}$$

Now let us find the discrepancy of values $M_{v^3} - M_{v^3}^{<}$, using (3.2.31), (3.2.32):

$$
\begin{aligned}
M_{v^3} &- M_{v^3}^{<} \\
&= \tau\left\{\frac{\partial}{\partial \mathbf{r}}\cdot\left[\overleftrightarrow{I}\,pv_0^2 + 7p\mathbf{v}_0\mathbf{v}_0 + 5\overleftrightarrow{I}\,\frac{p^2}{\rho}\right]\right. \\
&\quad - 5p\left[\left(\mathbf{v}_0\cdot\frac{\partial}{\partial \mathbf{r}}\right)\mathbf{v}_0 + \frac{1}{\rho}\frac{\partial p}{\partial \mathbf{r}}\right] - v_0^2\frac{\partial p}{\partial \mathbf{r}} \\
&\quad \left. - 5\mathbf{v}_0\left[\frac{\partial}{\partial \mathbf{r}}\cdot(p\mathbf{v}_0) + \frac{2}{3}p\frac{\partial}{\partial \mathbf{r}}\cdot\mathbf{v}_0\right] - 2\mathbf{v}_0\mathbf{v}_0\cdot\frac{\partial p}{\partial \mathbf{r}}\right\} \\
&= \tau\left\{\frac{\partial}{\partial \mathbf{r}}\cdot\left[\overleftrightarrow{I}\,pv_0^2 + 7p\mathbf{v}_0\mathbf{v}_0 + 5\overleftrightarrow{I}\,\frac{p^2}{\rho}\right] - 5p\left(\mathbf{v}_0\cdot\frac{\partial}{\partial \mathbf{r}}\right)\mathbf{v}_0\right. \\
&\quad \left. - 5\frac{p}{\rho}\frac{\partial p}{\partial \mathbf{r}} - v_0^2\frac{\partial p}{\partial \mathbf{r}} - 5\mathbf{v}_0\frac{\partial}{\partial \mathbf{r}}\cdot(p\mathbf{v}_0) - \frac{10}{3}\mathbf{v}_0 p\frac{\partial}{\partial \mathbf{r}}\cdot\mathbf{v}_0 - 2\mathbf{v}_0\mathbf{v}_0\cdot\frac{\partial p}{\partial \mathbf{r}}\right\} \\
&= \tau\left\{p\frac{\partial}{\partial \mathbf{r}}v_0^2 + 7p\frac{\partial}{\partial \mathbf{r}}\cdot(\mathbf{v}_0\mathbf{v}_0) + 7\left(\mathbf{v}_0\mathbf{v}_0\cdot\frac{\partial}{\partial \mathbf{r}}\right)p + 5p\frac{\partial}{\partial \mathbf{r}}\left(\frac{p}{\rho}\right)\right. \\
&\quad \left. - 5p\left(\mathbf{v}_0\cdot\frac{\partial}{\partial \mathbf{r}}\right)\mathbf{v}_0 - 5\mathbf{v}_0\frac{\partial}{\partial \mathbf{r}}\cdot(p\mathbf{v}_0) - \frac{10}{3}p\mathbf{v}_0\frac{\partial}{\partial \mathbf{r}}\cdot\mathbf{v}_0 - 2\mathbf{v}_0\mathbf{v}_0\cdot\frac{\partial p}{\partial \mathbf{r}}\right\} \\
&= \tau\left\{p\frac{\partial}{\partial \mathbf{r}}v_0^2 + 2p\left(\mathbf{v}_0\cdot\frac{\partial}{\partial \mathbf{r}}\right)\mathbf{v}_0 + 2p\mathbf{v}_0\left(\frac{\partial}{\partial \mathbf{r}}\cdot\mathbf{v}_0\right)\right. \\
&\quad \left. + 5p\frac{\partial}{\partial \mathbf{r}}\left(\frac{p}{\rho}\right) - \frac{10}{3}p\mathbf{v}_0\frac{\partial}{\partial \mathbf{r}}\cdot\mathbf{v}_0\right\} \\
&= \tau p\left\{\frac{\partial}{\partial \mathbf{r}}\left[5\frac{p}{\rho} + v_0^2\right] + 2\left(\mathbf{v}_0\cdot\frac{\partial}{\partial \mathbf{r}}\right)\mathbf{v}_0 - \frac{4}{3}\mathbf{v}_0\left(\frac{\partial}{\partial \mathbf{r}}\cdot\mathbf{v}_0\right)\right\}.
\end{aligned}
\tag{3.2.33}
$$

Then the chain of equalities (3.2.33) leads to result

$$
\begin{aligned}
M_{v^3} &- M_{v^3}^{<} \\
&= \tau p\left\{\frac{\partial}{\partial \mathbf{r}}\left[5\frac{p}{\rho} + v_0^2\right] + 2\left(\mathbf{v}_0\cdot\frac{\partial}{\partial \mathbf{r}}\right)\mathbf{v}_0 - \frac{4}{3}\mathbf{v}_0\left(\frac{\partial}{\partial \mathbf{r}}\cdot\mathbf{v}_0\right)\right\}.
\end{aligned}
\tag{3.2.34}
$$

Discrepancy (3.2.34) can be written in another form having transparent physical sense.

Introduce velocity distortion tensor $\overleftrightarrow{S}$ with components

$$
S_{ij} = \frac{1}{2}\left(\frac{\partial v_{0i}}{\partial r_j} + \frac{\partial v_{0j}}{\partial r_i}\right) - \frac{1}{3}\delta_{ij}\frac{\partial}{\partial \mathbf{r}}\cdot\mathbf{v}_0 \quad (i, j = 1, 2, 3).
\tag{3.2.35}
$$

Then j-component of vector equal to scalar product of vector $\mathbf{v}_0$ by tensor $\overleftrightarrow{S}$, has the form

$$\left(\mathbf{v}_0 \cdot \overleftrightarrow{S}\right)_j = \frac{1}{4}\sum_i \frac{\partial}{\partial r_j} v_{0i}^2 + \frac{1}{2}\sum_i \left(v_{0i}\frac{\partial}{\partial r_i}\right) v_{0j} - \frac{1}{3} v_{0j} \frac{\partial}{\partial \mathbf{r}} \cdot \mathbf{v}_0.$$

Then,

$$\mathbf{v}_0 \cdot \overleftrightarrow{S} = \frac{1}{4}\frac{\partial}{\partial \mathbf{r}} v_0^2 + \frac{1}{2}\left(\mathbf{v}_0 \frac{\partial}{\partial \mathbf{r}}\right)\mathbf{v}_0 - \frac{1}{3}\mathbf{v}_0 \frac{\partial}{\partial \mathbf{r}} \cdot \mathbf{v}_0. \tag{3.2.36}$$

Using (3.2.36), one obtains the form for discrepancy

$$M_{v^3} - M_{v^3}^{<} = 2\tau p\left[\frac{5}{2}\frac{\partial}{\partial \mathbf{r}}\left(\frac{p}{\rho}\right) + 2\mathbf{v}_0 \cdot \overleftrightarrow{S}\right]. \tag{3.2.37}$$

Now the possibility appears to write down the generalized Euler equations (GEuE) in the terminology of averaged values.

$$\frac{\partial \rho^a}{\partial t} + \frac{\partial}{\partial \mathbf{r}} \cdot (\rho \mathbf{v}_0)^a = 0, \tag{3.2.38}$$

$$\frac{\partial}{\partial t}(\rho \mathbf{v}_0)^a + \frac{\partial}{\partial \mathbf{r}} \cdot \left[p^a \overleftrightarrow{I} + \left(\rho \mathbf{v}_0 \mathbf{v}_0\right)^a\right] = \rho^a \mathbf{g}, \tag{3.2.39}$$

$$\frac{\partial}{\partial t}\left(3p + \rho v_0^2\right)^a + \frac{\partial}{\partial \mathbf{r}} \cdot \left[\mathbf{v}_0\left(\rho v_0^2 + 5p\right)\right] - 2\mathbf{g} \cdot (\rho \mathbf{v}_0)^a$$
$$= 2\frac{\partial}{\partial \mathbf{r}} \cdot \left\{\tau p\left[\frac{5}{2}\frac{\partial}{\partial \mathbf{r}}\left(\frac{p}{\rho}\right) + 2\mathbf{v}_0 \cdot \overleftrightarrow{S}\right]\right\}. \tag{3.2.40}$$

From the system of generalized Euler equations (3.2.38)–(3.2.40) we conclude the following:

(1) The formulation of the hydrodynamic equations in terms of average quantities is the objective of "classical" theory of turbulence. However, a rigorous approach based on the generalized Euler equations leads to a residual (with respect to true quantities) on the right of the equation of energy (3.2.40).
(2) The residual

$$\Pi^e = 2\frac{\partial}{\partial \mathbf{r}} \cdot \left\{\tau p\left[\frac{5}{2}\frac{\partial}{\partial \mathbf{r}}\frac{p}{\rho} + 2\mathbf{v}_0 \cdot \overleftrightarrow{S}\right]\right\} \tag{3.2.41}$$

in Eq. (3.2.40) turns out to be proportional to τp and hence to the viscosity (cf. Eq. (2.7.12)). If one puts

$$\Pi^e = 0, \tag{3.2.42}$$

then the set of Eqs. (3.2.38)–(3.2.40) reduces formally to the Euler equations for averaged quantities. It follows that the residual Π^e, which reflects the variation in space of the thermal-energy and shear-energy dissipation, stimulates the development of turbulence in the physical system under study.

(3) The so-called ‘soft’ boundary conditions commonly imposed at the output of the computational flow region follow from condition (3.2.42):

$$\frac{\partial}{\partial \mathbf{r}} \cdot \left\{ \mu \left[\frac{5}{2} \frac{\partial}{\partial \mathbf{r}} \frac{p}{\rho} + 2\mathbf{v}_0 \cdot \overset{\leftrightarrow}{S} \right] \right\} = 0. \tag{3.2.43}$$

(4) From position of kinetic theory the appearance of mentioned discrepancy is connected with approximation of distribution function (DF) with help of local Maxwellian function. This problem does not exist for generalized Enskog equations written for genuine DF. But situation is more complicated. Even if relation (3.2.42) is fulfilled, Eqs. (3.2.38)–(3.2.40) do not reduce exactly to the classical Euler equations even under condition (3.2.42) because the average of the product of hydrodynamic quantities is not equal to the product of their averages. Consequently, this system of equations contains more unknowns (namely, ρ^a, $(\rho\mathbf{v}_0)^a$, p^a, $(\rho\mathbf{v}_0\mathbf{v}_0)^a$, and $[\mathbf{v}_0(\rho v_0^2 + 5p)]^a$) than equations, thus presenting the typical problem of the classical theory of turbulence, which consists in closing the moment equations.

The theory presented here overcomes this problem by simply reverting to the formulation of the hydrodynamic equations in terms of the true quantities. And it is only in the case when turbulent fluctuations are completely absent or, equivalently, when the average product of hydrodynamic quantities is equal to the product of their averages, that we arrive at the classical form of the Euler and, of course, Navier–Stokes equations. Thus, the classical Euler and Navier–Stokes equations are not written for true quantities, and it is physically meaningless to employ the formal Reynolds procedure to try to “extract” small-scale fluctuations from these equations.

Let

$$\frac{\partial^{n_i} u_i}{\partial t^{n_i}} = F_i\left(t, x_1, \ldots, x_n, u_1, \ldots, u_N, \ldots, \frac{\partial^k u_j}{\partial t^{k_0} \partial x_1^{k_1} \ldots \partial x_n^{k_n}}, \ldots \right) \tag{3.2.44}$$

be a system of differential equations in N unknown functions $u_1, u_2, \ldots, u_N$ with partial derivatives with respect to independent variables $t, x_1, x_2, \ldots, x_n$, for which the following conditions fulfilled

$$k_0 + k_1 + \cdots + k_n = k \leqslant n_i, \quad k_0 < n_i, \quad i, j = 1, 2, \ldots, N. \tag{3.2.45}$$

For $t = t_0$ the “initial conditions” are formulated for all unknown functions and their derivatives

$$\frac{\partial^k u_i}{\partial t^k} = \varphi_i^{(k)}(x_1, x_2, \ldots, x_n) \quad (k = 0, 1, 2, \ldots, n_i - 1). \tag{3.2.46}$$

All functions $\varphi_i^{(k)}(x_1, x_2, \ldots, x_n)$ are given in the same domain G_0 of the space $(x_1, x_2, \ldots, x_n)$.

In this case the following Kovalevskaya theorem can be formulated for Cauchy problem:

If all functions F_i are analytical in a vicinity of point $(t_0, x_1^0, \ldots, x_n^0, \ldots, \varphi^0_{j,k_0,k_1,\ldots,k_n}, \ldots)$ and all functions $\varphi_i^{(k)}(x_1, x_2, \ldots, x_n)$ are analytical in vicinity of point $(x_1, x_2, \ldots, x_n)$, then Cauchy problem has analytical unique solution in class of analytical functions in vicinity of mentioned point $(x_1, x_2, \ldots, x_n)$.

The system of generalized hydrodynamic equations (cf. the generalized Euler equations) can be written in the form (3.2.44), satisfy the conditions (3.2.45) and – in reasonable assumptions concerning functions F_i – satisfy the conditions of Kovalevskaya theorem.

System of fluctuations of hydrodynamic values can be used for investigation of the flow stability. Basic guidelines of stability can be obtained on the basement of theory of differential equations with small parameter in front of derivative (Elsholz, 1955).

Let us consider the differential equation of the form

$$\tau \frac{\mathrm{d}A}{\mathrm{d}t} = f(A, t), \tag{3.2.47}$$

where τ is small parameter. If $\tau = 0$, from equation

$$f(A, t) = 0 \tag{3.2.48}$$

follows algebraic solution

$$A_i = \varphi_i(t), \tag{3.2.49}$$

where i is number of corresponding solution.

Construct the curve $\varphi_i(t)$ in coordinate system A, t and suppose that trajectory of $A(t)$ begins at a point placed under the curve $\varphi_i(t)$ where (for the sake of definiteness) $f > 0$. Then, because τ is a small parameter, the derivative $\mathrm{d}A/\mathrm{d}t$ is very big and integral curve going out from mentioned point practically parallel to A-axis, directs to algebraic solution $\varphi_i(t)$. After that this curve turns along t-axis because for $f = 0$ the derivative $\mathrm{d}A/\mathrm{d}t$ is equal to zero.

If outgoing point of trajectory is placed above the curve $A_i = \varphi_i(t)$ in domain where $f < 0$, the integral curve will come down to the curve $A_i = \varphi_i(t)$.

Then, in the considered scenario of behavior of integral curves and function $f(A, t) = 0$, there obviously exists the stability of solution of differential equation (3.2.47). In this case a criterion of stability of the form

$$\frac{\partial f}{\partial A} < 0, \quad A_i = \varphi_i(t), \tag{3.2.50}$$

can be introduced.

Table 3.1
Fluctuations of hydrodynamic quantities on the Kolmogorov scale in the framework of the generalized Euler equations

No.	Hydrodynamic value A	Fluctuation A^f
1	ρ	$\tau\left[\frac{\partial\rho}{\partial t}+\frac{\partial}{\partial\mathbf{r}}\cdot(\rho\mathbf{v}_0)\right]$
2	$\rho\mathbf{v}_0$	$\tau\left[\frac{\partial}{\partial t}(\rho\mathbf{v}_0)+\frac{\partial}{\partial\vec{r}}\cdot(\rho\mathbf{v}_0\mathbf{v}_0)+\frac{\partial p}{\partial\mathbf{r}}-\rho\mathbf{g}\right]$
3	$\mathbf{v}_0$	$\tau\left[\frac{\partial\mathbf{v}_0}{\partial t}+\left(\mathbf{v}_0\cdot\frac{\partial}{\partial\mathbf{r}}\right)\mathbf{v}_0+\frac{1}{\rho}\frac{\partial p}{\partial\mathbf{r}}-\mathbf{g}\right]$
4	$p\delta_{\alpha\beta}+\rho v_{0\alpha}v_{0\beta}$	$\tau\left[\frac{\partial}{\partial t}(p\delta_{\alpha\beta}+\rho v_{0\alpha}v_{0\beta})+\frac{\partial}{\partial r_\gamma}(pv_{0\alpha}\delta_{\beta\gamma}+pv_{0\beta}\delta_{\alpha\gamma}+pv_{0\gamma}\delta_{\alpha\beta}\right.$ $\left.+\rho v_{0\alpha}v_{0\beta}v_{0\gamma})-g_\beta\rho v_{0\alpha}-g_\alpha\rho v_{0\beta}\right]$
5	$3p+\rho v_0^2$	$\tau\left[\frac{\partial}{\partial t}(3p+\rho v_0^2)+\frac{\partial}{\partial\mathbf{r}}\cdot(\mathbf{v}_0(\rho v_0^2+5p))-2\mathbf{g}\cdot\rho\mathbf{v}_0\right]$
6	p	$\tau\left[\frac{\partial p}{\partial t}+\frac{\partial}{\partial\mathbf{r}}\cdot(p\mathbf{v}_0)+\frac{2}{3}p\frac{\partial}{\partial\mathbf{r}}\cdot\mathbf{v}_0\right]$
7	$\mathbf{v}_0(\rho v_0^2+5p)$	$\tau\left\{\frac{\partial}{\partial t}[\mathbf{v}_0(\rho v_0^2+5p)]+\frac{\partial}{\partial\mathbf{r}}\cdot\left[\rho v_0^2\mathbf{v}_0\mathbf{v}_0+\overleftrightarrow{I}pv_0^2+7p\mathbf{v}_0\mathbf{v}_0+5\overleftrightarrow{I}\frac{p^2}{\rho}\right]\right.$ $-2\rho\mathbf{v}_0\mathbf{v}_0\cdot\mathbf{g}-5p\overleftrightarrow{I}\cdot\mathbf{g}-\rho v_0^2\overleftrightarrow{I}\cdot\mathbf{g}$ $\left.-p\left[\frac{\partial}{\partial\mathbf{r}}\left(5\frac{p}{\rho}+v_0^2\right)+2\left(\mathbf{v}_0\cdot\frac{\partial}{\partial\mathbf{r}}\right)\mathbf{v}_0-\frac{4}{3}\mathbf{v}_0\left(\frac{\partial}{\partial\mathbf{r}}\cdot\mathbf{v}_0\right)\right]\right\}$

And vice versa, if function f is negative under the curve $A_i=\varphi_i(t)$ and positive above this curve, the solution (3.2.49) is not stable.

The formal possibility is existing of applications of these qualitative considerations for investigation of the problems of stability in the frame of GHE. Notice really that in one-dimensional non-stationary case the velocity fluctuation has the form (see Table 3.1):

$$\tau\frac{\mathrm{d}u}{\mathrm{d}t}=u-u^a-\tau\left(\frac{1}{\rho}\frac{\partial p}{\partial x}+u\frac{\partial u}{\partial x}\right). \tag{3.2.51}$$

The corresponding function f is written as

$$f=u-u^a-\tau\left(\frac{1}{\rho^a}\frac{\partial p^a}{\partial x}+u^a\frac{\partial u^a}{\partial x}\right), \tag{3.2.52}$$

taking into account that the genuine and averaged values differ about $\mathrm{O}(\tau)$. Obviously the sign of function f in vicinity of solution

$$u=u^a+\tau\left(\frac{1}{\rho^a}\frac{\partial p^a}{\partial x}+u^a\frac{\partial u^a}{\partial x}\right) \tag{3.2.53}$$

– and then stability or instability of solution of (3.2.47) – depends on behavior of averaged hydrodynamic values and therefore on solutions of concrete hydrodynamic problems.

But in all cases we can confirm the Heiseberg affirmation – stated in 1924 on the basement of the stability investigation of the Navier–Stokes equations (Heisenberg, 1924) – that small but finite liquid viscosity leads in the definite sense to the destabilizing influence on the flow in comparison with the case of ideal liquids.

3.3. Theory of turbulence and the generalized Enskog equations

Generalized hydrodynamic Enskog equations (GHEnE) (2.7.1)–(2.7.3) lead to the most general formulations of the micro scale (sub-grid or Kolmogorov) fluctuations. Let us consider the procedure of calculations of dependent fluctuations with help of independent fluctuations coming from the minor tensor velocity moments (containing in continuity equation (2.7.1)) to the senior moments contained in the energy equations. The general character of procedure corresponds to obtaining of related values for generalized Euler equations and, as in previous case, all found fluctuations are summarized in Table 3.2 where all independent fluctuations are underlined.

From continuity equation follows

$$\rho_\alpha^f = \tau_\alpha \left\{ \frac{\partial \rho_\alpha}{\partial t} + \frac{\partial}{\partial \mathbf{r}} \cdot (\rho_\alpha \bar{\mathbf{v}}_\alpha) \right\} \quad (\alpha = 1, \ldots, \eta), \tag{3.3.1}$$

and fluctuation of density of mixture

$$\rho^f = \sum_\alpha \tau_\alpha \left\{ \frac{\partial \rho_\alpha}{\partial t} + \frac{\partial}{\partial \mathbf{r}} \cdot (\rho_\alpha \bar{\mathbf{v}}_\alpha) \right\}. \tag{3.3.2}$$

If mean times of free path for species are not too different (see (1.3.59′)), $\tau_\alpha \cong \tau$, we have

$$\rho^f = \tau \left\{ \frac{\partial \rho}{\partial t} + \frac{\partial}{\partial \mathbf{r}} \cdot \rho \mathbf{v}_0 \right\}. \tag{3.3.3}$$

The first-order velocity tensor also can be found from continuity equation

$$(\rho_\alpha \bar{\mathbf{v}}_\alpha)^f = \tau_\alpha \left\{ \frac{\partial}{\partial t} (\rho_\alpha \bar{\mathbf{v}}_\alpha) + \frac{\partial}{\partial \vec{r}} \cdot (\rho_\alpha \overline{\mathbf{v}_\alpha \mathbf{v}_\alpha}) - \rho_\alpha \overline{\mathbf{F}}_\alpha \right\}, \tag{3.3.4}$$

and

$$(\rho \mathbf{v}_0)^f = \sum_\alpha \tau_\alpha \left\{ \frac{\partial}{\partial t} (\rho_\alpha \bar{\mathbf{v}}_\alpha) + \frac{\partial}{\partial \mathbf{r}} \cdot (\rho_\alpha \overline{\mathbf{v}_\alpha \mathbf{v}_\alpha}) - \rho_\alpha \overline{\mathbf{F}}_\alpha \right\}. \tag{3.3.5}$$

Table 3.2
Fluctuations of hydrodynamic quantities on the Kolmogorov scale in the framework of the generalized Enskog equations

No.	Hydrodynamic value A	Fluctuation A^f
1	ρ_α	$\tau_\alpha\left\{\frac{\partial\rho_\alpha}{\partial t}+\frac{\partial}{\partial\mathbf{r}}\cdot(\rho_\alpha\bar{\mathbf{v}}_\alpha)\right\}$
2	ρ	$\sum_\alpha\tau_\alpha\left\{\frac{\partial\rho_\alpha}{\partial t}+\frac{\partial}{\partial\mathbf{r}}\cdot(\rho_\alpha\bar{\mathbf{v}}_\alpha)\right\}$
3	$\rho_\alpha\bar{\mathbf{v}}_\alpha$	$\tau_\alpha\left\{\frac{\partial}{\partial t}(\rho_\alpha\bar{\mathbf{v}}_\alpha)+\frac{\partial}{\partial\mathbf{r}}\cdot(\rho_\alpha\overline{\mathbf{v}_\alpha\mathbf{v}_\alpha})-\rho_\alpha\overline{\mathbf{F}}_\alpha\right\}$
4	$\rho\mathbf{v}_0$	$\sum_\alpha\tau_\alpha\left\{\frac{\partial}{\partial t}(\rho_\alpha\bar{\mathbf{v}}_\alpha)+\frac{\partial}{\partial\mathbf{r}}\cdot(\rho_\alpha\overline{\mathbf{v}_\alpha\mathbf{v}_\alpha})-\rho_\alpha\overline{\mathbf{F}}_\alpha\right\}$
5	$\mathbf{v}_0$	$\frac{1}{\rho}\sum_\alpha\tau_\alpha\left\{\frac{\partial}{\partial t}(\rho_\alpha\bar{\mathbf{v}}_\alpha)+\frac{\partial}{\partial\mathbf{r}}\cdot(\rho_\alpha\overline{\mathbf{v}_\alpha\mathbf{v}_\alpha})-\rho_\alpha\overline{\mathbf{F}}_\alpha-\mathbf{v}_0\left[\frac{\partial\rho_\alpha}{\partial t}+\frac{\partial}{\partial\mathbf{r}}\cdot(\rho_\alpha\bar{\mathbf{v}}_\alpha)\right]\right\}$
6	$\bar{\mathbf{v}}_\alpha$	$\frac{1}{\rho_\alpha}\tau_\alpha\left\{\frac{\partial}{\partial t}(\rho_\alpha\bar{\mathbf{v}}_\alpha)+\frac{\partial}{\partial\mathbf{r}}\cdot(\rho_\alpha\overline{\mathbf{v}_\alpha\mathbf{v}_\alpha})-\rho_\alpha\overline{\mathbf{F}}_\alpha-\bar{\mathbf{v}}_\alpha\left[\frac{\partial\rho_\alpha}{\partial t}+\frac{\partial}{\partial\mathbf{r}}\cdot(\rho_\alpha\bar{\mathbf{v}}_\alpha)\right]\right\}$
7	$\rho_\alpha\overline{\mathbf{v}_\alpha\mathbf{v}_\alpha}$	$\tau_\alpha\left\{\frac{\partial}{\partial t}(\rho_\alpha\overline{\mathbf{v}_\alpha\mathbf{v}_\alpha})+\frac{\partial}{\partial\mathbf{r}}\cdot\rho_\alpha\overline{(\mathbf{v}_\alpha\mathbf{v}_\alpha)\mathbf{v}_\alpha}-\mathbf{F}_\alpha^{(1)}\rho_\alpha\bar{\mathbf{v}}_\alpha-\rho_\alpha\bar{\mathbf{v}}_\alpha\mathbf{F}_\alpha^{(1)}-\frac{q_\alpha}{m_\alpha}\rho_\alpha\overline{[\mathbf{v}_\alpha\times\mathbf{B}]\mathbf{v}_\alpha}-\frac{q_\alpha}{m_\alpha}\rho_\alpha\overline{\mathbf{v}_\alpha[\mathbf{v}_\alpha\times\mathbf{B}]}\right\}$
8	$\rho_\alpha\overline{v_\alpha^2}$	$\tau_\alpha\left\{\frac{\partial}{\partial t}(\rho_\alpha\overline{\mathbf{v}_\alpha^2})+\frac{\partial}{\partial\mathbf{r}}\cdot\rho_\alpha\overline{v_\alpha^2\mathbf{v}_\alpha}-2\rho_\alpha\overline{\mathbf{F}_\alpha\cdot\mathbf{v}_\alpha}\right\}$
9	$\varepsilon_\alpha n_\alpha$	$\tau_\alpha\left\{\frac{\partial}{\partial t}(\varepsilon_\alpha n_\alpha)+\frac{\partial}{\partial\mathbf{r}}\cdot(\varepsilon_\alpha n_\alpha\bar{\mathbf{v}}_\alpha)\right\}$
10	$\varepsilon_\alpha n_\alpha\bar{\mathbf{v}}_\alpha$	$\tau_\alpha\left\{\frac{\partial}{\partial t}(\varepsilon_\alpha n_\alpha\bar{\mathbf{v}}_\alpha)+\frac{\partial}{\partial\mathbf{r}}\cdot(\varepsilon_\alpha n_\alpha\overline{\mathbf{v}_\alpha\mathbf{v}_\alpha})-\varepsilon_\alpha n_\alpha\overline{\mathbf{F}}_\alpha\right\}$
11	$\rho_\alpha\mathbf{F}_\alpha^{(1)}\cdot\bar{\mathbf{v}}_\alpha$	$\tau_\alpha\mathbf{F}_\alpha^{(1)}\cdot\left\{\frac{\partial}{\partial t}(\rho_\alpha\bar{\mathbf{v}}_\alpha)+\frac{\partial}{\partial\mathbf{r}}\cdot\rho_\alpha\overline{\mathbf{v}_\alpha\mathbf{v}_\alpha}-\rho_\alpha\overline{\mathbf{F}}_\alpha\right\}$
12	$\rho_\alpha\overline{v_\alpha^2\mathbf{v}_\alpha}$	$\tau_\alpha\left\{\frac{\partial}{\partial t}(\rho_\alpha\overline{v_\alpha^2\mathbf{v}_\alpha})+\frac{\partial}{\partial\mathbf{r}}\cdot\rho_\alpha\overline{v_\alpha^2\mathbf{v}_\alpha\mathbf{v}_\alpha}-2\rho_\alpha\mathbf{F}_\alpha^{(1)}\cdot\overline{\mathbf{v}_\alpha\mathbf{v}_\alpha}-\rho_\alpha\overline{v_\alpha^2\mathbf{F}_\alpha}\right\}$
13	H_α	$\tau_\alpha\frac{\partial H_\alpha}{\partial t}$
14	H	$\sum_\alpha\tau_\alpha\frac{\partial H_\alpha}{\partial t}$

From Kolmogorov fluctuations Nos. 1 and 3 (see Table 3.2) find the fluctuations of hydrodynamic velocity and averaged velocity of species.

Because

$$\mathbf{v}_0^f=\frac{1}{\rho}\left[(\rho\mathbf{v}_0)^f-\rho^f\mathbf{v}_0\right], \tag{3.3.6}$$

one obtains

$$\mathbf{v}_0^f = \frac{1}{\rho}\sum_\alpha \tau_\alpha \left\{ \frac{\partial}{\partial t}(\rho_\alpha \bar{\mathbf{v}}_\alpha) + \frac{\partial}{\partial \vec{r}} \cdot (\rho_\alpha \overline{\mathbf{v}_\alpha \mathbf{v}_\alpha}) - \rho_\alpha \overline{\mathbf{F}}_\alpha - \mathbf{v}_0 \left[\frac{\partial \rho_\alpha}{\partial t} + \frac{\partial}{\partial \mathbf{r}} \cdot (\rho_\alpha \bar{\mathbf{v}}_\alpha) \right] \right\}. \tag{3.3.7}$$

Analogously

$$\bar{\mathbf{v}}_\alpha^f = \frac{1}{\rho_\alpha}\left[(\rho_\alpha \bar{\mathbf{v}}_\alpha)^f - \rho_\alpha^f \bar{\mathbf{v}}_\alpha \right], \tag{3.3.6$'$}$$

then from (3.3.1), (3.3.4) follows

$$\bar{\mathbf{v}}_\alpha^f = \frac{1}{\rho_\alpha}\tau_\alpha \left\{ \frac{\partial}{\partial t}(\rho_\alpha \bar{\mathbf{v}}_\alpha) + \frac{\partial}{\partial \mathbf{r}} \cdot (\rho_\alpha \overline{\mathbf{v}_\alpha \mathbf{v}_\alpha}) - \rho_\alpha \overline{\mathbf{F}}_\alpha - \bar{\mathbf{v}}_\alpha \left[\frac{\partial \rho_\alpha}{\partial t} + \frac{\partial}{\partial \mathbf{r}} \cdot (\rho_\alpha \bar{\mathbf{v}}_\alpha) \right] \right\}, \tag{3.3.8}$$

or

$$\bar{\mathbf{v}}_\alpha^f = \tau_\alpha \left\{ \frac{\partial \bar{\mathbf{v}}_\alpha}{\partial t} + \frac{1}{\rho_\alpha}\frac{\partial}{\partial \mathbf{r}} \cdot (\rho_\alpha \overline{\mathbf{v}_\alpha \mathbf{v}_\alpha}) - \overline{\mathbf{F}}_\alpha - \frac{\bar{\mathbf{v}}_\alpha}{\rho_\alpha}\frac{\partial}{\partial \mathbf{r}} \cdot (\rho_\alpha \mathbf{v}_\alpha) \right\}. \tag{3.3.9}$$

Fluctuations of the second-order tensor velocity moments $\rho_\alpha \overline{\mathbf{v}_\alpha \mathbf{v}_\alpha}$ are contained in momentum equation. As this takes place, the fluctuations of tensor moments of the zero-order and the first-order appearing in momentum equation coincide (in other words do not contradict) with corresponding fluctuations found from continuity equation.

Then the next independent fluctuation has the form

$$(\rho_\alpha \overline{\mathbf{v}_\alpha \mathbf{v}_\alpha})^f = \tau_\alpha \left\{ \frac{\partial}{\partial t}(\rho_\alpha \overline{\mathbf{v}_\alpha \mathbf{v}_\alpha}) + \frac{\partial}{\partial \mathbf{r}} \cdot \rho_\alpha \overline{(\mathbf{v}_\alpha \mathbf{v}_\alpha)\mathbf{v}_\alpha} - \mathbf{F}_\alpha^{(1)} \rho_\alpha \bar{\mathbf{v}}_\alpha - \rho_\alpha \bar{\mathbf{v}}_\alpha \mathbf{F}_\alpha^{(1)} - \frac{q_\alpha}{m_\alpha}\rho_\alpha \overline{[\mathbf{v}_\alpha \times \mathbf{B}]\mathbf{v}_\alpha} - \frac{q_\alpha}{m_\alpha}\rho_\alpha \overline{\mathbf{v}_\alpha [\mathbf{v}_\alpha \times \mathbf{B}]} \right\}. \tag{3.3.10}$$

The left side of energy equation contains fluctuation $(\rho_\alpha \overline{v_\alpha^2})^f$, which should be a consequence of (3.3.10). Such, indeed, is the case:

$$\overleftrightarrow{I} : (\rho_\alpha \overline{\mathbf{v}_\alpha \mathbf{v}_\alpha})^f = \left(\rho_\alpha \overline{v_\alpha^2}\right)^f = \tau_\alpha \left\{ \frac{\partial}{\partial t}\left(\rho_\alpha \overline{v_\alpha^2}\right) + \frac{\partial}{\partial \mathbf{r}} \cdot \rho_\alpha \overline{v_\alpha^2 \mathbf{v}_\alpha} - 2\mathbf{F}_\alpha^{(1)} \cdot \rho_\alpha \bar{\mathbf{v}}_\alpha \right\}. \tag{3.3.11}$$

The last terms of right side of (3.3.10) have no contribution in $(\rho_\alpha \overline{v_\alpha^2})^f$, because for example,

$$\overset{\leftrightarrow}{I}:\ \overline{\mathbf{v}_\alpha[\mathbf{v}_\alpha \times \mathbf{B}]} = \sum_{i=1}^{3} \overline{v_{\alpha i}[\mathbf{v}_\alpha \times \mathbf{B}]_i}$$
$$= v_{\alpha 1}(v_{\alpha 2}B_3 - v_{\alpha 3}B_2) - v_{\alpha 2}(v_{\alpha 1}B_3 - v_{\alpha 3}B_1)$$
$$+ v_{\alpha 3}(v_{\alpha 1}B_2 - v_{\alpha 2}B_1) = 0. \tag{3.3.12}$$

Then

$$\left(\rho_\alpha \overline{v_\alpha^2}\right)^f = \tau_\alpha \left\{ \frac{\partial}{\partial t}\left(\rho_\alpha \overline{v_\alpha^2}\right) + \frac{\partial}{\partial \mathbf{r}} \cdot \rho_\alpha \overline{v_\alpha^2 \mathbf{v}_\alpha} - 2\rho_\alpha \overline{\mathbf{F}_\alpha \mathbf{v}_\alpha} \right\}. \tag{3.3.13}$$

Fluctuation of internal energy (per unit of volume) of α-species has the form

$$(\varepsilon_\alpha n_\alpha)^f = \tau_\alpha \left\{ \frac{\partial}{\partial t}(\varepsilon_\alpha n_\alpha) + \frac{\partial}{\partial \mathbf{r}} \cdot (\varepsilon_\alpha n_\alpha \bar{\mathbf{v}}_\alpha) \right\}. \tag{3.3.14}$$

On the other hand,

$$(\varepsilon_\alpha n_\alpha)^f = \frac{(\varepsilon_\alpha n_\alpha \bar{\mathbf{v}}_\alpha)^f - \varepsilon_\alpha n_\alpha \bar{\mathbf{v}}_\alpha^f}{\bar{\mathbf{v}}_\alpha} \tag{3.3.15}$$

then

$$\bar{\mathbf{v}}_\alpha(\varepsilon_\alpha n_\alpha)^f = \tau_\alpha \left\{ \frac{\partial}{\partial t}(\varepsilon_\alpha n_\alpha \bar{\mathbf{v}}_\alpha) + \frac{\partial}{\partial \mathbf{r}} \cdot \varepsilon_\alpha n_\alpha \overline{\mathbf{v}_\alpha \mathbf{v}_\alpha} - \varepsilon_\alpha n_\alpha \mathbf{F}_\alpha \right.$$
$$- \frac{\varepsilon_\alpha n_\alpha}{\rho_\alpha}\left[\frac{\partial}{\partial t}(\rho_\alpha \bar{\mathbf{v}}_\alpha) + \frac{\partial}{\partial \mathbf{r}} \cdot (\rho_\alpha \overline{\mathbf{v}_\alpha \mathbf{v}_\alpha}) \right.$$
$$\left.\left. - \rho_\alpha \overline{\mathbf{F}}_\alpha - \bar{\mathbf{v}}_\alpha \frac{\partial \rho_\alpha}{\partial t} - \bar{\mathbf{v}}_\alpha \frac{\partial}{\partial \mathbf{r}} \cdot (\rho_\alpha \bar{\mathbf{v}}_\alpha) \right]\right\}$$
$$= \tau_\alpha \left\{ \bar{\mathbf{v}}_\alpha \frac{\partial}{\partial t}(\varepsilon_\alpha n_\alpha) + \frac{\varepsilon_\alpha n_\alpha}{\rho_\alpha} \bar{\mathbf{v}}_\alpha \frac{\partial}{\partial \mathbf{r}} \cdot (\rho_\alpha \bar{\mathbf{v}}_\alpha) \right\}$$
$$= \bar{\mathbf{v}}_\alpha \tau_\alpha \left\{ \frac{\partial}{\partial t}(\varepsilon_\alpha n_\alpha) + \varepsilon_\alpha \frac{\partial}{\partial \mathbf{r}} \cdot (n_\alpha \bar{\mathbf{v}}_\alpha) \right\}$$
$$= \bar{\mathbf{v}}_\alpha \tau_\alpha \left\{ \frac{\partial}{\partial t}(\varepsilon_\alpha n_\alpha) + \frac{\partial}{\partial \mathbf{r}} \cdot (\varepsilon_\alpha n_\alpha \bar{\mathbf{v}}_\alpha) \right\}. \tag{3.3.16}$$

This result agrees with relation (3.3.14). In (3.3.15), (3.3.16) the typical transformations directed on investigation of non-contradictions of found fluctuations of tensor moments are realized.

Finally, independent fluctuation of the third-order tensor moment contains the tensor moment of the fourth-order:

$$\left(\overline{\rho_\alpha \mathbf{v}_\alpha v_\alpha^2}\right)^f$$
$$= \tau_\alpha \left\{ \rho_\alpha \overline{\mathbf{v}_\alpha v_\alpha^2} + \frac{\partial}{\partial \vec{r}} \cdot \overline{v_\alpha^2 \mathbf{v}_\alpha \mathbf{v}_\alpha} - 2\rho_\alpha \mathbf{F}_\alpha^{(1)} \cdot \overline{\mathbf{v}_\alpha \mathbf{v}_\alpha} - \rho_\alpha \overline{v_\alpha^2 \mathbf{F}_\alpha} \right\}. \tag{3.3.17}$$

In Table 3.2 is introduced the fluctuation of the Boltzmann H-function.

Now we state that the most general boundary conditions for GHE should we written as follows (compare with (2.8.4)):

$$A_w^{f,\mathrm{ind}} = 0. \tag{3.3.18}$$

These conditions imply that all independent fluctuations $A^{f,\mathrm{ind}}$ indicated in Table 3.2, on the wall are equal to zero.

In the past many attempts were made at phenomenological refining of the Boltzmann equation (see, for example, Belotserkovski and Oparin, 2000, p. 69) for the purpose of introducing fluctuation terms into kinetic and hydrodynamic equations; however, these attempts were not rigorous to any extent.

The advent of generalized Boltzmann physical kinetics rendered invalid the statement of Belotserkovskii and Oparin (2000) to the effect that "the theory of turbulence remains the science of semi-empirical models on the kinetic level of description, as well".

In conclusion several remarks are necessary:

(1) By calculation of turbulent fluctuations contained in Table 3.2, the squared fluctuations were omitted as negligibly small, in other words were neglected by terms proportional to τ^2. It means that calculation of terms in curly brackets of Table 3.2 can be realized for averaged values.
(2) The application of the method of moments to the GBE leads to hydrodynamic equations, which include pulsation terms corresponding to the small-scale or sub-grid turbulence when, for small values of the Knudsen number, the mean time between collisions becomes proportional to viscosity. These fluctuations of hydrodynamic quantities are universal and tabulated. Every hydrodynamic value considered as a tensor velocity moment of the n-dimension, corresponds fluctuation containing senior tensor moment of the $(n+1)$-dimension.

In "classical" theory of turbulence this situation corresponds to the problem of closure of moment equations. Recall that this problem does not exist in the generalized Boltzmann physical kinetics; one needs only to return to equations written for genuine variables.

3.4. Generalized hydrodynamic equations and quantum mechanics

Fundamental equation of quantum mechanics – Schrödinger equation – can be written as a system of hydrodynamic equations without additional assumptions. It is interesting to discuss this fact from position of the GBKT.

Write down the Schrödinger equation

$$\mathrm{i}\hbar \frac{\partial \psi}{\partial t} = \widehat{H}\psi, \tag{3.4.1}$$

where non-relativistic Hamiltonian $\widehat{H}$ for a quantum mechanical system placed in electro-magnetic field is written as

$$\widehat{H} = \frac{(\hat{\vec{p}} - \mathrm{e}\hat{\vec{A}})^2}{2m} + \widehat{U}, \tag{3.4.2}$$

where **A** is vector potential, $\hat{\vec{A}}$ is corresponding operator, which reflects the magnet field influence on the quantum system. We use the standard notation: ψ – wave function, $\widehat{U}$ – operator of potential energy, $\hat{\vec{p}} = (\hbar/\mathrm{i})\nabla \equiv (\hbar/\mathrm{i})(\partial/\partial \mathbf{r}) \equiv (\hbar/\mathrm{i})\,\mathbf{grad}$ – momentum vector operator; m, e – mass and charge of particle, $h = 2\pi\hbar$ – Planck constant.

In these notations,

$$\widehat{H} = \frac{\hat{p}^2}{2m} + \widehat{U} - \frac{e}{m}(\hat{\vec{A}}, \hat{\vec{p}}) + \frac{e^2}{2m}\hat{A}^2. \tag{3.4.3}$$

Use complex form of wave function ψ

$$\psi(x, y, z, t) = \alpha(x, y, z, t)\, \mathrm{e}^{\mathrm{i}\beta(x,y,z,t)} \tag{3.4.4}$$

and introduce Laplacian Δ defined as usual

$$\Delta = \frac{\partial^2}{\partial x^2} + \frac{\partial^2}{\partial y^2} + \frac{\partial^2}{\partial z^2} = \frac{\partial}{\partial \mathbf{r}} \cdot \frac{\partial}{\partial \mathbf{r}}. \tag{3.4.5}$$

After differentiating (3.4.4) one obtains

$$\frac{1}{\psi}\frac{\partial \psi}{\partial t} = \frac{1}{\alpha}\frac{\partial \alpha}{\partial t} + \mathrm{i}\frac{\partial \beta}{\partial t}, \tag{3.4.6}$$

$$\frac{\mathbf{grad}\,\psi}{\psi} = \frac{1}{\alpha}\,\mathbf{grad}\,\alpha + \mathrm{i}\,\mathbf{grad}\,\beta, \tag{3.4.7}$$

$$\frac{\Delta\psi}{\psi} = \frac{\Delta\alpha}{\alpha} - (\mathbf{grad}\,\beta)^2 + \mathrm{i}\left(\Delta\beta + \frac{2}{\alpha}\,\mathbf{grad}\,\alpha \cdot \mathbf{grad}\,\beta\right). \tag{3.4.8}$$

Using relations (3.4.4), (3.4.6)–(3.4.8) and separating real and imaginary parts of complex numbers in Eq. (3.4.1), we find

$$\Delta\alpha - \alpha(\mathbf{grad}\,\beta)^2 - \frac{2m}{\hbar^2}\alpha\widetilde{U} - \frac{2m}{\hbar}\frac{\partial \beta}{\partial t}\alpha + \frac{2e}{\hbar}\alpha\mathbf{A} \cdot \mathbf{grad}\,\beta = 0, \tag{3.4.9}$$

$$\alpha\Delta\beta + 2\,\mathbf{grad}\,\alpha \cdot \mathbf{grad}\,\beta + \frac{2m}{\hbar}\frac{\partial \alpha}{\partial t} - \frac{2e}{\hbar}\mathbf{A} \cdot \mathbf{grad}\,\alpha = 0, \tag{3.4.10}$$

where

$$\widetilde{U} = U + \frac{e^2 A^2}{2m}. \tag{3.4.11}$$

By direct differentiating the following identity can be proved:

$$\frac{\partial}{\partial \mathbf{r}} \cdot \left(\alpha^2 \,\mathbf{grad}\, \beta\right) \equiv \mathrm{div}\left(\alpha^2 \,\mathbf{grad}\, \beta\right) = \alpha^2 \Delta\beta + 2\alpha \,\mathbf{grad}\, \alpha \cdot \mathbf{grad}\, \beta. \tag{3.4.12}$$

Then Eq. (3.4.10) is written as

$$\frac{\partial \alpha^2}{\partial t} + \mathrm{div}\left(\frac{\alpha^2 \hbar}{m} \,\mathbf{grad}\, \beta\right) = \frac{e}{m} \mathbf{A} \cdot \mathbf{grad}\, \alpha^2. \tag{3.4.13}$$

Introduce notations:

$$\rho = \alpha^2, \tag{3.4.14}$$

$$\mathbf{v} = \mathbf{grad}\left(\frac{\beta\hbar}{m}\right). \tag{3.4.15}$$

Following Madelung's idea we identify ρ, having the sense of probability density, with density of a hydrodynamic flow, and $\mathbf{v}$ with velocity of this flow.

Notice immediately, that existence of relation (3.4.15) means that the considered flow belongs to the class of potential flows with potential

$$\varphi = \frac{\beta\hbar}{m}. \tag{3.4.16}$$

Eq. (3.4.13) can be treated as hydrodynamic continuity equation with source term

$$R = \frac{e}{m} \mathbf{A} \cdot \mathbf{grad}\, \rho, \tag{3.4.17}$$

connected with fictitious rising and vanishing of hypothetical particles:

$$\frac{\partial \rho}{\partial t} + \mathrm{div}\, \rho \mathbf{v} = R. \tag{3.4.18}$$

Continuity Eq. (3.4.18) has a typical form for hydrodynamics of reacting gases (see, for example, Alekseev, 1982).

Transform now Eq. (3.4.9) to hydrodynamic form. With this aim divide left and right sides of (3.4.9) by $2\alpha m^2/\hbar^2$ and apply gradient operator to the terms of obtaining equation:

$$\frac{\hbar^2}{2m^2} \,\mathbf{grad}\, \frac{\Delta\alpha}{\alpha} - \frac{\hbar^2}{2m^2} \,\mathbf{grad}(\mathbf{grad}\, \beta)^2 - \frac{1}{m} \,\mathbf{grad}\, \widetilde{U} - \frac{\hbar}{m} \frac{\partial}{\partial t} \,\mathbf{grad}\, \beta + \frac{e\hbar}{m^2} \,\mathbf{grad}(\mathbf{A} \cdot \mathbf{grad}\, \beta) = 0. \tag{3.4.19}$$

Using definition of velocity (3.4.15), it is found

$$\frac{\partial \mathbf{v}}{\partial t} + \frac{1}{2}\,\mathbf{grad}\,v^2 = -\frac{1}{m}\,\mathbf{grad}\left(\widetilde{U} - e\mathbf{A}\cdot\mathbf{v} - \frac{\hbar^2}{2m}\frac{\Delta\alpha}{\alpha}\right). \tag{3.4.20}$$

Notice, that

$$\frac{\Delta\alpha}{\alpha} = \frac{\Delta\alpha^2}{2\alpha^2} - \frac{(\mathbf{grad}\,\alpha)^2}{\alpha^2}, \tag{3.4.21}$$

and – for potential flow –

$$\frac{1}{2}\,\mathbf{grad}\,v^2 = (\mathbf{v}\nabla)\mathbf{v}, \tag{3.4.22}$$

then Eq. (3.4.20) becomes

$$\frac{\partial \mathbf{v}}{\partial t} + (\mathbf{v}\nabla)\mathbf{v} = -\frac{1}{m}\nabla U^*, \tag{3.4.23}$$

where

$$U^* = \widetilde{U} - e\mathbf{A}\cdot\mathbf{v} - \frac{\hbar^2}{4m\rho}\left[\Delta\rho - \frac{(\mathbf{grad}\,\rho)^2}{2\rho}\right]. \tag{3.4.24}$$

The additive part of potential on the right side of Eq. (3.4.23) can be written in more compact form taking into account that

$$\frac{\hbar^2}{2m\sqrt{\rho}}\Delta\sqrt{\rho} = \frac{\hbar^2}{4m\rho}\left[\Delta\rho - \frac{(\mathbf{grad}\,\rho)^2}{2\rho}\right]. \tag{3.4.25}$$

Then effective potential energy U^* is the sum of potential energy $\widetilde{\widetilde{U}}$

$$\widetilde{\widetilde{U}} = \widetilde{U} - e\mathbf{A}\cdot\mathbf{v}, \tag{3.4.26}$$

which is not equal to zero in classical limit, and quantum part U_{qu},

$$U_{\text{qu}} = -\frac{\hbar^2}{2m\sqrt{\rho}}\Delta\sqrt{\rho} = -\frac{\hbar^2}{4m\rho}\left[\Delta\rho - \frac{(\mathbf{grad}\,\rho)^2}{2\rho}\right], \tag{3.4.27}$$

which contains squared Planck constant as coefficient. Usually transition to the classical description can be realized using $\hbar \to 0$. Obviously, in classical limit U_{qu} turns to zero.

Then

$$\begin{aligned} U^* &= \widetilde{\widetilde{U}} + U_{\text{qu}} = \widetilde{\widetilde{U}} - \frac{\hbar^2}{2m\sqrt{\rho}}\Delta\sqrt{\rho} \\ &= U - \frac{\hbar^2}{4m\rho}\left[\Delta\rho - \frac{(\mathbf{grad}\,\rho)^2}{2\rho}\right]. \end{aligned} \tag{3.4.28}$$

For this potential flow, hydrodynamic Cauchy integral exists:

$$\frac{\partial\varphi}{\partial t}+\frac{v^2}{2}+\frac{U^*}{m}=const. \tag{3.4.29}$$

Boundary and initial conditions for the hydrodynamic equations (3.4.18), (3.4.23) should be formulated separately taking into account the specific features of concrete quantum mechanics problem. One of these problems is considered as an example below.

Consider quantum mechanics analogue of stabilized flow. If potential U does not depend on time, wave function has the form

$$\psi=\alpha(x,y,z)\exp\left\{\mathrm{i}\left[f(x,y,z)-\frac{Et}{\hbar}\right]\right\}, \tag{3.4.30}$$

where E is total energy. Then from Eqs. (3.4.16), (3.4.29) follows Bernoulli equation

$$\frac{mv^2}{2}+U^*=E. \tag{3.4.31}$$

Incompressible flow corresponds – in quantum mechanics interpretation – to the motion of free particle.

Therefore, the Schrödinger equation can be treated in terms of potential flow of a compressible ideal liquid with rising and vanishing of fictitious particles, therewith the rate of "particles" formation is proportional to scalar product of vector potential by gradient of density. But vector potential $\mathbf{A}$ is proportional to electro-kinetic momentum $\mathbf{p}_{\mathrm{ek}}$ which is defined as

$$\mathbf{p}_{\mathrm{ek}}=\frac{1}{c}q\mathbf{A}, \tag{3.4.32}$$

where q is particle charge. Appearance of the additional electro-kinetic momentum leads to the change of probability density; quantum hydrodynamics reflects this fact by the rate of fictitious particles formation.

If magnetic field is absent, the usual continuity equation can be written

$$\frac{\partial\rho}{\partial t}+\operatorname{div}\rho\mathbf{v}=0, \tag{3.4.33}$$

and potential energy of the flow has the reduced form

$$U^*=U-\frac{\hbar^2}{4m\rho}\left[\Delta\rho-\frac{(\mathbf{grad}\,\rho)^2}{2\rho}\right]. \tag{3.4.34}$$

Cauchy and Bernoulli integrals should be used for a choice of energy levels.

The outlined theory can be applied for investigation of many quantum mechanics problems from position of hydrodynamics, and first of all in the context of numerical hydrodynamics. Difficulties, arising by numerical investigations of quantum mechanics problems, are well known. From this point of view it is interesting to apply well-developed methods of numerical hydrodynamics (see, for example, Richtmyer and Morton, 1967; Doering and Gibbon, 1995; Fletcher, 1987) for solution of quantum mechanics tasks.

Let us consider for example (Alekseev, 1982; Alexeev and Abakumov, 1982) the calculation of differential cross-sections by elastic scattering of electron bunch in spherically symmetric potential field of an atom ($\hbar = m = c = 1$).

$$U(r) = -\frac{2Z}{r}\sum_{i=1}^{n}\gamma_i\,\mathrm{e}^{-\lambda_i r}, \tag{3.4.35}$$

where n is a number of terms in potential function, γ_i, λ_i are constants calculated in Cox and Bonham (1967).

Numerical calculation of system of hydrodynamic equations (3.4.18), (3.4.23), (3.4.24) was realized in spherical coordinate system with taking into account the space symmetry of differential cross-section and initial and boundary conditions

$$\rho(r, \vartheta, 0) = 1, \tag{3.4.36}$$

$$v_r(r, \vartheta, 0) = \sqrt{2E_k}\sin\vartheta, \tag{3.4.37}$$

$$v_\vartheta(r, \vartheta, 0) = \sqrt{2E_k}\cos\vartheta, \tag{3.4.38}$$

$$U^*(r, \vartheta, 0) = 0, \tag{3.4.39}$$

$$\rho(0, \vartheta, t) = 0, \tag{3.4.40}$$

$$v_r(r_0, \vartheta, t) = \sqrt{2E_k}, \tag{3.4.41}$$

$$v_\vartheta(r_0, \vartheta, t) = 0, \tag{3.4.42}$$

$$U^*(r_0, \vartheta, t) = \frac{1}{4\rho r_0^2}\left[\frac{1}{2\rho}\left(\frac{\partial\rho}{\partial\vartheta}\right)^2 - \frac{1}{\sin\vartheta}\frac{\partial}{\partial\vartheta}\left(\sin\vartheta\frac{\partial\rho}{\partial\vartheta}\right)\right], \tag{3.4.43}$$

where E_k is kinetic energy of scattered electron bunch, r_0 is parameter of cutting of sphere of the atom action. Relation (3.4.43) is obtained with the help of asymptotic of scattered spherical wave.

Fig. 3.1 presents results of calculations of differential cross-section $\sigma(\vartheta)$ of electrons with energy 10 keV scattered by a krypton atom (solid curve). The results are in good coincidence with analogous results obtained by the method of partial waves (Abakumov and Vinogradov, 1981) (a dashed curve).

For comparison, the results the Spencer's theory with Moliere's screening are presented (dot-dashed curve in Figure 3.1) (Moliere, 1947; Spencer, 1955). This method of quantum hydrodynamics was applied later to calculations with taking into account external force fields and non-spherical atom potentials.

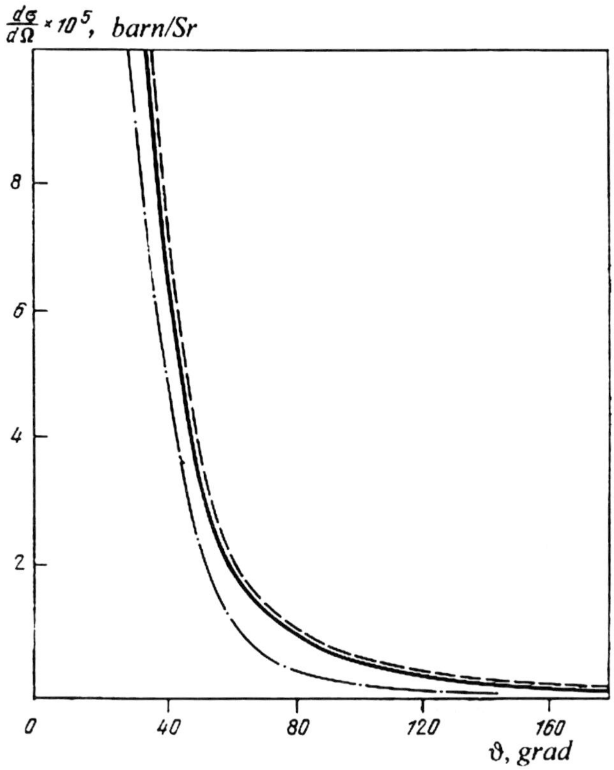

Fig. 3.1. Results of calculations of differential cross-section $\sigma(\vartheta)$ of electrons with energy 10 keV scattering by a krypton atom (a solid curve); The results obtained by the method of partial waves are shown by dashed curve, Molier's screening – by dot-dashed curve.

Flux of probability density

$$\mathbf{j} = \frac{i\hbar}{2m}\left(\psi\frac{\partial\psi^{\bullet}}{\partial\mathbf{r}} - \psi^{\bullet}\frac{\partial\psi}{\partial\mathbf{r}}\right) \tag{3.4.44}$$

also can be calculated in hydrodynamic terms using (3.4.4), (3.4.7), (3.4.14), (3.4.15):

$$\mathbf{j} = \rho\mathbf{v}. \tag{3.4.45}$$

If the quantum mechanics can be treated in terms of hydrodynamics, then the backward affirmation is also true. Hydrodynamics equations can be reduced (as minimum for ideal liquids) to Schrödinger equation, equations of quantum mechanics.

From this point of view no surprise comes in appearance of discrete structures in flow investigations like strange attractors (Doering and Gibbon, 1995). Quantum mechanics technique of quantization can be propagated in physics of continuum on the whole, and in particular in hydrodynamics. Chapter 8 contains examples of corresponding approach.

It is important to notice that Schrödinger equation is treated as Euler equation for ideal gas, more exactly as Euler equation for probabilistic liquid which is considered from the position of mechanics of continuum. It means that outgoing Schrödinger equation is a non-dissipative equation reflecting reversible processes in closed system. Schrödinger equation conserves its form by substitutions $t \to -t$, $\psi \to \psi^*$.

But for open systems, for example for interaction of atom with radiation, the situation is radically different. For construction of quantum mechanics of open systems the principles of the generalized hydrodynamics can be used.

Let us write down the generalized continuum equation – in the absence of magnetic field – in the form:

$$\frac{\partial}{\partial t}\left\{\rho - \tau_{\rm qu}\left[\frac{\partial \rho}{\partial t} + \frac{\partial}{\partial \mathbf{r}} \cdot (\rho \mathbf{v})\right]\right\} + \frac{\partial}{\partial \mathbf{r}} \cdot \left\{\rho \mathbf{v} - \tau_{\rm qu}\rho\left[\frac{\partial \mathbf{v}}{\partial t} + \frac{\partial}{\partial \mathbf{r}} \cdot \mathbf{v}\mathbf{v}\right.\right.$$

$$\left.\left. + \frac{1}{m}\frac{\partial}{\partial \mathbf{r}}\left(U - \frac{\hbar^2}{4m\rho}\left(\frac{\partial}{\partial \mathbf{r}} \cdot \frac{\partial \rho}{\partial \mathbf{r}} - \frac{1}{2\rho}\left(\frac{\partial \rho}{\partial \mathbf{r}}\right)^2\right)\right)\right]\right\} = 0, \tag{3.4.46}$$

where $\tau_{\rm qu}$ is a relaxation parameter reflecting the rate of interaction between quantum system and surrounding media. If $\tau_{\rm qu} \to \infty$ – and quantum system becomes isolating – Eq. (3.4.44) splits into two hydrodynamic equations corresponding to classical Schrödinger equation. If $\tau_{\rm qu} \to 0$, the remaining part corresponds only to quantum continuity equation related to the closed quantum system. This situation reflects the state of total thermodynamic equilibrium.

In general case of open quantum system, the energy equation should be used with consideration. Probably it forced us to introduce hyper-complex numbers description for wave function instead of complex ψ-function, real and imaginary parts of which lead only to two hydrodynamic equations: continuity and Euler momentum equations.

As we see the generalized hydrodynamic equations could be the theoretical basement of quantum mechanics of open systems.

CHAPTER 4

Physics of a Weakly Ionized Gas

Physics of weakly ionized gas has a very vast area of technical applications. Investigations of transport processes in gas discharge, quantum generators, magnetohydrodynamic generators and so on are based on physics of weakly ionized gases and, in particular, on the Boltzmann kinetic theory (BKT). Methods of physics of weakly ionized gas are significant not only for technical applications but also for fundamental physics, for example for physics of ionosphere and astrophysics.

The traditional area of application of the Boltzmann kinetic theory is the physics of a weakly ionized gas. It is interesting to see what the GBE yields in this case and how its results differ from those of the classical theory. To answer this fundamental question, let us consider the classical Lorentz formulation of the problem. We consider a spatially homogeneous, weakly ionized gas, for which it is assumed that collisions between charged particles may be ignored:

$$\nu_e \ll \delta \nu_{ea},$$

where ν_e is the collision rate between charged particles; ν_{ea} is the collision rate between charged and neutral particles, and δ is the relative amount of energy which a charged particle loses in one collision with a neutral particle. We assume that the magnetic field is either absent or has a static component B_z, and that the electric field is along the x-axis; all inelastic interactions are neglected. Topology of the external electromagnetic field and models of particle interactions will be discussed in the following for every considered case separately.

4.1. Relaxation of charged particles in "Maxwellian" gas and the hydrodynamic aspects of the theory

The classical BE in this case takes the form

$$\frac{Df_e}{Dt} \equiv \frac{\partial f_e}{\partial t} + \mathbf{F}_e \cdot \frac{\partial f_e}{\partial \mathbf{v}_e} = J_{ea}, \tag{4.1.1}$$

where $\mathbf{F}_e = q_e\mathbf{E}/m_e$ is a force acting on a unit mass of the charged particle, and q_e is the particle charge. The GBE is written in the following way:

$$\frac{Df_e}{Dt} - \frac{D}{Dt}\left(\tau_{ea}\frac{Df_e}{Dt}\right) = J_{ea}. \tag{4.1.2}$$

In Eq. (4.1.2), τ_{ea} is the mean time between the collisions of neutral and charged particles. As the GBE theory suggests, the collision integral can be taken in the Boltzmann form. Let us compare the results obtained in frames of BKT and generalized BKT.

Multiplying Eqs. (4.1.1) and (4.1.2) by the collision invariants m_e, $m_e\mathbf{v}_e$, $m_e v_e^2/2$ and then integrating them over the velocities, we arrive at the classical hydrodynamic equations (HEs) and the generalized hydrodynamic equations (GHEs), which assume a closed form providing we know how to evaluate the moments of the collision integrals involved. Note that in this case the following relation holds:

$$\int J_{ea} m_e \,d\mathbf{v}_e = 0 \tag{4.1.3}$$

owing to the law of conservation of mass in elastic nonrelativistic collisions. But the integrals

$$\int J_{ea} m_e \mathbf{v}_e \,d\mathbf{v}_e \quad \text{and} \quad \int J_{ea}\frac{m_e v_e^2}{2}\,d\mathbf{v}_e,$$

can be taken explicitly only for special models of particle interaction. Let us adopt the Maxwell model, in which the force F_{ea} of the intermolecular interaction depends on the inverse fifth power of the interparticle spacing:

$$F_{ea} = \frac{\chi_{ea}}{r^5}. \tag{4.1.4}$$

For this model, the integrals mentioned above are well known and the quantity τ_{ea} (hereinafter the subscript *ea* is dropped) was calculated to be

$$\tau = \left[2\pi 0.422\left(\frac{\chi(m_e + m_a)}{m_e m_a}\right)^{1/2} n_a\right]^{-1}. \tag{4.1.5}$$

Introducing the quantities

$$A = \frac{8\sqrt{\pi}}{3}\Gamma\left(\frac{5}{2}\right)0.422(m_e + m_a)^{-1}\left(\frac{\chi}{m_a M_e}\right)^{1/2} \tag{4.1.6}$$

and

$$M_a = \frac{m_a}{m_a + m_e}, \qquad M_e = \frac{m_e}{m_a + m_e}, \tag{4.1.7}$$

the rate ν of collisions between charged and neutral particles can be written in the form

$$\nu = n_a(m_a + m_e)A, \tag{4.1.8}$$

where $\tau = \nu^{-1}$, and n_a is the density number of neutral particles.

The continuity equations obtained from Eq. (4.1.2) yield the condition $n_e = const$, and the GBE in this case becomes

$$\frac{\partial f_e}{\partial t} + \mathbf{F}_e \cdot \frac{\partial f_e}{\partial \mathbf{v}_e} - \tau \left\{ \frac{\partial^2 f_e}{\partial t^2} + 2\mathbf{F}_e \cdot \frac{\partial^2 f_e}{\partial \mathbf{v}_e \partial t} + \frac{\partial^2 f_e}{\partial \mathbf{v}_e \partial \mathbf{v}_e} : \mathbf{F}_e \mathbf{F}_e \right\} = J_{ea}. \tag{4.1.9}$$

As before, a colon denotes the double scalar product of tensors.

We now introduce the drift velocity $\bar{v}_{ex}$ defined by the expression

$$\bar{v}_{ex} = \frac{1}{n_e} \int f_e v_{ex} \, \mathrm{d}\mathbf{v}_e. \tag{4.1.10}$$

Then the equation of motion entering the system of GHEs takes the form

$$\tau \frac{\mathrm{d}^2 \bar{v}_{ex}}{\mathrm{d}t^2} - \frac{\mathrm{d}\bar{v}_{ex}}{\mathrm{d}t} - A m_a n_a \bar{v}_{ex} + q_e E m_e^{-1} = 0. \tag{4.1.11}$$

In writing Eq. (4.1.11) we have used the result (Morse, 1963)

$$\int m_e v_{ex} J_{ea} \, \mathrm{d}\mathbf{v}_e = -A m_a n_a \bar{v}_{ea} n_a n_e.$$

The solution of Eq. (4.1.11) takes the form

$$\bar{v}_{ex}^{\mathrm{BAE}} = \left(\bar{v}_{ex}^0 - \frac{q_e E}{m_e m_a n_a A} \right) \mathrm{e}^{-(t/(2\tau))[\sqrt{4M_a+1}-1]} + \frac{q_e E}{m_e m_a n_a A}. \tag{4.1.12}$$

The superscript 0 here refers to the initial instant of time.

The problem of the time relaxation of Maxwell particles in an electric field is known to be amenable to a BE solution (Alekseev, 1982) giving for the drift velocity the result

$$\bar{v}_{ex}^{\mathrm{BE}} = \left(\bar{v}_{ea}^0 - \frac{q_e E}{m_e m_a n_a A} \right) \mathrm{e}^{-t A n_a m_a} + \frac{q_e E}{m_e m_a n_a A}. \tag{4.1.13}$$

For example, let us assume that $m_e \ll m_a$. Then, from Eqs. (4.1.12) and (4.1.13), it follows that

$$\bar{v}_{ex}^{\mathrm{GBE}} = \left(\bar{v}_{ex}^0 - F_{ex}\tau \right) \mathrm{e}^{-0.618t/\tau} + F_{ex}\tau, \tag{4.1.14}$$

$$\bar{v}_{ex}^{\mathrm{BE}} = \left(\bar{v}_{ex}^0 - F_{ex}\tau \right) \mathrm{e}^{-t/\tau} + F_{ex}\tau. \tag{4.1.15}$$

Thus, all other things being equal, the relaxation of the drift velocity $\bar{v}_{ex}$ in the framework of BKT proceeds faster than in the generalized BKT, whereas the steady-state drift velocities are found to be the same. We now turn our attention to the equation of energy and introduce the energy temperatures $\widehat{T}_e$ and $\widetilde{T}_e$ in accordance with the definitions:

$$\widehat{T}_e = \frac{m_e}{3n_e} \int f_e v_e^2 \, \mathrm{d}\mathbf{v}_e, \tag{4.1.16}$$

$$\widetilde{T}_e = \frac{m_e}{3n_e} \int f_e (\mathbf{v}_e - \bar{\mathbf{v}}_e)^2 \, \mathrm{d}\mathbf{v}_e. \tag{4.1.17}$$

Clearly, which of these temperatures is used is a matter of convenience, and in our case one obtains

$$\widehat{T}_e = \widetilde{T}_e + \frac{1}{3} m_e \bar{v}_{ex}^2. \tag{4.1.18}$$

We next evaluate the moments on the left-hand side of the kinetic equations. For example, the following relations hold true:

$$\frac{\partial}{\partial t} \int \frac{m_e v_e^2}{2} f_e \, \mathrm{d}\mathbf{v}_e = \frac{3}{2} n_e \frac{\partial \widehat{T}_e}{\partial t}, \tag{4.1.19}$$

$$\int \mathbf{F}_e \cdot \frac{\partial^2 f_e}{\partial \mathbf{v}_e \partial t} \frac{m_e v_e^2}{2} \, \mathrm{d}\mathbf{v}_e = -F_{ex} m_e n_e \frac{\partial \bar{v}_{ex}}{\partial t}, \tag{4.1.20}$$

$$\int \frac{m_e v_e^2}{2} \frac{\partial^2 f_e}{\partial \mathbf{v}_e \partial \mathbf{v}_e} : \mathbf{F}_e \mathbf{F}_e \, \mathrm{d}\mathbf{v}_e = F_{ex}^2 m_e n_e. \tag{4.1.21}$$

The corresponding integral on the right-hand side was calculated in Morse (1963) and is found to be

$$\int J_{ea} \frac{m_e v_e^2}{2} \, \mathrm{d}\mathbf{v}_e = -\frac{3(\widehat{T}_e - \widehat{T}_a)}{m_e + m_a} A m_e m_a n_e n_a. \tag{4.1.22}$$

We have then the following nonhomogeneous linear second-order differential equation

$$\begin{aligned} \frac{\mathrm{d}^2 \widetilde{T}_e}{\mathrm{d}t^2} - \frac{1}{\tau} \frac{\mathrm{d}\widetilde{T}_e}{\mathrm{d}t} - 2 \frac{\widehat{T}_e - \widehat{T}_a}{m_e + m_a} \frac{A}{\tau} m_a m_e n_a &= \frac{1}{3} \frac{m_e}{\tau} \frac{\mathrm{d}}{\mathrm{d}t} \bar{v}_{ex}^2 - \frac{2}{3} \frac{m_e}{\tau} F_{ex} \bar{v}_{ex} \\ - \frac{1}{3} m_e \frac{\mathrm{d}^2}{\mathrm{d}t^2} \bar{v}_{ex}^2 + \frac{4}{3} F_{ex} m_e \frac{\mathrm{d}\bar{v}_{ex}}{\mathrm{d}t} - \frac{2}{3} m_e F_{ex}^2. \end{aligned} \tag{4.1.23}$$

Omitting the straightforward but tedious algebra we arrive at the following results. For example, by setting $m_e \ll m_a$, the GBE yields

$$\begin{aligned} \widetilde{T}_e^{\mathrm{BAE}} = \widetilde{T}_a + C_2 \mathrm{e}^{-2m_e t/(m_a \tau)} - 2.157 m_e \tau F_{ex} \left(\bar{v}_{ex}^0 - F_{ex} \tau \right) \mathrm{e}^{-0.618t/\tau} \\ - \frac{1}{3} m_e \left(\bar{v}_{ex}^0 - F_{ex} \tau \right)^2 \mathrm{e}^{-1.236t/\tau} + \frac{2}{3} F_{ex}^2 \tau^2 m_a, \end{aligned} \tag{4.1.24}$$

where the following notation was used:

$$C_2 = \widetilde{T}_e^{\,0} - \widetilde{T}_a + 2.157 m_e \tau F_{ex} \left(\bar{v}_{ex}^0 - F_{ex}\tau \right) + \frac{1}{3} m_e \left(\bar{v}_{ex}^0 - F_{ex}\tau \right)^2 - \frac{2}{3} F_{ex}^2 \tau^2 m_a .$$

Similar BE results are as follows:

$$\widetilde{T}_e^{\,\mathrm{BE}} = \widetilde{T}_a + C_2\, \mathrm{e}^{-2m_e t/(m_a\tau)} - \frac{4}{3} m_e \tau F_{ex} \left(\bar{v}_{ex}^0 - F_{ex}\tau \right) \mathrm{e}^{-t/\tau} - \frac{1}{3} m_e \left(\bar{v}_{ex}^0 - F_{ex}\tau \right)^2 \mathrm{e}^{-2t/\tau} + \frac{1}{3} F_{ex}^2 \tau^2 m_a . \tag{4.1.25}$$

Here, the notation used is:

$$C_2 = \widetilde{T}_e^{\,0} - \widetilde{T}_a + \frac{4}{3} m_e \tau F_{ex} \left(\bar{v}_{ex}^0 - F_{ex}\tau \right) + \frac{1}{3} m_e \left(\bar{v}_{ex}^0 - F_{ex}\tau \right)^2 - \frac{1}{3} F_{ex}^2 \tau^2 m_a .$$

In the steady-state regime, the above solutions are related by the expression

$$\left(\widetilde{T}_e - \widetilde{T}_a \right)_{\mathrm{st}}^{\mathrm{GBE}} = 2 \left(\widetilde{T}_e - \widetilde{T}_a \right)_{\mathrm{st}}^{\mathrm{BE}} . \tag{4.1.26}$$

As before, the superscripts on the energy temperature differences in Eq. (4.1.26) refer to the type of the solution. We are now in a position to write down the solutions for the energy temperatures $\widehat{T}_e$; in the GBE scheme we have

$$\begin{aligned} \widehat{T}_e = \widehat{T}_a &+ \left[\widehat{T}_{ea}^0 - \frac{2}{3} F_{ex}\tau \frac{m_e}{M_a(2M_e - 1)} \sqrt{4M_a + 1} \left(\bar{v}_{ex}^0 - \frac{F_{ex}\tau}{M_a} \right) \right. \\ &\left. - \frac{\tau^2}{3} \frac{F_{ex}^2}{M_a^2} (1 + M_a)(m_e + m_a) \right] \mathrm{e}^{-(t/(2\tau))(\sqrt{8M_a M_e + 1} - 1)} \\ &+ \frac{2}{3} F_{ex}\tau \frac{m_e}{M_a(2M_e - 1)} \sqrt{4M_a + 1} \left(\bar{v}_{ex}^0 - \frac{F_{ex}\tau}{M_a} \right) \mathrm{e}^{-(t/(2\tau))(\sqrt{4M_a + 1} - 1)} \\ &+ \frac{\tau^2}{3} \frac{F_{ex}^2}{M_a^2} (m_e + m_a)(1 + M_a), \end{aligned} \tag{4.1.27}$$

with $\widehat{T}_{ea}^0 = \widehat{T}_e^{\,0} - \widehat{T}_a$.

In the framework of the classical Boltzmann equation, we find (Alekseev, 1982)

$$\widehat{T}_e = \widehat{T}_a + \left[\widehat{T}_{ea}^0 - \frac{2}{3} F_{ex}\tau \frac{m_e}{M_a(2M_e - 1)} \left(\bar{v}_{ex}^0 - \frac{F_{ex}\tau}{M_a} \right) - \frac{\tau^2}{3} \frac{F_{ex}^2}{M_a^2} (m_e + m_a) \right] \mathrm{e}^{-2M_a M_e t/\tau}$$

$$+\frac{2}{3}F_{ex}\tau\frac{m_e}{M_a(2M_e-1)}\left(\bar{v}_{ex}^0-\frac{F_{ex}\tau}{M_a}\right)e^{-M_a t/\tau}$$
$$+\frac{\tau^2}{3}\frac{F_{ex}^2}{M_a^2}(m_e+m_a). \tag{4.1.28}$$

Notice that the vanishing term $2M_e-1$ in the denominators in Eqs. (4.1.27) and (4.1.28) does not actually lead to singularities at $M_e=0.5$, because the corresponding terms cancel due to the exponential factors being equal. From Eqs. (4.1.27), (4.1.28) it follows that

$$\widehat{T}_{ea,\mathrm{st}}^{\mathrm{GBE}}=(1+M_a)\widehat{T}_{ea,\mathrm{st}}^{\mathrm{BE}}. \tag{4.1.29}$$

Thus, unlike drift velocity calculations, not only the GBE changes the trend of the relaxation curves but it also leads to different steady-state values of the energy temperatures. For a weakly ionized Lorentz gas, the effect of the self-consistent forces of electromagnetic origin can be neglected.

CBE for multicomponent reacting mixture of gases can be written in the form (1.3.71)

$$\frac{Df_\alpha}{Dt}-\frac{D}{Dt}\left(\tau_\alpha\frac{Df_\alpha}{Dt}\right)=J_\alpha^{\mathrm{st}}\quad(\alpha=1,\ldots,\eta), \tag{4.1.30}$$

where τ_α is mean time between collisions of αth particles ($\alpha=1,\ldots,\eta$) with particles of the η-component gas mixture.

Then, multiplying the GBE (4.1.30) by the collision invariants

$$m_\alpha,\quad m_\alpha\mathbf{v}_\alpha,\quad \frac{m_\alpha v_\alpha^2}{2}+\varepsilon_\alpha$$

(with ε_α being the internal energy of the particles of the component) and integrating with respect to the velocities we arrive at Enskog's system of generalized hydrodynamic equations (GHEs), in which only the effect of external forces is included:

– the continuity equation for the component α

$$\frac{\partial}{\partial t}\left\{\rho_\alpha-\tau_\alpha\left[\frac{\partial\rho_\alpha}{\partial t}+\frac{\partial}{\partial\mathbf{r}}\cdot(\rho_\alpha\bar{\mathbf{v}}_\alpha)\right]\right\}+\frac{\partial}{\partial\mathbf{r}}\cdot\left\{\rho_\alpha\bar{\mathbf{v}}_\alpha-\tau_\alpha\left[\frac{\partial}{\partial t}(\rho_\alpha\bar{\mathbf{v}}_\alpha)\right.\right.$$
$$\left.\left.+\frac{\partial}{\partial\mathbf{r}}\cdot\rho_\alpha\overline{\mathbf{v}_\alpha\mathbf{v}_\alpha}-\rho_\alpha\mathbf{F}_\alpha^{(1)}-\frac{q_\alpha}{m_\alpha}\rho_\alpha\mathbf{F}_\alpha\times\mathbf{B}\right]\right\}=R_\alpha, \tag{4.1.31}$$

– the equation of motion

$$\frac{\partial}{\partial t}\left\{\rho_\alpha\bar{\mathbf{v}}_\alpha-\tau_\alpha\left[\frac{\partial}{\partial t}(\rho_\alpha\bar{\mathbf{v}}_\alpha)+\frac{\partial}{\partial\mathbf{r}}\cdot\rho_\alpha\overline{\mathbf{v}_\alpha\mathbf{v}_\alpha}-\rho_\alpha\mathbf{F}_\alpha^{(1)}-\frac{q_\alpha}{m_\alpha}\rho_\alpha\bar{\mathbf{v}}_\alpha\times\mathbf{B}\right]\right\}$$

$$
\begin{aligned}
&-\left\{\mathbf{F}_\alpha^{(1)}\left[\rho_\alpha-\tau_\alpha\left(\frac{\partial\rho_\alpha}{\partial t}+\frac{\partial}{\partial\mathbf{r}}\cdot(\rho_\alpha\bar{\mathbf{v}}_\alpha)\right)\right]\right\}-\frac{q_\alpha}{m_\alpha}\left\{\rho_\alpha\bar{\mathbf{v}}_\alpha-\tau_\alpha\left[\frac{\partial}{\partial t}(\rho_\alpha\bar{\mathbf{v}}_\alpha)\right.\right.\\
&\left.\left.+\frac{\partial}{\partial\mathbf{r}}\cdot\rho_\alpha\overline{\mathbf{v}_\alpha\mathbf{v}_\alpha}-\rho_\alpha\mathbf{F}_\alpha^{(1)}-\frac{q_\alpha}{m_\alpha}\rho_\alpha\bar{\mathbf{v}}_\alpha\times\mathbf{B}\right]\right\}\times\mathbf{B}+\frac{\partial}{\partial\mathbf{r}}\cdot\left\{\rho_\alpha\overline{\mathbf{v}_\alpha\mathbf{v}_\alpha}\right.\\
&-\tau_\alpha\left[\frac{\partial}{\partial t}(\rho_\alpha\overline{\mathbf{v}_\alpha\mathbf{v}_\alpha})+\frac{\partial}{\partial\mathbf{r}}\cdot\rho_\alpha\overline{(\mathbf{v}_\alpha\mathbf{v}_\alpha)\mathbf{v}_\alpha}-\mathbf{F}_\alpha^{(1)}\rho_\alpha\bar{\mathbf{v}}_\alpha-\rho_\alpha\bar{\mathbf{v}}_\alpha\mathbf{F}_\alpha^{(1)}\right.\\
&\left.\left.-\frac{q_\alpha}{m_\alpha}\rho_\alpha\overline{[\mathbf{v}_\alpha\times\mathbf{B}]\mathbf{v}_\alpha}-\frac{q_\alpha}{m_\alpha}\rho_\alpha\overline{\mathbf{v}_\alpha[\mathbf{v}_\alpha\times\mathbf{B}]}\right]\right\}=\bar{\mathbf{J}}_{\alpha,\mathrm{mot}},
\end{aligned}
\tag{4.1.32}
$$

– the equation of energy

$$
\begin{aligned}
&\frac{\partial}{\partial t}\left\{\frac{\rho_\alpha\overline{v_\alpha^2}}{2}+\varepsilon_\alpha n_\alpha-\tau_\alpha\left[\frac{\partial}{\partial t}\left(\frac{\rho_\alpha\overline{v_\alpha^2}}{2}+\varepsilon_\alpha n_\alpha\right)+\frac{\partial}{\partial\mathbf{r}}\cdot\left(\frac{1}{2}\rho_\alpha\overline{v_\alpha^2\mathbf{v}_\alpha}+\varepsilon_\alpha n_\alpha\bar{\mathbf{v}}_\alpha\right)\right.\right.\\
&\left.\left.-\mathbf{F}_\alpha^{(1)}\cdot\rho_\alpha\bar{\mathbf{v}}_\alpha\right]\right\}+\frac{\partial}{\partial\mathbf{r}}\cdot\left\{\frac{1}{2}\rho_\alpha\overline{v_\alpha^2\mathbf{v}_\alpha}+\varepsilon_\alpha n_\alpha\bar{\mathbf{v}}_\alpha\right.\\
&-\tau_\alpha\left[\frac{\partial}{\partial t}\left(\frac{1}{2}\rho_\alpha\overline{v_\alpha^2\mathbf{v}_\alpha}+\varepsilon_\alpha n_\alpha\bar{\mathbf{v}}_\alpha\right)+\frac{\partial}{\partial\mathbf{r}}\cdot\left(\frac{1}{2}\rho_\alpha\overline{v_\alpha^2\mathbf{v}_\alpha\mathbf{v}_\alpha}+\varepsilon_\alpha n_\alpha\overline{\mathbf{v}_\alpha\mathbf{v}_\alpha}\right)\right.\\
&\left.\left.-\rho_\alpha\mathbf{F}_\alpha^{(1)}\cdot\overline{\mathbf{v}_\alpha\mathbf{v}_\alpha}-\frac{1}{2}\rho_\alpha\overline{v_\alpha^2\mathbf{F}_\alpha}-\varepsilon_\alpha n_\alpha\overline{\mathbf{F}}_\alpha\right]\right\}\\
&-\left\{\bar{\mathbf{v}}_\alpha\cdot\rho_\alpha\mathbf{F}_\alpha^{(1)}-\tau_\alpha\left[\mathbf{F}_\alpha^{(1)}\cdot\left(\frac{\partial}{\partial t}(\rho_\alpha\bar{\mathbf{v}}_\alpha)+\frac{\partial}{\partial\mathbf{r}}\cdot\rho_\alpha\overline{\mathbf{v}_\alpha\mathbf{v}_\alpha}-\rho_\alpha\mathbf{F}_\alpha^{(1)}\right.\right.\right.\\
&\left.\left.\left.-\frac{q_\alpha}{m_\alpha}\rho_\alpha\bar{\mathbf{v}}_\alpha\times\mathbf{B}\right)\right]\right\}=\bar{J}_{\alpha,\mathrm{en}},
\end{aligned}
\tag{4.1.33}
$$

where $\mathbf{v}_0$ is the hydrodynamical velocity, $\mathbf{B}$ is the magnetic induction, $\mathbf{F}_\alpha^{(1)}$ are external nonmagnetic forces per unit mass of the particle α, and q_α is the charge. The right-hand sides of Eqs. (4.1.32), (4.1.33) involve the integral relaxation terms $\bar{\mathbf{J}}_{\alpha,\mathrm{mot}}$ and $\bar{J}_{\alpha,\mathrm{en}}$, which, due to momentum and energy conservation laws, satisfy the relations

$$
\sum_{\alpha=1}^{\eta}\bar{\mathbf{J}}_{\alpha,\mathrm{mot}}=0,\qquad\sum_{\alpha=1}^{\eta}\bar{J}_{\alpha,\mathrm{en}}=0.
\tag{4.1.34}
$$

However, for the systems being far from equilibrium one has to introduce approximations for $\bar{\mathbf{J}}_{\alpha,\mathrm{mot}}$, $\bar{J}_{\alpha,\mathrm{en}}$. This can be done in a number of ways, including the Bhatnagar–Gross–Krook (BGK) method or its extensions (Bhatnagar, Gross and Krook, 1954; Shakhov, 1974).

The generalized Boltzmann equation and the system of GHEs can be used to study plasma in an electric field, in particular to understand the electron energy runaway effect (Golant, Zhilinski and Sakharov, 1977). We now proceed to apply the generalized Boltzmann kinetic theory to the classical problems of plasma physics.

4.2. Distribution function of the charged particles in the "Lorentz" gas

Calculating the distribution function for charged particles added as an impurity to a neutral gas in an external electric field is a classical problem in gas discharge physics, whose long history dates back to Pidduck's 1913 attempt to calculate the ion drift velocity in gases (Pidduck, 1913). We should also mention Compton's work concerned with computing the charge particle distribution function and its moments (Compton, 1916a; 1916b). Later on, Druyvesteyn (1930a, 1930b) and Davydov (1936) obtained analytical expressions for the distribution function and transport coefficients for the special case of elastic collisions. More recent work (note, in particular, the monograph of Ivanov, Lebedev and Polak (1981)) has been aimed principally at investigating the effect of inelastic collisions on the DF and transport processes within the BKT framework. It is important to note that the calculation of the DF depends heavily on what model of particle interaction is adopted – and hence ultimately on the collision cross sections involved. For example, the Davydov–Druyvesteyn distribution obtained under the assumption of a constant mean free path l for elastically colliding, charged gas particles significantly underpredict the number of "hot" particles on the tail of the DF and lead ultimately to unacceptable results when the theory is extended to calculating the kinetics of inelastic processes (Ivanov, Lebedev and Polak, 1981).

We apply the generalized Boltzmann equation

$$\mathbf{F}_e \cdot \frac{\partial f_e}{\partial \mathbf{v}_e} - \tau \frac{\partial^2 f_e}{\partial \mathbf{v}_e \partial \mathbf{v}_e} : \mathbf{F}_e \mathbf{F}_e = J_{ea}, \tag{4.2.1}$$

to consider charged particles in a steady state in a Lorentz gas subject to a stationary external electric field, where

$$\mathbf{F}_e = \frac{e\mathbf{E}}{m_e}.$$

The Boltzmann kinetic equation is usually solved by expanding the DF in a power series of zero-order solid spherical harmonics, i.e., of Legendre polynomials. The corresponding system of linked equations was obtained elsewhere (Ginzburg and Gurevich, 1960a, 1960b). The solution to the GBE (4.2.1) is conveniently sought as an expansion in terms of solid spherical harmonics:

$$f(\mathbf{v}_e) = f_0(v_e) + \mathbf{F}_e \cdot \mathbf{v}_e f_1(v_e) + \mathbf{F}_e \mathbf{F}_e : \mathbf{v}_e^0 \mathbf{v}_e f_2(v_e) + \cdots. \tag{4.2.2}$$

Here $\mathbf{v}_e^0 \mathbf{v}_e$ is the zero-trace tensor. For our further calculations in this section we assume that the force $\mathbf{F}_e$ is along the positive direction of a certain chosen coordinate axis. We now substitute expansion (4.2.2) into Eq. (4.2.1) and transform the corresponding terms; we have, for example, the following relations:

$$\mathbf{F}_e \cdot \frac{\partial}{\partial \mathbf{v}_e}(f_2 \mathbf{F}_e \mathbf{F}_e : \mathbf{v}_e \mathbf{v}_e) = \frac{(\mathbf{F}_e \cdot \mathbf{v}_e)^3}{v_e} \frac{\partial f_2}{\partial v_e} - \frac{1}{3} F_e^2 v_e (\mathbf{F}_e \cdot \mathbf{v}_e) \frac{\partial f_2}{\partial v_e} + \frac{4}{3} F_e^2 f_2 (\mathbf{v}_e \cdot \mathbf{F}_e), \tag{4.2.3}$$

$$\frac{\partial^2}{\partial \mathbf{v}_e \partial \mathbf{v}_e} : \left\{ f_2 \left(\mathbf{F}_e \mathbf{F}_e : \mathbf{v}_e^0 \mathbf{v}_e \right) \mathbf{F}_e \mathbf{F}_e \right\} = 5F_e^2 \frac{(\mathbf{F}_e \cdot \mathbf{v}_e)^2}{v_e} \frac{\partial f_2}{\partial v_e} + 2 f_2 F_e^4 + \frac{(\mathbf{F}_e \cdot \mathbf{v}_e)^4}{v_e^2} \frac{\partial^2 f_2}{\partial v_e^2} - \frac{(\mathbf{F}_e \cdot \mathbf{v}_e)^4}{v_e^3} \frac{\partial f_2}{\partial v_e} - \frac{1}{3} \sum_{i=1}^{3} F_{ei}^2 F_e^2 \frac{\partial^2}{\partial v_{ei}^2} \left(f_2 v_e^2 \right). \tag{4.2.4}$$

The left-hand side Λ of the generalized Boltzmann equation then takes the form

$$\begin{aligned}
\Lambda = {} & (\mathbf{F}_e \cdot \mathbf{v}_e) \frac{1}{v_e} \frac{\partial f_0}{\partial v_e} + F_e^2 f_1 + \frac{(\mathbf{F}_e \cdot \mathbf{v}_e)^2}{v_e} \frac{\partial f_1}{\partial v_e} + \frac{(\mathbf{F}_e \cdot \mathbf{v}_e)^3}{v_e} \frac{\partial f_2}{\partial v_e} \\
& - \frac{1}{3} F_e^2 v_e (\mathbf{F}_e \cdot \mathbf{v}_e) \frac{\partial f_2}{\partial v_e} + \frac{4}{3} F_e^2 f_2 (\mathbf{F}_e \cdot \mathbf{v}_e) \\
& - \tau \left\{ \frac{(\mathbf{F}_e \cdot \mathbf{v}_e)^2}{v_e} \frac{\partial}{\partial v_e} \left(\frac{1}{v_e} \frac{\partial f_0}{\partial v_e} \right) + \frac{F_e^2}{v_e} \frac{\partial f_0}{\partial v_e} + 3F_e^2 \frac{(\mathbf{F}_e \cdot \mathbf{v}_e)}{v_e} \frac{\partial f_1}{\partial v_e} \right. \\
& - \frac{(\mathbf{F}_e \cdot \mathbf{v}_e)^3}{v_e^3} \frac{\partial f_1}{\partial v_e} + \frac{(\mathbf{F}_e \cdot \mathbf{v}_e)^3}{v_e^2} \frac{\partial^2 f_1}{\partial v_e^2} + 4F_e^2 \frac{(\mathbf{F}_e \cdot \mathbf{v}_e)^2}{v_e} \frac{\partial f_2}{\partial v_e} \\
& + \frac{4}{3} f_2 F_e^2 + \frac{(\mathbf{F}_e \cdot \mathbf{v}_e)^4}{v_e^2} \frac{\partial^2 f_2}{\partial v_e^2} - \frac{1}{3} F_e^2 (\mathbf{F}_e \cdot \mathbf{v}_e)^2 \frac{\partial^2 f_2}{\partial v_e^2} \\
& \left. - \frac{(\mathbf{F}_e \cdot \mathbf{v}_e)^4}{v_e^3} \frac{\partial f_2}{\partial v_e} - \frac{1}{3} F_e^4 v_e \frac{\partial f_2}{\partial v_e} \right\}.
\end{aligned} \tag{4.2.5}$$

Denoting the angle between the vectors $\mathbf{F}_e$ and $\mathbf{v}_e$ by ϑ $(0 \leqslant \vartheta \leqslant \pi)$, multiplying the GBE by $\mathrm{d}\cos\vartheta$, and integrating over the entire range of angles, we arrive at

$$f_1 + \frac{1}{3} v_e \frac{\partial f_1}{\partial v_e} - \frac{\tau}{3} \left\{ \frac{2}{v_e} \frac{\partial f_0}{\partial v_e} + \frac{\partial^2 f_0}{\partial v_e^2} + \frac{12}{5} v_e \frac{\partial f_2}{\partial v_e} F_e^2 + 4 f_2 F_e^2 + \frac{4}{15} F_e^2 v_e^2 \frac{\partial^2 f_2}{\partial v_e^2} \right\} = \frac{1}{F_e^2} J_{ea}. \tag{4.2.6}$$

Multiplying the GBE by $\cos\vartheta\,\mathrm{d}\cos\vartheta$ and using a similar procedure,

$$\frac{\partial f_0}{\partial v_e} + \frac{4}{15}F_e^2 v_e^2 \frac{\partial f_2}{\partial v_e} + \frac{4}{3}F_e^2 v_e f_2 - \frac{3}{5}\tau F_e^2 \left\{ 4\frac{\partial f_1}{\partial v_e} + v_e \frac{\partial^2 f_1}{\partial v_e^2} \right\}$$
$$= -\frac{3}{2}\frac{1}{F_e} J_1. \tag{4.2.7}$$

As has been indicated, the collision terms J_{ea} and J_1 in the generalized Boltzmann kinetic theory (GBKT) can be taken in the form in which they are usually written in the Boltzmann equation. In the case we consider below, assuming that the change in the electron energy due to an elastic collision (approximately equal to $(m_e/m_a)^{1/2}\varepsilon$) is much less that the electron energy prior to the collision, in the Fokker–Planck approximation (see, for example, Chetverushkin, 1999) we have

$$J_{ea} = \frac{m_e}{m_a}\frac{\widehat{T}_a}{v_e^2}\frac{\partial}{\partial v_e}\left[v_e^3 \nu \left(\frac{f_0}{\widehat{T}} + \frac{1}{m_e v_e}\frac{\partial f_0}{\partial v_e} \right) \right], \tag{4.2.8}$$

$$J_1 = \frac{2}{3}F_e \frac{v_e^2}{\ell} f_1, \tag{4.2.9}$$

where $\widehat{T}_a$ is the energy temperature of the neutral gas $\widehat{T}_a = k_B T_a$, υ is the collision rate which generally depends on the velocity, and 1 is the mean free path for collisions of neutral and charged particles. It is relations (4.2.6)–(4.2.9) which provide the required basis for determining the DF and its moments. Traditionally, two limiting situations are considered in detail:

(1) a constant frequency rate, $\nu = const$, $\nu = \tau^{-1} = v_e \ell^{-1}$, and
(2) a constant mean free path between the collisions of charged and neutral particles, $\ell = const$.

We take up the former case first. Multiplying Eq. (4.2.6) through by $3v_e^2$ and using Eq. (4.2.8), we find, after some algebra, that

$$F_e^2 \frac{\mathrm{d}}{\mathrm{d}v_e}\left\{ v_e^3 f_1 - \tau v_e^2 \frac{\mathrm{d}f_0}{\mathrm{d}v_e} \right\} = 3\frac{\widehat{T}_a m_e}{\tau m_a}\frac{\mathrm{d}}{\mathrm{d}v_e}\left\{ v_e^3 \left(\frac{f_0}{\widehat{T}_a} + \frac{1}{m_e v_e}\frac{\mathrm{d}f_0}{\mathrm{d}v_e} \right) \right\} \tag{4.2.10}$$

or upon integration over v_e:

$$f_1 v_e = \left[\tau + \frac{3\widehat{T}_a}{F_e^2 m_a \tau} \right] \frac{\mathrm{d}f_0}{\mathrm{d}v_e} + \frac{3}{F_e^2}\frac{m_e}{m_a}\frac{v_e}{\tau} f_0, \tag{4.2.11}$$

because the constant of integration is zero due to the fact that both the left-hand and the right-hand sides of Eq. (4.2.11) vanish for $v_e = 0$. Eq. (4.2.11) was obtained under the condition (which will also be used in the following analysis) that small terms proportional to f_2 may be dropped. Substituting Eq. (4.2.11) into Eq. (4.2.9) and making use of the result produced to eliminate f_1 from Eq. (4.2.7), we arrive at the following equation in f_0:

$$v_e^2\left(\tau+\frac{3\widehat{T}_a}{F_e^2 m_a \tau}\right)\frac{\mathrm{d}^3 f_0}{\mathrm{d}v_e^3}+v_e\left(2\tau+\frac{3m_e}{F_e^2 m_a \tau}v_e^2+\frac{6\widehat{T}_a}{F_e^2 m_a \tau}\right)\frac{\mathrm{d}^2 f_0}{\mathrm{d}v_e^2}$$
$$+\left(-2\tau-\frac{10}{3\tau F_e^2}v_e^2-\frac{5\widehat{T}_a}{\tau^3 F_e^4 m_a}v_e^2+\frac{12m_e}{F_e^2 m_a \tau}v_e^2-\frac{6\widehat{T}_a}{F_e^2 m_a \tau}\right)\frac{\mathrm{d}f_0}{\mathrm{d}v_e}$$
$$-\frac{5m_e}{F_e^4\tau^3 m_a}v_e^3 f_0=0. \tag{4.2.12}$$

To solve Eq. (4.2.12), three boundary conditions are needed. These are in fact quite obvious. Indeed, for $v_e=0$, we can specify a certain value of f_0, determined only by the normalization of the function. From Eq. (4.2.12) it is also seen that $f_0'=0$ for $v_e=0$. Finally, dividing the above equation by v_e^3 we find that $f_0\to 0$ for $v_e\to\infty$.

Thus, Eq. (4.2.12) is easily solved by, for example, the sweep method. To do this, it is convenient to first bring the equation to the dimensionless form by introducing the following dimensionless quantities labelled with arcs over the symbols:

$$\check{v}_e=\frac{v_e}{F_e\tau},\qquad \check{\varepsilon}=\frac{m_e F_e^2\tau^2}{\widehat{T}_a},\qquad \check{f}_0=\frac{f_0}{f_0(v_e=0)}. \tag{4.2.13}$$

The procedure is realized to yield

$$\check{v}_e^2\left[1+\frac{3m_e}{m_a\check{\varepsilon}}\right]\frac{\mathrm{d}^3\check{f}_0}{\mathrm{d}\check{v}_e^3}+\check{v}_e\left[2+\frac{6m_e}{m_a\check{\varepsilon}}+\frac{3m_e}{m_a}\check{v}_e^2\right]\frac{\mathrm{d}^2\check{f}_0}{\mathrm{d}\check{v}_e^2}$$
$$-\left[2+\frac{6m_e}{m_a\check{\varepsilon}}+\check{v}_e^2\left(\frac{10}{3}+\frac{5m_e}{m_a\check{\varepsilon}}-12\frac{m_e}{m_a}\right)\right]\frac{\mathrm{d}\check{f}_0}{\mathrm{d}\check{v}_e}-5\frac{m_e}{m_a}\check{v}_e^3\check{f}_0=0. \tag{4.2.14}$$

Let us define the energy temperature of charged particles as follows:

$$\widehat{T}_e=\frac{1}{3n_e}\int f_e m_e v_e^2\,\mathrm{d}v_e\cong\frac{1}{3n_e}\int f_0 m_e v_e^2\,\mathrm{d}v_e. \tag{4.2.15}$$

This means, for example, that, in terms of definitions (4.2.13), the Maxwellian function $\check{f}_M$ has the form

$$\check{f}_M=\mathrm{e}^{-(\widehat{T}_a\check{\varepsilon}/(2\widehat{T}_e))\check{v}_e^2}. \tag{4.2.16}$$

Let us examine the asymptotics of the function f_0 at large velocities v_e. From Eq. (4.2.14) it follows that for $v_e\to\infty$ the equation

$$\frac{\mathrm{d}^2 f_0}{\mathrm{d}v_e^2}-\frac{5}{3F_e^2\tau^2}f_0=0, \tag{4.2.17}$$

holds, which has the solution

$$f_0 \sim \mathrm{e}^{-\sqrt{5/3}\, v_e/(F_e \tau)}. \tag{4.2.18}$$

Note that, in the limiting case we are considering, the classical solution of the Boltzmann equation (Smirnov, 1985) leads to a Maxwellian distribution function with a temperature $\widehat{T}_e$, different from the neutral gas temperature $\widehat{T}_a$. Thus, the solution of the GBE results in a large number of "hot" charged particles in the tail of the distribution function. Of course, the moments of the distribution function – the temperature $\widehat{T}_e$ and the drift velocity $\bar{v}_{ex}$ – can be found by properly integrating the DF after the solution of Eq. (4.2.14) has been found. There is no need to do this, however. Indeed, multiplying Eq. (4.2.14) by v_e and integrating term-by-term we obtain

$$\left(\frac{3\widehat{T}_a}{m_a \tau^2 F_e^2} + 2\right) \int_0^\infty f_0 v_e^2 \, \mathrm{d}v_e = \frac{m_e}{m_a \tau^2 F_e^2} \int_0^\infty f_0 v_e^4 \, \mathrm{d}v_e. \tag{4.2.19}$$

Assuming that

$$\int f \, \mathrm{d}\mathbf{v}_e \cong \int f_0 \, \mathrm{d}\mathbf{v}_e = 4\pi \int_0^\infty f_0 v_e^2 \, \mathrm{d}v_e = n_e, \tag{4.2.20}$$

as was done in Eq. (4.2.15), it is found that

$$\widehat{T}_e = \widehat{T}_a + \frac{2}{3} m_a \tau^2 F_e^2. \tag{4.2.21}$$

In a similar way, without explicitly solving Eqs. (4.2.11) and (4.2.12), we can determine the drift velocity. To accomplish this, we multiply Eq. (4.2.11) term-wise by v_e^3 and integrate the resulting expression to yield

$$\int_0^\infty f_1 v_e^4 \, \mathrm{d}v_e = \left[\tau + \frac{3\widehat{T}_a}{F_e^2 m_a \tau}\right] \int_0^\infty v_e^3 \frac{\partial f_0}{\partial v_e} \, \mathrm{d}v_e + \frac{3}{F_e^2} \frac{m_e}{m_a \tau} \int_0^\infty f_0 v_e^4 \, \mathrm{d}v_e, \tag{4.2.22}$$

leading to

$$\bar{v}_{ex} = \frac{3(\widehat{T}_e - \widehat{T}_a)}{m_a \tau F_e} - \tau F_e, \tag{4.2.23}$$

because, by definition, the following relations hold true:

$$\bar{v}_{ex} = \frac{1}{n_e} \int f v_{ex} \, \mathrm{d}\mathbf{v}_e = \frac{4\pi F_e}{3 n_e} \int_0^\infty f_1 v_e^4 \, \mathrm{d}v_e. \tag{4.2.24}$$

Using expressions (4.2,21) and (4.2.23), we achieve the result sought:

$$\bar{\upsilon}_{ex} = \tau F_e. \tag{4.2.25}$$

Comparing relations (4.2.21) and (4.2.25) with known classical results (Smirnov, 1985, p. 108) suggests that in the limiting case $\upsilon = const$ the drift velocity remains unchanged and that $\widehat{T}_e$ increases (the classical analogue of Eq. (4.2.21) contains the numerical coefficient 1/2 instead of 2/3). In concluding the discussion of this limiting case, we present the corresponding form of Eq. (4.2.14) $m_e \ll m_a$ for $\check{\varepsilon} \gtrsim 1$:

$$\check{\upsilon}_e^2 \frac{\mathrm{d}^3 \check{f}_0}{\mathrm{d}\check{\upsilon}_e^3} + \left(2 + 3\frac{m_e}{m_a}\check{\upsilon}_e^2\right)\check{\upsilon}_e \frac{\mathrm{d}^2 \check{f}_0}{\mathrm{d}\check{\upsilon}_e^2} - \left(2 + \frac{10}{3}\check{\upsilon}_e^2\right)\frac{\mathrm{d}\check{f}_0}{\mathrm{d}\check{\upsilon}_e} - 5\frac{m_e}{m_a}\check{\upsilon}_e^3 \check{f}_0 = 0. \tag{4.2.26}$$

As a check on the correctness of the above results, note that if $F_e \equiv 0$ then Eq. (4.2.11) leads, as it should, to the Maxwellian distribution function f_{0M}:

$$\frac{\mathrm{d}f_0}{\mathrm{d}\upsilon_e} = -\frac{m_e \upsilon_e}{\widehat{T}} f_0, \tag{4.2.27}$$

$$f_0 = C\,\mathrm{e}^{-m_e \upsilon_e^2/(2\widehat{T})}. \tag{4.2.28}$$

We proceed now to the second limiting case, $\ell = const$. In this case, the analogue of Eq. (4.2.11) is as follows

$$f_1 \upsilon_e = \left[\tau + \frac{3\widehat{T}\upsilon_e}{F_e^2 m_a \ell}\right]\frac{\mathrm{d}f_0}{\mathrm{d}\upsilon_e} + \frac{3}{F_e^2}\frac{m_e}{m_a}\frac{\upsilon_e^2}{\ell} f_0. \tag{4.2.29}$$

By the same procedure used in the limiting case $\nu = const$, we arrive at the following equation in f_0:

$$\upsilon_e^2\left[\tau + \frac{3\widehat{T}_a \upsilon_e}{F_e^2 m_a \ell}\right]\frac{\mathrm{d}^3 f_0}{\mathrm{d}\upsilon_e^3} + \left[2\tau + \frac{12\widehat{T}_a}{m_a \ell F_e^2}\upsilon_e + \frac{3m_e \upsilon_e^3}{m_a \ell F_e^2}\right]\upsilon_e \frac{\mathrm{d}^2 f_0}{\mathrm{d}\upsilon_e^2} + \left[18\frac{m_e}{m_a \ell F_e^2}\upsilon_e^3 - 2\tau - \frac{5\upsilon_e^2}{3\tau F_e^2} - \frac{5}{3F_e^2 \ell}\upsilon_e^3 - \frac{5\widehat{T}_a}{\tau F_e^4 m_a \ell^2}\upsilon_e^4\right]\frac{\mathrm{d}f_0}{\mathrm{d}\upsilon_e} + \upsilon_e^2\left[12\frac{m_e}{F_e^2 m_a \ell} - 5\frac{m_e}{\tau F_e^4 m_a \ell^2}\upsilon_e^3\right] f_0 = 0. \tag{4.2.30}$$

Again, it is easily seen by multiplying Eq. (4.2.30) term-wise by F_e^4 that the vanishing external force F_e leads to Eq. (4.2.27) and then, upon integration, to the Maxwellian distribution function (4.2.28). The boundary conditions for Eq. (4.2.30) are as follows: f_0 is specified for $\upsilon_e = 0$ in accordance with the chosen normalization; for $\upsilon_e = 0$, as Eq. (4.2.30) suggests, $f_0' = 0$; and, finally, $f_0 \to 0$, when $\upsilon_e \to \infty$.

The last result becomes evident after dividing Eq. (4.2.30) through by v_e^5. In order to numerically integrate Eq. (4.2.30), it is convenient to bring it to the dimensionless form by using the dimensionless quantities

$$\breve{v}_e = \frac{v_e}{\ell/\tau}, \qquad \breve{\varepsilon} = \frac{m_e F_e^2 \tau^2}{\widehat{T}_a}, \qquad \breve{A} = \frac{F_e^2 \tau^4}{\ell^2}, \tag{4.2.31}$$

to give the ordinary differential equation

$$\begin{aligned} &\breve{v}_e^2 \breve{A}\left[1 + 3\frac{m_e}{m_a}\frac{\breve{v}_e}{\breve{\varepsilon}}\right]\frac{\mathrm{d}^3 \breve{f}_0}{\mathrm{d}\breve{v}_e^3} + \left[\breve{A}\left(2 + 12\frac{m_e}{m_a}\frac{\breve{v}_e}{\breve{\varepsilon}}\right) + 3\frac{m_e}{m_a}\breve{v}_e^3\right]\breve{v}_e \frac{\mathrm{d}^2 \breve{f}_0}{\mathrm{d}\breve{v}_e^2} \\ &+ \left[-2\breve{A} - \frac{5}{3}\breve{v}_e^2 - \left(\frac{5}{3} - 18\frac{m_e}{m_a}\right)\breve{v}_e^3 - 5\frac{m_e}{m_a}\frac{\breve{v}_e^4}{\breve{\varepsilon}}\right]\frac{\mathrm{d}\breve{f}_0}{\mathrm{d}\breve{v}_e} \\ &+ \breve{v}_e^2 \frac{m_e}{m_a}\left[12 - 5\frac{\breve{v}_e^3}{\breve{A}}\right]\breve{f}_0 = 0, \end{aligned} \tag{4.2.32}$$

with the boundary conditions

$$\breve{f}_0(0) = 1, \qquad \breve{f}_0'(0) = 0, \qquad \breve{f}_0(\infty) = 0. \tag{4.2.33}$$

Here the term-by-term integration no longer leads to elegant results like Eqs. (4.2.21) and (4.2.25). We can, however, give a useful formula for computing the drift velocity $\bar{v}_{ex}$, which is obtained from Eq. (4.2.32) by multiplying it by v_e^3 and then integrating, to yield

$$\begin{aligned} \bar{v}_{ex} = \frac{4\pi F_e}{3n_e}\Bigg\{ &\frac{18}{5}\frac{\widehat{T}_a \tau}{m_a}\int_0^\infty f_0 \,\mathrm{d}v_e + C\frac{6}{5}\tau^2 \ell F_e^2 + 2\ell \int_0^\infty f_0 v_e \,\mathrm{d}v_e \\ &- \frac{9}{10}\frac{m_e}{\pi m_a}\tau n_e\Bigg\}, \end{aligned} \tag{4.2.34}$$

with $C = f_0\ (v_e = 0)$.

Although Eq. (4.2.34) can of course be used only after numerically integrating Eq. (4.2.33), it is of interest to note that, unlike Eq. (4.2.25), the drift velocity is a nonlinear function of F_e in this limiting case. Let us consider here some numerical results for the distribution function of charged particles in an external electric field, produced when employing the generalized Boltzmann equation. Numerical integration of corresponding differential equations was realized by the three-diagonal method of Gauss elimination techniques for the differential second-order equation (see Appendix 4).

In Figure 4.1, the dimensionless distribution function $\breve{f}_0$ is plotted versus the dimensionless velocity $\breve{v}_e$ for $\breve{\varepsilon} = 10^{-3}$ and $\tau = const$. The curve 1 corresponds to the Maxwellian distribution function, and the curve 2 to the distribution function obtained using the GBE. As $\breve{\varepsilon}$ is decreasing, the two distributions approach each other. Note that the function $\breve{f}_0^{\mathrm{GBE}}$ lies above the Maxwellian function.

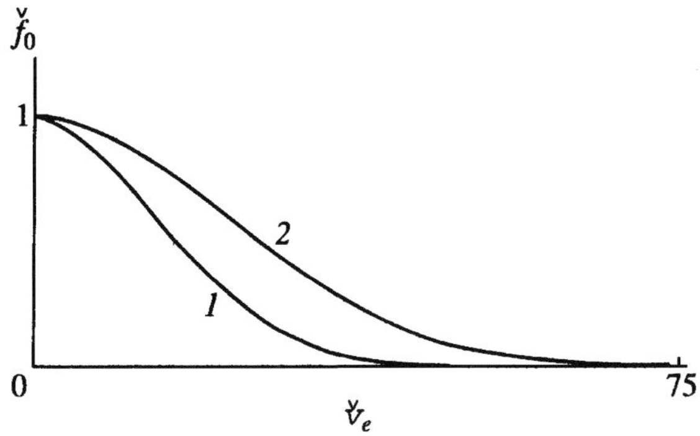

Fig. 4.1. Dependence of $\check{f}_0$ on $\check{v}_e$ for $\tau = const$: 1, Maxwellian distribution function; 2, $\check{f}_0^{\mathrm{GBE}}$.

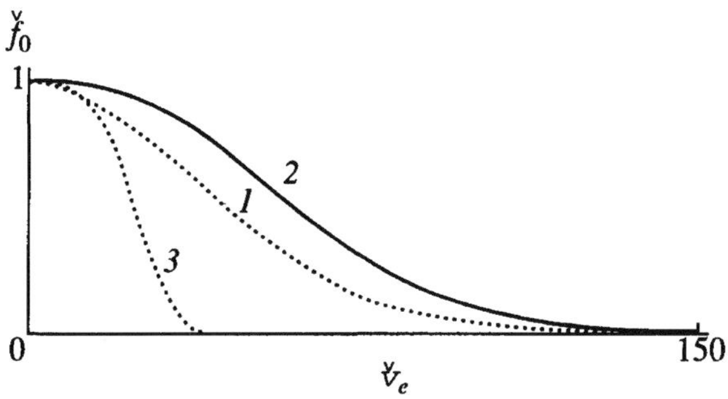

Fig. 4.2. Dependence of $\check{f}_0$ on $\check{v}_e$ for $\ell = const$, $\check{\varepsilon} = 10^{-2}$, $\check{A} = 1$: 1, Maxwellian distribution function; 2, $\check{f}_0^{\mathrm{GBE}}$; 3, Druyvesteyn distribution function.

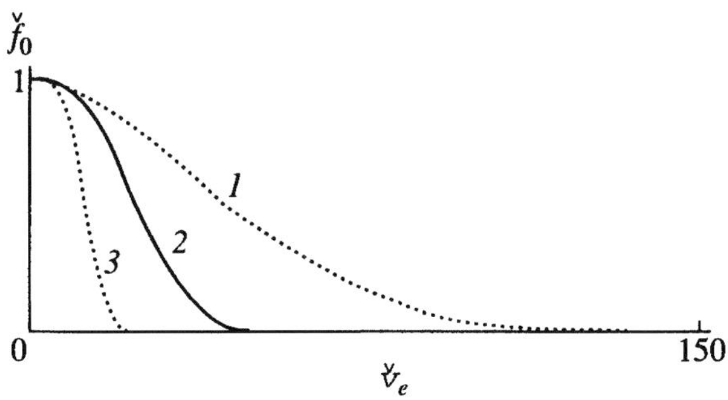

Fig. 4.3. Dependence of $\check{f}_0$ on $\check{v}_e$ for $\ell = const$, $\check{\varepsilon} = 10^{-2}$, $\check{A} = 10^{-1}$: 1, Maxwellian distribution function; 2, $\check{f}_0^{\mathrm{GBE}}$; 3, Druyvesteyn distribution function.

Figures 4.2 and 4.3 present $\check{f}_0$, calculated in the case of $\ell = const$ under the conditions $\check{\varepsilon} = 10^{-2}$, $\check{A} = 1$, and $\check{\varepsilon} = 10^{-2}$, $\check{A} = 10^{-1}$, respectively. The curves 1, 2, and 3 in Figures 4.2 and 4.3 correspond to the Maxwellian distribution function, the generalized Boltzmann equation, and the Druyvesteyn distribution function, respectively. It is interesting to note that the distribution function $\check{f}_0^{\mathrm{GBE}}$ may lie between the Maxwell and Druyvesteyn functions, as well as above the two. In practical computations, to reemphasize, the distribution functions can be normalized to the number density of the charged particles involved.

4.3. Charged particles in alternating electric field

As another example of the application of the GBE, let us consider the time evolution of the DF of charged particles moving in an alternating electric field. In this problem, only elastic collisions will be considered; the GBE takes the form

$$\left(\frac{\partial f_e}{\partial t} + \mathbf{F}_e \cdot \frac{\partial f_e}{\partial \mathbf{v}_e}\right)\left(1 - \frac{\partial \tau}{\partial t}\right) - \tau\left\{\frac{\partial^2 f_e}{\partial t^2} + 2\mathbf{F}_e \cdot \frac{\partial^2 f_e}{\partial \mathbf{v}_e \partial t}\right.$$
$$\left. + \frac{\partial \mathbf{F}_e}{\partial t} \cdot \frac{\partial f_e}{\partial \mathbf{v}_e} + \frac{\partial^2 f_e}{\partial \mathbf{v}_e \partial \mathbf{v}_e} : \mathbf{F}_e\mathbf{F}_e\right\} = J_{ea}. \tag{4.3.1}$$

If we make use of the expansion

$$f_e(\mathbf{v}_e, t) = f_0(v_e, t) + \mathbf{F}_e \cdot \mathbf{v}_e f_1(v_e, t) + \mathbf{F}_e\mathbf{F}_e : \mathbf{v}_e^0 \mathbf{v}_e f_2(v_e, t), \tag{4.3.2}$$

then utilizing a procedure analogous to that described above we arrive at the following equations in the functions f_0 and f_1:

$$\left(1 - \frac{\partial \tau}{\partial t}\right)\left[\frac{\partial f_0}{\partial t} + F_e^2 f_1 + \frac{1}{3}F_e^2 v_e \frac{\partial f_1}{\partial v_e}\right] - \tau\left[\frac{\partial^2 f_0}{\partial t^2} + \frac{1}{3}F_e^2 \frac{\partial^2 f_0}{\partial v_e^2}\right.$$
$$+ \frac{2}{3}F_e^2 \frac{1}{v_e}\frac{\partial f_0}{\partial v_e} + \frac{1}{2}f_1 \frac{\partial F_e^2}{\partial t} + 2F_e^2 \frac{\partial f_1}{\partial t} + \frac{1}{2}\frac{\partial F_e^2}{\partial t} v_e \frac{\partial f_1}{\partial v_e}$$
$$\left. + \frac{2}{3}F_e^2 v_e \frac{\partial^2 f_1}{\partial v_e \partial t}\right] = J_{ea}, \tag{4.3.3}$$

$$\left(1 - \frac{\partial \tau}{\partial t}\right)\left[F_e \frac{\partial f_0}{\partial v_e} + v_e \frac{\partial F_e}{\partial t} f_1 + F_e v_e \frac{\partial f_1}{\partial t}\right]$$
$$- \tau\left[2F_e \frac{\partial^2 f_0}{\partial v_e \partial t} + \frac{\partial F_e}{\partial t}\frac{\partial f_0}{\partial v_e} + 2v_e \frac{\partial F_e}{\partial t}\frac{\partial f_1}{\partial t} + v_e \frac{\partial^2 F_e}{\partial t^2} f_1 + F_e v_e \frac{\partial^2 f_1}{\partial t^2}\right.$$
$$\left. + \frac{12}{5}F_e^3 \frac{\partial f_1}{\partial v_e} + \frac{3}{5}F_e^3 v_e \frac{\partial^2 f_1}{\partial v_e^2}\right] = -\frac{3}{2}J_1, \tag{4.3.4}$$

where, for example,

$$J_1 = \frac{2}{3}F_e f_1 \frac{v_e^2}{\ell}.$$

Now consider a case in which the mean time between collisions (and then $\nu = v_e/\ell$ which is used for J_{ea} and J_1 approximations), is independent of the velocity. It proves possible to determine the distribution function moments $\bar{v}_{ex}$ and $\widehat{T}_e$ without directly solving Eqs. (4.3.3) and (4.3.4). Multiply Eq. (4.3.4) term-wise by v_e^3 and integrate the resulting expression over all absolute velocities. Using the additional conditions

$$\int_0^\infty f_1 v_e^4 \, \mathrm{d}v_e = \frac{3n_e}{4\pi F_e}\bar{v}_{ex}, \qquad \int_0^\infty f_0 v_e^2 \, \mathrm{d}v_e = \frac{n_e}{4\pi},$$

$$\int_0^\infty f_0 v_e^4 \, \mathrm{d}v_e = \frac{3}{2} n_e \frac{\widehat{T}_e}{m_e},$$

we derive the following equation:

$$\tau \frac{\mathrm{d}^2 \bar{v}_{ex}}{\mathrm{d}t^2} - \frac{\mathrm{d}\bar{v}_{ex}}{\mathrm{d}t} - \frac{1}{\tau}\bar{v}_{ex} + F_e - \tau \frac{\mathrm{d}F_e}{\mathrm{d}t} = 0. \tag{4.3.5}$$

Suppose that the time dependence of the external force can be represented as $F_e = (eE_0/m_e)\cos\omega t$, where the frequency ω is related to the external electric field strength. The solution of the nonhomogeneous differential equation (4.3.5) is

$$\begin{aligned}\bar{v}_{ex}^{\mathrm{GBE}} &= C_1 \,\mathrm{e}^{-at/\tau} \\ &\quad + \frac{b\tau}{(\omega\tau)^4 + 3(\omega\tau)^2 + 1}\left[\cos\omega t + \omega\tau\left(2 + \omega^2\tau^2\right)\sin\omega t\right],\end{aligned} \tag{4.3.6}$$

where $b = eE_0/m_e$, and C_1 is the constant of integration which is determined by the initial conditions of the problem. The classical result which can be obtained from the BE for the quasi-stationary case is given by

$$\bar{v}_{ex}^{\mathrm{BE}} = \frac{b\tau}{(\omega\tau)^2 + 1}[\cos\omega t + \omega t \sin\omega t]. \tag{4.3.7}$$

The introduction of the mobility K usually defined through the expression

$$\bar{v}_{ex} = K\frac{m_e}{e}F_e$$

would serve no purpose due to singularities that can appear for $F_e = 0$.

We now turn our attention to Eq. (4.3.3). We multiply this equation term-wise by v_e^4 and integrate the resulting expression over all v_e:

$$\begin{aligned}&\frac{\mathrm{d}^2\widehat{T}_e}{\mathrm{d}t^2} - \frac{1}{\tau}\frac{\mathrm{d}\widehat{T}_e}{\mathrm{d}t} - \frac{2m_e}{m_a\tau^2}\left(\widehat{T}_e - \widehat{T}_a\right) \\ &\quad = -\frac{2}{3}m_e\frac{F_e}{\tau}\bar{v}_{ex} - \frac{2}{3}m_e F_e^2 + \frac{2}{3}m_e\bar{v}_{ex}\frac{\mathrm{d}F_e}{\mathrm{d}t} + \frac{4}{3}m_e F_e\frac{\mathrm{d}\bar{v}_{ex}}{\mathrm{d}t},\end{aligned} \tag{4.3.8}$$

where $F_e = b \sin \omega t$ and consequently,

$$\bar{v}_{ex} = C_1 \,\mathrm{e}^{-at/\tau} + \frac{b\tau}{(\omega\tau)^4 + 3(\omega\tau)^2 + 1}\big\{\sin \omega t - \big(2 + (\omega\tau)^2\big)\omega\tau \cos \omega t\big\}. \tag{4.3.9}$$

The differential equation (4.3.8) integrates in the finite form to the following expression:

$$\begin{aligned}
\widehat{T}_{ea} &= C_1 \,\mathrm{e}^{-dt/\tau} \\
&\quad + \mathrm{e}^{-at/\tau} \frac{\tau^2 Z}{[(d+a+1)^2 + \omega^2\tau^2][(d-a)^2 + \omega^2\tau^2]} \big\{\sin \omega t \big[\omega^2\tau^2 \big(2\sqrt{5} \\
&\quad - 2(d+a+1) + \omega^2\tau^2(\sqrt{5} - 1 - d - a)\big) + (d-a)\big(2\omega^2\tau^2 + \omega^4\tau^4 \\
&\quad + \omega^2\tau^2\sqrt{5}(d+a+1) + 2\sqrt{5}(d+a+1)\big)\big] \\
&\quad + \cos \omega t \big[(d-a)\omega\tau \big(2\sqrt{5} + \omega^2\tau^2(\sqrt{5} - 1 - d - a) - 2(d+a+1)\big) \\
&\quad - \omega\tau \big(2\omega^2\tau^2 + \omega^4\tau^4 + \omega^2\tau^2\sqrt{5}(d+a+1) + 2\sqrt{5}(d+a+1)\big)\big]\big\} \\
&\quad + Z\tau^2 \bigg\{ \frac{\sin^2 \omega t}{2(d^2 + 4\omega^2\tau^2)[(d+1)^2 + 4\omega^2\tau^2]} \big[4d(d+1) \\
&\quad - 2d(d+1)\omega^2\tau^2 - 4\omega^2\tau^2 - 2d(d+1)\omega^4\tau^4 - 28\omega^4\tau^4 - 16\omega^6\tau^6\big] \\
&\quad + \cos^2 \omega t \frac{\omega^2\tau^2(2 + \omega^2\tau^2) d(d+1)}{[(d+1)^2 + 4\omega^2\tau^2][d^2 + 4\omega^2\tau^2]} \\
&\quad - \omega\tau \cos \omega t \sin \omega t \frac{\omega^2\tau^2(d^2 + d + 14) + 4d + 4}{(d^2 + 4\omega^2\tau^2)((d+1)^2 + 4\omega^2\tau^2)} \\
&\quad + \omega^2\tau^2\big[9d^2 + 9d + 4 + \omega^2\tau^2\big(12d^2 + 10d + 18\big) \\
&\quad + 4\omega^4\tau^4\big(d^2 + d + 2\big)\big]\big[d(d+1)\big((d+1) + 4\tau^2\omega^2\big) \\
&\quad \times \big(d^2 + 4\tau^2\omega^2\big)\big]^{-1} \bigg\},
\end{aligned} \tag{4.3.10}$$

where the notation used is:

$$a = \frac{\sqrt{5} - 1}{2}, \qquad d = \frac{1}{2}\left[\sqrt{1 + 8\frac{m_e}{m_a}} - 1\right] \cong 2\frac{m_e}{m_a} \ll 1,$$

$$Z = \frac{2}{3} \frac{m_e b^2}{\omega^4\tau^4 + 3\omega^2\tau^2 - 1}.$$

In the quasi-stationary limiting case, under the condition $\omega\tau \gg 1$, one finds

$$\widehat{T}_{ea} = -\tau^2 Z \frac{\omega^2\tau^2}{2} \sin^2 \omega t + \tau^2 Z \frac{\omega^2\tau^2}{2d}, \tag{4.3.11}$$

or, taking into account that the time average

$$\overline{\sin^2 \omega t} = \frac{1}{2} \ll d^{-1},$$

we arrive at

$$\widehat{T}_e = \frac{m_a}{6}\left(\frac{eE_0}{m_e\omega}\right)^2. \tag{4.3.12}$$

Thus, in the limiting case $l = const$, the following equality holds true:

$$\widehat{T}_e^{\mathrm{BE}} = \widehat{T}_e^{\mathrm{GBE}}. \tag{4.3.13}$$

In the opposite limit of $\omega\tau \ll d \ll 1$, one has

$$\widehat{T}_{ea} = \tau^2 Z \frac{2}{d} \sin^2 \omega t$$

or, computing the average over the time, we obtain

$$\widehat{T}_{ea} = \frac{m_a \tau^2}{3}\left(\frac{eE_0}{m_e}\right)^2, \tag{4.3.14}$$

i.e., a result which corresponds to the solution of the classical Boltzmann equation (Smirnov, 1985).

4.4. Conductivity of a weakly ionized gas in crossed electric and magnetic fields

In this section, the conductivity of a weakly ionized gas subject to crossed magnetic and (alternating) electric field will be examined using the GBE with the BGK approximation for the elastic collision integral. The BGK-approximated kinetic equation takes the form

$$\frac{\partial f_e}{\partial t} + \mathbf{F}_e \cdot \frac{\partial f_e}{\partial \mathbf{v}_e} - \tau\left[\frac{\partial^2 f_e}{\partial t^2} + 2\mathbf{F}_e \cdot \frac{\partial^2 f_e}{\partial \mathbf{v}_e \partial t} + \frac{\partial \mathbf{F}_e}{\partial t} \cdot \frac{\partial f_e}{\partial \mathbf{v}_e}\right.$$
$$\left. + \frac{\partial^2 f_e}{\partial \mathbf{v}_e \partial \mathbf{v}_e} : \mathbf{F}_e\mathbf{F}_e + \frac{\partial f_e}{\partial \mathbf{v}_e}\mathbf{F}_e : \frac{\partial}{\partial \mathbf{v}_e}\mathbf{F}_e\right] = -\frac{f_e - f_e^{(0)}}{\tau}, \tag{4.4.1}$$

where $\mathbf{F}_e = \mathbf{F}_e^{(1)} + \mathbf{F}_B$ is the Lorentz force which, in our case, includes the effect of the alternating electric field

$$\mathbf{F}_e^{(1)} = \frac{e\mathbf{E}^0}{m_e} \mathrm{e}^{\mathrm{i}\omega t},$$

directed along the x-axis, and of the static magnetic field, whose induction is along the z-axis. The equation of motion (4.1.32) reduces to the form

$$\frac{\partial}{\partial t}\left\{\bar{\mathbf{v}}_e - \tau\left[\frac{\partial \bar{\mathbf{v}}_e}{\partial t} - \frac{e\mathbf{E}^0}{m_e}\mathrm{e}^{\mathrm{i}\omega t} - \frac{e}{m_e}\bar{\mathbf{v}}\times\mathbf{B}\right]\right\} - \frac{e\mathbf{E}^0}{m_e}\mathrm{e}^{\mathrm{i}\omega t}$$
$$- \frac{e}{m_e}\left\{\bar{\mathbf{v}}_e - \tau\left[\frac{\partial \bar{\mathbf{v}}_e}{\partial t} - \frac{e\mathbf{E}^0}{m_e}\mathrm{e}^{\mathrm{i}\omega t} - \frac{e}{m_e}\bar{\mathbf{v}}\times\mathbf{B}\right]\right\}\times\mathbf{B} = -\frac{\bar{\mathbf{v}}_e}{\tau}. \tag{4.4.2}$$

The components of the drift velocity $\bar{\mathbf{v}}_e$ along the axes x and y are determined by the following set of equations ($\bar{v}_{ez} = 0$):

$$\frac{\partial}{\partial t}\left\{\bar{v}_{ex} - \tau\left[\frac{\partial \bar{v}_{ex}}{\partial t} - \frac{eE^0}{m_e}\mathrm{e}^{\mathrm{i}\omega t} - \frac{e}{m_e}\bar{v}_{ey}B\right]\right\} - \frac{eE^0}{m_e}\mathrm{e}^{\mathrm{i}\omega t}$$
$$- \frac{e}{m_e}\left\{B\bar{v}_{ey} - \tau B\frac{\partial \bar{v}_{ey}}{\partial t}\right\} = -\frac{\bar{v}_{ex}}{\tau} - \frac{e^2\tau}{m_e^2}B^2\bar{v}_{ex}, \tag{4.4.3}$$

$$\frac{\partial}{\partial t}\left\{\bar{v}_{ey} - \tau\left[\frac{\partial \bar{v}_{ey}}{\partial t} + \frac{e}{m_e}\bar{v}_{ex}B\right]\right\} - \frac{e}{m_e}\left\{-\bar{v}_{ex}B + \tau B\frac{\partial \bar{v}_{ex}}{\partial t} - \frac{eE^0}{m_e}\mathrm{e}^{\mathrm{i}\omega t}\tau B\right\}$$
$$= -\frac{\bar{v}_{ey}}{\tau} - \frac{e^2\tau}{m_e^2}B^2\bar{v}_{ey}. \tag{4.4.4}$$

The solution to Eq. (4.4.2) is naturally sought in the form $\bar{v} = \bar{v}^0\,\mathrm{e}^{\mathrm{i}\omega t}$, thus leading to the following system of algebraic equations

$$\bar{v}_{ex}^0\left[\mathrm{i}\omega + \tau\left(\omega^2 + \omega_B^2\right) + \frac{1}{\tau}\right] = \frac{eE^0}{m_e} + \bar{v}_{ey}^0\omega_B[1 - 2\mathrm{i}\omega\tau] - \mathrm{i}\omega\tau\frac{eE^0}{m_e}, \tag{4.4.5}$$

$$\bar{v}_{ey}^0\left[\mathrm{i}\omega + \tau\left(\omega^2 + \omega_B^2\right) + \frac{1}{\tau}\right] = \bar{v}_{ex}^0\omega_B[2\mathrm{i}\omega\tau - 1] - \omega_B\tau\frac{eE^0}{m_e}, \tag{4.4.6}$$

where $\omega_B = eB/m_e$.

From these equations it is not difficult to find the components $\bar{v}_{ex}^0$ and $\bar{v}_{ey}^0$ of the drift velocity, and hence to determine the components of the electrical conductivity tensor. In our case, the complex conductivity σ_x assumes the form

$$\sigma_x = \sigma_0\frac{1 + 2\omega^2\tau^2 + \mathrm{i}(\omega_B^2\tau^2 - \omega^2\tau^2)\omega\tau}{1+\omega^2\tau^2+\omega^4\tau^4+3\omega_B^2\tau^2+\omega_B^4\tau^4-2\omega^2\tau^2\omega_B^2\tau^2+2\mathrm{i}\omega\tau(1+\omega^2\tau^2-\omega_B^2\tau^2)}, \tag{4.4.7}$$

where we have used the notation:

$$\sigma_x = \frac{en_e\bar{v}_{ex}^0}{E^0}, \qquad \sigma_0 = \frac{n_e e^2\tau}{m_e}.$$

Separating the real part of σ_x now yields

$$\begin{aligned}\mathrm{Re}\,\sigma_x &= \sigma_0\left\{1+3\omega^2\tau^2+\omega^4\tau^4+\omega_B^2\tau^2\left[3+6\omega^2\tau^2+\omega_B^2\tau^2\right]\right\}\\ &\times\left\{\left[1+\omega^2\tau^2+\omega^4\tau^4+\omega_B^2\tau^2(3+\omega_B^2\tau^2-2\omega^2\tau^2)\right]^2\right.\\ &\left.+4\omega^2\tau^2[1+\omega^2\tau^2-\omega_B^2\tau^2]^2\right\}^{-1}. \qquad (4.4.8)\end{aligned}$$

The Boltzmann theory, as is known, leads to the following results

$$\mathrm{Re}\,\sigma_x = \sigma_0\frac{1+\omega^2\tau^2+\omega_B^2\tau^2}{1+2\omega^2\tau^2+\omega^4\tau^4+\omega_B^2\tau^2[2-2\omega^2\tau^2+\omega_B^2\tau^2]}, \qquad (4.4.9)$$

$$\sigma_x = \sigma_0\frac{1+\mathrm{i}\omega\tau}{1+(\omega_B^2-\omega^2)\tau^2+2\mathrm{i}\omega\tau}. \qquad (4.4.10)$$

For the traditionally considered limiting cases, Eqs. (4.4.7)–(4.4.10) give the following results (Alekseev, 1995a):

(a) $\omega = 0$, a constant electric field:

$$\mathrm{Re}\,\sigma_x^{\mathrm{GBE}} = \sigma_0\frac{1}{1+3\omega_B^2\tau^2+\omega_B^4\tau^4}, \qquad (4.4.11)$$

$$\mathrm{Re}\,\sigma_x^{\mathrm{BE}} = \sigma_0\frac{1}{1+\omega_B^2\tau^2}; \qquad (4.4.12)$$

(b) $\omega_B = 0$, no magnetic field:

$$\begin{aligned}\mathrm{Re}\,\sigma_x^{\mathrm{GBE}} &= \sigma_0\frac{1+3\omega^2\tau^2+\omega^4\tau^4}{[1+\omega^2\tau^2+\omega^4\tau^4]^2+4\omega^2\tau^2[1+\omega^2\tau^2]^2}\\ &= \frac{1}{1+3\omega^2\tau^2+\omega^4\tau^4}, \qquad (4.4.13)\end{aligned}$$

$$\mathrm{Re}\,\sigma_x^{\mathrm{BE}} = \sigma_0\frac{1}{1+\omega^2\tau^2}; \qquad (4.4.14)$$

(c) $\omega = \omega_B$,

$$\mathrm{Re}\,\sigma_x^{\mathrm{GBE}} = \sigma_0\frac{1+6\omega^2\tau^2+8\omega^4\tau^4}{1+12\omega^2\tau^2+16\omega^4\tau^4}, \qquad (4.4.15)$$

$$\mathrm{Re}\,\sigma_x^{\mathrm{BE}} = \sigma_0\frac{1+2\omega^2\tau^2}{1+4\omega^2\tau^2}; \qquad (4.4.16)$$

(d) $\omega = \omega_B$, $\omega\tau \gg 1$, the cyclotron resonance condition:

$$\mathrm{Re}\,\sigma_x^{\mathrm{GBE}} = \mathrm{Re}\,\sigma_x^{\mathrm{BE}} = \frac{1}{2}\sigma_0. \qquad (4.4.17)$$

Finally, from the system of Eqs. (4.4.5), (4.4.6) the drift velocity along the y-axis is found as

$$\bar{v}_{ey}^{0} = \frac{eE^{0}}{m_e}\omega_B\tau^2\big[\big(\omega^4-\omega_B^4\big)\tau^4 - 3\big(\omega^2+\omega_B^2\big)\tau^2 - 2 + \mathrm{i}\omega\tau\big(3\omega^2\tau^2+\omega_B^2\tau^2\big)\big]D^{-1}, \tag{4.4.18}$$

where

$$D = 1 + \big(\omega^6+\omega_B^6\big)\tau^6 + 4\omega_B^2\tau^2 + 4\omega^2\tau^2\omega_B^2\tau^2 - \omega^4\tau^4\omega_B^2\tau^2 + 4\omega_B^4\tau^4 - \omega^2\tau^2\omega_B^4\tau^4 + \mathrm{i}\omega\tau\big[3\omega^4\tau^4 - 5\omega^2\tau^2 + 3 + 3\omega_B^2\tau^2 - 2\omega^2\tau^2\omega_B^2\tau^2 - \omega_B^4\tau^4\big].$$

Notice that the BE implies that

$$\bar{v}_{ey}^{0} = -\frac{eE^{0}}{m_e}\omega_B\tau^2\frac{1}{1+(\omega_B^2-\omega^2)\tau^2+2\mathrm{i}\omega\tau}. \tag{4.4.19}$$

In particular, it follows that

(a) $\omega = 0$,

$$\bar{v}_{ey}^{0\ \mathrm{GBE}} = -\frac{eE^{0}}{m_e}\omega_B\tau^2\frac{\omega_B^4\tau^4+3\omega_B^2\tau^2+2}{\omega_B^6\tau^6+4\omega_B^4\tau^4+4\omega_B^2\tau^2+1}, \tag{4.4.20}$$

$$\bar{v}_{ey}^{0\ \mathrm{BE}} = -\frac{eE^{0}}{m_e}\omega_B\tau^2\frac{1}{\omega_B^2\tau^2+1}; \tag{4.4.21}$$

(b) $\omega = \omega_B$,

$$\mathrm{Re}\,\bar{v}_{ey}^{0\ \mathrm{GBE}} = -\frac{eE^{0}}{m_e}\omega\tau^2\frac{2+14\omega^2\tau^2+28\omega^4\tau^4+56\omega^6\tau^6}{[1+4\omega^2\tau^2+8\omega^4\tau^4]^2+\omega^2\tau^2[3-2\omega^2\tau^2]^2}, \tag{4.4.22}$$

$$\mathrm{Re}\,\bar{v}_{ey}^{0\ \mathrm{BE}} = -\frac{eE^{0}}{m_e}\omega\tau^2\frac{1}{1+4\omega^2\tau^2}. \tag{4.4.23}$$

The calculations show that while the BE and GBE results may happen to be identical, they may also be significantly different, both qualitatively and quantitatively. The question of exactly how significantly can only be answered through the solution of concrete problems. In particular, the generalized Boltzmann equation has been applied successfully to transport processes in a partially ionized gas of inelastic colliding particles (Alekseev, Lebedev and Michailov, 1997).

Table 4.1
Calculation of $\mathrm{Re}\,\sigma_x^{\mathrm{GBE}}/\mathrm{Re}\,\sigma_x^{\mathrm{BE}}$ for $\omega = 0$

$\omega_B\tau$	$\mathrm{Re}\,\sigma_x^{\mathrm{GBE}}/\mathrm{Re}\,\sigma_x^{\mathrm{BE}}$
0	1
1	0.4
2	0.172
3	0.0917
4	0.0557
5	0.0371
6	0.0263
7	0.0196
8	0.0152
9	0.0120
10	0.0098

Table 4.1 contains the calculation of ratio of conductivities $\mathrm{Re}\,\sigma_x^{\mathrm{GBE}}/\mathrm{Re}\,\sigma_x^{\mathrm{BE}}$ for the case "a" ($\omega = 0$, a constant electric field):

$$\frac{\mathrm{Re}\,\sigma_x^{\mathrm{GBE}}}{\mathrm{Re}\,\sigma_x^{\mathrm{BE}}} = \frac{1+\omega_B^2\tau^2}{1+3\omega_B^2\tau^2+\omega_B^4\tau^4}. \tag{4.4.24}$$

From Eqs. (4.4.13), (4.4.14) it follows that for the case "b" (no magnetic field), the ratio of conductivities can be written as

$$\frac{\mathrm{Re}\,\sigma_x^{\mathrm{GBE}}}{\mathrm{Re}\,\sigma_x^{\mathrm{BE}}} = \frac{1+\omega^2\tau^2}{1+3\omega^2\tau^2+\omega^4\tau^4}, \tag{4.4.25}$$

and dependence $\mathrm{Re}\,\sigma_x^{\mathrm{GBE}}/\mathrm{Re}\,\sigma_x^{\mathrm{BE}}$ as a function of $\omega\tau$ has the same character reflected in Table 4.1.

Table 4.2 corresponds to the case "b", the coincidence of frequencies $\omega = \omega_B$.

In Tables 4.3, 4.4 the calculation of ratio of drift velocities $\mathrm{Re}\,\bar{v}_{ey}^{0\ \mathrm{GBE}}/\mathrm{Re}\,\bar{v}_{ey}^{0\ \mathrm{BE}}$ across magnetic field is presented $\mathrm{Re}\,\bar{v}_{ey}^{0\ \mathrm{GBE}}/\mathrm{Re}\,\bar{v}_{ey}^{0\ \mathrm{BE}}$ in constant electric field ($\omega = 0$) and in electromagnetic field by the condition $\omega = \omega_B$. In these cases, correspondingly,

$$\frac{\bar{v}_{ey}^{0\ \mathrm{GBE}}}{\bar{v}_{ey}^{0\ \mathrm{BE}}} = \frac{\omega_B^6\tau^6+4\omega_B^4\tau^4+5\omega_B^2\tau^2+2}{\omega_B^6\tau^6+4\omega_B^4\tau^4+4\omega_B^2\tau^2+1}, \tag{4.4.26}$$

$$\frac{\mathrm{Re}\,\bar{v}_{ey}^{0\ \mathrm{GBE}}}{\mathrm{Re}\,\bar{v}_{ey}^{0\ \mathrm{BE}}} = \frac{(2+14\omega^2\tau^2+28\omega^4\tau^4+56\omega^6\tau^6)(1+4\omega^2\tau^2)}{(1+4\omega^2\tau^2+8\omega^4\tau^4)^2+\omega^2\tau^2(3-2\omega^2\tau^2)^2}. \tag{4.4.27}$$

The drift velocity $\bar{v}_{ey}^0$ defines diffusion of charged particles in the perpendicular direction to vectors of intensity of electric field and magnetic induction and then the Hall effect. Usually experimental data in the theory of Hall effect are presented with the help

Table 4.2
Calculation of $\mathrm{Re}\,\sigma_x^{\mathrm{GBE}}/\mathrm{Re}_x^{\mathrm{BE}}$ for the case $\omega = \omega_B$

$\omega\tau$	$\mathrm{Re}\,\sigma_x^{\mathrm{GBE}}/\mathrm{Re}\,\sigma_x^{\mathrm{BE}}$	$\omega\tau$	$\mathrm{Re}\,\sigma_x^{\mathrm{GBE}}/\mathrm{Re}\,\sigma_x^{\mathrm{BE}}$
0	1	8	0.996
0.1	0.964	9	0.997
0.2	0.937	10	0.9975
0.3	0.837	11	0.998
0.4	0.807	12	0.9983
0.5	0.8	13	0.9985
0.6	0.805	14	0.9987
0.7	0.817	15	0.9989
0.8	0.832	16	0.999
0.9	0.847	17	0.9991
1	0.862	18	0.9992
2	0.948	19	0.9993
3	0.974	20	0.9994
4	0.985	30	0.9997
5	0.990	40	0.9998
6	0.993	50	0.9999
7	0.995		

Table 4.3
Calculation of $\bar{v}_{ey}^{0\ \mathrm{GBE}}/\bar{v}_{ey}^{0\ \mathrm{BE}}$ in constant electric field ($\omega = 0$)

$\omega_B\tau$	$\bar{v}_{ey}^{0\ \mathrm{GBE}}/\bar{v}_{ey}^{0\ \mathrm{BE}}$	$\omega_B\tau$	$\bar{v}_{ey}^{0\ \mathrm{GBE}}/\bar{v}_{ey}^{0\ \mathrm{BE}}$
0.1	1.971	2	1.034
0.2	1.891	3	1.009
0.3	1.782	4	1.003
0.4	1.664	5	1.001
0.5	1.552	6	1.0007
0.6	1.453	7	1.0004
0.7	1.369	8	1.0002
0.8	1.300	9	1.0001
0.9	1.245	10	1.0001
1	1.2		

of Hall constant. Let us introduce the Hall constant using Eq. (4.4.21). After multiplying both parts of Eq. (4.4.21) by $n_e e$ and using current density $J_{ey} = n_e e v_{ey}^0$ (where e is particle charge), we have

$$J_{ey}\frac{m_e}{e^2 n}\omega_B\left(1+\frac{1}{\omega_B^2\tau^2}\right) = E^0. \tag{4.4.28}$$

Taking into account that $\omega_B = eB/m_e$, from Eq. (4.4.28) can be found

$$E^0 = -\frac{1}{ne}\left(1+\frac{1}{\omega_B^2\tau^2}\right)J_{ey}B. \tag{4.4.29}$$

Table 4.4
Calculation of $\operatorname{Re}\bar{v}_{ey}^{0\ \mathrm{GBE}}/\operatorname{Re}\bar{v}_{ey}^{0\ \mathrm{BE}}$ for the case $\omega=\omega_B$

$\omega\tau$	$\operatorname{Re}\bar{v}_{ey}^{0\ \mathrm{GBE}}/\operatorname{Re}\bar{v}_{ey}^{0\ \mathrm{BE}}$	$\omega\tau$	$\operatorname{Re}\bar{v}_{ey}^{0\ \mathrm{GBE}}/\operatorname{Re}\bar{v}_{ey}^{0\ \mathrm{BE}}$
0.1	1.901	4	3.437
0.2	1.763	5	3.458
0.3	1.747	6	3.471
0.4	1.868	7	3.478
0.5	2	8	3.483
0.6	2.321	9	3.487
0.7	2.542	10	3.489
0.8	2.716	20	3.497
0.9	2.846	30	3.499
1	2.941	40	3.4993
2	3.291	50	3.4996
3	3.393		

But the Hall constant is the coefficient of proportionality between the potential difference and production of current density and magnetic induction, therefore

$$R^{\mathrm{BE}}=\frac{1}{ne}\left(1+\frac{1}{\omega_B^2\tau^2}\right). \tag{4.4.30}$$

The sign of R is defined by the sign of the charge e. Obviously, in this case

$$R^{\mathrm{GBE}}=\frac{1}{ne}\frac{\omega_B^6\tau^6+4\omega_B^6\tau^6+4\omega_B^2\tau^2+1}{\omega_B^2\tau^2(\omega_B^4\tau^4+3\omega_B^2\tau^2+2)}. \tag{4.4.31}$$

From relations (4.4.20), (4.4.21), (4.4.26), (4.4.30), (4.4.31) follow

$$\frac{\bar{v}_{ey}^{0\ \mathrm{GBE}}}{\bar{v}_{ey}^{0\ \mathrm{BE}}}=\frac{R^{\mathrm{BE}}}{R^{\mathrm{GBE}}}. \tag{4.4.32}$$

Therefore, for the case "a", Table 4.3 defines also the ratio of the Hall coefficients R^{BE} and R^{GBE}, obtained in frames of BKT and GBKT.

By the formulated assumptions, the following conclusions can be drawn from Tables 4.1–4.4:

(1) In constant electric field, the decrease of conductivity with increase of $\omega_B\tau$ is being realized in the frame of GBKT much faster than it follows from BKT (more significant effect of confining the ionized gas in magnetic field).
(2) In the absence of magnetic field, the increase of $\omega\tau$ in alternating electric field in the frame of GBKT leads to more significant decrease of conductivity than it follows from BKT.
(3) In the case of equality of frequencies $\omega=\omega_B$, the ratio of conductivities varies non-monotonically with increase of $\omega\tau$.

(4) Transverse drift velocities in electromagnetic fields are larger than it follows from BKT.

The calculations show that while the BE and GBE results may happen to be identical, they may also be significantly different, both qualitatively and quantitatively. The question of exactly how significantly can only be answered through the solution of concrete problems.

CHAPTER 5

Kinetic Coefficients in the Theory of the Generalized Kinetic Equations

5.1. Linearization of the generalized Boltzmann equation

Let us use the dimensionless form of the generalized Boltzmann equation (GBE) convenient for investigation of transport processes in hydrodynamic limit. Define for GBE

$$\frac{Df_\alpha}{Dt} - \frac{D}{Dt}\left(\tau_\alpha \frac{Df_\alpha}{Dt}\right) = \sum_{j=1}^{\eta} \int \left[f'_\alpha f'_j - f_\alpha f_j\right] g_{\alpha j} b \, \mathrm{d}b \, \mathrm{d}\varphi \, \mathrm{d}\mathbf{v}_j, \tag{5.1.1}$$

where

$$\frac{D}{Dt} = \frac{\partial}{\partial t} + \mathbf{v}_\alpha \cdot \frac{\partial}{\partial \mathbf{r}} + \mathbf{F}_\alpha \cdot \frac{\partial}{\partial \mathbf{v}_\alpha}, \tag{5.1.2}$$

are the corresponding scales. Namely, for radius-vector $\mathbf{r}$ the characteristic hydrodynamic length L is used as scale ($M_r = L$); for velocity $\mathbf{v}_\alpha$ the scale is $M_{v_\alpha} = v_{0\lambda,\alpha}$ ($M_{v_\alpha} = \sqrt{\overline{v_\alpha^2}}$), which is consistent with root-mean-square velocity for particles belonging to species α ("α-particles"). In accordance with definition,

$$\overline{v_\alpha^2} = \frac{1}{n_\alpha} \int f_\alpha v_\alpha^2 \, \mathrm{d}\mathbf{v}_\alpha.$$

For distribution function f_α the scale introduced is compatible with its normalization

$$M_{f_\alpha} = \frac{n_{\alpha 0}}{v_{0\lambda,\alpha}^3}, \tag{5.1.3}$$

where $n_{\alpha 0}$ is characteristic value of numerical concentration for α-particles. The scale $G_{\alpha j}$ for module of relative velocity $g_{\alpha j}$ is defined as

$$M_{g_{\alpha j}} = \frac{1}{n_\alpha n_j} \int |\mathbf{v}_\alpha - \mathbf{v}_j| f_\alpha f_j \, \mathrm{d}\mathbf{v}_\alpha \, \mathrm{d}\mathbf{v}_j = G_{\alpha j}, \tag{5.1.4}$$

and the scale M_b for impact parameter b of encountering particles α and j,

$$M_b = \sigma_{\alpha j}\sqrt[4]{1+\frac{m_\alpha}{m_j}}, \quad \sigma_{\alpha j} = \frac{\sigma_\alpha + \sigma_j}{2}. \tag{5.1.5}$$

The rest of scales: $M_{F_\alpha} = F_{0\alpha}$, $M_{\tau_\alpha} = l_\alpha / v_{0\lambda,\alpha}$, where l_α is the mean free path for α-particles, $M_\varphi = \pi$. Rewrite now Eq. (5.1.1) in dimensionless form, marking off the dimensionless values by sign ^.

$$\begin{aligned}
&\frac{v_{0\lambda,\alpha}}{L}\frac{\partial \hat{f}_\alpha}{\partial \hat{t}} + \frac{v_{0\lambda,\alpha}}{L}\hat{\mathbf{v}}_\alpha \cdot \frac{\partial \hat{f}_\alpha}{\partial \hat{\mathbf{r}}} + \frac{F_{0\alpha}}{v_{0\lambda,\alpha}}\widehat{\mathbf{F}}_\alpha \cdot \frac{\partial \hat{f}_\alpha}{\partial \hat{\mathbf{v}}_\alpha} \\
&\quad - \left\{\left(\frac{v_{0\lambda,\alpha}}{L}\frac{\partial}{\partial \hat{t}} + \frac{v_{0\lambda,\alpha}}{L}\hat{\mathbf{v}}_\alpha \cdot \frac{\partial}{\partial \hat{\mathbf{r}}} + \frac{F_{0\alpha}}{v_{0\lambda,\alpha}}\widehat{\mathbf{F}}_\alpha \cdot \frac{\partial}{\partial \hat{\mathbf{v}}_\alpha}\right)\right. \\
&\quad \left.\times\left[\frac{\hat{\tau}_\alpha l_\alpha}{v_{0\lambda,\alpha}}\left(\frac{v_{0\lambda,\alpha}}{L}\frac{\partial \hat{f}_\alpha}{\partial \hat{t}} + \frac{v_{0\lambda,\alpha}}{L}\hat{\mathbf{v}}_\alpha \cdot \frac{\partial \hat{f}_\alpha}{\partial \hat{\mathbf{r}}} + \frac{F_{0\alpha}}{v_{0\lambda,\alpha}}\widehat{\mathbf{F}}_\alpha \cdot \frac{\partial \hat{f}_\alpha}{\partial \hat{\mathbf{v}}}\right)\right]\right\} \\
&= \pi \sum_{j=1}^{\eta} n_{j0}\sigma_{\alpha j}^2 G_{\alpha j}\int \left(\hat{f}'_\alpha \hat{f}'_j - \hat{f}_\alpha \hat{f}_j\right)\hat{g}_{\alpha j}\hat{b}\,\mathrm{d}\hat{b}\,\mathrm{d}\hat{\varphi}\,\mathrm{d}\hat{\mathbf{v}}_j\sqrt{1+\frac{m_\alpha}{m_j}}.
\end{aligned} \tag{5.1.6}$$

After dividing Eq. (5.1.6) term-by-term into $v_{0\lambda,\alpha}/L$, we find

$$\begin{aligned}
&\frac{\partial \hat{f}_\alpha}{\partial \hat{t}} + \hat{\mathbf{v}}_\alpha \cdot \frac{\partial \hat{f}_\alpha}{\partial \hat{\mathbf{r}}} + \frac{F_{0\alpha}L}{v_{0\lambda,\alpha}^2}\widehat{\mathbf{F}}_\alpha \cdot \frac{\partial \hat{f}_\alpha}{\partial \hat{\mathbf{v}}_\alpha} - \left\{\left(\frac{\partial}{\partial \hat{t}} + \hat{\mathbf{v}} \cdot \frac{\partial}{\partial \hat{\mathbf{r}}} + \frac{F_{0\alpha}L}{v_{0\lambda,\alpha}^2}\widehat{\mathbf{F}} \cdot \frac{\partial}{\partial \hat{\mathbf{v}}_\alpha}\right)\right. \\
&\quad \left.\times\left[\hat{\tau}_\alpha \frac{l_\alpha}{L}\left(\frac{\partial \hat{f}_\alpha}{\partial \hat{t}} + \hat{\mathbf{v}}_\alpha \cdot \frac{\partial \hat{f}_\alpha}{\partial \hat{\mathbf{r}}} + \frac{F_{0\alpha}L}{v_{0\lambda,\alpha}^2}\widehat{\mathbf{F}}_\alpha \cdot \frac{\partial \hat{f}_\alpha}{\partial \hat{\mathbf{v}}_\alpha}\right)\right]\right\} \\
&= \frac{\pi L}{v_{0\lambda,\alpha}}\sum_{j=1}^{\eta} n_{j0}\sqrt{1+\frac{m_\alpha}{m_j}}\,\sigma_{\alpha j}^2 G_{\alpha j}\int \left(\hat{f}'_\alpha \hat{f}'_j - \hat{f}_\alpha \hat{f}_j\right)\hat{g}_{\alpha j}\hat{b}\,\mathrm{d}\hat{b}\,\mathrm{d}\hat{\varphi}\,\mathrm{d}\hat{\mathbf{v}}_j.
\end{aligned} \tag{5.1.7}$$

For the hard spheres model in multi-component mixture of gases the following relation is valid:

$$l_\alpha^{-1} = \pi \sum_{j=1}^{\eta} n_j \sigma_{\alpha j}^2 \sqrt{1+\frac{m_\alpha}{m_j}}. \tag{5.1.8}$$

Let us introduce Knudsen numbers

$$Kn_{\alpha j}^{-1} = \frac{L}{l_{\alpha j}} = \pi L n_j \sigma_{\alpha j}^2 \sqrt{1+\frac{m_\alpha}{m_j}}, \tag{5.1.9}$$

$$Kn_\alpha^{-1} = \frac{L}{l_\alpha} = \pi L \sum_{j=1}^{\eta} n_j \sigma_{\alpha j}^2 \sqrt{1+\frac{m_\alpha}{m_j}} = \sum_{j=1}^{\eta} Kn_{\alpha j}^{-1}, \tag{5.1.10}$$

and rewrite Eq. (5.1.7) in the form

$$\frac{D\hat{f}_\alpha}{D\hat{t}} - \frac{D}{D\hat{t}}\left[\hat{\tau}_\alpha Kn_\alpha \frac{D\hat{f}_\alpha}{D\hat{t}}\right] = \sum_j Kn_{\alpha j}^{-1} \frac{G_{\alpha j}}{v_{0\lambda,\alpha}} \int \left[\hat{f}'_\alpha \hat{f}'_j - \hat{f}_\alpha \hat{f}_j\right] \hat{g}_{\alpha j} \hat{b}\, \mathrm{d}\hat{b}\, \mathrm{d}\hat{\varphi}\, \mathrm{d}\hat{\mathbf{v}}_j, \tag{5.1.11}$$

where

$$\frac{D}{D\hat{t}} = \frac{\partial}{\partial \hat{t}} + \hat{\mathbf{v}}_\alpha \cdot \frac{\partial}{\partial \hat{\mathbf{r}}} + \gamma \widehat{\mathbf{F}}_\alpha \cdot \frac{\partial}{\partial \hat{\mathbf{v}}_\alpha}, \quad \gamma = \frac{F_{0\alpha}}{v_{0\lambda,\alpha}^2/L}. \tag{5.1.12}$$

For a mixture of neutral gases, the molecules of which are not strongly different in mass, the scale of relative velocity $G_{\alpha j}$ can be adopted as $v_{0\lambda,\alpha}$. In hydrodynamic limit, when Knudsen numbers $Kn_{\alpha j}$ $(\alpha, j = 1, \dots, \eta)$ are small (or, which is the same, the mean free path between collisions much smaller than characteristic hydrodynamic length L), the perturbation method of the GBE solution can be developed using expansion of distribution function f_α in a power series in Knudsen numbers. Namely, introduce now the small parameter

$$\varepsilon = \max\{Kn_{\alpha j}\} \ll 1 \quad (\alpha, j = 1, \dots, \eta), \tag{5.1.13}$$

in this case $Kn_\alpha^{-1} \geqslant \eta\varepsilon^{-1}$, then

$$Kn_\alpha \leqslant \frac{\varepsilon}{\eta}, \tag{5.1.14}$$

where η is a number of species in the gas mixture. The mentioned expansion for f_α has the form

$$f_\alpha = \sum_k f_\alpha^{(k)} \varepsilon^k. \tag{5.1.15}$$

Now it is obvious that for obtaining a successive approach the model equation with the large parameter ε^{-1} in the right integral part of equation can be used

$$\frac{Df_\alpha}{Dt} - \frac{D}{Dt}\left(\tau_\alpha \frac{Df_\alpha}{Dt}\right) = \frac{1}{\varepsilon} \sum_{j=1}^{\eta} (f'_\alpha f'_j - f_\alpha f_j) g_{\alpha j} b\, \mathrm{d}b\, \mathrm{d}\varphi\, \mathrm{d}\mathbf{v}_j. \tag{5.1.16}$$

If inequality (5.1.13) is not valid for all components of mixture, further modification of the Chapman–Enskog method for solution of Eq. (5.1.16) needs a correction which can lead to multi-velocities hydrodynamics.

Equating the coefficients of ε^{-1} on both sides of Eq. (5.1.16) leads to the integral relation

$$\sum_{j=1}^{\eta}\int\left(f_{\alpha}^{(0)\prime}f_{j}^{(0)\prime}-f_{\alpha}^{(0)}f_{j}^{(0)}\right)g_{\alpha j}b\,\mathrm{d}b\,\mathrm{d}\varphi\,\mathrm{d}\mathbf{v}_{j}=0, \tag{5.1.17}$$

where $f_{\alpha}^{(0)}$ is a distribution function (DF) corresponding to the state of local equilibrium. Then to the Maxwellian DF

$$f_{\alpha}^{(0)}=n_{\alpha}\left(\frac{m_{\alpha}}{2\pi k_{B}T}\right)^{3/2}\mathrm{e}^{-m_{\alpha}V_{\alpha}^{2}/(2k_{B}T)}, \tag{5.1.18}$$

in Eq. (5.1.18) $\mathbf{V}_{\alpha}$ is thermal velocity of α-component particles, T – temperature of gas mixture. The following approach corresponds to equating the coefficients in ε^{0} of Eq. (5.1.16). In this case two alternative methods exist:

1. The second term on the left-hand side of Eq. (5.1.11), containing the second substantive derivation, is proportional to Knudsen number (therefore to ε), and formally speaking this term is out of the Navier–Stokes approximation. In this case, the linear equation which forms the basis for obtaining kinetic coefficients becomes the standard form

$$\frac{Df_{\alpha}^{(0)}}{Dt}=\sum_{j}\int\left[f_{\alpha}^{(0)\prime}f_{j}^{(1)\prime}+f_{j}^{(0)\prime}f_{\alpha}^{(1)\prime}-f_{\alpha}^{(0)}f_{j}^{(1)}-f_{\alpha}^{(1)}f_{j}^{(0)}\right]\mathrm{d}\omega, \tag{5.1.19}$$

where $\mathrm{d}\omega\equiv g_{\alpha j}b\,\mathrm{d}b\,\mathrm{d}\varphi\,\mathrm{d}\mathbf{v}_{j}$.

Application to Eq. (5.1.19) of the Chapman–Enskog procedure allows to obtain classical expressions for kinetic coefficients (Chapman and Cowling, 1952; Hirschfelder, Curtiss and Bird, 1954). Differences from known results of the Chapman–Enskog method appear on the level of obtaining $f_{\alpha}^{(2)}$, i.e., in the generalized Barnett approach.

2. In the formulated first method in essence the turbulent fluctuations of kinetic coefficients on the Kolmogorov micro-scale (or statistical fluctuations of transport coefficients) are ignored. Taking into account this effect leads to equation

$$\begin{aligned}-\frac{D}{Dt}\left(\tau_{\alpha}^{(0)}\left(\frac{Df_{\alpha}^{(0)}}{Dt}\right)_{\tau=0}\right)+\frac{Df_{\alpha}^{(0)}}{Dt}\\=\sum_{j}\int\left[f_{\alpha}^{(0)\prime}f_{j}^{(1)\prime}+f_{j}^{(0)\prime}f_{\alpha}^{(1)\prime}-f_{\alpha}^{(0)}f_{j}^{(1)}-f_{j}^{(0)}f_{\alpha}^{(1)}\right]\mathrm{d}\omega.\end{aligned} \tag{5.1.20}$$

It is important to notice that the additional term on the left-hand side of the linearized equation does not exceed the limit of approximation and can be calculated using only Maxwellian DF. In particular, $\tau_{\alpha}^{(0)}$ is mean time between collisions of α-particles, calculated for the state of local thermodynamic equilibrium. Moreover, introduction of this term leaves the description on the level of linear Fredholm equations.

For the proof of this affirmation we need to show that arising additional integral equation satisfies the corresponding solubility conditions. With this aim we consider for simplicity one-component gas and obtain the explicit form of $Df_\alpha^{(0)}/Dt$ with taking τ-terms into account.

The next relation is valid in independent variables **r**, **V**, t (Chapman and Cowling, 1952):

$$\frac{Df^{(0)}}{Dt} \equiv \frac{\breve{D}f^{(0)}}{Dt} + \mathbf{V}\cdot\frac{\partial f^{(0)}}{\partial \mathbf{r}} + \left(\mathbf{F} - \frac{D\mathbf{v}_0}{Dt}\right)\cdot\frac{\partial f^{(0)}}{\partial \mathbf{V}} - \frac{\partial f^{(0)}}{\partial \mathbf{V}}\mathbf{V} : \frac{\partial}{\partial \mathbf{r}}\mathbf{v}_0, \tag{5.1.21}$$

where operator is introduced

$$\frac{\breve{D}}{Dt} \equiv \frac{\partial}{\partial t} + \mathbf{v}_0\cdot\frac{\partial}{\partial \mathbf{r}}. \tag{5.1.22}$$

Generalized hydrodynamic Euler equations are written as:

$$\frac{\partial n}{\partial t} = -\frac{\partial}{\partial \mathbf{r}}\cdot(n\mathbf{v}_0) + \int \frac{D}{Dt}\left(\tau^{(0)}\frac{Df^{(0)}}{Dt}\right)d\mathbf{v}, \tag{5.1.23}$$

$$\frac{\partial \mathbf{v}_0}{\partial t} = -\left(\mathbf{v}_0\cdot\frac{\partial}{\partial \mathbf{r}}\right)\mathbf{v}_0 + \mathbf{F} - \frac{1}{\rho}\frac{\partial p}{\partial \mathbf{r}} + \frac{1}{n}\int \mathbf{V}\frac{D}{Dt}\left(\tau^{(0)}\frac{Df^{(0)}}{Dt}\right)d\mathbf{v}, \tag{5.1.24}$$

$$\frac{\partial}{\partial t}\left(\frac{3}{2}nk_BT\right) = -\mathbf{v}_0\cdot\frac{\partial}{\partial \mathbf{r}}\left(\frac{3}{2}nk_BT\right) - \frac{3}{2}nk_BT\frac{\partial}{\partial \mathbf{r}}\cdot\mathbf{v}_0 - p\frac{\partial}{\partial \mathbf{r}}\cdot\mathbf{v}_0 + \int \frac{mV^2}{2}\frac{D}{Dt}\left(\tau^{(0)}\frac{Df^{(0)}}{Dt}\right)d\mathbf{v}. \tag{5.1.25}$$

Substantial derivative $Df^{(0)}/Dt$ in integral terms (5.1.23)–(5.1.25) should be calculated in Euler approach, because terms proportional to $\tau^{(0)2}$ are not taken into account. In (5.1.20) this fact is reflected by introduction of symbolic condition $\tau = 0$ for $[Df^{(0)}/Dt]_{\tau=0}$. Rewrite hydrodynamic equations using operator $\breve{D}/Dt$ from (5.1.22):

$$\frac{\breve{D}n}{Dt} = -n\frac{\partial}{\partial \mathbf{r}}\cdot\mathbf{v}_0 + \int \frac{D}{Dt}\left(\tau^{(0)}\frac{Df^{(0)}}{Dt}\right)d\mathbf{v}, \tag{5.1.26}$$

$$\frac{\breve{D}\mathbf{v}_0}{Dt} = \mathbf{F} - \frac{1}{\rho}\frac{\partial p}{\partial \mathbf{r}} + \frac{1}{n}\int \mathbf{V}\frac{D}{Dt}\left(\tau^{(0)}\frac{Df^{(0)}}{Dt}\right)d\mathbf{v}, \tag{5.1.27}$$

$$\frac{\breve{D}T}{Dt} = -\frac{2}{3}T\frac{\partial}{\partial \mathbf{r}}\cdot\mathbf{v}_0 - \frac{T}{n}\int \frac{D}{Dt}\left(\tau^{(0)}\frac{Df^{(0)}}{Dt}\right)d\mathbf{v} + \frac{2}{3k_Bn}\int \frac{mV^2}{2}\frac{D}{Dt}\left(\tau^{(0)}\frac{Df^{(0)}}{Dt}\right)d\mathbf{v}. \tag{5.1.28}$$

Transform relation (5.1.21)

$$\frac{Df^{(0)}}{Dt} = f^{(0)}\left\{\frac{\breve{D}\ln f^{(0)}}{Dt} + \mathbf{V}\cdot\frac{\partial \ln f^{(0)}}{\partial \mathbf{r}} + \left(\mathbf{F} - \frac{\breve{D}\mathbf{v}_0}{Dt}\right)\cdot\frac{\partial \ln f^{(0)}}{\partial \mathbf{V}}\right.$$
$$\left. - \frac{\partial \ln f^{(0)}}{\partial \mathbf{V}}\mathbf{V} : \frac{\partial}{\partial \mathbf{r}}\mathbf{v}_0\right\}, \tag{5.1.29}$$

and use set of Eqs. (5.1.26)–(5.1.28) resulting in the relation

$$\frac{\breve{D}\ln f^{(0)}}{Dt} = -\frac{mV^2}{3k_BT}\frac{\partial}{\partial \mathbf{r}}\cdot\mathbf{v}_0 + \left(\frac{5}{2n} - \frac{mV^2}{2p}\right)\int \frac{D}{Dt}\left(\tau^{(0)}\frac{Df^{(0)}}{Dt}\right)\mathrm{d}\mathbf{v}$$
$$+ \left(\frac{mV^2}{3nk_B^2T^2} - \frac{1}{p}\right)\int \frac{mV^2}{2}\frac{D}{Dt}\left(\tau^{(0)}\frac{Df^{(0)}}{Dt}\right)\mathrm{d}\mathbf{v}. \tag{5.1.30}$$

We use (5.1.30) for further transformation of (5.1.29):

$$\frac{Df^{(0)}}{Dt} = f^{(0)}\left\{\left(\frac{mV^2}{2k_BT} - \frac{5}{2}\right)\mathbf{V}\cdot\frac{\partial \ln T}{\partial \mathbf{r}} + \frac{m}{k_BT}\mathbf{V}^0\mathbf{V} : \frac{\partial}{\partial \mathbf{r}}\mathbf{v}_0\right.$$
$$+ \left(\frac{5}{2n} - \frac{mV^2}{2p}\right)\int \frac{D}{Dt}\left(\tau^{(0)}\frac{Df^{(0)}}{Dt}\right)\mathrm{d}\mathbf{v}$$
$$+ \frac{m\mathbf{V}}{p}\cdot\int \mathbf{V}\frac{D}{Dt}\left(\tau^{(0)}\frac{Df^{(0)}}{Dt}\right)\mathrm{d}\mathbf{v}$$
$$\left. + \frac{1}{p}\left(\frac{mV^2}{3k_BT} - 1\right)\int \frac{mV^2}{2}\frac{D}{Dt}\left(\tau^{(0)}\frac{Df^{(0)}}{Dt}\right)\mathrm{d}\mathbf{v}\right\}. \tag{5.1.31}$$

Thence, in view of (5.1.31), the integral equation (5.1.20) is written as:

$$f^{(0)}\left\{\left(\frac{mV^2}{2k_BT} - \frac{5}{2}\right)\mathbf{V}\cdot\frac{\partial \ln T}{\partial \mathbf{r}} + \frac{m}{k_BT}\mathbf{V}^0\mathbf{V} : \frac{\partial}{\partial \mathbf{r}}\mathbf{v}_0\right.$$
$$+ \left(\frac{5}{2n} - \frac{mV^2}{2p}\right)\int \frac{D}{Dt}\left(\tau^{(0)}\left[\frac{Df^{(0)}}{Dt}\right]_{\tau=0}\right)\mathrm{d}\mathbf{v}$$
$$+ \frac{m\mathbf{V}}{p}\cdot\int \mathbf{V}\frac{D}{Dt}\left(\tau^{(0)}\left[\frac{Df^{(0)}}{Dt}\right]_{\tau=0}\right)\mathrm{d}\mathbf{v}$$
$$\left. + \frac{1}{p}\left(\frac{mV^2}{3k_BT} - 1\right)\int \frac{mV^2}{2}\frac{D}{Dt}\left(\tau^{(0)}\left[\frac{Df^{(0)}}{Dt}\right]_{\tau=0}\right)\mathrm{d}\mathbf{v}\right\}$$
$$- \frac{D}{Dt}\left(\tau^{(0)}\left[\frac{Df^{(0)}}{Dt}\right]_{\tau=0}\right)$$
$$= \int f^{(0)}f_1^{(0)}(\psi_1' + \psi' - \psi - \psi_1)gb\,\mathrm{d}b\,\mathrm{d}\varphi\,\mathrm{d}\mathbf{v}_1, \tag{5.1.32}$$

where

$$f = f^{(0)}(1+\psi), \qquad f^{(1)} = \psi f^{(0)}. \tag{5.1.33}$$

The solution ψ of Eq. (5.1.32) is the sum of two functions

$$\psi = \psi^E + \psi^\tau, \tag{5.1.34}$$

corresponding to equations

$$f^{(0)}\left\{\left(\frac{mV^2}{2k_BT} - \frac{5}{2}\right)\mathbf{V}\cdot\frac{\partial \ln T}{\partial \mathbf{r}} + \frac{m}{k_BT}\mathbf{V}^0\mathbf{V} : \frac{\partial}{\partial \mathbf{r}}\mathbf{v}_0\right\}$$
$$= \int f^{(0)} f_1^{(0)} \left(\psi_1'^E + \psi'^E - \psi_1^E - \psi^E\right) \mathrm{d}\omega, \tag{5.1.35}$$

$$f^{(0)}\left\{\left(\frac{5}{2n} - \frac{mV^2}{2p}\right)\int \frac{D}{Dt}\left(\tau^{(0)}\left[\frac{Df^{(0)}}{Dt}\right]_{\tau=0}\right)\mathrm{d}\mathbf{v}\right.$$
$$+ \frac{m\mathbf{V}}{p}\cdot\int \mathbf{V}\frac{D}{Dt}\left(\tau^{(0)}\left[\frac{Df^{(0)}}{Dt}\right]_{\tau=0}\right)\mathrm{d}\mathbf{v}$$
$$\left. + \frac{1}{p}\left(\frac{mV^2}{3k_BT} - 1\right)\int \frac{mV^2}{2}\frac{D}{Dt}\left(\tau^{(0)}\left[\frac{Df^{(0)}}{Dt}\right]_{\tau=0}\right)\mathrm{d}\mathbf{v}\right\}$$
$$- \frac{D}{Dt}\left(\tau^{(0)}\left[\frac{Df^{(0)}}{Dt}\right]_{\tau=0}\right)$$
$$= \int f^{(0)} f_1^{(0)} \left(\psi_1'^\tau + \psi'^\tau - \psi^\tau - \psi_1^\tau\right) \mathrm{d}\omega. \tag{5.1.36}$$

Eq. (5.1.35) is the classical Enskog integral equation, the solubility conditions for this equation are satisfied, and its solution is known (Chapman and Cowling, 1952).

Let us investigate the solubility of integral equation (5.1.36). With this aim we multiply both (integral and differential) parts of Eq. (5.1.36) by additive invariants ψ^i ($i = 1, \mathbf{V}, V^2/2$) and integrate with respect to $\mathbf{v}$. Right-hand sides of these relations are equal to zero because of conservation laws of mass, momentum and energy for encountering particles. Now we show that integrated left-hand sides of these relations are also equal to zero, therefore the solubility conditions for this Fredholm equation are fulfilled.

(1) $\psi^{(1)} = 1$.

For this invariant we have

$$\int f^{(0)}\left(\frac{5}{2n} - \frac{mV^2}{2p}\right)\mathrm{d}\mathbf{v}\int \frac{D}{Dt}\left(\tau^{(0)}\left[\frac{Df^{(0)}}{Dt}\right]_{\tau=0}\right)\mathrm{d}\mathbf{v}$$
$$= \int \frac{D}{Dt}\left(\tau^{(0)}\left[\frac{Df^{(0)}}{Dt}\right]_{\tau=0}\right)\mathrm{d}\mathbf{v}. \tag{5.1.37}$$

The second integrated term in (5.1.36) is equal to zero because the integrand is an odd function in thermal velocity, the third term is also equal to zero because of equality

$$\int \left(\frac{mV^2}{3kT} - 1 \right) f^{(0)} \, \mathrm{d}\mathbf{v} = 0. \tag{5.1.38}$$

Notice that expression (5.1.37) and the rest of the terms in (5.1.36) (integrated in $\mathbf{v}$) on the left side cancel each other.

(2) $\psi^{(2)} = \mathbf{V}$.

Define

$$R_i = \int V_i \frac{D}{Dt} \left(\tau^{(0)} \left[\frac{Df^{(0)}}{Dt} \right]_{\tau=0} \right) \mathrm{d}\mathbf{v} \quad (i = 1, 2, 3),$$

then

$$\frac{m}{p} \int f^{(0)} V_i (\mathbf{V} \cdot \mathbf{R}) \, \mathrm{d}\mathbf{v} - R_i = 0. \tag{5.1.39}$$

Relation (5.1.39) leads to accomplishing of solubility condition for invariant $\psi^{(2)}$.

(3) $\psi^{(3)} = mV^2/2$.

$$\begin{aligned}
&\int \frac{mV^2}{2} \left(\frac{5}{2n} - \frac{mV^2}{2p} \right) f^{(0)} \, \mathrm{d}\mathbf{v} \int \frac{D}{Dt} \left(\tau^{(0)} \left[\frac{Df^{(0)}}{Dt} \right]_{\tau=0} \right) \mathrm{d}\mathbf{v} \\
&\quad + \frac{1}{p} \int \left(\frac{mV^2}{3k_B T} - 1 \right) \frac{mV^2}{2} f^{(0)} \, \mathrm{d}\mathbf{v} \int \frac{mV^2}{2} \frac{D}{Dt} \left(\tau^{(0)} \left[\frac{Df^{(0)}}{Dt} \right]_{\tau=0} \right) \mathrm{d}\mathbf{v} \\
&\quad - \int \frac{mV^2}{2} \frac{D}{Dt} \left(\tau^{(0)} \left[\frac{Df^{(0)}}{Dt} \right]_{\tau=0} \right) \mathrm{d}\mathbf{v} = 0,
\end{aligned} \tag{5.1.40}$$

because

$$\int \frac{mV^2}{2} \left(\frac{5}{2n} - \frac{mV^2}{2p} \right) f^{(0)} \, \mathrm{d}\mathbf{v} = 0,$$

$$\frac{1}{p} \int \left(\frac{mV^2}{3k_B T} - 1 \right) \frac{mV^2}{2} f^{(0)} \, \mathrm{d}\mathbf{v} = 1.$$

Then solubility conditions are fulfilled for Eq. (5.1.36). For inscription of the part ψ^{τ} of solution we need the explicit form of operator

$$\begin{aligned}
&\frac{D}{Dt} \left(\tau^{(0)} \left[\frac{Df^{(0)}}{Dt} \right]_{\tau=0} \right) \\
&\quad = \left\{ \frac{\breve{D}}{Dt} + \mathbf{V} \cdot \frac{\partial}{\partial \mathbf{r}} + \left(\mathbf{F} - \frac{\breve{D}\mathbf{v}_0}{Dt} \right) \cdot \frac{\partial}{\partial \mathbf{V}} - \frac{\partial}{\partial \mathbf{V}} \mathbf{V} : \frac{\partial}{\partial \mathbf{r}} \mathbf{v}_0 \right\} \\
&\qquad \times \left\{ \tau^{(0)} \left[\left(W^2 - \frac{5}{2} \right) \mathbf{V} \cdot \frac{\partial \ln T}{\partial \mathbf{r}} + 2\mathbf{W}^0 \mathbf{W} : \frac{\partial}{\partial \mathbf{r}} \mathbf{v}_0 \right] \right\}.
\end{aligned} \tag{5.1.41}$$

Transformation of (5.1.41) leads to result

$$
\begin{aligned}
&\frac{D}{Dt}\left(\tau\left[\frac{Df^{(0)}}{Dt}\right]_{\tau=0}\right)\\
&=\left(W^2-\frac{5}{2}\right)\frac{\breve{D}}{Dt}\left[\tau^{(0)}\frac{\partial\ln T}{\partial\mathbf{r}}\right]+2\mathbf{W}^0\mathbf{W}:\frac{\breve{D}}{Dt}\left[\tau^{(0)}\frac{\partial}{\partial\mathbf{r}}\mathbf{v}_0\right]\\
&\quad-\tau^{(0)}W^2\left(\frac{\partial\ln T}{\partial\mathbf{r}}\cdot\mathbf{V}\right)\frac{\partial\ln T}{\partial t}-\tau^{(0)}W^2\left[\mathbf{v}_0\cdot\frac{\partial\ln T}{\partial\mathbf{r}}\right]\left[\mathbf{V}\cdot\frac{\partial\ln T}{\partial\mathbf{r}}\right]\\
&\quad-2\frac{\breve{D}\ln T}{Dt}\tau^{(0)}\mathbf{W}^0\mathbf{W}:\frac{\partial}{\partial\mathbf{r}}\mathbf{v}_0+\tau^{(0)}\left(\mathbf{F}-\frac{\breve{D}\mathbf{v}_0}{Dt}\right)\cdot\frac{\partial\ln T}{\partial\mathbf{r}}\left(W^2-\frac{5}{2}\right)\\
&\quad+2\tau^{(0)}\left(\mathbf{F}-\frac{\breve{D}\mathbf{v}_0}{Dt}\right)\frac{\partial\ln T}{\partial\mathbf{r}}:\mathbf{W}\mathbf{W}+2\sqrt{\frac{m}{2k_BT}}\left(\mathbf{F}-\frac{\breve{D}\mathbf{v}_0}{Dt}\right)\mathbf{W}:\frac{\partial}{\partial\mathbf{r}}\mathbf{v}_0\\
&\quad+2\sqrt{\frac{m}{2k_BT}}\mathbf{W}\left(\mathbf{F}-\frac{\breve{D}\mathbf{v}_0}{Dt}\right):\frac{\partial}{\partial\mathbf{r}}\mathbf{v}_0-\tau^{(0)}\left(W^2-\frac{5}{2}\right)\frac{\partial\ln T}{\partial\mathbf{r}}\mathbf{V}:\frac{\partial}{\partial\mathbf{r}}\mathbf{v}_0\\
&\quad-2\tau^{(0)}\left(\frac{\partial\ln T}{\partial\mathbf{r}}\cdot\mathbf{W}\right)\mathbf{W}\mathbf{V}:\frac{\partial}{\partial\mathbf{r}}\mathbf{v}_0-2\tau^{(0)}\left(\mathbf{W}\cdot\frac{\partial}{\partial\mathbf{r}}\right)\mathbf{v}_0\cdot\left(\mathbf{W}\cdot\frac{\partial}{\partial\mathbf{r}}\right)\mathbf{v}_0\\
&\quad-2\tau^{(0)}\left[\left(\mathbf{W}\cdot\frac{\partial}{\partial\mathbf{r}}\right)\mathbf{v}_0\cdot\frac{\partial}{\partial\mathbf{r}}\right](\mathbf{v}_0\cdot\mathbf{W}).
\end{aligned}
\tag{5.1.42}
$$

Taking into account the explicit form of $Df^{(0)}/Dt$ with terms proportional to $\tau^{(0)}$ (see also Appendix 2 and corresponding generalized hydrodynamic equations), we state, that the structure of the left-hand side of Eq. (5.1.36) is so complicated that in many cases it is more convenient to use self-consistent numerical schemes of solution of GHE and Eq. (5.1.36) (defining kinetic coefficients) or to use variants of the generalized Grad's method. Nevertheless, we formulate approximate method of Eq. (5.1.36) solution making possible to find explicit forms of the fluctuation terms for kinetic coefficients.

5.2. Approximate modified Chapman–Enskog method

We intend to develop in the theory of GBE the approximate modified Chapman–Enskog method. With this aim let us rewrite Eq. (5.1.32) in the form

$$
\begin{aligned}
&f^{(0)}\left\{\left(\frac{mV^2}{2k_BT}-\frac{5}{2}\right)\left[\mathbf{V}\cdot\frac{\partial\ln T}{\partial\mathbf{r}}-\frac{1}{n}\int\frac{D}{Dt}\left(\tau^{(0)}\left(\frac{Df^{(0)}}{Dt}\right)_{\tau=0}\right)\mathrm{d}\mathbf{v}\right]\right.\\
&\quad+\frac{m}{k_BT}\mathbf{V}\mathbf{V}:\frac{\partial}{\partial\mathbf{r}}\mathbf{v}_0-\frac{mV^2}{3k_BT}\left[\frac{\partial}{\partial\mathbf{r}}\cdot\mathbf{v}_0\right.\\
&\quad\left.-\frac{1}{p}\int\frac{mV^2}{2}\frac{D}{Dt}\left(\tau^{(0)}\left(\frac{Df^{(0)}}{Dt}\right)_{\tau=0}\right)\mathrm{d}\mathbf{v}\right]
\end{aligned}
$$

$$
\begin{aligned}
&+\frac{m}{p}\left[\mathbf{V}\cdot\int \mathbf{V}\frac{D}{Dt}\left(\tau^{(0)}\left(\frac{Df^{(0)}}{Dt}\right)_{\tau=0}\right)\mathrm{d}\mathbf{v}\right.\\
&\left.\left.-\int\frac{V^2}{2}\frac{D}{Dt}\left(\tau^{(0)}\left(\frac{Df^{(0)}}{Dt}\right)_{\tau=0}\right)\mathrm{d}\mathbf{v}\right]\right\}-\frac{D}{Dt}\left(\tau^{(0)}\left(\frac{Df^{(0)}}{Dt}\right)_{\tau=0}\right)\\
&=\int f^{(0)}f_1^{(0)}(\psi_1'+\psi'-\psi-\psi_1)\,\mathrm{d}\omega.
\end{aligned}
\tag{5.2.1}
$$

In the following calculations of the kinetic coefficient's perturbations we neglect the small influence of τ-terms in square brackets of (5.2.1). It means that we intend to obtain solution of the shortened approximate integral equation

$$
\begin{aligned}
&\left(\frac{Df^{(0)}}{Dt}\right)_{\tau=0}-\frac{D}{Dt}\left(\tau^{(0)}\left(\frac{Df^{(0)}}{Dt}\right)_{\tau=0}\right)\\
&=\int f^{(0)}f_1^{(0)}(\psi_1'+\psi'-\psi-\psi_1)\,\mathrm{d}\omega,
\end{aligned}
\tag{5.2.2}
$$

where the substantial derivative $(Df^{(0)}/Dt)_{\tau=0}$ exists. Let us consider more general case of multi-component mixture of gases while neglecting the influence of external forces $\mathbf{F}_\alpha$. We use also approximation (1.3.86) $\tau_\alpha=\tau$. It leads to some simplification (Alexeev, 1988, 1992, 1994) of the following calculations, which nevertheless remain complicated. Then,

$$
\begin{aligned}
&\left(\frac{Df_\alpha^{(0)}}{Dt}\right)_{\tau=0}-\frac{D}{Dt}\left(\tau^{(0)}\left(\frac{Df_\alpha^{(0)}}{Dt}\right)_{\tau=0}\right)\\
&=\sum_{j=1}^{\eta}\int\left[f_\alpha^{(0)\prime}f_j^{(1)\prime}+f_\alpha^{(1)\prime}f_j^{(0)\prime}-f_\alpha^{(0)}f_j^{(1)}-f_\alpha^{(1)}f_j^{(0)}\right]\mathrm{d}\omega,\\
&\mathrm{d}\omega=g_{\alpha j}b\,\mathrm{d}b\,\mathrm{d}\varphi\,\mathrm{d}\mathbf{v}_j.
\end{aligned}
\tag{5.2.3}
$$

Suppose ψ_α is the solution of equation

$$
\left[\frac{Df_\alpha^{(0)}}{Dt}\right]_{\tau=0}=\sum_j\int f_\alpha^{(0)}f_j^{(0)}(\psi_\alpha'+\psi_j'-\psi_\alpha-\psi_j)\,\mathrm{d}\omega,
\tag{5.2.4}
$$

$$
\frac{D}{Dt}=\frac{\partial}{\partial t}+\mathbf{v}_\alpha\cdot\frac{\partial}{\partial\mathbf{r}}.
\tag{5.2.5}
$$

We intend to find the solution of the linearized GBE (5.2.3) in the form

$$
f_\alpha^{(1)}=f_\alpha^{(0)}\psi_\alpha-\frac{D}{Dt}\left[\tau^{(0)}f_\alpha^{(0)}\psi_\alpha\right].
\tag{5.2.6}
$$

After substituting (5.2.6) into (5.2.3):

$$\left[\frac{Df_\alpha^{(0)}}{Dt}\right]_{\tau=0} - \frac{D}{Dt}\left[\tau^{(0)}\left(\frac{Df_\alpha^{(0)}}{Dt}\right)_{\tau=0}\right]$$
$$= \sum_{j=1}^{\eta} \int f_\alpha^{(0)} f_j^{(0)} (\psi'_\alpha + \psi'_j - \psi_\alpha - \psi_j)\, \mathrm{d}\omega$$
$$- \sum_{j=1}^{\eta} \int \left\{ f_\alpha^{(0)\prime} \frac{D'}{Dt}\left(\tau^{(0)} f_j^{(0)\prime} \psi'_j\right) + f_j^{(0)\prime} \frac{D'}{Dt}\left(\tau^{(0)} f_\alpha^{(0)\prime} \psi'_\alpha\right)\right.$$
$$\left. - f_\alpha^{(0)} \frac{D}{Dt}\left(\tau^{(0)} f_j^{(0)} \psi_j\right) - f_j^{(0)} \frac{D}{Dt}\left(\tau^{(0)} f_\alpha^{(0)} \psi_\alpha\right)\right\} \mathrm{d}\omega, \tag{5.2.7}$$

where

$$\frac{D'}{Dt} = \frac{\partial}{\partial t} + \mathbf{V}'_\alpha \cdot \frac{\partial}{\partial \mathbf{r}}. \tag{5.2.8}$$

Eq. (5.2.7) is being simplified with the help of (5.2.4):

$$\frac{D}{Dt}\left[\tau^{(0)}\left(\frac{Df_\alpha^{(0)}}{Dt}\right)_{\tau=0}\right]$$
$$= \sum_j \int \left\{ f_\alpha^{(0)\prime} \frac{D'}{Dt}\left(\tau^{(0)} f_j^{(0)\prime} \psi'_j\right) + f_j^{(0)\prime} \frac{D'}{Dt}\left(\tau^{(0)} f_\alpha^{(0)\prime} \psi'_\alpha\right)\right.$$
$$\left. - f_\alpha^{(0)} \frac{D}{Dt}\left(\tau^{(0)} f_j^{(0)} \psi_j\right) - f_j^{(0)} \frac{D}{Dt}\left(\tau^{(0)} f_\alpha^{(0)} \psi_\alpha\right)\right\} \mathrm{d}\omega. \tag{5.2.9}$$

Maxwellian functions in front of substantial derivatives on right-hand side of (5.2.9) can be introduced in round squares after signing the derivatives with accuracy $\mathrm{O}[\psi^2]$. For example, let us consider the term

$$f_\alpha^{(0)} \frac{D}{Dt}\left(\tau^{(0)} f_j^{(0)} \psi_j\right) = \frac{D}{Dt}\left(\tau^{(0)} f_\alpha^{(0)} f_j^{(0)} \psi_j\right) - \tau^{(0)} f_j^{(0)} \psi_j \frac{Df_\alpha^{(0)}}{Dt}. \tag{5.2.10}$$

The form (5.2.10) is obtained with the help of relation

$$\frac{D}{Dt} \equiv \frac{\partial}{\partial t} + \mathbf{v}_\alpha \cdot \frac{\partial}{\partial \mathbf{r}} = \frac{\partial}{\partial t} + \frac{\partial}{\partial \mathbf{r}} \cdot \mathbf{v}_\alpha, \tag{5.2.11}$$

which is valid, because $\mathbf{r}, \mathbf{v}_\alpha, t$ are independent variables. But from (5.2.4) follows that $Df_\alpha^{(0)}/Dt \sim \psi$, and therefore the formulated affirmation.

Then

$$\frac{D}{Dt}\left(\tau^{(0)}\left[\frac{Df_\alpha^{(0)}}{Dt}\right]_{\tau=0}\right)$$
$$=\sum_{j=1}^{\eta}\int\left\{\frac{D'}{Dt}\left(f_\alpha^{(0)\prime}f_j^{(0)\prime}\tau^{(0)}\psi_j'\right)+\frac{D'}{Dt}\left(f_\alpha^{(0)\prime}f_j^{(0)\prime}\tau^{(0)}\psi_\alpha'\right)\right.$$
$$\left.-\frac{D}{Dt}\left(f_\alpha^{(0)}f_j^{(0)}\tau^{(0)}\psi_j\right)-\frac{D}{Dt}\left(f_\alpha^{(0)}f_j^{(0)}\tau^{(0)}\psi_\alpha\right)\right\}\mathrm{d}\omega. \tag{5.2.12}$$

This linear equation can be split into the following integral equations.

$$\frac{\partial}{\partial t}\left(\tau^{(0)}\left[\frac{Df_\alpha^{(0)}}{Dt}\right]_{\tau=0}\right)$$
$$=\frac{\partial}{\partial t}\left[\tau^{(0)}\sum_{j=1}^{\eta}\int f_\alpha^{(0)}f_j^{(0)}(\psi_\alpha'+\psi_j'-\psi_\alpha-\psi_j)\,\mathrm{d}\omega\right], \tag{5.2.13}$$

$$\frac{\partial}{\partial \mathbf{r}}\cdot\left(\mathbf{v}_0\tau^{(0)}\left[\frac{Df_\alpha^{(0)}}{Dt}\right]_{\tau=0}\right)$$
$$=\frac{\partial}{\partial \mathbf{r}}\cdot\left[\mathbf{v}_0\tau^{(0)}\sum_{j=1}^{\eta}\int f_\alpha^{(0)}f_j^{(0)}(\psi_\alpha'+\psi_j'-\psi_\alpha-\psi_j)\,\mathrm{d}\omega\right], \tag{5.2.14}$$

$$\frac{\partial}{\partial \mathbf{r}}\cdot\left(\mathbf{V}_\alpha\tau^{(0)}\left[\frac{Df_\alpha^{(0)}}{Dt}\right]_{\tau=0}\right)$$
$$=\frac{\partial}{\partial \mathbf{r}}\cdot\left[\tau^{(0)}\sum_{j=1}^{\eta}\int f_\alpha^{(0)}f_j^{(0)}\left(\psi_\alpha'\mathbf{V}_\alpha'-\psi_j'\mathbf{V}_j'-\psi_\alpha\mathbf{V}_\alpha-\psi_j\mathbf{V}_j'\right)\mathrm{d}\omega\right]. \tag{5.2.15}$$

Obviously, Eqs. (5.2.4) and (5.2.13), (5.2.14) are consistent. Let us show that Eq. (5.2.15) can be satisfied identically by solution of Eq. (5.2.15) in frame of Enskog's moment method; in other words, it can be stated by use the of moment method for solution of equation

$$\mathbf{V}_\alpha\left[\frac{Df_\alpha^{(0)}}{Dt}\right]_{\tau=0}$$
$$=\sum_j\int f_\alpha^{(0)}f_j^{(0)}\left(\psi_\alpha'\mathbf{V}_\alpha'-\psi_j'\mathbf{V}_j'-\psi_\alpha'\mathbf{V}_\alpha'-\psi_j'\mathbf{V}_j'\right)\mathrm{d}\omega, \tag{5.2.16}$$

where ψ_α is the Enskog solution, the moment equations corresponding to the first approximation, are satisfied identically. In variables $\mathbf{r}$, $\mathbf{V}_\alpha$, t after exclusion of time derivatives using hydrodynamic equations (Chapman and Cowling, 1952; Hirschfelder, Curtiss and Bird, 1954), the left-hand side of equation takes the form (5.2.17)

$$\mathbf{V}_\alpha\left[\frac{Df_\alpha^{(0)}}{Dt}\right]_{\tau=0} = f_\alpha^{(0)}\mathbf{V}_\alpha\left[\frac{n}{n_\alpha}(\mathbf{V}_\alpha\cdot\mathbf{d}_\alpha)+2\mathbf{W}_\alpha^0\mathbf{W}_\alpha:\frac{\partial}{\partial\mathbf{r}}\mathbf{v}_0 - \left(\frac{5}{2}-W_\alpha^2\right)\left(\mathbf{V}_\alpha\cdot\frac{\partial\ln T}{\partial\mathbf{r}}\right)\right], \tag{5.2.17}$$

where

$$\mathbf{d}_\alpha = \frac{\partial}{\partial\mathbf{r}}\left(\frac{n_\alpha}{n}\right)+\left(\frac{n_\alpha}{n}-\frac{\rho_\alpha}{\rho}\right)\frac{\partial\ln p}{\partial\mathbf{r}},\qquad \mathbf{W}_\alpha=\sqrt{\frac{m}{2k_BT}}\,\mathbf{V}_\alpha. \tag{5.2.18}$$

ψ_α is linear function of derivatives and can be written as:

$$\psi_\alpha = -\mathbf{A}_\alpha\cdot\frac{\partial\ln T}{\partial\mathbf{r}} - \overleftrightarrow{B}_\alpha : \frac{\partial}{\partial\mathbf{r}}\mathbf{v}_0 + n\sum_j \mathbf{C}_\alpha^{(j)}\cdot\mathbf{d}_j, \tag{5.2.19}$$

$$\mathbf{C}_\alpha^{(j)} = C_\alpha^{(j)}(W_\alpha)\mathbf{W}_\alpha,\quad \mathbf{A}_\alpha = A_\alpha(W_\alpha)\mathbf{W}_\alpha,\qquad \overleftrightarrow{B}_\alpha = \mathbf{W}_\alpha^0\mathbf{W}_\alpha B_\alpha(W_\alpha). \tag{5.2.20}$$

Because of (5.2.19), Eq. (5.2.16) is split into three equations

$$\frac{1}{n_\alpha}V_{\alpha i}f_\alpha^{(0)}(\delta_{\alpha n}-\delta_{\alpha k})\mathbf{V}_\alpha = \sum_j\int\big[V'_{\alpha i}\mathbf{C}_\alpha^{(n)\prime}+V'_{ji}\mathbf{C}_j^{(n)\prime}-V'_{\alpha i}\mathbf{C}_\alpha^{(k)\prime}-V'_{ji}\mathbf{C}_j^{(k)\prime}-V_{\alpha i}\mathbf{C}_\alpha^{(n)} - V_{ji}\mathbf{C}_j^{(n)}+V_{\alpha i}\mathbf{C}_\alpha^{(k)}+V_{ji}\mathbf{C}_j^{(k)}\big]f_\alpha^{(0)}f_j^{(0)}\,\mathrm{d}\omega\quad (i=1,2,3), \tag{5.2.21}$$

$$2V_{\alpha i}f_\alpha^{(0)}\mathbf{W}_\alpha^0\mathbf{W}_\alpha = -\sum_j\int\big[V'_{\alpha i}\overleftrightarrow{B}'_\alpha+V_{ji}\overleftrightarrow{B}'_j-V_{\alpha i}\overleftrightarrow{B}_\alpha-V_{ji}\overleftrightarrow{B}_j\big]f_\alpha^{(0)}f_j^{(0)}\,\mathrm{d}\omega \tag{5.2.22}$$

and

$$V_{\alpha i}f_\alpha^{(0)}\left(\frac{5}{2}-W_\alpha^2\right)\mathbf{V}_\alpha = \sum_j\int\big[V'_{\alpha i}\mathbf{A}'_\alpha+V'_{ji}\mathbf{A}'_j-V_{\alpha i}\mathbf{A}_\alpha-V_{ji}\mathbf{A}_j\big]f_\alpha^{(0)}f_j^{(0)}\,\mathrm{d}\omega. \tag{5.2.23}$$

By solution of Eq. (5.2.4) as probe functions $t_\alpha^{(n,k)}$ (which correspond to functions $\mathbf{A}_\alpha, \overleftrightarrow{B}_\alpha, \mathbf{C}_\alpha^{(n)} - \mathbf{C}_\alpha^{(k)}$), the finite linear combinations of Sonine polynomials can be used ($\overleftrightarrow{W}_\alpha = \mathbf{W}_\alpha, \mathbf{W}_\alpha^0\mathbf{W}_\alpha, \mathbf{W}_\alpha$ relevant to (5.2.21)–(5.2.23)):

$$\overleftrightarrow{t}_\alpha^{(n,k)} = \overleftrightarrow{W}_\alpha \sum_{m=0}^{\xi} t_{\alpha m}^{(n,k)} S_n^{(m)}\big(W_\alpha^2\big), \quad m = 0, 1, 2, \ldots. \tag{5.2.24}$$

After multiplication of each side of Eqs. (5.2.21)–(5.2.23) by Sonine polynomials – correspondingly $\mathbf{W}_\alpha S_{3/2}^{(m)}(W_\alpha^2)$, $\mathbf{W}_\alpha^0\mathbf{W}_\alpha S_{5/2}^{(m)}(W_\alpha^2)$, $\mathbf{W}_\alpha S_{3/2}^{(m)}(W_\alpha^2)$ – the following integrals appear on the left-hand sides of equations:

$$I_1 = \frac{1}{n_\alpha}\int f_\alpha^{(0)} V_{\alpha i}(\delta_{\alpha n} - \delta_{\alpha k})\mathbf{V}_\alpha \cdot \mathbf{W}_\alpha S_{3/2}^{(m)}\big(W_\alpha^2\big)\,\mathrm{d}\mathbf{V}_\alpha, \tag{5.2.25}$$

$$I_2 = 2\int f_\alpha^{(0)} V_{\alpha i} f_\alpha^{(0)} \mathbf{W}_\alpha^0\mathbf{W}_\alpha : \mathbf{W}_\alpha^0\mathbf{W}_\alpha S_{5/2}^{(m)}\big(W_\alpha^2\big)\,\mathrm{d}\mathbf{V}_\alpha, \tag{5.2.26}$$

$$I_3 = \int f_\alpha^{(0)} V_{\alpha i}\left(\frac{5}{2} - W_\alpha^2\right)\mathbf{V}_\alpha \cdot \mathbf{W}_\alpha S_{3/2}^{(m)}\big(W_\alpha^2\big)\,\mathrm{d}\mathbf{V}_\alpha \tag{5.2.27}$$

$$(i = 1, \ldots, 3).$$

Integrands in (5.2.25)–(5.2.27) are odd functions of velocity components and then integrals I_1, I_2, I_3 over all $\mathbf{V}$-space are equal to zero. This affirmation is obvious for integrals I_1 and I_3. For proof that $I_2 = 0$, it is sufficient to notice that

$$\mathbf{W}_\alpha^0\mathbf{W}_\alpha : \mathbf{W}_\alpha^0\mathbf{W}_\alpha = \frac{2}{3}W_\alpha^4. \tag{5.2.28}$$

Let us consider now the right-hand sides of moment equations. By scalar multiplication of right-hand sides of Eqs. (5.2.21) and (5.2.23) by $\mathbf{W}_\alpha S_{3/2}^{(m)}(W_\alpha^2)$ and by following integration over all $\mathbf{V}_\alpha$, the bracket expressions appear of the type

$$\begin{aligned}&\big[V_{\alpha i}\mathbf{W}_\alpha S_{3/2}^{(p)}\big(W_\alpha^2\big), \mathbf{W}_j S_{3/2}^{(q)}\big(W_j^2\big)\big]_{\alpha j}\\ &\quad = -\frac{1}{n_\alpha n_j}\int f_\alpha^{(0)} f_j^{(0)} S_{3/2}^{(q)}\big(W_j^2\big)\mathbf{W}_j \cdot \big[V'_{\alpha i}\mathbf{W}'_\alpha S_{3/2}^{(p)}\big(W'^2_\alpha\big)\\ &\qquad - V_{\alpha i}\mathbf{W}_\alpha S_{3/2}^{(p)}\big(W_\alpha^2\big)\big] g_{\alpha j} b\,\mathrm{d}b\,\mathrm{d}\varphi\,\mathrm{d}\mathbf{v}_\alpha\,\mathrm{d}\mathbf{v}_j,\end{aligned} \tag{5.2.29}$$

$$\begin{aligned}&\big[V_{\alpha i}\mathbf{W}_\alpha S_{3/2}^{(p)}\big(W_\alpha^2\big), \mathbf{W}_\alpha S_{3/2}^{(q)}\big(W_\alpha^2\big)\big]_{\alpha j}\\ &\quad = -\frac{1}{n_\alpha n_j}\int f_\alpha^{(0)} f_j^{(0)} S_{3/2}^{(q)}\big(W_\alpha^2\big)\mathbf{W}_\alpha \cdot \big[V'_{\alpha i}\mathbf{W}'_\alpha S_{3/2}^{(p)}\big(W'^2_\alpha\big)\\ &\qquad - V_{\alpha i}\mathbf{W}_\alpha S_{3/2}^{(p)}\big(W_\alpha^2\big)\big] g_{\alpha j} b\,\mathrm{d}b\,\mathrm{d}\varphi\,\mathrm{d}\mathbf{v}_\alpha\,\mathrm{d}\mathbf{v}_j.\end{aligned} \tag{5.2.30}$$

Expression (5.2.29) is a coefficient of $s^p t^q$ in expansion of function

$$J = \frac{1}{n_\alpha n_j} \int f_\alpha^{(0)} f_j^{(0)} (1-s)^{-5/2} (1-t)^{-5/2} \left\{ \exp\left(-\frac{W_\alpha^2 s}{1-s} - \frac{W_j^2 t}{1-t} \right) \right.$$
$$\times \mathbf{W}_\alpha \cdot \mathbf{W}_j V_{\alpha i} - \exp\left(-\frac{W_\alpha'^2 s}{1-s} - \frac{W_j^2 t}{1-t} \right) \mathbf{W}'_\alpha \cdot \mathbf{W}'_j V'_{\alpha i} \Big\}$$
$$\times g_{\alpha j} b \, \mathrm{d}b \, \mathrm{d}\varphi \, \mathrm{d}\mathbf{v}_\alpha \, \mathrm{d}\mathbf{v}_j. \tag{5.2.31}$$

Let us use now in relation (5.2.31) other variables, $\mathbf{G}_0$, $\mathbf{g}_{\alpha j}$, using formulae

$$\mathbf{V}_\alpha = \mathbf{G}_0 - M_j \mathbf{g}_{j\alpha}, \tag{5.2.32}$$
$$\mathbf{V}_j = \mathbf{G}_0 + M_\alpha \mathbf{g}_{j\alpha}, \tag{5.2.33}$$

where $\mathbf{G}_0$ is mass center velocity of particles α, relative coordinate axes j moving with the mean mass velocity $\mathbf{v}_0$ of gas ($M_j = m_j/m_0$, $m_0 = m_\alpha + m_j$). In this coordinate system,

$$\frac{1}{2} m_\alpha V_\alpha^2 + \frac{1}{2} m_j V_j^2 = \frac{1}{2} m_0 \left(G_0^2 + M_\alpha M_j g_{\alpha j}^2 \right). \tag{5.2.34}$$

Introduce notation

$$\widetilde{G}_0 = G_0 \sqrt{\frac{m_0}{2k_B T}}, \qquad \tilde{g}_{\alpha j} = g_{\alpha j} \sqrt{\frac{m_0 M_\alpha M_j}{2k_B T}}. \tag{5.2.35}$$

Now we use in formula (5.2.31) the integration over variables $\mathbf{G}_0$, $\mathbf{g}_{\alpha j}$. Jacobian of transformation can be written as

$$\frac{\partial(\widetilde{G}_0, \tilde{g}_{j\alpha})}{\partial(\mathbf{v}_\alpha, \mathbf{v}_j)} = \frac{(m_\alpha m_j)^{3/2}}{(2k_B T)^3}. \tag{5.2.36}$$

Formula (5.2.31) in variables $\widetilde{G}_0$, $\tilde{g}_{\alpha j}$ has the form

$$J = (1-s)^{-5/2} (1-t)^{-5/2} \pi^{-3} \int \mathrm{e}^{-\widetilde{G}_0^2 - \tilde{g}_{\alpha j}^2} \left(\mathrm{e}^{-\widetilde{S} W_\alpha^2} \mathbf{W}_\alpha V_{\alpha i} \right.$$
$$\left. - \mathrm{e}^{-\widetilde{S} W_\alpha'^2} \mathbf{W}'_\alpha V'_{\alpha i} \right) \mathbf{W}_j \, \mathrm{e}^{-\widetilde{T} W_j^2} g_{\alpha j} b \, \mathrm{d}b \, \mathrm{d}\varphi \, \mathrm{d}\widetilde{\mathbf{G}}_0 \, \mathrm{d}\tilde{\mathbf{g}}_{j\alpha}, \tag{5.2.37}$$

where

$$\widetilde{S} = \frac{s}{1-s}, \qquad \widetilde{T} = \frac{t}{1-t}.$$

Calculate

$${}^1 H_{\alpha j, i} = \int \exp\left(-\widetilde{G}_0^2 - \tilde{g}_{\alpha j}^2 - \widetilde{S} W_\alpha'^2 - \widetilde{T} W_j^2 \right) \mathbf{W}'_\alpha \cdot \mathbf{W}_j V'_{\alpha i} \, \mathrm{d}\widetilde{\mathbf{G}}_0. \tag{5.2.38}$$

With this aim introduce a new variable

$$\mathbf{c} = \widetilde{\mathbf{G}}_0 + \frac{1}{i_{\alpha j}} \sqrt{M_\alpha M_j} \left(\widetilde{T} \tilde{\mathbf{g}}_{j\alpha} - \widetilde{S} \tilde{\mathbf{g}}'_{j\alpha} \right), \tag{5.2.39}$$

where

$$i_{\alpha j} = \frac{1 - sM_j - tM_\alpha}{(1-s)(1-t)}. \tag{5.2.40}$$

Then exponential index in (5.2.37) has the form

$$\widetilde{G}_0^2 + \tilde{g}_{\alpha j}^2 + \widetilde{S} W_\alpha'^2 + \widetilde{T} W_j^2 = i_{\alpha j} c^2 + j_{\alpha j} \tilde{g}_{\alpha j}^2, \tag{5.2.41}$$

where

$$j_{\alpha j} = \frac{1 - 2M_\alpha M_j st(1 - \cos\chi)}{1 - sM_j - tM_\alpha}, \quad \chi = \mathbf{g}_{j\alpha}, \tilde{\mathbf{g}}'_{j\alpha}. \tag{5.2.42}$$

Notice, that

$$\begin{aligned}
\mathbf{W}'_\alpha \cdot \mathbf{W}_j &= \left(\sqrt{M_\alpha}\, \widetilde{\mathbf{G}}_0 - \sqrt{M_j}\, \tilde{\mathbf{g}}_{j\alpha} \right) \cdot \left(\sqrt{M_j}\, \widetilde{\mathbf{G}}_0 + \sqrt{M_\alpha}\, \tilde{\mathbf{g}}'_{j\alpha} \right) \\
&= \left(\sqrt{M_\alpha}\, \mathbf{c} - \sqrt{M_j}\, \mathbf{c}_\alpha \right) \cdot \left(\sqrt{M_j}\, \mathbf{c} - \sqrt{M_\alpha}\, \mathbf{c}_j \right) \\
&= \sqrt{M_\alpha M_j}\, c^2 - \mathbf{c} \cdot (M_j \mathbf{c}_\alpha + M_\alpha \mathbf{c}_j) + \sqrt{M_\alpha M_j}\, \mathbf{c}_\alpha \cdot \mathbf{c}_j,
\end{aligned} \tag{5.2.43}$$

$$W'_{\alpha i} = \sqrt{M_\alpha}\, c_i - \sqrt{M_j}\, c_{\alpha i} \quad (i = 1, 2, 3). \tag{5.2.44}$$

Therefore

$$\begin{aligned}
\mathbf{W}'_\alpha \cdot \mathbf{W}_j W'_{\alpha i} = {} & M_\alpha \sqrt{M_j}\, c^2 c_i - \sqrt{M_\alpha}\, c_i \mathbf{c} \cdot (M_j \mathbf{c}_\alpha + M_\alpha \mathbf{c}_j) \\
& + M_\alpha \sqrt{M_j}\, c_i \mathbf{c}_\alpha \cdot \mathbf{c}_j - M_j \sqrt{M_\alpha}\, c_{\alpha i} c^2 \\
& + \sqrt{M_j}\, c_{\alpha i} \mathbf{c} \cdot (M_j \mathbf{c}_\alpha + M_\alpha \mathbf{c}_j) \\
& - M_j \sqrt{M_\alpha}\, c_{\alpha i} \mathbf{c}_\alpha \cdot \mathbf{c}_j,
\end{aligned} \tag{5.2.45}$$

where

$$\mathbf{c}_\alpha = \frac{M_\alpha}{i_{\alpha j}} \left(\widetilde{T} \tilde{\mathbf{g}}_{j\alpha} - \widetilde{S} \tilde{\mathbf{g}}'_{j\alpha} \right) + \tilde{\mathbf{g}}'_{j\alpha}, \tag{5.2.46}$$

$$\mathbf{c}_j = \frac{M_j}{i_{\alpha j}} \left(\widetilde{T} \tilde{\mathbf{g}}'_{j\alpha} - \widetilde{S} \tilde{\mathbf{g}}'_{j\alpha} \right) - \tilde{\mathbf{g}}'_{j\alpha}. \tag{5.2.47}$$

By integration over all $\mathbf{c}$ in $\mathbf{c}$-space, all integrals containing the odd "c" in integrand will be equal to zero. It means that after substitution of (5.2.45) into integrand of (5.2.38), the first, second and fifth terms in right-hand side of (5.2.45) should be omitted. As a result we have

$$
\begin{aligned}
{}^1H_{\alpha j,i} = 4\pi\sqrt{\frac{2k_BT}{m_\alpha}}\int_0^\infty \mathrm{e}^{-i_{\alpha j}c^2}c^2\Big[-\frac{1}{3}c^2\sqrt{M_\alpha}\,(M_jc_{\alpha i}+M_\alpha c_{ji})\\
- M_j\sqrt{M_\alpha}\,c_{\alpha i}c^2 - M_j\sqrt{M_\alpha}\,c_{\alpha i}\mathbf{c}_\alpha\cdot\mathbf{c}_j\Big]\mathrm{d}c\,\mathrm{e}^{-j_{\alpha j}\tilde{g}^2_{\alpha j}}. \qquad (5.2.48)
\end{aligned}
$$

Carry the integration over "c" in (5.2.48):

$$
\begin{aligned}
{}^1H_{\alpha j,i} = \frac{\pi^{3/2}}{2}\sqrt{\frac{2k_BT}{m_0}}\,\{(M_jc_{\alpha i}+M_\alpha c_{ji})i_{\alpha j}^{-5/2} - 3M_jc_{\alpha i}i_{\alpha j}^{-5/2}\\
- 2M_j\mathbf{c}_\alpha\cdot\mathbf{c}_jc_{\alpha i}i_{\alpha j}^{-3/2}\}\,\mathrm{e}^{-j_{\alpha j}\tilde{g}^2_{\alpha j}},
\end{aligned}
$$

or

$$
\begin{aligned}
{}^1H_{\alpha j,i} = -\frac{\pi^{3/2}}{2}\sqrt{\frac{2k_BT}{m_0}}\,\{2M_jc_{\alpha i}+2M_j\mathbf{c}_\alpha\cdot\mathbf{c}_jc_{\alpha i}i_{\alpha j}\\
- M_\alpha c_{ji}\}\,\mathrm{e}^{-j_{\alpha j}\tilde{g}^2_{\alpha j}}i_{\alpha j}^{-5/2}, \quad i=1,2,3. \qquad (5.2.49)
\end{aligned}
$$

Using also

$$
i_{\alpha j}\mathbf{c}_\alpha\cdot\mathbf{c}_j = \tilde{g}^2_{\alpha j}(1-j_{\alpha j}-\cos\chi), \qquad (5.2.50)
$$

we find

$$
\begin{aligned}
{}^1H_{\alpha j,i} = -\frac{\pi^{3/2}}{2}\sqrt{\frac{2k_BT}{m_0}}\,\{2M_jc_{\alpha i}+2M_jc_{\alpha i}\tilde{g}^2_{\alpha j}(1-j_{\alpha j}-\cos\chi)\\
- M_\alpha c_{ji}\}\,\mathrm{e}^{-j_{\alpha j}\tilde{g}^2_{\alpha j}}i_{\alpha j}^{-5/2}. \qquad (5.2.51)
\end{aligned}
$$

After substitution of $c_{\alpha i}, c_{ji}$ from relations (5.2.46), (5.2.47) into integral $\int {}^1H_{\alpha j,i}\,\mathrm{d}\tilde{\mathbf{g}}_{j\alpha}$, the following integrals appear in calculations

$$
J_1 = \int \mathrm{e}^{-j_{\alpha j}\tilde{g}^2_{\alpha j}}Z_1(\tilde{g}^2_{\alpha j})\tilde{g}_{\alpha j,i}\,\mathrm{d}\tilde{\mathbf{g}}_{\alpha j}, \qquad (5.2.52)
$$

$$
J_2 = \int \mathrm{e}^{-j_{\alpha j}\tilde{g}^2_{\alpha j}}Z_2(\tilde{g}^2_{\alpha j})\tilde{g}'_{\alpha j,i}\,\mathrm{d}\tilde{\mathbf{g}}_{\alpha j} \quad (i=1,2,3). \qquad (5.2.53)
$$

The first integral mentioned above is known to be equal to zero as having the odd $\tilde{g}_{\alpha j,i}$ in integrand. Consider now integral J_2. Since

$$\mathrm{d}\widetilde{\mathbf{G}}_0\,\mathrm{d}\tilde{\mathbf{g}}_{\alpha j} = \mathrm{d}\mathbf{v}_\alpha\,\mathrm{d}\mathbf{v}_j\left[\frac{(m_\alpha m_j)^{1/2}}{2k_BT}\right]^3$$

and

$$\mathrm{d}\mathbf{v}_\alpha\,\mathrm{d}\mathbf{v}_j = \mathrm{d}\mathbf{v}'_\alpha\,\mathrm{d}\mathbf{v}'_j,$$

then

$$\mathrm{d}\widetilde{\mathbf{G}}'_0\,\mathrm{d}\tilde{\mathbf{g}}'_{\alpha j} = \mathrm{d}\mathbf{v}_\alpha\,\mathrm{d}\mathbf{v}_j\left[\frac{(m_\alpha m_j)^{1/2}}{2k_BT}\right]^3. \tag{5.2.54}$$

The mass center velocity is not changing during collision of particles also as a module of relative velocity by elastic collisions, $\widetilde{\mathbf{G}}_0 = \widetilde{\mathbf{G}}'_0$, $\tilde{g}_{\alpha j} = \tilde{g}'_{\alpha j}$. Then the use of variables associated with backward collisions leads to relation

$$\mathrm{d}\widetilde{\mathbf{G}}_0\,\mathrm{d}\tilde{\mathbf{g}}'_{\alpha j} = \mathrm{d}\mathbf{v}_\alpha\,\mathrm{d}\mathbf{v}_j\left[\frac{(m_\alpha m_j)^{1/2}}{2k_BT}\right]^3, \tag{5.2.55}$$

and to possible transformation of integral J_2:

$$\int \mathrm{e}^{-j_{\alpha j}\tilde{g}'^2_{\alpha j}} Z_2\big(\tilde{g}'^2_{\alpha j}\big)\tilde{g}'_{\alpha j,i}\,\mathrm{d}\tilde{\mathbf{g}}'_{\alpha j},$$

wich also turns into zero because the integrand contains odd function. Calculate now

$${}^2H_{\alpha j,i} = \int \exp\big(-\widetilde{G}_0^2 - \tilde{g}^2_{\alpha j} - \widetilde{S}W^2_\alpha - \widetilde{T}W^2_j\big)\mathbf{W}_\alpha\cdot\mathbf{W}_jV_{\alpha i}\,\mathrm{d}\widetilde{\mathbf{G}}_0. \tag{5.2.56}$$

Notice, that

$$\mathbf{W}_\alpha = \sqrt{M_\alpha}\,\mathbf{c} - \sqrt{M_j}\,\mathbf{c}_\alpha + \sqrt{M_j}\,\big(\tilde{\mathbf{g}}'_{j\alpha} - \tilde{\mathbf{g}}_{j\alpha}\big).$$

Using also

$$\mathbf{W}_j = \sqrt{M_j}\,\mathbf{c} - \sqrt{M_\alpha}\,\mathbf{c}_j,$$

all combinations of velocities we need for calculation of ${}^2H_{\alpha j,i}$ in (5.2.56) can be found. But we need not do it because (Chapman and Cowling, 1952) arbitrary function of $\mathbf{W}'_\alpha$ and $\mathbf{W}_j$ can be transformed into corresponding function of $\mathbf{W}_\alpha$ and $\mathbf{W}_j$, if we suppose $\chi = 0$ (see (5.2.51)). Then all considerations which led to conditions $J_1 = 0$, $J_2 = 0$ are valid and the bracket expression in (5.2.29) turns into zero.

By calculation of bracket expression in (5.2.30), the following function appears:

$$H_{\alpha j,i}(\chi) = \int \exp\left(-\widetilde{G}_0^2 - \tilde{g}_{\alpha j}^2 - \widetilde{S}W_\alpha'^2 - \widetilde{T}W_\alpha^2\right)\mathbf{W}_\alpha \cdot \mathbf{W}_\alpha' V_{\alpha i}' \,\mathrm{d}\widetilde{G}_0, \quad (5.2.57)$$

because the bracket expression in (5.2.30) is the coefficient of $s^p t^q$ in expansion of function

$$N = \frac{1}{n_\alpha n_j} \int f_\alpha^{(0)} f_j^{(0)} (1-s)^{-5/2}(1-t)^{-5/2}\left\{\mathrm{e}^{-W_\alpha^2 \widetilde{S}} W_\alpha^2 V_{\alpha i} - \mathrm{e}^{-W_\alpha'^2 \widetilde{S}} \mathbf{W}_\alpha' \cdot \mathbf{W}_\alpha V_{\alpha i}\right\} \mathrm{e}^{-W_\alpha^2 \widetilde{T}} g_{\alpha j} b \,\mathrm{d}b \,\mathrm{d}\varphi \,\mathrm{d}\mathbf{v}_\alpha \,\mathrm{d}\mathbf{v}_j. \quad (5.2.58)$$

Really, in variables $\widetilde{\mathbf{G}}_0$, $\tilde{\mathbf{g}}_{\alpha j}$ function N takes the form

$$N = (1-s)^{-5/2}(1-t)^{-5/2}\pi^{-3} \times \int \left[H_{\alpha j}(0) - H_\alpha(\chi)\right] g_{\alpha j} b \,\mathrm{d}b \,\mathrm{d}\varphi \,\mathrm{d}\tilde{\mathbf{g}}_{j\alpha}, \quad (5.2.59)$$

where

$$H_{\alpha j,i}(0) = \int \exp\left(-\widetilde{G}_0^2 - \tilde{g}_{\alpha j}^2 - SW_\alpha^2 - TW_\alpha^2\right) W_\alpha^2 V_{\alpha i} \,\mathrm{d}\widetilde{G}_0.$$

Let us go from variables $\widetilde{\mathbf{G}}_0$, $\tilde{\mathbf{g}}_{j\alpha}$ to variables $\mathbf{c}$, $\tilde{\mathbf{g}}_{j\alpha}$ (see, for example, Alekseev, 1982, p. 167), supposing that

$$\mathbf{c} = \mathbf{G}_0 + d_1 \tilde{\mathbf{g}}_{j\alpha} + d_2 \tilde{\mathbf{g}}_{j\alpha}'. \quad (5.2.60)$$

The values d_1, d_2 are chosen in a special way, i.e., from condition

$$\widetilde{G}_0^2 + \tilde{g}_{\alpha j} + SW_\alpha'^2 + TW_\alpha^2 = i_{\alpha j} c^2 + j_{\alpha j} \tilde{g}_{\alpha j}^2. \quad (5.2.61)$$

This means that the integration over variable $\mathbf{c}$ may be easily realized. Not difficult to find that if

$$d_1 = -\sqrt{\frac{M_\alpha M_j}{i_{\alpha j}}}\,\widetilde{T}, \qquad d_2 = -\sqrt{\frac{M_\alpha M_j}{i_{\alpha j}}}\,\widetilde{S},$$

then

$$i_{\alpha j} = \frac{1 - M_j(s+t) + (M_j - M_\alpha)st}{(1-s)(1-t)}, \quad (5.2.62)$$

$$j_{\alpha j} = \frac{1 - st(M_\alpha^2 + M_j^2 + 2M_\alpha M_j \cos\chi)}{1 - M_j(s+t) + (M_j - M_\alpha)st}, \quad (5.2.63)$$

and in this case

$$\mathbf{c} = \widetilde{\mathbf{G}} - \frac{\sqrt{M_\alpha M_j}\,[s(1-t)\tilde{\mathbf{g}}'_{j\alpha} + t(1-s)\tilde{\mathbf{g}}_{j\alpha}]}{1 - M_j(s+t) + (M_j - M_\alpha)st}, \tag{5.2.64}$$

or

$$\mathbf{c} = \widetilde{\mathbf{G}}_0 - \frac{1}{i_{\alpha j}}\sqrt{M_\alpha M_j}\left(\widetilde{S}\tilde{\mathbf{g}}'_{j\alpha} + \widetilde{T}\tilde{\mathbf{g}}_{j\alpha}\right). \tag{5.2.65}$$

Notice, that

$$\mathbf{W}_\alpha = \sqrt{M_\alpha}\,\widetilde{\mathbf{G}}_0 - \sqrt{M_j}\,\tilde{\mathbf{g}}_{j\alpha},$$
$$\mathbf{W}'_\alpha = \sqrt{M_\alpha}\,\widetilde{\mathbf{G}}_0 - \sqrt{M_j}\,\tilde{\mathbf{g}}'_{j\alpha}.$$

Then

$$\mathbf{W}_\alpha = \sqrt{M_\alpha}\left(\mathbf{c} + \frac{1}{i_{\alpha j}}\sqrt{M_\alpha M_j}\left(\widetilde{S}\tilde{\mathbf{g}}'_{j\alpha} + \widetilde{T}\tilde{\mathbf{g}}_{j\alpha}\right)\right) - \sqrt{M_j}\,\tilde{\mathbf{g}}_{j\alpha},$$
$$\mathbf{W}'_\alpha = \sqrt{M_\alpha}\left(\mathbf{c} + \frac{1}{i_{\alpha j}}\sqrt{M_\alpha M_j}\left(\widetilde{S}\tilde{\mathbf{g}}'_{j\alpha} + \widetilde{T}\tilde{\mathbf{g}}_{j\alpha}\right)\right) - \sqrt{M_j}\,\tilde{\mathbf{g}}'_{j\alpha},$$
$$\mathbf{W}_\alpha = \sqrt{M_\alpha}\,\mathbf{c} + \sqrt{M_j}\,\mathbf{c}_\alpha, \tag{5.2.66}$$
$$\mathbf{W}'_\alpha = \sqrt{M_\alpha}\,\mathbf{c} + \sqrt{M_j}\,\mathbf{c}_j, \tag{5.2.67}$$

where in similar manner as in (5.2.46), (5.2.47) the next notation is introduced

$$\mathbf{c}_\alpha = \frac{M_\alpha}{i_{\alpha j}}\left(\widetilde{T}\tilde{\mathbf{g}}_{j\alpha} + \widetilde{S}\tilde{\mathbf{g}}'_{j\alpha}\right) - \tilde{\mathbf{g}}_{j\alpha}, \tag{5.2.68}$$

$$\mathbf{c}_j = \frac{M_\alpha}{i_{\alpha j}}\left(\widetilde{T}\tilde{\mathbf{g}}_{j\alpha} + \widetilde{S}\tilde{\mathbf{g}}'_{j\alpha}\right) - \tilde{\mathbf{g}}'_{j\alpha}. \tag{5.2.69}$$

Using also (5.2.66), (5.2.67), we find

$$\begin{aligned}\mathbf{W}_\alpha \cdot \mathbf{W}'_\alpha W_{\alpha i} &= \left(M_\alpha c^2 + \sqrt{M_\alpha M_j}\,\mathbf{c}\cdot\mathbf{c}_j + \sqrt{M_\alpha M_j}\,\mathbf{c}\cdot\mathbf{c}_\alpha + M_j\mathbf{c}_\alpha\cdot\mathbf{c}_j\right)\\ &\quad\times\left(\sqrt{M_\alpha}\,c_i + \sqrt{M_j}\,c_{ji}\right)\\ &= \cdots + M_\alpha\sqrt{M_j}\,c_i\mathbf{c}\cdot\mathbf{c}_j + M_\alpha\sqrt{M_j}\,c_i\mathbf{c}\cdot\mathbf{c}_\alpha\\ &\quad + \sqrt{M_j}\,M_\alpha c^2 c_{ji} + \sqrt{M_j}\,M_j c_{ji}\mathbf{c}_\alpha\cdot\mathbf{c}_j.\end{aligned} \tag{5.2.70}$$

In the right-hand side of relation (5.2.70) the terms which contain the odd powers of velocity components **c** are omitted. Analogously, by (5.2.48) we have

$$H_{\alpha j,i}(\chi) = 4\pi\sqrt{\frac{2k_BT}{m_\alpha}}\int_0^\infty \mathrm{e}^{-i_{\alpha j}c^2}c^2\bigg[\frac{1}{3}M_\alpha\sqrt{M_j}\,c^2(c_{ji}+c_{\alpha i})$$
$$+ M_\alpha\sqrt{M_j}\,c^2c_{ji} + M_j\sqrt{M_j}\,c_{ji}\mathbf{c}_\alpha\cdot\mathbf{c}_j\bigg]\mathrm{d}c\,\mathrm{e}^{-j_{\alpha j}\tilde{g}^2_{\alpha j}}, \qquad (5.2.71)$$

and after integration

$$H_{\alpha j,i}(\chi) = \frac{\pi^{3/2}}{2}\sqrt{\frac{2k_BT}{m_\alpha}}\big\{M_\alpha\sqrt{M_j}\,(c_{ji}+c_{\alpha i})i_{\alpha j}^{-5/2}$$
$$+ 3c_{ji}M_\alpha\sqrt{M_j}\,i_{\alpha j}^{-5/2} + 2M_j\sqrt{M_j}\,c_{ji}\mathbf{c}_\alpha\cdot\mathbf{c}_j i_{\alpha j}^{-3/2}\big\}\,\mathrm{e}^{-j_{\alpha j}\tilde{g}^2_{\alpha j}},$$

or

$$H_{\alpha j,i}(\chi) = i_{\alpha j}^{-5/2}\frac{\pi^{3/2}}{2}\sqrt{\frac{2k_BT}{m_0}}\sqrt{\frac{M_j}{M_\alpha}}\big\{M_\alpha(c_{\alpha i}+4c_{ji})$$
$$+ 2M_jc_{ji}\mathbf{c}_\alpha\cdot\mathbf{c}_j i_{\alpha j}\big\}\,\mathrm{e}^{-j_{\alpha j}\tilde{g}^2_{\alpha j}}. \qquad (5.2.72)$$

Further practically word for word all considerations can be repeated which led expressions like (5.2.52), (5.2.53) equal to zero as odd integrand functions. Then bracket expressions of the type (5.2.30) turn into zero.

Finally notice, that Eq. (5.2.22) leads to bracket expressions of the kind

$$\big[V_{\alpha i}\mathbf{W}^0_\alpha\mathbf{W}_\alpha S^{(p)}_{5/2}(W^2_\alpha), \mathbf{W}^0_j\mathbf{W}_j S^{(q)}_{5/2}(W^2_j)\big]_{\alpha j}$$
$$= -\frac{1}{n_\alpha n_j}\int f^{(0)}_\alpha f^{(0)}_j\mathbf{W}^0_j\mathbf{W}_j S^{(q)}_{5/2}(W^2_j) : \big[V'_{\alpha i}\mathbf{W}'^0_\alpha\mathbf{W}'_\alpha S^{(p)}_{5/2}(W'^2_\alpha)$$
$$- V_{\alpha i}\mathbf{W}^0_\alpha\mathbf{W}_\alpha S^{(p)}_{5/2}(W^2_\alpha)\big]g_{\alpha j}b\,\mathrm{d}b\,\mathrm{d}\varphi\,\mathrm{d}\mathbf{v}_\alpha\,\mathrm{d}\mathbf{v}_j, \qquad (5.2.73)$$

$$\big[V_{\alpha i}\mathbf{W}^0_\alpha\mathbf{W}_\alpha S^{(p)}_{5/2}(W^2_\alpha), \mathbf{W}^0_\alpha\mathbf{W}_\alpha S^{(q)}_{5/2}(W^2_\alpha)\big]_{\alpha j}$$
$$= -\frac{1}{n_\alpha n_j}\int f^{(0)}_\alpha f^{(0)}_j\mathbf{W}^0_\alpha\mathbf{W}_\alpha S^{(q)}_{5/2}(W^2_\alpha) : \big[V'_{\alpha i}\mathbf{W}'^0_\alpha\mathbf{W}'_\alpha S^{(p)}_{5/2}(W'^2_\alpha)$$
$$- V_{\alpha i}\mathbf{W}^0_\alpha\mathbf{W}_\alpha S^{(p)}_{5/2}(W^2_\alpha)\big]g_{\alpha j}b\,\mathrm{d}b\,\mathrm{d}\varphi\,\mathrm{d}\mathbf{v}_\alpha\,\mathrm{d}\mathbf{v}_j. \qquad (5.2.74)$$

As a result of analogous calculations it can be shown that bracket integrals (5.2.73), (5.2.74) are equal to zero like (5.2.29), (5.2.30).

5.3. Kinetic coefficient calculation with taking into account the statistical fluctuations

In the frame of developed approximate method we need to use the following form of distribution function for calculation of kinetic coefficients and hydrodynamic fluxes:

$$f_\alpha = f_\alpha^{(0)} + f_\alpha^{(0)}\psi_\alpha - \frac{D}{Dt}\left[\tau^{(0)} f_\alpha^{(0)}\psi_\alpha\right], \tag{5.3.1}$$

where

$$\psi_\alpha = -A_\alpha \mathbf{W}_\alpha - B_\alpha \mathbf{W}_\alpha^0 \mathbf{W}_\alpha : \frac{\partial}{\partial \vec{r}}\mathbf{v}_0 + n\sum_j C_\alpha^{(j)} \mathbf{W}_\alpha \cdot \mathbf{d}_j. \tag{5.3.2}$$

Let us calculate diffusive fluxes of α-species ($\alpha = 1, \ldots, \eta$).

$$\begin{aligned} \mathbf{J}_\alpha &= \rho_\alpha \overline{\mathbf{V}}_\alpha \\ &= m_\alpha \int \mathbf{V}_\alpha f_\alpha^{(0)}\psi_\alpha \,\mathrm{d}\mathbf{V}_\alpha - m_\alpha \int \mathbf{V}_\alpha \frac{D}{Dt}\left[\tau^{(0)} f_\alpha^{(0)}\psi_\alpha\right]\mathrm{d}\mathbf{V}_\alpha, \end{aligned} \tag{5.3.3}$$

where

$$\frac{D}{Dt} = \frac{\partial}{\partial t} + \mathbf{v}_\alpha \cdot \frac{\partial}{\partial \mathbf{r}}. \tag{5.3.4}$$

Let us denote

$$\mathbf{J}_\alpha^\wedge = m_\alpha \int \mathbf{V}_\alpha f_\alpha^{(0)}\psi_\alpha \,\mathrm{d}\mathbf{V}_\alpha, \tag{5.3.5}$$

$$\mathbf{J}_\alpha^\tau = -m_\alpha \int \mathbf{V}_\alpha \frac{D}{Dt}\left[\tau^{(0)} f_\alpha^{(0)}\psi_\alpha\right]\mathrm{d}\mathbf{V}_\alpha. \tag{5.3.6}$$

Obviously diffusive flux $\mathbf{J}_\alpha^\wedge$ can be obtained from the Chapman–Enskog expression (Hirschfelder, Curtiss and Bird, 1954)

$$\mathbf{J}_\alpha^\wedge = \frac{m_\alpha n^2}{\rho}\sum_{j=1}^{\eta} m_j D_{\alpha j}\mathbf{d}_j - D_\alpha^T \frac{\partial \ln T}{\partial \mathbf{r}}, \tag{5.3.7}$$

where $D_{\alpha j}$ D_α^T are coefficients of diffusion and thermo-diffusion, correspondingly. For calculation of $\mathbf{J}_\alpha^\tau$ we need to substitute (5.3.2) into integral (5.3.6) and use the additional conditions which – as in Enskog theory – have the form

$$\int f_\alpha^{(1)} \,\mathrm{d}\mathbf{v}_\alpha = 0, \tag{5.3.8}$$

$$\sum_{\alpha=1}^{\eta} m_\alpha \int \mathbf{v}_\alpha f_\alpha^{(1)} \, \mathrm{d}\mathbf{v}_\alpha = 0, \tag{5.3.9}$$

$$\sum_{\alpha=1}^{\eta} \int V_\alpha^2 f_\alpha^{(1)} \, \mathrm{d}\mathbf{v}_\alpha = 0, \tag{5.3.10}$$

because $f_\alpha^{(1)}$ corresponds to classical solution of Enskog equation in the first approximation $f_\alpha^{(1)} = f_\alpha^{(0)} \psi_\alpha$.

Then one supposes that numerical density of component, mean mass velocity of gas mixture and temperature are defined by Maxwellian distribution function. Mentioned transformations lead to result

$$\mathbf{J}_\alpha^\tau = -\frac{\partial}{\partial t}\left(\tau^{(0)} \mathbf{J}_\alpha^\wedge\right) - \tau^{(0)} \left(\mathbf{J}_\alpha^\wedge \cdot \frac{\partial}{\partial \mathbf{r}}\right) \mathbf{v}_0 - \frac{\partial}{\partial \mathbf{r}} \cdot \left(\mathbf{v}_0 \tau^{(0)} \mathbf{J}_\alpha^\wedge\right)$$
$$- \frac{\partial}{\partial \mathbf{r}} \cdot \left[\tau^{(0)} m_\alpha \int \mathbf{V}_\alpha \mathbf{V}_\alpha f_\alpha^{(0)} \, \mathrm{d}\mathbf{V}_\alpha\right]. \tag{5.3.11}$$

Calculation of integral on the right-hand side of (5.3.11) is realized by the use of (5.3.2):

$$\int \mathbf{V}_\alpha \mathbf{V}_\alpha f_\alpha^{(1)} \, \mathrm{d}\mathbf{V}_\alpha = -\frac{m_\alpha}{15kT} \overleftrightarrow{S} \int B_\alpha(W_\alpha) V_\alpha^4 f_\alpha^{(0)} \, \mathrm{d}\mathbf{V}_\alpha, \tag{5.3.12}$$

where $\overleftrightarrow{S}$ is tensor, defined by expression ($i, k = 1, 2, 3$)

$$S_{ik} = \frac{1}{2}\left(\frac{\partial v_{0i}}{\partial r_k} + \frac{\partial v_{0k}}{\partial r_i}\right) - \frac{1}{3}\delta_{ik} \frac{\partial}{\partial \mathbf{r}} \cdot \mathbf{v}_0. \tag{5.3.13}$$

Introduce dynamical viscosity of α-component,

$$\mu_\alpha = \frac{1}{15} \frac{m_\alpha^2}{2k_B T} \int B_\alpha(W_\alpha) V_\alpha^4 f_\alpha^{(0)} \, \mathrm{d}\mathbf{V}_\alpha. \tag{5.3.14}$$

Then (5.3.11) can be transformed:

$$\mathbf{J}_\alpha^\tau = -\tau^{(0)} \left(\mathbf{J}_\alpha^\wedge \cdot \frac{\partial}{\partial \mathbf{r}}\right) \mathbf{v}_0 - \frac{\partial}{\partial t}\left(\tau^{(0)} \mathbf{J}_\alpha^\wedge\right) - \frac{\partial}{\partial \mathbf{r}} \cdot \left(\tau^{(0)} \mathbf{v}_0 \mathbf{J}_\alpha^\wedge\right)$$
$$+ \frac{\partial}{\partial \mathbf{r}} \cdot 2\tau^{(0)} \mu_\alpha \overleftrightarrow{S}. \tag{5.3.15}$$

Using condition

$$\sum_{\alpha=1}^{\eta} \mathbf{J}_\alpha^\wedge = 0, \tag{5.3.16}$$

we see that

$$\sum_{\alpha} \mathbf{J}_{\alpha}^{\tau} = 2\frac{\partial}{\partial \mathbf{r}} \cdot \left(\tau^{(0)} \mu \overleftrightarrow{S}\right), \tag{5.3.17}$$

where μ is viscosity of gas mixture

$$\mu = \sum_{\alpha} \mu_{\alpha}.$$

Therefore inconsistency in (5.3.17) appear which it is not equal to zero, and is proportional to squared viscosity (see (1.3.66)) or – for dimensionless form of equation – to squared Knudsen number. The origin of this inconsistency has to do with the use of additional conditions in the form (5.3.8)–(5.3.10), obtained for shortened classical solution $f_{\alpha} = f_{\alpha}^{(0)}(1 + \psi_{\alpha})$. Practically this inconsistency is related with generalized Barnett approximation, but nevertheless the sum of diffusive fluxes is turned into zero for another definition of fluxes $\mathbf{J}_{\alpha}$ for generalized Navier–Stokes approximation

$$\mathbf{J}_{\alpha} = \mathbf{J}_{\alpha}^{\wedge} - \tau^{(0)}\left(\mathbf{J}_{\alpha}^{\wedge} \cdot \frac{\partial}{\partial \mathbf{r}}\right)\mathbf{v}_0 - \frac{\partial}{\partial t}\left(\tau^{(0)}\mathbf{J}_{\alpha}^{\wedge}\right) - \frac{\partial}{\partial \mathbf{r}} \cdot \left(\tau^{(0)}\mathbf{v}_0\mathbf{J}_{\alpha}^{\wedge}\right), \tag{5.3.18}$$

introducing the corresponding inconsistent terms in fluctuation parts of the generalized equations.

Consider now pressure tensor

$$\begin{aligned}\overleftrightarrow{P} = {} & m_{\alpha}\int \mathbf{V}_{\alpha}\mathbf{V}_{\alpha} f_{\alpha}^{(0)}\,\mathrm{d}\mathbf{V}_{\alpha} + m_{\alpha}\int \mathbf{V}_{\alpha}\mathbf{V}_{\alpha} f_{\alpha}^{(0)}\psi_{\alpha}\,\mathrm{d}\mathbf{V}_{\alpha} \\ & - m_{\alpha}\int \mathbf{V}_{\alpha}\mathbf{V}_{\alpha}\frac{\partial}{\partial t}\left(\tau^{(0)} f_{\alpha}^{(0)}\psi_{\alpha}\right)\mathrm{d}\mathbf{V}_{\alpha} \\ & - m_{\alpha}\int \mathbf{V}_{\alpha}\mathbf{V}_{\alpha}\frac{\partial}{\partial \mathbf{r}} \cdot \left(\tau^{(0)}\mathbf{V}_{\alpha} f_{\alpha}^{(0)}\psi_{\alpha}\right)\mathrm{d}\mathbf{V}_{\alpha}.\end{aligned} \tag{5.3.19}$$

Let us now pass over cumbersome calculations and present their result (see also Alexeev, 1994):

$$\begin{aligned}\overleftrightarrow{P}_{\alpha} = {} & p_{\alpha}\overleftrightarrow{I} - 2\mu\overleftrightarrow{S} - \mathbf{J}_{\alpha}^{\wedge}\tau^{(0)}\frac{\partial \mathbf{v}_0}{\partial t} - \tau^{(0)}\frac{\partial \mathbf{v}_0}{\partial t}\mathbf{J}_{\alpha}^{\wedge} + 2\frac{\partial}{\partial t}\left(\tau^{(0)}\mu_{\alpha}\overleftrightarrow{S}\right) \\ & - 2\mathbf{v}_0\frac{\partial}{\partial \mathbf{r}} \cdot \left(\tau^{(0)}\mu_{\alpha}\overleftrightarrow{S}\right) + 2\tau^{(0)}\mu_{\alpha}\left[\overleftrightarrow{S} \cdot \frac{\partial}{\partial \mathbf{r}}\right]\mathbf{v}_0 + 2\left[\frac{\partial}{\partial \mathbf{r}} \cdot \left(\tau^{(0)}\mathbf{v}_0\right]\mu_{\alpha}\overleftrightarrow{S}\right) \\ & - \mathbf{J}_{\alpha}^{\wedge}\tau^{(0)}\left(\mathbf{v}_0 \cdot \frac{\partial}{\partial \mathbf{r}}\right)\mathbf{v}_0 - \left[\left(\mathbf{v}_0 \cdot \frac{\partial}{\partial \mathbf{r}}\right)\mathbf{v}_0\right]\mathbf{J}_{\alpha}^{\wedge}\tau^{(0)} + \overleftrightarrow{T}_{\alpha} \\ & - \overleftrightarrow{D}_{\alpha} - {}^{v_0}\overleftrightarrow{H}_{\alpha}{}^{v_0}.\end{aligned} \tag{5.3.20}$$

The cross-disposition of brackets in (5.3.20) underlines the order of tensor calculations:

$$\left[\frac{\partial}{\partial \mathbf{r}} \cdot \left(\tau^{(0)}\mathbf{v}_0\right]\mu_\alpha \overleftrightarrow{S}\right) \equiv \left(\mathbf{v}_0 \cdot \frac{\partial}{\partial \mathbf{r}}\right)\mu_\alpha \overleftrightarrow{S} + \mu_\alpha \overleftrightarrow{S} \frac{\partial}{\partial \mathbf{r}} \cdot \mathbf{v}_0; \tag{5.3.21}$$

$\overleftrightarrow{T}_\alpha$, $\overleftrightarrow{D}_\alpha$, ${}^{v_0}\overleftrightarrow{H}_\alpha$ are tensors with components of the form $(k, \ell = 1, 2, 3)$

$$\begin{aligned} T_{k\ell,\alpha} &= \frac{\partial}{\partial r_k}\left(\tau^{(0)} K_\alpha \frac{\partial T}{\partial r_\ell}\right) + \frac{\partial}{\partial r_\ell}\left(\tau^{(0)} K_\alpha \frac{\partial T}{\partial r_k}\right) \\ &\quad + \delta_{k\ell} \sum_{i=1}^{3} \frac{\partial}{\partial r_i}\left(\tau^{(0)} K_\alpha \frac{\partial T}{\partial r_i}\right), \quad K_\alpha = \frac{k_B}{m_\alpha} D_\alpha^T + \frac{2}{5}\lambda'_\alpha, \end{aligned} \tag{5.3.22}$$

$$\begin{aligned} D_{k\ell,\alpha} &= \frac{\partial}{\partial r_k}\left(\tau^{(0)} \mathcal{L}_{\alpha\ell}\right) + \frac{\partial}{\partial r_\ell}\left(\tau^{(0)} \mathcal{L}_{\alpha k}\right) + \delta_{k\ell} \sum_{i=1}^{3} \frac{\partial}{\partial r_i}\left(\tau^{(0)} \mathcal{L}_{\alpha i}\right), \\ \mathcal{L}_{\alpha k} &= \frac{p}{\rho} n \sum_\beta m_\beta \left(D_{\alpha\beta} - D^1_{\alpha\beta}\right) d_{\beta k}, \end{aligned} \tag{5.3.23}$$

$${}^{v_0}H_{k\ell,\alpha} = -2\sum_i \frac{\partial}{\partial r_i}\left\{\tau^{(0)} v_{0k}\mu_\alpha \left[\frac{1}{2}\left(\frac{\partial v_{0\ell}}{\partial r_i} + \frac{\partial v_{0i}}{\partial r_\ell}\right) - \frac{1}{3}\delta_{i\ell}\frac{\partial}{\partial r_i} \cdot \mathbf{v}_0\right]\right\}. \tag{5.3.24}$$

Eq. (5.3.22) contains the Boltzmann's constant k_B and the coefficient of thermal conduction λ'_α (Hirschfelder, Curtiss and Bird, 1954)

$$\lambda'_\alpha = -\frac{5}{4} k_B n \alpha \sqrt{\frac{2kT}{m_\alpha}}\, a_{\alpha 1}, \tag{5.3.25}$$

where $a_{\alpha 1}$ is a coefficient in expansion

$$A_\alpha = \sum_{m=0}^{\xi} a_{\alpha m} S^{(m)}_{3/2}\left(W_\alpha^2\right). \tag{5.3.26}$$

Relation (5.3.23), apart from usual "zero" diffusive coefficient contains

$$D_{\alpha\beta} = \frac{\rho n_\alpha}{2nm_\beta}\sqrt{\frac{2k_B T}{m_\alpha}}\, C^{(\beta,\alpha)}_{\alpha 0}, \tag{5.3.27}$$

also the "first" diffusive coefficient defined by relation

$$D^1_{\alpha\beta} = \frac{\rho n_\alpha}{2nm_\beta}\sqrt{\frac{2k_B T}{m_\alpha}}\, C^{(\beta,\alpha)}_{\alpha 1}. \tag{5.3.28}$$

Pressure tensor of gas mixture $\overset{\leftrightarrow}{P}$ forms by summation over all components of gas mixture

$$\overset{\leftrightarrow}{P} = p\overset{\leftrightarrow}{I} - 2\eta\overset{\leftrightarrow}{S} + 2\frac{\partial}{\partial t}\left(\tau^{(0)}\mu\overset{\leftrightarrow}{S}\right) - 2\mathbf{v}_0 \cdot \left(\tau^{(0)}\mu\overset{\leftrightarrow}{S}\right) + 2\tau^{(0)}\mu\left[\overset{\leftrightarrow}{S} \cdot \frac{\partial}{\partial \mathbf{r}}\right]\mathbf{v}_0$$
$$+ \overset{\leftrightarrow}{T} - \overset{\leftrightarrow}{D} - {}^{v_0}\overset{\leftrightarrow}{H}, \tag{5.3.29}$$

where

$$\overset{\leftrightarrow}{T} = \sum_\alpha \overset{\leftrightarrow}{T}_\alpha; \qquad \overset{\leftrightarrow}{D} = \sum_\alpha \overset{\leftrightarrow}{D}_\alpha; \qquad {}^{v_0}\overset{\leftrightarrow}{H} = \sum_\alpha {}^{v_0}\overset{\leftrightarrow}{H}_\alpha. \tag{5.3.30}$$

To use hydrodynamic equations in the generalized Navier–Stokes approximation we need to calculate the thermal flux

$$\mathbf{q}_\alpha = \mathbf{q}_\alpha^\wedge - \frac{m_\alpha}{2}\int \mathbf{V}_\alpha V_\alpha^2 \frac{\partial}{\partial t}\left(\tau^{(0)} f_\alpha^{(0)} \psi_\alpha\right) \mathrm{d}\mathbf{V}_\alpha$$
$$- \frac{m_\alpha}{2}\int \mathbf{V}_\alpha V_\alpha^2 \frac{\partial}{\partial \mathbf{r}} \cdot \left(\mathbf{v}_\alpha \tau^{(0)} f_\alpha^{(0)} \psi_\alpha\right) \mathrm{d}\mathbf{V}_\alpha. \tag{5.3.31}$$

Calculations lead to result:

$$\mathbf{q}_\alpha = \mathbf{q}_\alpha^\wedge - \frac{\partial}{\partial t}\left(\tau^{(0)}\mathbf{q}_\alpha^\wedge\right) + \frac{1}{2}v_0^2\tau^{(0)}\mathbf{J}_\alpha^\wedge \frac{\partial}{\partial \mathbf{r}} \cdot \mathbf{v}_0 + \frac{1}{2}v_0^2\left(\mathbf{v}_0 \cdot \frac{\partial}{\partial \mathbf{r}}\right)\left(\tau^{(0)}\mathbf{J}_\alpha^\wedge\right)$$
$$+ \frac{1}{2}\mathbf{J}_\alpha^\wedge \tau^{(0)}\left(\mathbf{v}_0 \cdot \frac{\partial}{\partial \mathbf{r}}\right)v_0^2 + 2\tau^{(0)}\mu_\alpha\left(\mathbf{v}_0 \cdot \frac{\partial}{\partial \mathbf{r}}\right)\left(\mathbf{v}_0 \cdot \overset{\leftrightarrow}{S}\right)$$
$$+ 2\tau^{(0)}\mu_\alpha\left(\mathbf{v}_0 \cdot \overset{\leftrightarrow}{S}\right) \cdot \frac{\partial}{\partial \mathbf{r}}\mathbf{v}_0 - \frac{\partial}{\partial \mathbf{r}} \cdot \left(\tau^{(0)}\mu_\alpha \overset{\leftrightarrow}{S} v_0^2\right) + v_0^2 \frac{\partial}{\partial \mathbf{r}} \cdot \left(\tau^{(0)}\mu_\alpha \overset{\leftrightarrow}{S}\right)$$
$$- \frac{\partial}{\partial \mathbf{r}} \cdot \left[\frac{7p_\alpha}{\rho_\alpha}\tau^{(0)}\left(\mu_\alpha^1 - \mu_\alpha\right)\overset{\leftrightarrow}{S}\right] + \tau^{(0)}K_\alpha\left[\frac{\partial}{\partial \mathbf{r}}\mathbf{v}_0 \cdot \frac{\partial T}{\partial \mathbf{r}} + \frac{\partial T}{\partial \mathbf{r}} \cdot \frac{\partial}{\partial \mathbf{r}}\mathbf{v}_0\right.$$
$$\left. + \frac{\partial T}{\partial \mathbf{r}}\frac{\partial}{\partial \mathbf{r}} \cdot \mathbf{v}_0\right] - \tau^{(0)}\frac{pn}{\rho}\left[\frac{\partial}{\partial \mathbf{r}}\mathbf{v}_0 \cdot \sum_\beta m_\beta\left(D_{\alpha\beta} - D_{\alpha\beta}^1\right)\mathbf{d}_\beta\right.$$
$$\left. + \sum_\beta m_\beta\left(D_{\alpha\beta} - D_{\alpha\beta}^1\right)\mathbf{d}_\beta \cdot \frac{\partial}{\partial \mathbf{r}}\mathbf{v}_0 + \sum_\beta m_\beta\left(D_{\alpha\beta} - D_{\alpha\beta}^1\right)\mathbf{d}_\beta \frac{\partial}{\partial \mathbf{r}} \cdot \mathbf{v}_0\right]$$
$$- \frac{5}{2}\frac{p}{\rho}\tau^{(0)}\sum_\beta m_\beta\left(D_{\alpha\beta} - D_{\alpha\beta}^1\right)\left(\mathbf{d}_\beta \cdot \frac{\partial}{\partial \mathbf{r}}\right)\mathbf{v}_0$$
$$+ \frac{5}{2}\tau^{(0)}K_\alpha\left(\frac{\partial T}{\partial \mathbf{r}} \cdot \frac{\partial}{\partial \mathbf{r}}\right)\mathbf{v}_0. \tag{5.3.32}$$

Eq. (5.3.32) contains left- and right scalar tensor products of diada $\frac{\partial}{\partial \mathbf{r}}\mathbf{v}_0$.

The "first" viscosity coefficient introduced in (5.3.32) is

$$\mu_\alpha^1 = \frac{1}{2}k_B T n_\alpha b_{\alpha 1}. \tag{5.3.33}$$

After summation over all components of the gas mixture we obtain heat flux for the mixture:

$$
\begin{aligned}
\mathbf{q} = {} & \mathbf{q}^{\wedge} - \frac{\partial}{\partial t}\left(\tau^{(0)}\mathbf{q}^{\wedge}\right) + 2\tau^{(0)}\mu\left(\mathbf{v}_0 \cdot \frac{\partial}{\partial \mathbf{r}}\right)\left(\mathbf{v}_0 \cdot \overleftrightarrow{S}\right) \\
& + 2\tau^{(0)}\mu\left(\mathbf{v}_0 \cdot \overleftrightarrow{S}\right) \cdot \frac{\partial}{\partial \mathbf{r}}\mathbf{v}_0 - \frac{\partial}{\partial \mathbf{r}} \cdot \left(\tau^{(0)}\overleftrightarrow{S} v_0^2\right) + v_0^2 \frac{\partial}{\partial \mathbf{r}} \cdot \left(\tau^{(0)}\mu\overleftrightarrow{S}\right) \\
& - \frac{\partial}{\partial \mathbf{r}} \cdot \left[7\overleftrightarrow{S}\tau^{(0)}\sum_{\alpha}\frac{p_\alpha}{\rho_\alpha}\left(\mu_\alpha^1 - \mu_\alpha\right)\right] + \tau^{(0)}K\left[\frac{\partial}{\partial \mathbf{r}}\mathbf{v}_0 \cdot \frac{\partial T}{\partial \mathbf{r}} + \frac{\partial T}{\partial \mathbf{r}} \cdot \frac{\partial}{\partial \mathbf{r}}\mathbf{v}_0\right. \\
& \left. + \frac{\partial T}{\partial \mathbf{r}}\frac{\partial}{\partial \mathbf{r}} \cdot \mathbf{v}_0\right] - \tau^{(0)}\frac{pn}{\rho}\left[\frac{\partial}{\partial \mathbf{r}}\mathbf{v}_0 \cdot \sum_{\alpha\beta} m_\beta\left(D_{\alpha\beta} - D_{\alpha\beta}^1\right)\mathbf{d}_\beta\right. \\
& \left. + \sum_{\alpha\beta} m_\beta\left(D_{\alpha\beta} - D_{\alpha\beta}^1\right)\mathbf{d}_\beta \cdot \frac{\partial}{\partial \mathbf{r}}\mathbf{v}_0 + \sum_{\alpha\beta} m_\beta\left(D_{\alpha\beta} - D_{\alpha\beta}^1\right)\mathbf{d}_\beta\frac{\partial}{\partial \mathbf{r}} \cdot \mathbf{v}_0\right] \\
& - \frac{5}{2}\tau^{(0)}\frac{p}{\rho}\sum_{\alpha\beta} m_\beta\left(D_{\alpha\beta} - D_{\alpha\beta}^1\right)\left(\mathbf{d}_\beta \cdot \frac{\partial}{\partial \mathbf{r}}\right)\mathbf{v}_0 \\
& + \frac{5}{2}K\tau^{(0)}\left(\frac{\partial T}{\partial \mathbf{r}} \cdot \frac{\partial}{\partial \mathbf{r}}\right)\mathbf{v}_0.
\end{aligned} \tag{5.3.34}
$$

As is evident from the foregoing, the heat flux – in its fluctuation part – depends also on coefficients of diffusion and viscosity in considered physical system. It remains only to calculate integrals, corresponding to average values in the generalized Navier–Stokes approximation. These integrals are given in Appendix 3. We can state that developed theory of calculation of kinetic coefficients leads to the appearance of additional τ-terms which contain time and space fluctuations including cross effects of different transport mechanisms.

CHAPTER 6

Some Applications of the Generalized Boltzmann Physical Kinetics

The classical Boltzmann equation (BE) is valid only on characteristic scales related to the hydrodynamic time of flow and average time between particle collisions. It was previously established that the inclusion of the third possible scale, the time of collisions between particles, results in the emergence of additional terms of the Boltzmann equation, which are of the same order of magnitude as the other terms in the equation (cf., Alexeev, 1994, 1995c, 2000b). It means that additional terms (in comparison with classical BE) in GBE are proportional to Knudsen number and cannot be considered small values in region $Kn \gtrsim 1$. But even if $Kn \ll 1$, the corresponding terms generally speaking cannot be omitted because these terms contain small coefficients in front of senior derivatives, which can lead to effects of "boundary layer". The solution of the problem using the generalized Boltzmann equation (GBE) results in the change, to some extent, in the results in the field of traditional applications of GBE and classical hydrodynamic theory, which is a direct consequence of the kinetic theory. In treating some problems, the results obtained by the generalized theory coincide with experimental data much more precisely than similar results of the classical theory and sometimes describe the processes for which classical theories are simply invalid.

In the following we intend to demonstrate the possibilities of the generalized Boltzmann kinetics solving some problems in area of traditional application of BE: transport processes in weakly ionized gas with inelastic collisions; investigation of the sound wave propagation; calculation of the shock wave structure.

6.1. Investigation of the generalized Boltzmann equation for electron energy distribution in a constant electric field with due regard for inelastic collisions

We begin with numerical solutions of the generalized Boltzmann equation, which are carried out for electrons of a weakly ionized plasma in a constant electric field with due regard for inelastic collisions for the case of constant collision frequency between electrons and heavy particles ($\nu_{ea} = 1/\tau = const$). For this purpose, a set of cross sections of collisions between electrons and heavy particles is used, applied in simulation of processes in a hydrogen plasma. The results are compared with those obtained by means of numerical solution of the "traditional" Boltzmann equation.

Until now the results related to the GBE as a rule have been obtained using analytical methods. These methods impose considerable limitations on the application of the equation for description of real objects and, in particular, make the detailed account of inelastic interactions impossible. It is these interactions that are important in solving numerous problems of physics and plasma chemistry. The aim of this study was to construct an algorithm for numerical solution of GBE for plasma electrons in a constant electric field with due regard for inelastic collisions between electrons and heavy particles.

In view of this, a nontrivial problem is that on the set of cross sections used in solution of the equation. Strictly speaking, this should be a set corresponding to a certain atom (molecule) obtained by calculation or experimentally. In practice, such a rigorous approach cannot be realized because of the fact that far from all the cross sections are known even for well-studied particles such as inert gases and hydrogen. Therefore, sets of cross sections are used which were constructed from the known ones in such a way that the calculated drift velocity and the first Townsend coefficient should coincide with experimental quantities in a wide range of values of reduced electric field E/N. In so doing, the cross sections entering the set may be changed in magnitude and shifted along the axis of electron energy (as compared to initial values). Thus, it is only a full set of cross sections that has the physical meaning, because it yields the electron distribution function in energies, corresponding to the gas being described. No cross-section may be added to or excluded from this set. The rate coefficients of the processes involving electrons are calculated with the help of electron energy distribution and cross sections of these processes.

All known solutions of the BE are obtained using such sets of cross sections. In this study, we restricted ourselves to the construction of the algorithm of numerical solution of the GBE for the case of constant collision frequency between electrons and heavy particles ($\nu_{ea} = const$). As is known, this condition is valid for helium and hydrogen (Shkarovsky, Johnstone and Bachinsky, 1969; Golant, 1959). In solving the GBE, we used the set of cross sections for molecular hydrogen (Gal'tsev et al., 1979; Lebedev and Epstein, 1995).

The second objective of this study was to compare the results obtained by numerical solution of the GBE and BE for one and the same set of cross sections. The results of such comparison are of extraordinary importance in defining the trend of further treatment of the GBE. Indeed, if both equations yield coinciding results, this may be indicative of the fact that, for the problems being treated, additional terms appearing in the GBE make no contribution to the results. This conclusion must simply be tested for various sets of cross sections.

If the results are different, the role of additional terms is considerable. However, the physical interpretation of the results is hardly possible at this stage, because, as follows from the foregoing, the used set of cross sections corresponds to the BE rather than to GBE. Therefore, the further stage should be construction of a set of cross sections for the GBE (in the way described above) and physical interpretation of the obtained solutions. This is an independent problem falling beyond the scope of this study.

Detailed derivation of the equations for numerical solution of the GBE is given below. The generalized Boltzmann equation for plasma electron component is written in the form

$$\frac{Df_e}{Dt} - \frac{D}{Dt}\left(\tau \frac{Df_e}{Dt}\right) = J_{ea}, \tag{6.1.1}$$

where

$$\frac{D}{Dt} = \frac{\partial}{\partial t} + \mathbf{v}_e \cdot \frac{\partial}{\partial \mathbf{r}} + \mathbf{F}_e \cdot \frac{\partial}{\partial \mathbf{v}_e};$$

$\tau = 1/\overline{N\sigma(v)v}$ is the average time between collisions of electrons and heavy particles (the bar designates averaging over the electron distribution in energies); and J_{ea} is the collision integral for electrons, including elastic and inelastic processes. For comparison, the classical Boltzmann equation is given,

$$\frac{Df_e}{Dt} = J_{ea}. \tag{6.1.2}$$

Consider the case of spatially homogeneous weakly ionized gas, assuming that collisions between charged particles may be ignored. This means that

$$\nu_{ee} \ll \delta \nu_{ea}. \tag{6.1.3}$$

Here, ν_{ee} is the frequency of collisions between electrons; $\nu_{ea} = \overline{N\sigma_{ea}(v)v}$ is the frequency of collisions between electrons and neutral particles; N is the concentration of heavy particles; σ_{ea} is the cross section of electron collisions with atoms (molecules); and δ is the fraction of energy lost by a electron in a single collision against a neutral particle. It is also assumed that the external electric field is stationary, and there is no magnetic field. In this case, Eq. (6.1.1) is written in the form

$$\mathbf{F}_e \cdot \frac{\partial f_e}{\partial \mathbf{v}_e} - \tau \frac{\partial^2 f_e}{\partial \mathbf{v}_e \partial \mathbf{v}_e} : \mathbf{F}_e \mathbf{F}_e = J_{ea}, \tag{6.1.4}$$

where $\mathbf{F}_e = e\mathbf{E}/m_e$. Solution of (6.1.4) for the energy distribution of electrons is found using the traditional perturbation method with the help of expansion,

$$f_e(v_e) = f_0(v_e) + \mathbf{F}_e \cdot \mathbf{v}_e f_1(v_e) + \cdots. \tag{6.1.5}$$

In the case of weak anisotropy, we may restrict our consideration to the first two terms of energy expansion, and, with the elastic electron collisions alone taken into account, the set of equation with respect to $f_0(v)$ and $f_1(v)$ has the form (4.2.6), (4.2.7)

$$f_1 + \frac{1}{3} v_e \frac{\partial f_1}{\partial v_e} - \frac{\tau}{3}\left\{\frac{2}{v_e}\frac{\partial f_0}{\partial v_e} + \frac{\partial^2 f_0}{\partial v_e^2}\right\} = \frac{1}{F_e^2} J_0, \tag{6.1.6}$$

$$\frac{\partial f_0}{\partial v_e} - \frac{3}{5}\tau F_e^2 \left\{ 4\frac{\partial f_1}{\partial v_e} + v_e \frac{\partial^2 f_1}{\partial v_e^2} \right\} = -\frac{3}{2}\frac{1}{F_e} J_1. \tag{6.1.7}$$

In Eqs. (6.1.6) and (6.1.7), J_0, J_1 are the collision integrals, which should be presented in the form conventional for the Boltzmann equation (4.2.8), (4.2.9),

$$J_0 = \frac{m_e \widehat{T}_a}{m_a v_e^2} \frac{\partial}{\partial v_e} \left[v_e^3 \nu_{ea}^{(\mathrm{el})} \left(\frac{f_0}{\widehat{T}_a} + \frac{1}{m_e v_e} \frac{\partial f_0}{\partial v_e} \right) \right], \tag{6.1.8}$$

$$J_1 = \frac{2}{3} F_e \frac{v_e^2}{\ell} f_1, \tag{6.1.9}$$

where $\nu_{ea}^{(\mathrm{el})}$ is the frequency of elastic collisions; ℓ is the mean free path of electrons; and $\widehat{T}_a = k_B T$ is the gas temperature in energy units. Further transformation of Eqs. (6.1.6) and (6.1.7) makes it possible to derive the equation for the case of $\nu_{ea} = const$

$$\begin{aligned} &\breve{v}_e^2 \left[1 + \frac{3m_e}{\breve{\varepsilon} m_a} \right] \breve{f}_0''' + \breve{v}_e \left[2 + \frac{6m_e}{\breve{\varepsilon} m_a} + \frac{3m_e \breve{v}_e^2}{m_a} \right] \breve{f}_0'' \\ &\quad - \left[2 + \frac{6m_e}{\breve{\varepsilon} m_a} + \breve{v}_e^2 \left(\frac{10}{3} + \frac{5m_e}{m_a} - 12\frac{m_e}{m_a} \right) \right] \breve{f}_0' - 5\frac{m_e}{m_a} \breve{v}_e^3 \breve{f}_0 = 0, \end{aligned} \tag{6.1.10}$$

where

$$\breve{v}_e = \frac{v_e}{\tau F_e}, \qquad \breve{\varepsilon} = \frac{m_e \tau^2 F_e^2}{\widehat{T}_a}, \qquad \breve{f}_0 = \frac{f_0}{f_0(v_e = 0)}.$$

The next stage of transformation of the generalized Boltzmann equation is the inclusion of real cross sections of elastic and inelastic collisions. The Boltzmann equation with due regard for inelastic collisions is written in the form (Lukovnikov and Fetisov, 1969)

$$\mathbf{F}_e \cdot \frac{\partial f_e}{\partial \mathbf{v}_e} = J_{ea}^{(\mathrm{el})} + \sum_m J_{ea}^{(m)} + \sum_n J_i^{(n)} + J_{ea}^{(\mathrm{r})}. \tag{6.1.11}$$

Here, $J_{ea}^{(\mathrm{el})}$, $J_{ea}^{(m)}$, $J_i^{(n)}$, $J_{ea}^{(\mathrm{r})}$ are the terms of the collision integral, taking into account elastic collisions between electrons and neutral particles, excitation of the mth level, n-fold ionization, and recombination, respectively. We introduce the following notation for the inelastic part of the collision integral:

$$J_{ea}^{(\mathrm{inel})} = \sum_m J_{ea}^{(m)} + \sum_n J_i^{(n)} + J_{ea}^{(\mathrm{r})}. \tag{6.1.12}$$

Now, write the generalized Boltzmann equation with due regard for inelastic collisions

$$\mathbf{F}_e \cdot \frac{\partial f_e}{\partial \mathbf{v}_e} - \tau \frac{\partial^2 f_e}{\partial \mathbf{v}_e \partial \mathbf{v}_e} : \mathbf{F}_e \mathbf{F}_e = J_{ea}^{(\mathrm{el})} + J_{ea}^{(\mathrm{inel})}, \tag{6.1.13}$$

which is reduced, with the help of Eq. (6.1.5), to the following set of equations:

$$f_1 + \frac{1}{3} v_e \frac{\partial f_1}{\partial v_e} - \frac{\tau}{3} \left\{ \frac{2}{v_e} \frac{\partial f_0}{\partial v_e} + \frac{\partial^2 f_0}{\partial v_e^2} \right\} = \frac{1}{F_e^2} J_0^{(\mathrm{el})} + \frac{1}{F_e^2} J_{ea}^{(\mathrm{inel})}(f_0), \qquad (6.1.14)$$

$$\frac{\partial f_0}{\partial v_e} - \frac{3}{5} \tau F_e^2 \left\{ 4 \frac{\partial f_1}{\partial v_e} + v_e \frac{\partial^2 f_1}{\partial v_e^2} \right\} = -\frac{3}{2} \frac{1}{F_e} \left[J_1^{(\mathrm{el})} + J_{ea}^{(\mathrm{inel})}(f_1) \right], \qquad (6.1.15)$$

where J_0 and J_1, are defined by Eqs. (6.1.8) and (6.1.9), respectively.

We dwell into more detail on the choice of the inelastic part of the collision integral. We will use the following approximation for the component parts of $J_{ea}^{(\mathrm{inel})}$ (the first term on the right-hand side allows for electron energy losses in collisions against heavy particles in direct processes, and the second one for the energy return to electrons upon collisions of the second kind) (Gal'tsev et al., 1979; Lebedev and Epstein, 1995):

$$J_{ea}^{(m)} = -\nu_e^{(m)}(v_e) f(v_e) + \nu_e^{(m)}\left(v_e^{(m)}\right) \frac{v_e^{(m)}}{v_e} f\left(v_e^{(m)}\right),$$

$$J_i^{(n)}(v_e) = -\nu_i^{(n)}(v_e) f(v_e) + \eta \nu_i^{(n)}\left(v_e^{(n)}\right) \frac{v_e^{(n)}}{v_e} f\left(v_e^{(n)}\right),$$

$$J_{ea}^{(\mathrm{r})} = -\nu_r(v_e) f(v_e), \qquad (6.1.16)$$

where

$$\nu_e^{(m)} = N v_e \sigma_e^{(m)}(v_e), \qquad \nu_i^{(n)} = N v_e \sigma_i^{(n)}(v_e);$$

$v_e^{(m)}$, $v_i^{(n)}$ are the velocities of electrons defined by the equations

$$v_e^{(m)2} = v_e^2 + \frac{2\varepsilon_e^{(m)}}{m_e}, \qquad v_i^{(n)2} = \eta v_e^2 + \frac{2\varepsilon_i^{(n)}}{m_e}. \qquad (6.1.17)$$

Here, $\varepsilon_e^{(m)}$, $\varepsilon_i^{(n)}$, $\sigma_e^{(m)}$, $\sigma_i^{(n)}$ are the threshold energies and cross sections of the mth and nth processes, respectively, which are defined by the equilibrium between the primary and secondary electrons $\eta \approx 1$. For a weakly ionized plasma, the terms $J_{ea}^{(n)}$, $J_{ea}^{(\mathrm{r})}$ in the collision integral may be ignored (Karoulina and Lebedev, 1988, 1992; Lukovnikov and Fetisov, 1969).

We will now turn to the transformation of Eqs. (6.1.14) and (6.1.15) for the case $\tau = const$. It follows from Eqs. (6.1.14) and (6.1.15) that

$$F_e^2 \left\{ v_e^3 f_1 - \tau v_e^2 \frac{\partial f_0}{\partial v_e} \right\}'$$
$$= 3\widehat{T}_a \left\{ v_e^3 \nu_{ea}^{(\mathrm{el})} \left(\frac{f_0}{\widehat{T}_a} + \frac{1}{m_e v_e} \frac{\partial f_0}{\partial v_e} \right) \right\}' + 3 v_e^3 J_{ea}^{(\mathrm{inel})}(f_0), \qquad (6.1.18)$$

$$\frac{\partial f_0}{\partial v_e} - \frac{3}{5}\tau F_e^2 \left\{ 4\frac{\partial f_1}{\partial v_e} + v_e \frac{\partial^2 f_1}{\partial v_e^2} \right\} = -\nu_{ea} v_e f_1, \tag{6.1.19}$$

where $\nu_{ea} = \nu_{ea}^{(\mathrm{el})} + \nu_{ea}^{(\mathrm{inel})}$.

Relations (6.1.18) and (6.1.19) make it possible to exclude f_1 from the set of equations, because an explicit representation for f_1, may be obtained from Eq. (6.1.18).

It follows directly from Eq. (6.1.18) that

$$\begin{aligned} \{v_e^3 f_1\}' = {} & 3\widehat{T}_a \frac{m_e}{F_e^2 m_a} \left\{ v_e^3 \nu_{ea}^{(\mathrm{el})} \left(\frac{f_0}{\widehat{T}_a} + \frac{1}{m_e v_e} \frac{\partial f_0}{\partial v_e} \right) \right\}' + \left(\tau v_e^2 \frac{\partial f_0}{\partial v_e} \right)' \\ & + 3\frac{v_e^2}{F_e^2} J_{ea}^{(\mathrm{inel})}(f_0). \end{aligned} \tag{6.1.20}$$

On performing the integration of Eq. (6.1.20) in a way similar to that used for the expression for the case of elastic collisions in Alekseev (1995a), we derive

$$\begin{aligned} v_e^3 f_1 = {} & 3\widehat{T}_a \frac{m_e}{F_e^2 m_a} v_e^3 \nu_{ea}^{(\mathrm{el})} \left(\frac{f_0}{\widehat{T}_a} + \frac{1}{m_e v_e} \frac{\partial f_0}{\partial v_e} \right) + \tau v_e^2 \frac{\partial f_0}{\partial v_e} \\ & + 3\frac{1}{F_e^2} \int_0^{v_e} v_e^2 J_{ea}^{(\mathrm{inel})}(f_0)\, \mathrm{d}v_e. \end{aligned} \tag{6.1.21}$$

It follows directly from Eq. (6.1.21) that

$$\begin{aligned} f_1 = {} & 3\widehat{T}_a \frac{m_e}{F_e^2 m_a} \nu_{ea}^{(\mathrm{el})} \left(\frac{f_0}{\widehat{T}_a} + \frac{1}{m_e v_e} \frac{\partial f_0}{\partial v_e} \right) + \frac{\tau}{v_e} \frac{\partial f_0}{\partial v_e} \\ & + \frac{3}{F_e^2 v_e^3} \int_0^{v_e} v_e^2 J_{ea}^{(\mathrm{inel})}(f_0)\, \mathrm{d}v_e. \end{aligned} \tag{6.1.22}$$

We perform double differentiation of Eq. (6.1.22) to derive

$$\begin{aligned} \frac{\partial f_1}{\partial v_e} = {} & 3\widehat{T}_a \frac{m_e}{F_e^2 m_a} \frac{\partial \nu_{ea}^{(\mathrm{el})}}{\partial v_e} \left(\frac{f_0}{\widehat{T}_a} + \frac{1}{m_e v_e} \frac{\partial f_0}{\partial v_e} \right) - \frac{\tau}{v_e^2} \frac{\partial f_0}{\partial v_e} \\ & + 3\widehat{T}_a \frac{m_e}{F_e^2 m_a} \nu_{ea}^{(\mathrm{el})} \left(\frac{1}{\widehat{T}_a} \frac{\partial f_0}{\partial v_e} + \frac{1}{m_e v_e} \frac{\partial^2 f_0}{\partial v_e^2} - \frac{1}{m_e v_e^2} \frac{\partial f_0}{\partial v_e} \right) + \frac{\tau}{v_e} \frac{\partial^2 f_0}{\partial v_e^2} \\ & - \frac{9}{F_e^2 v_e^4} \int_0^{v_e} v_e^2 J_{ea}^{(\mathrm{inel})}(f_0)\, \mathrm{d}v_e + \frac{3}{F_e^2 v_e} J_{ea}^{(\mathrm{inel})}(f_0), \end{aligned} \tag{6.1.23}$$

$$\begin{aligned} \frac{\partial^2 f_1}{\partial v_e^2} = {} & \frac{3m_e}{F_e^2 m_a} \frac{\partial^2 \nu_{ea}^{(\mathrm{el})}}{\partial v_e^2} f_0 + \left(\frac{6m_e}{F_e^2 m_a} \frac{\partial \nu_{ea}^{(\mathrm{el})}}{\partial v_e} - \frac{6\widehat{T}_a}{F_e^2 m_a v_e^2} \frac{\partial \nu_{ea}^{(\mathrm{el})}}{\partial v_e} + \frac{2\tau}{v_e^3} \right. \\ & \left. + \frac{6\widehat{T}_a}{F_e^2 m_a v_e^3} \nu_{ea}^{(\mathrm{el})} + \frac{3\widehat{T}_a}{F_e^2 m_a v_e} \frac{\partial^2 \nu_{ea}^{(\mathrm{el})}}{\partial v_e^2} \right) \frac{\partial f_0}{\partial v_e} \end{aligned}$$

$$
\begin{aligned}
&+\left(\frac{6\widehat{T}_a}{F_e^2 m_a v_e}\frac{\partial \nu_{ea}^{(\mathrm{el})}}{\partial v_e}+\frac{3m_e}{F_e^2 m_a}\nu_{ea}^{(\mathrm{el})}-\frac{6\widehat{T}_a}{F_e^2 m_a v_e^2}\nu_{ea}^{(\mathrm{el})}-\frac{2\tau}{v_e^2}\right)\frac{\partial^2 f_0}{\partial v_e^2}\\
&+\left(\frac{\tau}{v_e}+\frac{3\widehat{T}_a}{F_e^2 m_a v_e}\nu_{ea}^{(\mathrm{el})}\right)\frac{\partial^3 f_0}{\partial v_e^3}+\frac{36}{F_e^2 v_e^5}\int_0^{v_e} v_e^2 J_{ea}^{(\mathrm{inel})}(f_0)\,\mathrm{d}v_e\\
&-\frac{12}{F_e^2 v_e^2}J_{ea}^{(\mathrm{inel})}(f_0)+\frac{3}{F_e^2 v_e}\frac{\partial J_{ea}^{(\mathrm{inel})}(f_0)}{\partial v_e},
\end{aligned}
\tag{6.1.24}
$$

accordingly,

$$
J_{ea}^{(\mathrm{inel})}(f_0)=\sum_\alpha\left[\nu_{ea}^{(\mathrm{inel})}\left(v_e+v_e^\alpha\right)\frac{v_e^{(m)}}{v_e}f_0\left(v_e+v_e^\alpha\right)-\nu_{ea}^{(\mathrm{inel})}(v_e)f_0(v_e)\right],
$$

where $\nu_{ea}^{(\mathrm{el})}$ is the frequency of elastic collisions; $\nu_{ea}^{(\mathrm{inel})}$ is the frequency of inelastic collisions; v_e^α is the threshold of the process a; and $v_e^{(m)}$ is calculated by Eq. (6.1.17).

The resultant equation is considerably simplified if collisions of the second kind are ignored, and the integral of inelastic collisions is written in the form

$$
J_{ea}^{(\mathrm{inel})}(f_e)=-\sum_\alpha \nu_{ea_\alpha}^{(\mathrm{inel})}f_e(v_e)\equiv-\nu_{ea}^{(\mathrm{inel})}f_e(v_e).
$$

On substituting Eqs. (6.1.22)–(6.1.24) into (6.1.19), we derive

$$
\begin{aligned}
&f_0'''\left(\tau+\frac{3\widehat{T}_a\nu_a^{(\mathrm{el})}}{F_e^2 m_a}\right)+f_0''\left(2\frac{\tau}{v_e}+\frac{6\widehat{T}_a}{F_e^2 m_a}\frac{\partial\nu_{ea}^{(\mathrm{el})}}{\partial v_e}+\frac{3m_e v_e}{F_e^2 m_a}\nu_{ea}^{(\mathrm{el})}\right.\\
&\left.+\frac{6\widehat{T}_a}{F_e^2 m_a v_e}\nu_{ea}^{(\mathrm{el})}\right)+f_0'\left(-\frac{2\tau}{v_e^2}-\frac{6\widehat{T}_a}{F_e^2 m_a v_e^2}\nu_{ea}^{(\mathrm{el})}\right.\\
&+\frac{6m_e}{F_e^2 m_a}\frac{\partial\nu_{ea}^{(\mathrm{el})}}{\partial v_e}\left(\frac{\widehat{T}_a}{m_a v_e}+v_e\right)+12\frac{m_e}{F_e^2 m_a}\nu_{ea}^{(\mathrm{el})}-\frac{5}{3F_e^2}\left(\nu_{ea}^{(\mathrm{el})}+\nu_{ea}^{(\mathrm{inel})}\right)\\
&\left.+\frac{3\widehat{T}_a}{F_e^2 m_a}\frac{\partial^2\nu_{ea}^{(\mathrm{el})}}{\partial v_e^2}-\frac{3}{F_e^2}\nu_{ea}^{(\mathrm{inel})}-\frac{5}{3}\frac{1}{\tau F_e^2}-5\left(\nu_{ea}^{(\mathrm{el})}+\nu_{ea}^{(\mathrm{inel})}\right)\frac{\widehat{T}_a\nu_{ea}^{(\mathrm{el})}}{F_e^4\tau m_a}\right)\\
&+f_0\left(12\frac{m_e}{F_e^2 m_a}\frac{\partial\nu_{ea}^{(\mathrm{el})}}{\partial v_e}+3\frac{m_e}{F_e^2 m_a}\frac{\partial^2\nu_{ea}^{(\mathrm{el})}}{\partial v_e^2}v_e-\frac{3}{F_e^2}\frac{\partial\nu_{ea}^{(\mathrm{inel})}}{\partial v_e}\right.\\
&\left.-5\left(\nu_{ea}^{(\mathrm{el})}+\nu_{ea}^{(\mathrm{inel})}\right)v_e\frac{m_e\nu_{ea}^{(\mathrm{el})}}{F_e^2\tau m_a}\right)\\
&+5\frac{\nu_{ea}^{(\mathrm{el})}+\nu_{ea}^{(\mathrm{inel})}}{F_e^4\tau v_e^2}\int_0^{v_e} v_e^2\nu_{ea}^{(\mathrm{inel})}f_0\,\mathrm{d}v_e=0.
\end{aligned}
\tag{6.1.25}
$$

In dimensionless form, the equation for the isotropic part of energy distribution of electrons may be presented in the form

$$\begin{aligned}
&\breve{f}_0'''\left(1+3\frac{\breve{m}}{\breve{\varepsilon}}\breve{\nu}_{ea}^{(\mathrm{el})}\right)\breve{v}_e^2+\breve{f}_0''\left(2+6\frac{\breve{m}}{\breve{\varepsilon}}\frac{\partial\breve{\nu}_{ea}^{(\mathrm{el})}}{\partial\breve{v}_e}\breve{v}_e+3\breve{m}\breve{v}_e^2\breve{\nu}_{ea}^{(\mathrm{el})}+6\frac{\breve{m}}{\breve{\varepsilon}}\breve{\nu}_{ea}^{(\mathrm{el})}\right)\breve{v}_e\\
&+\breve{f}_0'\left(-2-6\frac{\breve{m}}{\breve{\varepsilon}}\breve{\nu}_{ea}^{(\mathrm{el})}+6\breve{m}\left(\frac{\breve{v}_e}{\breve{\varepsilon}}+\breve{v}_e^3\right)\frac{\partial\breve{\nu}_{ea}^{(\mathrm{el})}}{\partial\breve{v}_e}\right.\\
&-\breve{v}_e^2\left(\frac{5}{3}+\frac{5}{3}\breve{\nu}_{ea}^{(\mathrm{el})}+\frac{14}{3}\breve{\nu}_{ea}^{(\mathrm{inel})}-12\breve{m}\breve{\nu}_{ea}^{(\mathrm{el})}-3\frac{\breve{m}}{\breve{\varepsilon}}\frac{\partial^2\breve{\nu}_{ea}^{(\mathrm{el})}}{\partial\breve{v}_e^2}\right.\\
&\left.\left.+5\left(\breve{\nu}_{ea}^{(\mathrm{el})}+\breve{\nu}_{ea}^{(\mathrm{inel})}\right)\frac{\breve{m}}{\breve{\varepsilon}}\breve{\nu}_{ea}^{(\mathrm{el})}\right)\right)+\breve{f}_0\left(12\breve{m}\breve{v}_e^2\frac{\partial\breve{\nu}_{ea}^{(\mathrm{el})}}{\partial\breve{v}_e}\right.\\
&\left.+3\breve{m}\breve{v}_e^3\frac{\partial^2\breve{\nu}_{ea}^{(\mathrm{el})}}{\partial\breve{v}_e^2}-3\breve{v}_e^2\frac{\partial\breve{\nu}_{ea}^{(\mathrm{inel})}}{\partial\breve{v}_e}-5\left(\breve{\nu}_{ea}^{(\mathrm{el})}+\breve{\nu}_{ea}^{(\mathrm{inel})}\right)\breve{\nu}_{ea}^{(\mathrm{inel})}\breve{v}_{ea}^3\breve{m}\right)\\
&+5\left(\breve{\nu}_{ea}^{(\mathrm{el})}+\breve{\nu}_{ea}^{(\mathrm{inel})}\right)\int_0^{\breve{v}_e}\breve{v}_e^2\breve{\nu}_{ea}^{(\mathrm{inel})}f_0\,\mathrm{d}\breve{v}_e=0,
\end{aligned}\tag{6.1.26}$$

where

$$\begin{aligned}
&\breve{m}=\frac{m_e}{m_a},\qquad \breve{v}_e=\frac{v_e}{\tau F_e},\qquad \breve{\varepsilon}=\frac{m_eF_e^2\tau}{\widehat{T}_a},\\
&\breve{\nu}_{ea}^{(\mathrm{el})}=\frac{\nu_{ea}^{(\mathrm{el})}}{1/\tau},\qquad \breve{\nu}_{ea}^{(\mathrm{inel})}=\frac{\nu_{ea}^{(\mathrm{inel})}}{1/\tau},\qquad \breve{f}_0=\frac{f_0}{f_0(v_e=0)}.
\end{aligned}\tag{6.1.27}$$

Note that Eq. (6.1.26) transforms to Eq. (6.1.10) if elastic collisions are ignored ($\breve{\nu}_{ea}^{(\mathrm{el})}\to 1$ and $\breve{\nu}_{ea}^{(\mathrm{inel})}\to 0$, respectively).

The boundary conditions for the solution of Eq. (6.1.26) have the form

$$\breve{f}_0(\breve{v}_e=0)=1,\qquad \left(\frac{\partial\breve{f}_0}{\partial\breve{v}_e}\right)_{\breve{v}_e=0}=0,\qquad \breve{f}_0(\breve{v}_e=\infty)=0.\tag{6.1.28}$$

Eq. (6.1.26) was solved by the iterative three-diagonal method of Gauss elimination technique for the differential second-order equation (see Appendix 4) using a PC i486. The calculation is time depended on the parameter E/N and was within the limits of 10–30 minutes (Alekseev, Lebedev and Michailov, 1997).

As was pointed out in the introduction, at this stage of the investigation into the capabilities and prospects of application of the GBE, it is necessary to determine the ratio of energy distributions of electrons calculated by the GBE and traditional BE. In the latter case, the algorithm of solution of the BE and computer code were used that were described in detail in Lebedev and Epstein (1995) and Karoulina and Lebedev (1988, 1992). The equation was solved for the isotropic part of energy distribution of electrons obtained from the BE in binomial expansion in spherical harmonics of the distribution function.

Note once again that the distribution functions of electrons (DFE) were found for conditions under which the electron–electron collisions and collisions of the second kind may be ignored. Besides the DFE, some moments of the energy distribution are considered below, such as the average electron energy ($\bar{\varepsilon}$) (Lebedev and Epstein, 1995; Karoulina and Lebedev, 1988, 1992)

$$\bar{\varepsilon} = \int_0^\infty \varepsilon^{3/2} f_0(\varepsilon)\, \mathrm{d}\varepsilon, \tag{6.1.29}$$

electron diffusion coefficient (D)

$$D = \frac{4\pi}{3} \int_0^\infty \frac{v_e^4}{\sum_\alpha \nu_{ea}^\alpha} f_0(v)\, \mathrm{d}v, \tag{6.1.30}$$

and the ratio of mobility of electrons (drift velocity) calculated by the GBE to the mobility calculated by the BE ($\mu_{\mathrm{GBE}}/\mu_{\mathrm{BE}}$). The electron mobility was defined with due regard for normalization of the distribution function (6.1.29) and expansion in spherical harmonics (6.1.5) by the formula

$$\mu_{\mathrm{GBE}} = \frac{4\pi}{3} \frac{e}{m_e} \int_0^\infty v_e^4 f_1(v_e)\, \mathrm{d}v_e. \tag{6.1.31}$$

Results are given that were derived within the generalized Boltzmann kinetic theory (GBE) and Boltzmann kinetic theory (BE), respectively.

Figure 6.1 shows the characteristic form of the electron distribution functions in hydrogen plasma obtained by the BE (curve 1) and GBE (curve 2). The energy distributions differ considerably, and, as was pointed out in the introduction, this is indicative of the fact that the role of additional terms in the GBE is great. It is of interest to compare

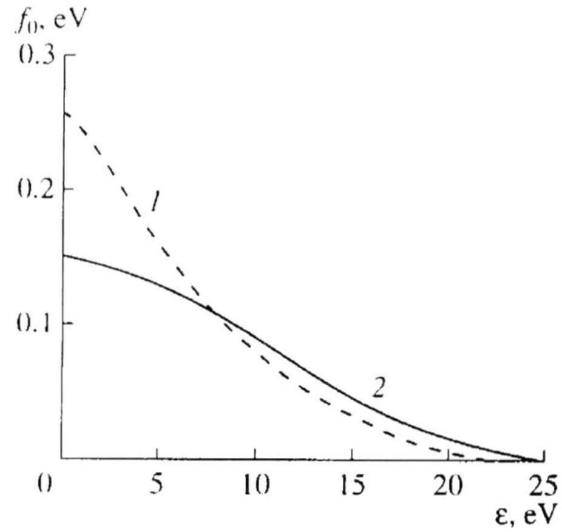

Fig. 6.1. Electron distribution function in energies for $T = 1000$ °C, $p = 1$ torr, $E = 500$ V/m. (1) The solution of the Boltzmann equation; (2) the solution of the generalized Boltzmann equation.

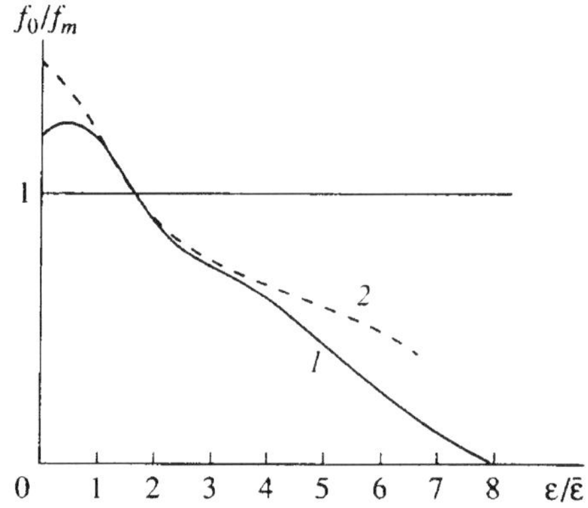

Fig. 6.2. The ratio of the distribution functions to Maxwellian distribution (MDF). (1) The ratio of solution of the Boltzmann equation to MDF; (2) the ratio of solution of the generalized Boltzmann equation to MDF.

Table 6.1

$(E/N)10^{16}$ (V/cm^2)	Average energy (eV)		Diffusion coefficient $D \times 10^{-6}$ (cm^2/s)		Mobility $\mu \times 10^{-6}$ (cm^2/(V·s)	
	GBE	BE	GBE	BE	GBE	BE
2.07	1.85	1.17	1.67	1.24	1.36	1.46
3.11	2.88	1.80	2.29	1.59	1.23	1.25
4.14	3.91	2.59	2.97	2.07	1.14	1.15
5.18	4.81	3.27	3.59	2.50	1.13	1.12
6.21	5.56	3.78	4.11	2.84	1.12	1.10
7.25	6.19	4.21	4.56	3.11	1.11	1.09
8.02	6.59	4.48	4.84	3.35	1.10	1.08
9.06	7.05	4.78	5.18	3.55	1.10	1.07
10.10	7.45	5.06	5.46	3.73	1.09	1.07
11.10	7.79	5.31	5.69	3.90	1.09	1.06
12.20	8.09	5.54	5.93	4.06	1.09	1.05
13.20	8.35	5.75	6.12	4.20	1.08	1.05
14.20	8.58	5.94	6.29	4.33	1.09	1.04

the shapes of energy distributions. Because the average energies of electrons calculated by the DFEs using two different methods are different, it is reasonable to perform comparison with the help of the functions f_0/f_m (f_m is the Maxwellian DFE at the same average electron energy as that defined by f_0). The abscissa is the quantity $\varepsilon/\bar{\varepsilon}$ rather than the electron energy. Thus, by comparison, it is possible to exclude from consideration the differences caused by different average electron energies. Such curves are shown in Figure 6.2. Table 6.1 and Figures 6.3 and 6.4 give the dependencies of $\bar{\varepsilon}$, D and $\mu_{\mathrm{GBE}}/\mu_{\mathrm{BE}}$ on the parameter E/N.

The problems related to the used set of cross sections (see the introduction) apart, and assuming this set to be model, we can draw the following conclusions. First, for one and the same value of E/N, the quantities $\bar{\varepsilon}$ and D calculated by the GBE are greater than those calculated using the BE. Second, although the ratio of f_0 to the Maxwellian DFE

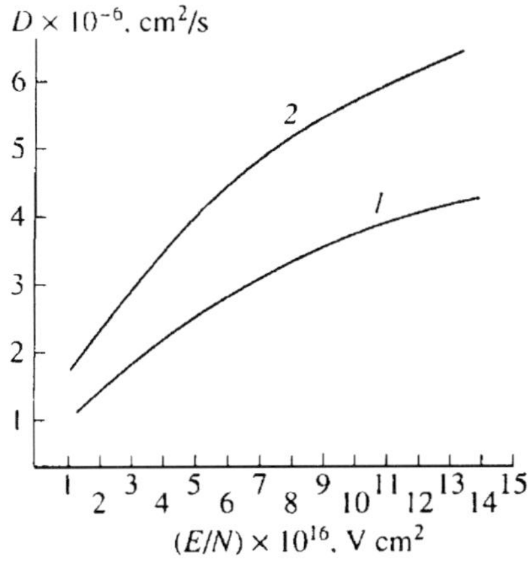

Fig. 6.3. Dependence of the diffusion coefficient on the parameter E/N. (1) The Boltzmann equation; (2) the generalized Boltzmann equation.

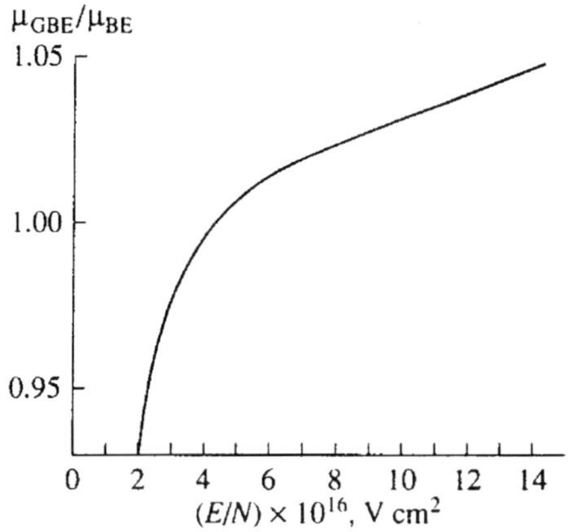

Fig. 6.4. Dependence of the ratio of electron mobilities on the parameter E/N.

is similar (Figure 6.2), the difference from the Maxwellian DFE is greater in the region of low energies and smaller in the region of high energies for the DFE calculated using the GBE rather than the BE. This means that the rate coefficients of high-threshold processes are higher in the first case than in the second one, and additional terms in the GBE result in an increase of the energy stored in electron gas per electron.

These results are also indicative of the fact that the calculated values of the mobility and ionization coefficient are different from the measured ones, and the used set of cross sections should be changed as applied to the GBE. This should be the main trend of further studies into numerical solution of the GBE. Other trends include the construction of algorithms of solution of the GBE for the case of arbitrary dependence of the collision frequency υ_{ea} on the energy (velocity) of electrons, as well as treatment of the behavior of electron gas in variable electric fields. An analytical approach to solution of the latter problem is developed in Alekseev (1995b).

6.2. Sound propagation studied with the generalized equations of fluid dynamics

The propagation of sound is a classical problem in kinetic and hydrodynamic theories. Let an infinite plate oscillate in a gas with the frequency ω in the direction of its normal. Put $a = \omega\tau$, where τ is the mean time between collisions. For a Boltzmann gas of hard spheres one has

$$p\tau = \Pi\mu. \tag{6.2.1}$$

In addition to the static pressure p and dynamic viscosity μ, the hydrodynamic relation (6.2.1) contains the parameter Π, which is equal to 0.786 if the distribution function is expanded in terms of Sonine polynomials, and 0.8 in the first-order (Maxwell) approximation.

The parameter a may be linked to the Reynolds number analogue

$$r = \frac{\Pi}{a} = \frac{p}{\omega\mu}. \tag{6.2.2}$$

For large enough values of r, classical hydrodynamics works quite satisfactorily. In linear acoustics, the attenuation of sound tends to zero as $r \to \infty$, and the velocity of sound is given by

$$c_0^2 = \gamma\frac{p_0}{\rho_0}, \tag{6.2.3}$$

where ρ_0 is the density of the unperturbed gas and

$$\gamma = \chi^{-1} = \frac{C_p}{C_V} \tag{6.2.4}$$

is the ratio of the heat capacity at constant pressure to that at constant volume.

Complications arise when $r \sim 1$ and especially in the limit as $r \to 0$. The Euler equations do not "feel" that the situation has changed and yield a constant velocity of sound and zero attenuation over the entire range of r. The Navier–Stokes equation leads to an entirely unreasonable prediction that the attenuation tends to zero after having reached a maximum at $r \sim 1$, and that the speed of sound tends to infinity as $r \to 0$. Therefore the problem of sound propagation at small r numbers requires a kinetic theory treatment. Without entering into a detailed discussion of the methods mentioned (see, e.g., Cercignani, 1975), it should be admitted that the situation as a whole is unsatisfactory in this field.

In particular, the increased number of moments employed when solving the Boltzmann equation by moment methods gives a poorer agreement with experimental data.

One commonly speaks of the "critical Reynolds number" r_{cr}, below which it is impossible to obtain a plane-wave solution. For each particular type of model or moment equations there exists a unique number r_{cr}, thus revealing the purely mathematical – rather than physical – nature of the effect observed.

Let us apply the generalized equations of fluid dynamics to the propagation of sound waves in a monatomic gas. In linear acoustics, density and temperature perturbations are written as

$$\rho = \rho_0(1+s), \tag{6.2.5}$$

$$T = T_0(1+\eta), \tag{6.2.6}$$

respectively, and the solution of the generalized hydrodynamic equations is taken in the form

$$s = \bar{s}\exp(\mathrm{i}\omega t - k'x), \tag{6.2.7}$$

$$\eta = \bar{\eta}\exp(\mathrm{i}\omega t - k'x), \tag{6.2.8}$$

$$v = \bar{v}\exp(\mathrm{i}\omega t - k'x), \tag{6.2.9}$$

with v the hydrodynamic velocity, and k' the complex wave number.

We now write down the system of non-stationary generalized Euler equations in a one-dimensional case (cf. (2.7.52)–(2.7.54)):

$$\begin{aligned}&\frac{\partial}{\partial t}\left\{\rho - \tau\left[\frac{\partial \rho}{\partial t} + \frac{\partial}{\partial t}(\rho v)\right]\right\}\\&\quad + \frac{\partial}{\partial x}\left\{\rho v - \tau\left[\frac{\partial}{\partial t}(\rho v) + \frac{\partial}{\partial x}(\rho v^2) + \frac{\partial p}{\partial x}\right]\right\} = 0,\end{aligned} \tag{6.2.10}$$

$$\begin{aligned}&\frac{\partial}{\partial t}\left\{\rho v - \tau\left[\frac{\partial}{\partial t}(\rho v) + \frac{\partial}{\partial x}(\rho v^2) + \frac{\partial p}{\partial x}\right]\right\}\\&\quad + \frac{\partial}{\partial x}\left\{\rho v^2 + p - \tau\left[\frac{\partial}{\partial t}(\rho v^2 + p) + \frac{\partial}{\partial x}(\rho v^3 + 3pv)\right]\right\} = 0,\end{aligned} \tag{6.2.11}$$

$$\begin{aligned}&\frac{\partial}{\partial t}\left\{\rho v^2 + 3p - \tau\left[\frac{\partial}{\partial t}(\rho v^2 + 3p) + \frac{\partial}{\partial x}(\rho v^3 + 5pv)\right]\right\}\\&\quad + \frac{\partial}{\partial x}\left\{\rho v^3 + 5pv - \tau\left[\frac{\partial}{\partial t}(\rho v^3 + 5pv)\right.\right.\\&\quad \left.\left. + \frac{\partial}{\partial x}\left(\rho v^4 + 8pv^2 + 5\frac{p^2}{\rho}\right)\right]\right\} = 0.\end{aligned} \tag{6.2.12}$$

From the equation of state

$$p = \rho RT, \tag{6.2.13}$$

in which R is the universal gas constant and which is valid for the Maxwellian distribution function, it follows that

$$p = p_0(1 + s + \eta). \tag{6.2.14}$$

On carrying out the linearization, Eqs. (6.2.10)–(6.2.12) reduce to

$$\frac{\partial s}{\partial t} + \frac{\partial v}{\partial x} - \tau\left[\frac{\partial^2 s}{\partial t^2} + \frac{p_0}{\rho_0}\frac{\partial^2}{\partial x^2}(s + \eta) + 2\frac{\partial^2 v}{\partial t \partial x}\right] = 0, \tag{6.2.15}$$

$$\frac{\partial v}{\partial t} + \frac{p_0}{\rho_0}\frac{\partial}{\partial x}(s + \eta) - \tau\left[\frac{\partial^2 v}{\partial t^2} + 2\frac{p_0}{\rho_0}\frac{\partial^2}{\partial t \partial x}(s + \eta) + 3\frac{p_0}{\rho_0}\frac{\partial^2 v}{\partial x^2}\right] = 0, \tag{6.2.16}$$

$$3\frac{\partial}{\partial t}(s + \eta) + 5\frac{\partial v}{\partial x} - \tau\left[3\frac{\partial^2}{\partial t^2}(s + \eta) + 5\frac{p_0}{\rho_0}\frac{\partial^2}{\partial x^2}(s + 2\eta) + 10\frac{\partial^2 v}{\partial t \partial x}\right] = 0. \tag{6.2.17}$$

To consider in somewhat more detail the linearization process, we use the continuity equation (6.2.10) as an example. Let us rewrite it as

$$\frac{\partial \rho}{\partial t} + \frac{\partial}{\partial x}(\rho v) - \tau\left[\frac{\partial^2 \rho}{\partial t^2} + \frac{\partial^2}{\partial x^2}\left(p + \rho v^2\right) + 2\frac{\partial^2}{\partial t \partial x}(\rho v) - \left(\frac{\partial \rho}{\partial t} + \frac{\partial}{\partial x}(\rho v)\right)\frac{\partial \ln \tau}{\partial x} - \left(\frac{\partial}{\partial t}(\rho v) + \frac{\partial}{\partial x}\left(p + \rho v^2\right)\right)\frac{\partial \ln \tau}{\partial x}\right] = 0. \tag{6.2.18}$$

Because for the one-component ("simple") gas of hard spheres

$$\tau = \frac{1}{\sqrt{2}\pi\sigma^2}\sqrt{\frac{\pi m}{2k_B T}} = \frac{const}{n\sqrt{T}}, \tag{6.2.19}$$

we have

$$\frac{\partial \ln \tau}{\partial t} = -\frac{1}{\rho}\frac{\partial \rho}{\partial t} - \frac{1}{2T}\frac{\partial T}{\partial t}, \tag{6.2.20}$$

$$\frac{\partial \ln \tau}{\partial x} = -\frac{1}{\rho}\frac{\partial \rho}{\partial x} - \frac{1}{2T}\frac{\partial T}{\partial x}. \tag{6.2.21}$$

Therefore, in linear acoustics, terms containing derivatives of $\ln \tau$ contribute nothing to the first order equations.

Using the representations (6.2.7)–(6.2.9) we now arrive at a system of equations

$$\mathrm{i}\omega s - k'v - \tau\left[-\omega^2 s + \frac{p_0}{\rho_0}k'^2(s + \eta) - 2\mathrm{i}\omega k' v\right] = 0, \tag{6.2.22}$$

$$i\omega v-\frac{p_0}{\rho_0}k'(s+\eta)-\tau\left[-\omega^2 v-2i\omega\frac{p_0}{\rho_0}k'(s+\eta)+3\frac{p_0}{\rho_0}k'^2 v\right]=0, \quad (6.2.23)$$

$$3i\omega(s+\eta)-5k'v$$
$$-\tau\left[-3\omega^2(s+\eta)+5\frac{p_0}{\rho_0}k'^2(s+2\eta)-10i\omega k'v\right]=0. \quad (6.2.24)$$

If the rank of matrix equals the number of equations then the system of homogeneous algebraic equations (6.2.22)–(6.2.24) has only a trivial solution The requirement that there be a nonzero solution to this system is that its determinant must be zero:

$$\begin{vmatrix} i\omega+\omega^2\tau-\tau\frac{p_0}{\rho_0}k'^2 & -\tau\frac{p_0}{\rho_0}k' & -k+2i\omega\tau k' \\ -\frac{p_0}{\rho_0}k'+2i\omega\tau\frac{p_0}{\rho_0}k' & -\frac{p_0}{\rho_0}k'+2i\omega\tau\frac{p_0}{\rho_0}k' & i\omega+\omega^2\tau-3\tau\frac{p_0}{\rho_0}k'^2 \\ 3i\omega+3\omega^2\tau-5\tau\frac{p_0}{\rho_0}k'^2 & 3i\omega+3\omega^2\tau-10\tau\frac{p_0}{\rho_0}k'^2 & -5k'+10i\omega\tau k' \end{vmatrix}=0$$

giving an algebraic equation of the sixth order in the wave number k':

$$5\tau^3\frac{p_0^3}{\rho_0^3}k'^6-\frac{p_0^2}{\rho_0^2}\left[5i\omega\tau^2+5\omega^2\tau^3+\frac{5}{3}\tau\right]k'^4$$
$$+\frac{p_0}{\rho_0}\left[\frac{5}{3}i\omega+2\omega^2\tau-\frac{2}{3}i\omega^3\tau^2-\frac{1}{3}\omega^4\tau^3\right]k'^2$$
$$+i\omega^3+3\omega^4\tau-3i\omega^5\tau^2-\omega^6\tau^3=0. \quad (6.2.25)$$

This equation reduces to the dimensionless form

$$3a^3\chi^2\hat{k}^6-\left[3ia^2+3a^3+a\right]\chi\hat{k}^4+\left[i+\frac{6}{5}a-\frac{2}{5}ia^2-\frac{1}{5}a^3\right]\hat{k}^2$$
$$+\frac{3}{5\chi}i+\frac{9}{5\chi}a-\frac{9}{5\chi}a^2-\frac{9}{5\chi}a^3=0, \quad (6.2.26)$$

where we have introduced the dimensionless wave number $\hat{k}=k'c_0/\omega$ with the characteristic velocity

$$c_0=\sqrt{\gamma\frac{p_0}{\rho_0}}. \quad (6.2.27)$$

The separation of the real from imaginary part in Eq. (6.2.26), according to the equality

$$\hat{k}=\alpha+i\beta, \quad (6.2.28)$$

now yields the system of equations for α and β:

$$
\begin{aligned}
&3a^3\chi^2(\alpha^2-\beta^2)(\alpha^4+\beta^4-14\alpha^2\beta^2)+12\alpha\beta a^2(\alpha^2-\beta^2)\chi \\
&\quad-(3a^2+a)\chi(\alpha^4+\beta^4-6\alpha^2\beta^2)+(\alpha^2-\beta^2)\left(\frac{6}{5}a-\frac{1}{5}a^3\right) \\
&\quad-2\alpha\beta\left(1-\frac{2}{5}a^2\right)+\frac{9}{5\chi}-\frac{3}{5\chi}a^3=0,
\end{aligned}
\tag{6.2.29}
$$

$$
\begin{aligned}
&6a^3\chi^2\alpha\beta(3\alpha^4+3\beta^4-10\alpha^2\beta^2)-3a^2\chi(\alpha^4+\beta^4-6\alpha^2\beta^2) \\
&\quad-4(3a^2+a)\chi\alpha\beta(\alpha^2-\beta^2)+(\alpha^2-\beta^2)\left(1-\frac{1}{5}a^2\right) \\
&\quad+2\alpha\beta\left(\frac{6}{5}a-\frac{1}{5}a^3\right)+\frac{3}{5\chi}-\frac{9}{5\chi}a^2=0.
\end{aligned}
\tag{6.2.30}
$$

From Eq. (6.2.7) it follows

$$
s=\bar{s}\exp\left(-\omega\frac{\alpha}{c_0}x\right)\exp\left[\mathrm{i}\omega\left(t-\frac{\beta}{c_0}x\right)\right],
$$

showing that the factor α characterizes the attenuation of sound and that $\beta=c_0/c$ is the ratio of the classical Eulerian speed of sound to its calculated value.

Let us now consider two asymptotic solutions to Eqs. (6.2.29) and (6.2.30).

(1) If $a=\omega\tau\to 0$, then in the limiting case $a=0$ it follows from Eqs. (6.2.29) and (6.2.30) that

$$
\hat{k}^2=-\frac{3}{5}\chi^{-1}. \tag{6.2.31}
$$

Using the value

$$
\chi=\frac{5}{3} \tag{6.2.32}
$$

for a monatomic gas (see Eq. (6.2.4)), one is led to the classical Euler limit

$$
k'=\pm\mathrm{i}\frac{\omega}{c_0}. \tag{6.2.33}
$$

The wave number k' proves to be imaginary; for the density perturbation, for example, the solution is written as

$$
s=\bar{s}\exp\left[\mathrm{i}\omega\left(t\pm\frac{x}{c_0}\right)\right]. \tag{6.2.34}
$$

Thus, in the classical Euler description sound is undamped and its velocity remains constant and equal to c_0 (cf. Eq. (6.2.27)). In other words, for oscillations travelling in the positive x direction one obtains

$$\alpha = 0, \qquad \beta = 1. \tag{6.2.35}$$

(2) If $a \to \infty$, then it follows from Eq. (6.2.26) that

$$\hat{k}^6 - \frac{5}{3}\hat{k}^4 - \frac{5}{27}\hat{k}^2 = \frac{25}{27}, \tag{6.2.36}$$

or (see Eqs. (6.2.26) and (6.2.30))

$$\begin{aligned} &-x\left(x^2 - 12y\right) - \frac{5}{3}x^2 + \frac{20}{3}y + \frac{5}{27}x = \frac{25}{27}, \\ &3x^2 - 4y + \frac{10}{3}x = \frac{5}{27}, \end{aligned} \tag{6.2.37}$$

where

$$\beta^2 - \alpha^2 = x, \qquad \alpha^2\beta^2 = y. \tag{6.2.38}$$

From Eq. (6.2.37), one finds

$$x^3 + \frac{5}{3}x^2 + \frac{35}{54}x - \frac{25}{162} = 0, \tag{6.2.39}$$

and the asymptotic solution then follows readily as

$$\alpha = 0.509, \qquad \beta = 0.650. \tag{6.2.40}$$

Figures 6.5 and 6.6 plot the dimensionless velocity of sound, β, and the dimensionless attenuation rate, α, as calculated by the generalized Euler equations (Alexeev, 1994; Alekseev, Poddoskin and Mikhailov, 1990), the Navier–Stokes equations, and the generalized Navier–Stokes equations (Alekseev, 1990a, 1990b), and compares the results with the experimental data of Greenspan (1956), and Meyer and Sessler (1957).

Detailed generalized Navier–Stokes calculations are given elsewhere (Alekseev, 1990a, 1990b), they are rather cumbersome and we present here main steps of transformations. Linearization of the generalized hydrodynamic equations (GHE) with taking into account (6.2.7)–(6.2.9) lead to following equations:

– continuity equation

$$\begin{aligned} &\left[\mathrm{i} + a - \chi a\hat{k}^2\right]s + \left[-\chi a\hat{k}^2 - \frac{9}{2}\frac{\chi^2}{\Pi}a^3\hat{k}^4\right]\eta \\ &\quad + \left[-\frac{\hat{k}}{c_0} + 2\mathrm{i}a\frac{\hat{k}}{c_0} - \frac{8}{3}\chi\frac{a^2}{c_0\Pi}\hat{k}^3 + 4\mathrm{i}\frac{\chi a^3}{c_0\Pi}\hat{k}^3\right]v = 0, \end{aligned} \tag{6.2.41}$$

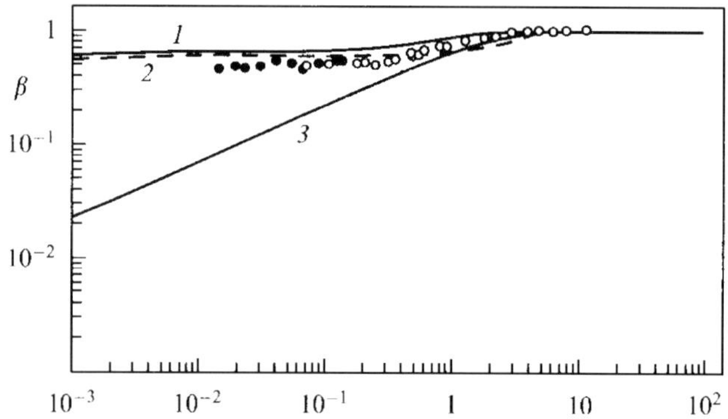

Fig. 6.5. Comparison of observed (symbols) and calculated (lines) dimensionless velocity of sound $\beta = c_0/c$ as a function of the Reynolds number analogue r. 1, generalized Euler equation; 2, generalized Navier–Stokes equation; 3, Navier–Stokes equation. Open circles, data by Greenspan; filled circles, data by Meyer and Sessler.

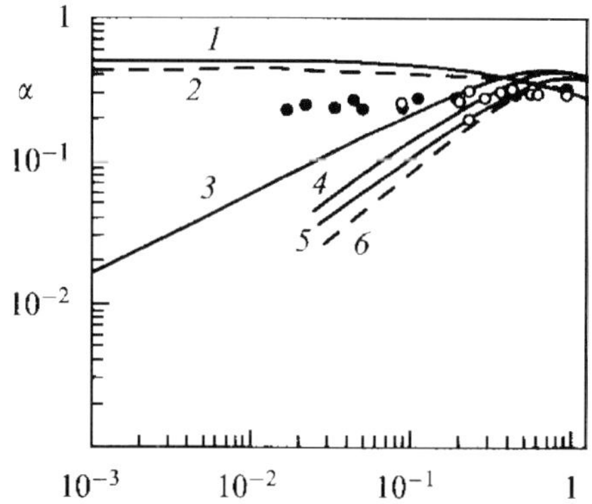

Fig. 6.6. Comparison of observed (symbols) and calculated (lines) dimensionless attenuation rate α as a function of the Reynolds number analogue r: 1, generalized Euler equation; 2, generalized Navier–Stokes equation; 3, Navier–Stokes equation; 4, Bumett equation; 5, super-Burnett equation; 6, moment equations (the number of moments $N = 105$). Open circles, data by Greenspan; filled circles, data by Meyer and Sessler.

– equation of motion

$$[-\chi\hat{k} + 2\mathrm{i}a\chi\hat{k}]s + \left[-\chi\hat{k} - 9\frac{\chi^2}{\Pi}a^2\hat{k}^3 + \frac{27}{2}\mathrm{i}\frac{\chi^2}{\Pi}a^3\hat{k}^3 + 2\mathrm{i}\chi a\hat{k}\right]\eta + \left[\frac{i}{c_0} - \frac{4}{3}\frac{\chi a}{\Pi c_0}\hat{k}^2 + 4\mathrm{i}\frac{\chi a^2}{\Pi c_0}\hat{k}^2 + \frac{a}{c_0} - 3\chi\frac{a}{c_0}\hat{k}^2 - 8\chi^2\frac{a^3}{\Pi c_0}A\hat{k}^4 + \frac{8}{3}\frac{\chi a^3}{\Pi c_0}\hat{k}^2\right]v = 0, \qquad (6.2.42)$$

– energy equation

$$\begin{aligned}
&[-5\chi a\hat{k}^2+3\mathrm{i}+3a]s+\left[3\mathrm{i}-\frac{15}{2}\frac{\chi}{\Pi}a\hat{k}^2+\frac{45}{2}\mathrm{i}\frac{\chi}{\Pi}a^2\hat{k}^2+3a\right.\\
&\quad\left.-10\chi a\hat{k}^2-126\frac{\chi^2}{\Pi}a^3B\hat{k}^4+15\frac{\chi}{\Pi}a^3\hat{k}^2\right]\eta\\
&\quad+\left[-5\frac{\hat{k}}{c_0}-28\frac{\chi a^2}{\Pi c_0}A\hat{k}^3+\frac{112}{3}\frac{\chi a^2}{\Pi c_0}\mathrm{i}a^3A\hat{k}^3+10\mathrm{i}\frac{a}{c_0}\hat{k}\right]v=0. \qquad (6.2.43)
\end{aligned}$$

Here

$$A=1-\frac{\mu^1}{\mu},\qquad B=1-\frac{\lambda^1}{2\lambda}, \qquad (6.2.44)$$

μ^1, λ^1 – additional coefficients of viscosity and thermal conductivity (5.3.33), (5.3.25), which are calculated – in comparison to usual coefficients viscosity μ and thermal conductivity λ, using the first and second coefficients of Sonine polynomial expansions, for example $\mu^1=pb_1/2$.

As a result, GHE lead to the dispersion equation of the tenth power:

$$\begin{aligned}
&\{1008AB-180A\}\frac{\chi^5}{\Pi^2}a^7\hat{k}^{10}+\left\{-\frac{1008}{\Pi^2}a^7AB+\frac{156}{\Pi^2}a^7A+\frac{672}{\Pi^2}a^7B\right.\\
&\quad-\frac{210}{\Pi^2}a^7-\frac{66}{\Pi^2}a^5A-\frac{168}{\Pi^2}a^5B+\frac{40}{\Pi}a^5A+\frac{378}{\Pi}a^5B+\frac{90}{\Pi^2}a^5-\frac{135}{2\Pi}a^5\\
&\quad\left.+\mathrm{i}\frac{a^6}{\Pi^2}[-1008AB+222A+672B-270]\right\}\chi^4\hat{k}^8\\
&\quad+\left\{-\frac{384}{\Pi^2}a^7A+\frac{336}{\Pi^2}a^7B+\frac{46}{\Pi^2}a^7+\frac{726}{\Pi^2}a^5A-\frac{672}{\Pi^2}a^5B-\frac{80}{\Pi}a^5A\right.\\
&\quad-\frac{36}{\Pi^2}a^5-\frac{137}{6\Pi}a^5-\frac{10}{\Pi^2}a^3+\frac{95}{6\Pi}a^3+15a^3-\frac{126}{\Pi}a^3B\\
&\quad+\mathrm{i}\left[-\frac{918}{\Pi^2}a^6A+\frac{840}{\Pi^2}a^6B+\frac{88}{\Pi^2}a^6+\frac{192}{\Pi^2}a^4A-\frac{168}{\Pi^2}a^4B\right.\\
&\quad\left.\left.-\frac{80}{\Pi}a^4A+\frac{36}{\Pi^2}a^4-\frac{527}{6\Pi}a^4\right]\right\}\chi^3\hat{k}^6+\left\{-\frac{40}{\Pi^2}a^7-\frac{152}{3\Pi}a^5A\right.\\
&\quad+\frac{126}{\Pi}a^5B+\frac{250}{\Pi^2}a^5-\frac{245}{6\Pi}a^5+\frac{124}{3\Pi}a^3A-\frac{126}{\Pi}a^3B+\frac{337}{6\Pi}a^3\\
&\quad-\frac{70}{\Pi^2}a^3-15a^3-\frac{15}{2\Pi}a-5a+\mathrm{i}\left[-\frac{160}{\Pi^2}a^6-\frac{120}{\Pi}a^4A+\frac{252}{\Pi}a^4B\right.\\
&\quad\left.\left.+\frac{190}{\Pi^2}a^4-\frac{232}{3\Pi}a^4+\frac{103}{6\Pi}a^2-\frac{28}{\Pi}a^2A-15a^2\right]\right\}\chi^2\hat{k}^4+\left\{-\frac{23}{\Pi}a^5\right.
\end{aligned}$$

$$+\frac{207}{2\Pi}a^3 - a^3 - \frac{23}{2\Pi}a + 6a + \mathrm{i}\left[-\frac{161}{2\Pi}a^4 + \frac{115}{2\Pi}a^2 - 2a^2 + 5\right]\Bigg\}\chi\hat{k}^2$$

$$+\left\{-3a^3 + 9a + \mathrm{i}\left[-9a^2 + 3\right]\right\} = 0. \tag{6.2.45}$$

Dispersion equation (6.2.45) has two obvious asymptotic solutions

(a) $a \to 0$ ($r \to \infty$, see also (6.2.2)). Eq. (6.2.45) is transforming to (6.2.31)

$$5\chi\hat{k}^2 + 3 = 0,$$

(b) $a \to \infty$ ($r \to 0$). From Eq. (6.2.45) it can be found:

$$\{336AB - 60A\}\chi^3\hat{k}^6 + \{-336AB + 52A + 224B - 70\}\chi^2\hat{k}^4$$

$$+\left\{-128A + 112B + \frac{46}{3}\right\}\chi\hat{k}^2 - \frac{40}{3} = 0. \tag{6.2.46}$$

The models of hard spheres $A = 0.9415$, $B = 0.956$ and solution of Eq. (6.2.46) lead to (compare with (6.2.40))

$$\alpha = 0.428, \qquad \beta = 0.564. \tag{6.2.47}$$

Introduction of Lennard–Jones model of particle interaction instead of the hard spheres model does not lead to significant influence on results of calculations. For example, in the limit $r \to 0$ ($T = 300$ K) calculations lead practically to the same attenuation rate and to $\beta = 0.55$.

We now proceed to discuss the numerical results obtained from the generalized hydrodynamical equations and to make comparisons with available published data. We will also consider some aspects of the method of moments as applied to the Boltzmann equations used in the sound propagation problem. We will base our analysis on the results of Sirovich and Thurber (1965), presented in Figures 6.7–6.9 with necessary notational changes.

Figure 6.7 compares the numerically calculated dimensionless velocity of sound $\beta = c_0/c$ and dimensionless attenuation rate α as functions of the Reynolds number analogue r for the eight- and eleven-moment models using the interaction potential of Maxwellian molecules and that of the hard sphere models. Notice that the two models yield very close results.

Figures 6.8 and 6.9 plot similar data for hard spheres and Maxwellian molecules and compare them with the experimental data of Greenspan (1956) and Meyer and Sessler (1957) mononatomic gases. Also shown are Navier–Stokes results and those obtained by Pekeris, Alterman and Finkelstein (1960, 1962) in their unprecedently voluminous computations using the 105- and 483-moment models.

A comparison of the experimental data with the theoretical results obtained with the Boltzmann equations suggests what appears at first sight to be a paradoxical conclusion – the more accurate a theoretical model, the worse its agreement with experiment. We see indeed that the hard sphere model, of Sirovich and Thurber (1965) works better with

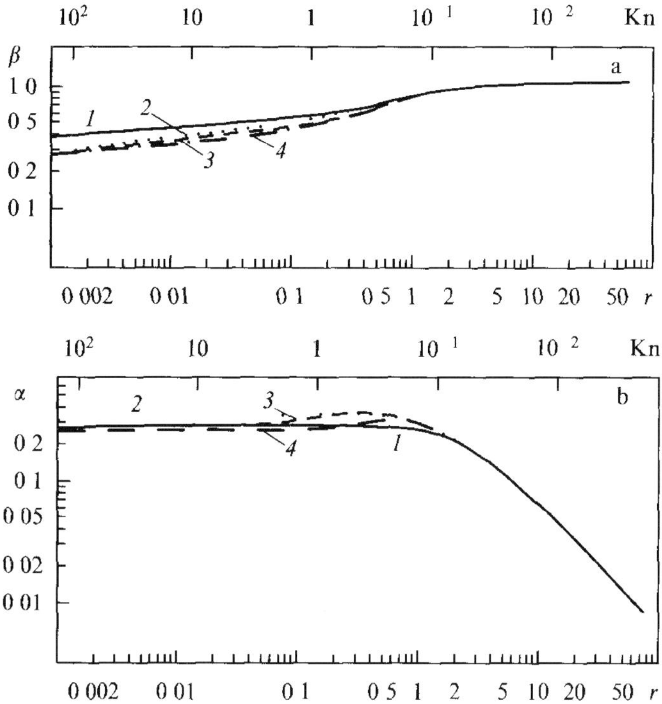

Fig. 6.7. Comparison of the velocity of sound (a) and the attenuation rate (b) as calculated from the Boltzmann equation for two models: 1 – 11-moment hard sphere model, 2 – 11-moment model of Maxwellian molecules, 3 – 8-moment hard sphere model, 4 – 8-moment model of Maxwellian molecules.

eight moments than with eleven, and that the 105-moment results of Pekeris, Alterman and Finkelstein (1960, 1962) are much poorer.

Considering the weak correlation between the molecular interaction model and the velocity of sound and attenuation rate calculations, the results of the 483-moment computations for the Maxwellian molecules within the range $r < 1$ should be viewed as simply catastrophic. Note that as the number of the moments used increases, the "critical" number r decreases (Sirovich and Thurber, 1965), apparently raising hopes for a better final result.

A similar situation exists with regard to the hydrodynamic results: while the Navier–Stokes equations are totally invalid for $r < 1$, "corrected" models (e.g., the Burnett equation) are even less successful. It can be argued that, paradoxically though it may seem, the best classical theory approach is to employ the Euler equation which, although yielding zero attenuation and a constant velocity of sound, at least does not involve divergences or nonphysical "critical" points.

Viewed in the context of the generalized Boltzmann kinetic theory, this effect has a very clear origin. Let us introduce the Knudsen number as the ratio l/l_ω of the mean free path of hard-sphere particles to the wavelength

$$l_\omega = 2\pi \frac{c}{\omega}. \tag{6.2.48}$$

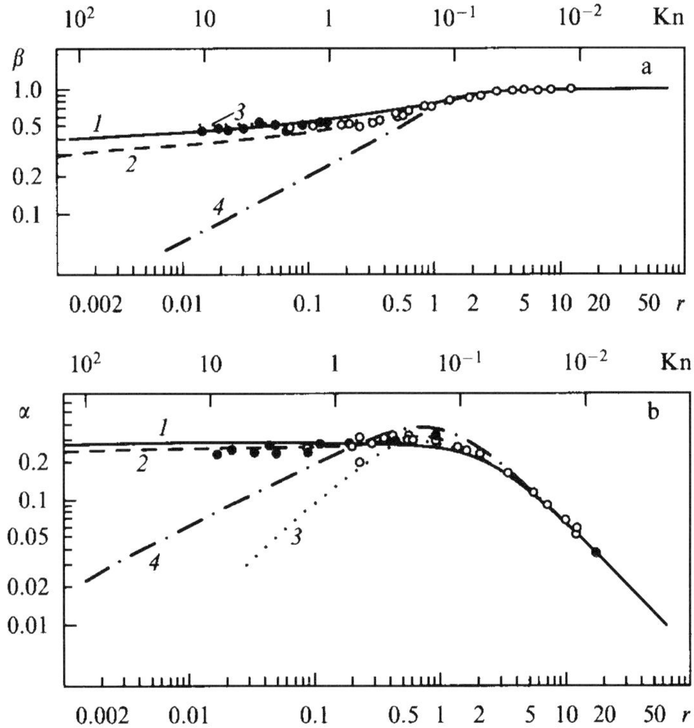

Fig. 6.8. Comparison of observed (symbols) and calculated (lines) velocities of sound (a) and attenuation rates (b) as calculated from the Boltzmann equation for the hard sphere model: 1 – 11-moment model, 2 – 8-moment model, 3 – 105-moment model, 4 – Navier–Stokes equation. Open circles – data by Greenspan filled circles data by Meyer and Sessler.

Since in the hard-sphere model

$$l = \frac{1}{\sqrt{2}\pi n\sigma^2}, \qquad \mu = \frac{5}{16}\frac{\sqrt{mk_B T}}{\sqrt{\pi}\sigma^2}, \tag{6.2.49}$$

where σ is the particle diameter, it follows from Eqs. (6.2.2) and (6.2.3) that

$$Kn = \frac{l}{l_\omega} = \frac{8}{5\pi\sqrt{2\pi\gamma}}\frac{1}{r}. \tag{6.2.50}$$

In this case, the ratio of the heat capacities $\gamma = C_p/C_V$ at constant pressure and constant volume is $5/3$, and using Eq. (6.2.49) we can recalculate the scale of the Reynolds number analogue r to an equivalent scale of the Knudsen number as shown in Figures 6.7–6.9.

Thus, discrepancies between the experiment and the "revised" theoretical models based on the Boltzmann equations start to appear for Knudsen number values $Kn \sim 1$. This is to be expected because the additional terms in the kinetic equation of the gener-

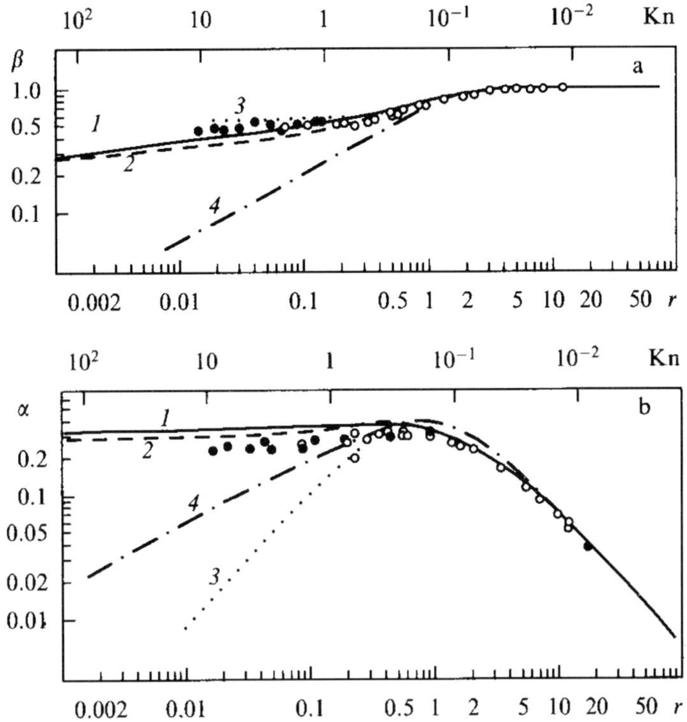

Fig. 6.9. Comparison of observed (symbols) and calculated (lines) velocities of sound (a) and attenuation rates (b) as calculated from the Boltzmann equation for the model of Maxwelhan molecules: 1 – 11-moment model, 2 – 8-moment model, 3 – 483-moment model, 4 – Navier–Stokes equation. Open circles – data by Greenspan, filled circles – data by Meyer and Sessler.

alized Boltzmann theory first become comparable in magnitude and then start to dominate the terms on the left-hand side of the Boltzmann equations as the Knudsen number increases. This means, in particular, that neither the Burnett equations nor, less still, super-Burnett equations hold promise for higher Knudsen number computations.

The generalized Boltzmann equation performs much better. The generalized Euler equations and Navier–Stokes equations give quite satisfactory agreement with the experimental data over the entire range of Knudsen numbers, including the asymptotic regions. The generalized Navier–Stokes equations fit the experimental points better than the generalized Euler equations. Another important point about this result is that it is obtained from the hydrodynamic equations, this raises hopes for a "through" computation of hydrodynamic flows including shock layers, shock waves, and intermediate Knudsen numbers, thus eliminating the necessity of coupling the hydrodynamic and free-molecular solutions.

Coupling problems of this kind are discussed widely in the scientific literature (see, e.g., Longo, Preziosi and Bellomo, 1992; Bourgat et al., 1992). In the next section we will see that the generalized hydrodynamical equations make it possible to perform accurate through computations via the shock wave or, in other words, to examine the structure of the shock wave.

6.3. Shock wave structure examined with the generalized equations of fluid dynamics

Let us consider the structure of the shock wave in a monatomic gas based on the solution of the generalized hydrodynamic equations. The solution of the usual gas-dynamic equations in this case is given by discontinuous density, velocity, and temperature functions interrelated by the Rankine–Hugoniot equations.

This classical problem of kinetic theory has long since become a kind of a testing ground for approximate kinetic theories, as well as for methods of solving the Boltzmann equation. Note, that although the solution of this problem has also been obtained with the Navier–Stokes equations, but the conditions for the applicability of the Navier–Stokes equations (small variations of hydrodynamic quantities over the molecular mean free path) are of course not fulfilled – at Mach numbers not too close to unity – and a qualitative description of the transition layer is the most that seems achievable. In this case the viscous terms in the Navier–Stokes equations play the same role as the artificial viscosity terms introduced into the Euler equations in shock wave calculations.

The generalized Euler equations in the one-dimensional steady case are (see Eqs. (2.7.52)–(2.7.54))

$$\frac{\mathrm{d}}{\mathrm{d}x}\left(\rho v_0 - \tau^{(0)}\frac{\mathrm{d}}{\mathrm{d}x}(\rho v_0 + p)\right) = 0, \tag{6.3.1}$$

$$\frac{\mathrm{d}}{\mathrm{d}x}\left(\rho v_0^2 + p - \tau^{(0)}\frac{\mathrm{d}}{\mathrm{d}x}(\rho v_0^3 + 3pv_0)\right) = 0, \tag{6.3.2}$$

$$\frac{\mathrm{d}}{\mathrm{d}x}\left(\rho v_0^3 + 5pv_0 - \tau^{(0)}\frac{\mathrm{d}}{\mathrm{d}x}\left(\rho v_0^4 + 8pv_0^2 + 5\frac{p^2}{\rho}\right)\right) = 0. \tag{6.3.3}$$

Recall that in the hard-sphere model $\tau^{(0)}p = 0.8\mu$ in the first-order approximation within the framework of the Enskog method.

Eqs. (6.3.1)–(6.3.3) in the one-dimensional steady case are readily integrated once to yield

$$\rho v_0 = \tau^{(0)}\frac{\mathrm{d}}{\mathrm{d}x}(p + \rho v_0^2) + C_1, \tag{6.3.4}$$

$$p + \rho v_0^2 = \tau^{(0)}\frac{\mathrm{d}}{\mathrm{d}x}(v_0(3p + \rho v_0^2)) + C_2, \tag{6.3.5}$$

$$v_0(5p + \rho v_0^2) = \tau^{(0)}\frac{\mathrm{d}}{\mathrm{d}x}\left(8pv_0^2 + 5\frac{p^2}{\rho} + \rho v_0^4\right) + C_3. \tag{6.3.6}$$

To integrate Eqs. (6.3.1)–(6.3.3), it is necessary to specify two boundary conditions for the hydrodynamical velocity, density, and pressure. These are the so-called Hugoniot

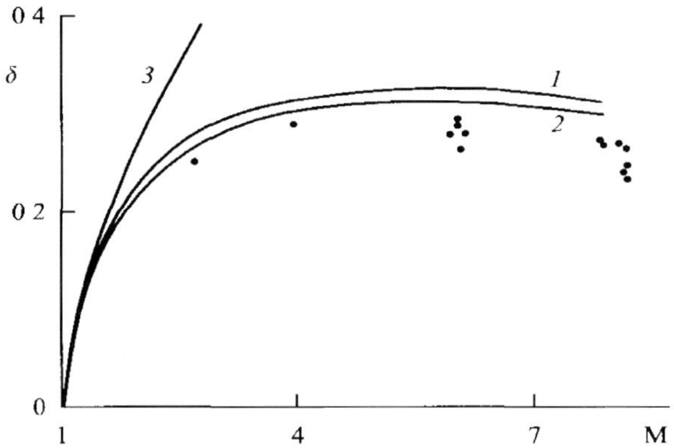

Fig. 6.10. Comparison of observed (experimental points by Schmidt, 1969) and calculated (lines) dimensionless shock wave width $\bar{\delta}$ as a function of the Mach number M: 1 – generalized Euler equations, 2 – generalized Navier–Stokes equations, 3 – Navier–Stokes equations.

conditions. The constants C_1, C_2 and C_3 for Eqs. (6.3.4)–(6.3.6) are determined by the conditions before the shock wave

$$(\rho v_0)_\mathrm{b} = C_1, \tag{6.3.7}$$

$$\left(p + \rho v_0^2\right)_\mathrm{b} = C_2, \tag{6.3.8}$$

$$\left[v_0\left(5p + \rho v_0^2\right)\right]_\mathrm{b} = C_3, \tag{6.3.9}$$

where the subscript "b" refers to the flow before the shock wave.

However, the numerical integration of Eqs. (6.3.4)–(6.3.6) is complicated by the necessity of satisfying the boundary conditions at the opposite ends of the integration region. A simpler approach in this case is to solve the boundary value problem directly, by applying the three-diagonal method of Gauss elimination technique for the differential second-order equation (see Appendix 4).

This is exactly what Polev and the present author did in 1988 (Alekseev and Polev, 1990).

Let us define the width of the shock wave by the relation

$$d = \frac{\rho_\mathrm{a} - \rho_\mathrm{b}}{(\mathrm{d}\rho/\mathrm{d}x)_\mathrm{max}}, \tag{6.3.10}$$

where the subscript "a" refers to the flow parameters after the shock wave, and $(\mathrm{d}\rho/\mathrm{d}x)_\mathrm{max}$ corresponds to the maximum value of the density gradient in the shock wave.

Let us next define the dimensionless shock wave width

$$\bar{\delta} = \frac{l_\mathrm{b}}{d}, \tag{6.3.11}$$

where λ_b is the mean free path in the region before the shock. For the hard-sphere model

$$l_b = \frac{m}{\sqrt{2}\pi\rho_b\sigma^2}, \tag{6.3.12}$$

or, using Eqs. (6.2.3) and (6.2.49),

$$l_b = \frac{16}{5}\sqrt{\frac{5}{6\pi}}\frac{\mu_b}{c_b\rho_b}, \tag{6.3.13}$$

where is the velocity of sound before the shock as calculated in the Euler approximation. It is also useful to define the dimensionless density

$$\bar{\rho} = \frac{\rho - \rho_b}{\rho_a - \rho_b}. \tag{6.3.14}$$

Figure 6.10 plots the dimensionless shock wave width $\bar{\delta}$ as a function of the Mach number M. The theoretical curves 1, 2, and 3 (computed with the generalized Euler equations, the generalized Navier–Stokes equations, and the ordinary Navier–Stokes equations (Alekseev and Polev, 1990; Alexeev and Chikhaoui, 1993), respectively) are compared to the experimental data of Schmidt (1969). Curves 1 and 2 lie somewhat above the experimental points, the generalized Navier–Stokes calculations giving a better fit. Notice that the Navier–Stokes results (curve 3) become unsatisfactory for $M > 1.6$.

The use of the Grad method also has proved unsatisfactory. Grad himself used the thirteen-moment approximation to determine the shock wave structure (Grad, 1949). He found that a solution to this problem does not exist for $M > 1.65$, and Holway later established (Holway, 1964) that the Grad series for the distribution function in the Boltzmann equation diverges for $M > 1.85$.

It is important to note that our results on shock wave structure were obtained in the framework of the generalized hydrodynamic equations, leading to the expectation that these equations may be used effectively in through calculations for arbitrary Mach and Knudsen numbers.

CHAPTER 7

Numerical Simulation of Vortex Gas Flow Using the Generalized Euler Equations

7.1. Unsteady flow of a compressible gas in a cavity

To demonstrate the research potential of numerical vortex-flow simulation using the generalized hydrodynamic equations, we examine the two-dimensional unsteady flow of a compressible gas. We begin our consideration with calculations of flow in a rectangular cavity and after that the mathematical modeling of flows in channels of different forms will be delivered.

We examine the two-dimensional unsteady flow of a compressible gas in a cavity (Alexeev and Mikhailov, 1999; Fedoseyev and Alexeev, 1998a). The problem to be solved is the following. Consider a flow over a flat surface and suppose there suddenly appears – as a result of some mechanical action, for example – a certain cavity of square cross section, whose length is much longer than the side of the square (Figure 7.1). It is assumed that at some instant of time gas suddenly starts to move along the segment OL of the axis x with the velocity V_s which is subsequently maintained constant.

The system of the generalized Euler equations for a two-dimensional, unsteady and nonisothermic flow of compressible gas is written in the following way:

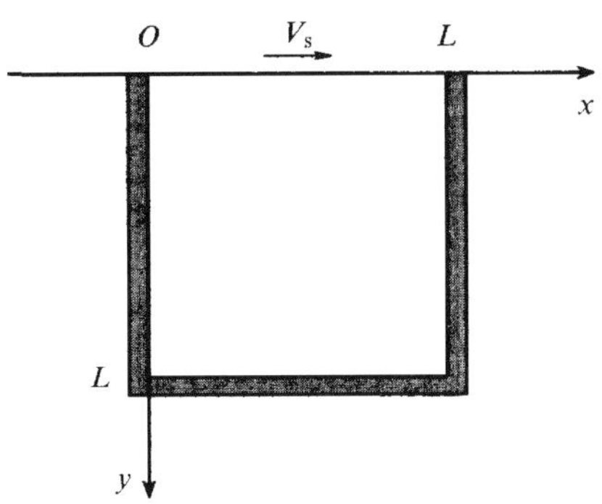

Fig. 7.1. Unsteady flow of compressible gas in a cavity.

$$\frac{\partial}{\partial t}\left\{\rho-\tau\left[\frac{\partial\rho}{\partial t}+\frac{\partial}{\partial x}(\rho u)+\frac{\partial}{\partial y}(\rho v)\right]\right\}$$
$$+\frac{\partial}{\partial x}\left\{\rho u-\tau\left[\frac{\partial}{\partial t}(\rho u)+\frac{\partial}{\partial x}\left(\rho u^2\right)+\frac{\partial}{\partial y}(\rho u v)+\frac{\partial p}{\partial x}\right]\right\}$$
$$+\frac{\partial}{\partial y}\left\{\rho v-\tau\left[\frac{\partial}{\partial t}(\rho v)+\frac{\partial}{\partial x}(\rho u v)+\frac{\partial}{\partial y}\left(\rho v^2\right)+\frac{\partial p}{\partial y}\right]\right\}=0, \qquad (7.1.1)$$

$$\frac{\partial}{\partial t}\left\{\rho u-\tau\left[\frac{\partial}{\partial t}(\rho u)+\frac{\partial}{\partial x}\left(\rho u^2\right)+\frac{\partial}{\partial y}(\rho u v)\frac{\partial p}{\partial x}\right]\right\}$$
$$+\frac{\partial}{\partial x}\left\{p+\rho u^2-\tau\left[\frac{\partial}{\partial t}\left(p+\rho u^2\right)+3\frac{\partial}{\partial x}(p u)+\frac{\partial}{\partial x}\left(\rho u^3\right)\right.\right.$$
$$\left.\left.+\frac{\partial}{\partial y}\left(\rho u^2 v\right)+2\frac{\partial}{\partial y}(p v)\right]\right\}+\frac{\partial}{\partial y}\left\{\rho u v-\tau\left[\frac{\partial}{\partial t}(\rho u v)+\frac{\partial}{\partial x}\left(\rho u^2 v\right)\right.\right.$$
$$\left.\left.+\frac{\partial}{\partial y}(p u)+\frac{\partial}{\partial y}\left(\rho u v^2\right)\right]\right\}=0, \qquad (7.1.2)$$

$$\frac{\partial}{\partial t}\left\{\rho v-\tau\left[\frac{\partial}{\partial t}(\rho v)+\frac{\partial}{\partial x}(\rho u v)+\frac{\partial}{\partial y}\left(\rho v^2\right)\frac{\partial p}{\partial y}\right]\right\}$$
$$+\frac{\partial}{\partial x}\left\{\rho u v-\tau\left[\frac{\partial}{\partial t}(\rho u v)+\frac{\partial}{\partial x}(p v)+\frac{\partial}{\partial x}\left(\rho u^2 v\right)+\frac{\partial}{\partial y}\left(\rho u v^2\right)\right]\right\}$$
$$+\frac{\partial}{\partial y}\left\{p+\rho v^2-\tau\left[\frac{\partial}{\partial t}\left(p+\rho v^2\right)+3\frac{\partial}{\partial y}(p v)+\frac{\partial}{\partial y}\left(\rho v^3\right)\right.\right.$$
$$\left.\left.+\frac{\partial}{\partial x}\left(\rho u v^2\right)+2\frac{\partial}{\partial x}(p u)\right]\right\}=0, \qquad (7.1.3)$$

$$\frac{\partial}{\partial t}\left\{\rho v_0^2+3p-\tau\left[\frac{\partial}{\partial t}\left(\rho v_0^2+3p\right)+\frac{\partial}{\partial x}\left(u\left(\rho v_0^2+5p\right)\right)\right.\right.$$
$$\left.\left.+\frac{\partial}{\partial y}\left(v\left(\rho v_0^2+5p\right)\right)\right]\right\}+\frac{\partial}{\partial x}\left\{u\left(\rho v_0^2+5p\right)-\tau\left[\frac{\partial}{\partial t}\left(u\left(\rho v_0^2+5p\right)\right)\right.\right.$$
$$\left.\left.+\frac{\partial}{\partial x}\left(u^2\rho v_0^2+p v_0^2+7 p u^2+5\frac{p^2}{\rho}\right)+\frac{\partial}{\partial y}\left(u v\rho v_0^2+7 p u v\right)\right]\right\}$$
$$+\frac{\partial}{\partial y}\left\{v\left(\rho v_0^2+5p\right)-\tau\left[\frac{\partial}{\partial t}\left(v\left(\rho v_0^2+5p\right)\right)+\frac{\partial}{\partial x}\left(\rho u v v_0^2+7 p u v\right)\right.\right.$$
$$\left.\left.+\frac{\partial}{\partial y}\left(v^2\rho v_0^2+p v_0^2+7 p v^2+5\frac{p^2}{\rho}\right)\right]\right\}=0. \qquad (7.1.4)$$

Here, mean time between collision is equal

$$\tau = \frac{\Pi \mu}{p},$$

where the factor Π allows for the model of interaction of gas particles. In the so-called "elementary" kinetic theory, $\Pi = 1$; $v_0^2 = u^2 + v^2$, and $\mathbf{v}_0$ is the hydrodynamical flow velocity with components u and v.

The system of generalized equations (7.1.1)–(7.1.4) was made dimensionless by the use of dimensionless variables $\hat{p} = p/p_\infty$, $\hat{\rho} = \rho/\rho_\infty$, $\hat{u} = u/V_s$, $\hat{v} = v/V_s$, and $\hat{t} = tV_s/L$.

$$\begin{aligned}
&\frac{\partial}{\partial \hat{t}}\left\{\hat{\rho} - \Pi\frac{\hat{\mu}}{\hat{p}}\frac{Eu^{-1}}{Re}\left[\frac{\partial \hat{\rho}}{\partial \hat{t}} + \frac{\partial}{\partial \hat{x}}(\hat{\rho}\hat{u}) + \frac{\partial}{\partial \hat{y}}(\hat{\rho}\hat{v})\right]\right\}\\
&\quad + \frac{\partial}{\partial \hat{x}}\left\{\hat{\rho}\hat{u} - \Pi\frac{\hat{\mu}}{\hat{p}}\frac{Eu^{-1}}{Re}\left[\frac{\partial}{\partial \hat{t}}(\hat{\rho}\hat{u}) + \frac{\partial}{\partial \hat{x}}\left(\hat{\rho}\hat{u}^2\right) + \frac{\partial}{\partial \hat{y}}(\hat{\rho}\hat{u}\hat{v}) + Eu\frac{\partial \hat{p}}{\partial \hat{x}}\right]\right\}\\
&\quad + \frac{\partial}{\partial \hat{y}}\left\{\hat{\rho}\hat{v} - \Pi\frac{\hat{\mu}}{\hat{p}}\frac{Eu^{-1}}{Re}\left[\frac{\partial}{\partial \hat{t}}(\hat{\rho}\hat{v}) + \frac{\partial}{\partial \hat{x}}(\hat{\rho}\hat{u}\hat{v})\right.\right.\\
&\quad \left.\left. + \frac{\partial}{\partial \hat{y}}\left(\hat{\rho}\hat{v}^2\right) + Eu\frac{\partial \hat{p}}{\partial \hat{y}}\right]\right\} = 0,
\end{aligned} \tag{7.1.5}$$

$$\begin{aligned}
&\frac{\partial}{\partial \hat{t}}\left\{\hat{\rho}\hat{u} - \Pi\frac{\hat{\mu}}{\hat{p}}\frac{Eu^{-1}}{Re}\left[\frac{\partial}{\partial \hat{t}}(\hat{\rho}\hat{u}) + \frac{\partial}{\partial \hat{x}}\left(\hat{\rho}\hat{u}^2\right) + \frac{\partial}{\partial \hat{y}}(\hat{\rho}\hat{u}\hat{v}) + Eu\frac{\partial \hat{p}}{\partial \hat{x}}\right]\right\}\\
&\quad + \frac{\partial}{\partial \hat{x}}\left\{Eu\hat{p} + \hat{\rho}\hat{u}^2 - \Pi\frac{\hat{\mu}}{\hat{p}}\frac{Eu^{-1}}{Re}\left[\frac{\partial}{\partial \hat{t}}\left(Eu\hat{p} + \hat{\rho}\hat{u}^2\right) + 3Eu\frac{\partial}{\partial \hat{x}}(\hat{p}\hat{u})\right.\right.\\
&\quad \left.\left. + \frac{\partial}{\partial \hat{x}}\left(\hat{\rho}\hat{u}^3\right) + \frac{\partial}{\partial \hat{y}}\left(\hat{\rho}\hat{u}^2\hat{v}\right) + 2Eu\frac{\partial}{\partial \hat{y}}(\hat{p}\hat{v})\right]\right\}\\
&\quad + \frac{\partial}{\partial \hat{y}}\left\{\hat{\rho}\hat{u}\hat{v} - \Pi\frac{\hat{\mu}}{\hat{p}}\frac{Eu^{-1}}{Re}\left[\frac{\partial}{\partial \hat{t}}(\hat{\rho}\hat{u}\hat{v}) + \frac{\partial}{\partial \hat{x}}\left(\hat{\rho}\hat{u}^2\hat{v}\right) + Eu\frac{\partial}{\partial y}(\hat{p}\hat{u})\right.\right.\\
&\quad \left.\left. + \frac{\partial}{\partial y}\left(\hat{\rho}\hat{u}\hat{v}^2\right)\right]\right\} = 0,
\end{aligned} \tag{7.1.6}$$

$$\begin{aligned}
&\frac{\partial}{\partial \hat{t}}\left\{\hat{\rho}\hat{v} - \Pi\frac{\hat{\mu}}{\hat{p}}\frac{Eu^{-1}}{Re}\left[\frac{\partial}{\partial \hat{t}}(\hat{\rho}\hat{v}) + \frac{\partial}{\partial \hat{x}}(\hat{\rho}\hat{u}\hat{v}) + \frac{\partial}{\partial \hat{y}}\left(\hat{\rho}\hat{v}^2\right) + Eu\frac{\partial \hat{p}}{\partial \hat{y}}\right]\right\}\\
&\quad + \frac{\partial}{\partial \hat{x}}\left\{\hat{\rho}\hat{u}\hat{v} - \Pi\frac{\hat{\mu}}{\hat{p}}\frac{Eu^{-1}}{Re}\left[\frac{\partial}{\partial \hat{t}}(\hat{\rho}\hat{u}\hat{v}) + Eu\frac{\partial}{\partial \hat{x}}(\hat{p}\hat{v}) + \frac{\partial}{\partial \hat{x}}\left(\hat{\rho}\hat{u}^2\hat{v}\right)\right.\right.\\
&\quad \left.\left. + \frac{\partial}{\partial \hat{y}}\left(\hat{\rho}\hat{u}\hat{v}^2\right)\right]\right\} + \frac{\partial}{\partial \hat{y}}\left\{Eu\hat{p} + \hat{\rho}\hat{v}^2 - \Pi\frac{\hat{\mu}}{\hat{p}}\frac{Eu^{-1}}{Re}\left[\frac{\partial}{\partial \hat{t}}\left(Eu\hat{p} + \hat{\rho}\hat{v}^2\right)\right.\right.\\
&\quad \left.\left. + 3Eu\frac{\partial}{\partial \hat{y}}(\hat{p}\hat{v}) + \frac{\partial}{\partial y}\left(\hat{\rho}\hat{v}^3\right) + \frac{\partial}{\partial \hat{x}}\left(\hat{\rho}\hat{u}\hat{v}^2\right) + 2Eu\frac{\partial}{\partial \hat{x}}(\hat{p}\hat{u})\right]\right\} = 0,
\end{aligned} \tag{7.1.7}$$

$$
\begin{aligned}
&\frac{\partial}{\partial \hat{t}}\left\{\hat{\rho}\hat{v}_0^2+3\hat{p}-\Pi\frac{\hat{\mu}}{\hat{p}}\frac{Eu^{-1}}{Re}\left[\frac{\partial}{\partial \hat{t}}\left(\hat{\rho}\hat{v}_0^2+3Eu\hat{p}\right)+\frac{\partial}{\partial \hat{x}}\left(\hat{u}\left(\hat{\rho}\hat{v}_0^2+5Eu\hat{p}\right)\right)\right.\right.\\
&\quad\left.\left.+\frac{\partial}{\partial \hat{y}}\left(\hat{v}\left(\hat{\rho}\hat{v}_0^2+5Eu\hat{p}\right)\right)\right]\right\}\\
&\quad+\frac{\partial}{\partial \hat{x}}\left\{\hat{u}\left(\hat{\rho}\hat{v}_0^2+5Eu\hat{p}\right)-\Pi\frac{\hat{\mu}}{\hat{p}}\frac{Eu^{-1}}{Re}\left[\frac{\partial}{\partial \hat{t}}\left(\hat{u}\left(\hat{\rho}\hat{v}_0^2+5Eu\hat{p}\right)\right)\right.\right.\\
&\quad+\frac{\partial}{\partial \hat{x}}\left(\hat{u}^2\hat{\rho}\hat{v}_0^2+Eu\hat{p}\hat{v}_0^2+7Eu\hat{p}\hat{u}^2+5Eu\frac{\hat{p}^2}{\hat{\rho}}\right)\\
&\quad\left.\left.+\frac{\partial}{\partial \hat{y}}\left(\hat{u}\hat{v}\hat{\rho}\hat{v}_0^2+7Eu\hat{p}\hat{u}\hat{v}\right)\right]\right\}\\
&\quad+\frac{\partial}{\partial \hat{y}}\left\{\hat{v}\left(\hat{\rho}\hat{v}_0^2+5Eu\hat{p}\right)-\Pi\frac{\hat{\mu}}{\hat{p}}\frac{Eu^{-1}}{Re}\left[\frac{\partial}{\partial \hat{t}}\left(\hat{v}\left(\hat{\rho}\hat{v}_0^2+5Eu\hat{p}\right)\right)\right.\right.\\
&\quad+\frac{\partial}{\partial \hat{x}}\left(\hat{\rho}\hat{u}\hat{v}\hat{v}_0^2+7Eu\hat{p}\hat{u}\hat{v}\right)\\
&\quad\left.\left.+\frac{\partial}{\partial \hat{y}}\left(\hat{v}^2\hat{\rho}\hat{v}_0^2+Eu\hat{p}\hat{v}_0^2+7Eu\hat{p}\hat{v}^2+5Eu^2\frac{\hat{p}^2}{\hat{\rho}}\right)\right]\right\}=0. \qquad (7.1.8)
\end{aligned}
$$

The effect of the force of gravity is neglected, so that there are two similarity criteria in the picture: $Eu = p_\infty/(\rho_\infty V_s^2)$ and $Re = LV_s\rho_\infty/\mu_\infty$, where V_s is the velocity of the external flow. The parameter Π corresponds to the first-order approximation in the hard-sphere model: $\Pi = 0.8$.

These systems of dimensional hydrodynamic equations (7.1.1)–(7.1.4) and corresponding dimensionless equations (7.1.5), (7.1.6) will be applied in following two-dimensional flow calculations in this chapter. The difference consists mainly in geometry of flow boundaries and boundary conditions.

The initial conditions ($t = 0$) are as follows

$$
\begin{aligned}
&\rho=\rho_\infty, \qquad p=p_\infty, \qquad v=0,\\
&u=V_s \quad \text{for } y=0, \qquad u=0 \quad \text{for } y>0,\\
&\frac{\partial u}{\partial t}=0, \quad \frac{\partial v}{\partial t}=0, \quad \frac{\partial \rho}{\partial t}=0, \quad \frac{\partial p}{\partial t}=0.
\end{aligned}
$$

The boundary conditions to be satisfied are

$$
\begin{aligned}
&u(x,0)=V_s, \qquad v(x,0)=0,\\
&u(x,L)=0, \quad v(x,L)=0 \quad \text{for } x\in[0,L],\\
&u(0,y)=0, \qquad v(0,y)=0,\\
&u(L,y)=0, \quad v(L,y)=0 \quad \text{for } y\in[0,L],\\
&\left[\frac{\partial \rho}{\partial x}\right]_{x=0}=0, \quad \left[\frac{\partial \rho}{\partial x}\right]_{x=L}=0 \quad \text{for } y\in[0,L],
\end{aligned}
$$

$$\left[\frac{\partial \rho}{\partial y}\right]_{y=0} = 0, \quad \left[\frac{\partial \rho}{\partial y}\right]_{y=L} = 0 \quad \text{for } x \in [0, L],$$

$$\left[\frac{\partial p}{\partial x}\right]_{x=0} = 0, \quad \left[\frac{\partial p}{\partial x}\right]_{x=L} = 0 \quad \text{for } y \in [0, L],$$

$$\left[\frac{\partial p}{\partial y}\right]_{y=0} = 0, \quad \left[\frac{\partial p}{\partial y}\right]_{y=L} = 0 \quad \text{for } x \in [0, L].$$

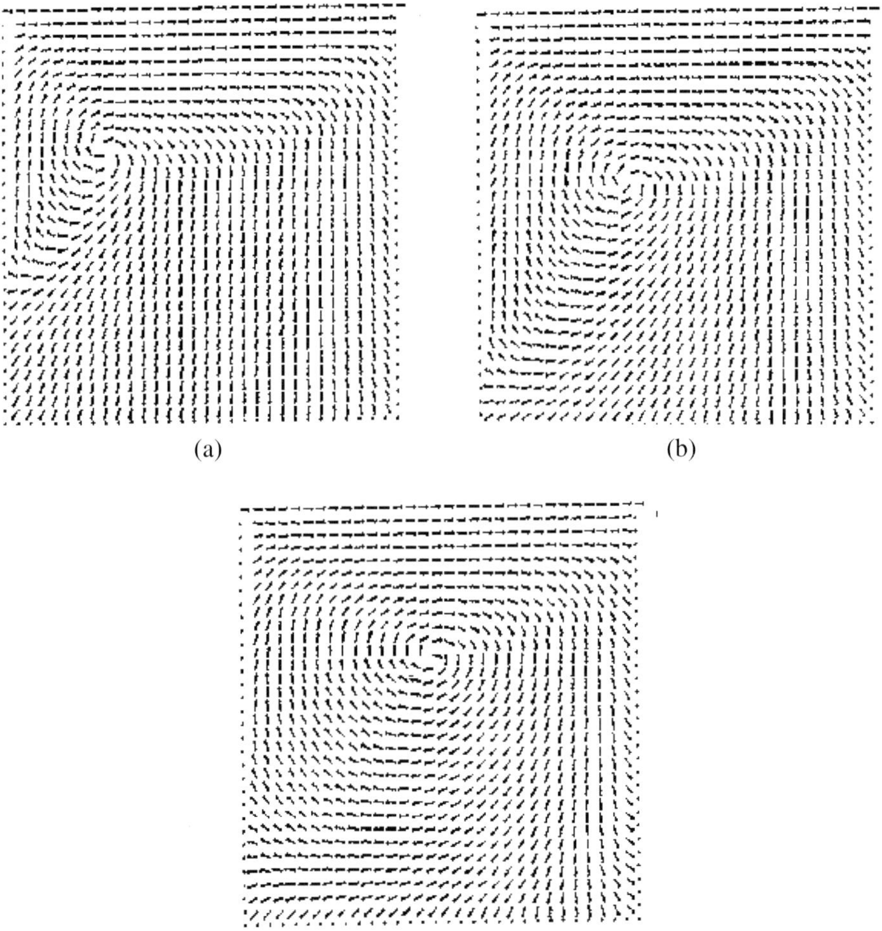

(a) (b)

(c)

Fig. 7.2. A flow of gas in the space at $Re = 6.24$ and $Eu = 2.33$ for the moments of time: (a) $\hat{t} = 0.20$, (b) $\hat{t} = 0.50$, (c) $\hat{t} = 1.40$. The calculations are performed by the GHE.

These boundary conditions imply that there is no slip, no leakage of compressible gas through the wall, and no heat transfer at the cavity wall – a good enough model to demonstrate the potential of the generalized hydrodynamical equations. The computations performed covered a wide range of Reynolds numbers. Many calculated results, including those for other types of flow (along a heated cylinder and over a step) may be found elsewhere (Fedoseyev and Alexeev, 1998a, 1998b), and in what follows only some characteristic results will be given.

Along with the program described above, calculations using the generalized Euler equations and the Navier–Stokes equations were carried out simultaneously. While the cavity flowfield solutions obtained by the different approaches are qualitatively different for the drastically unsteady regime, they start getting closer for sufficiently long times.

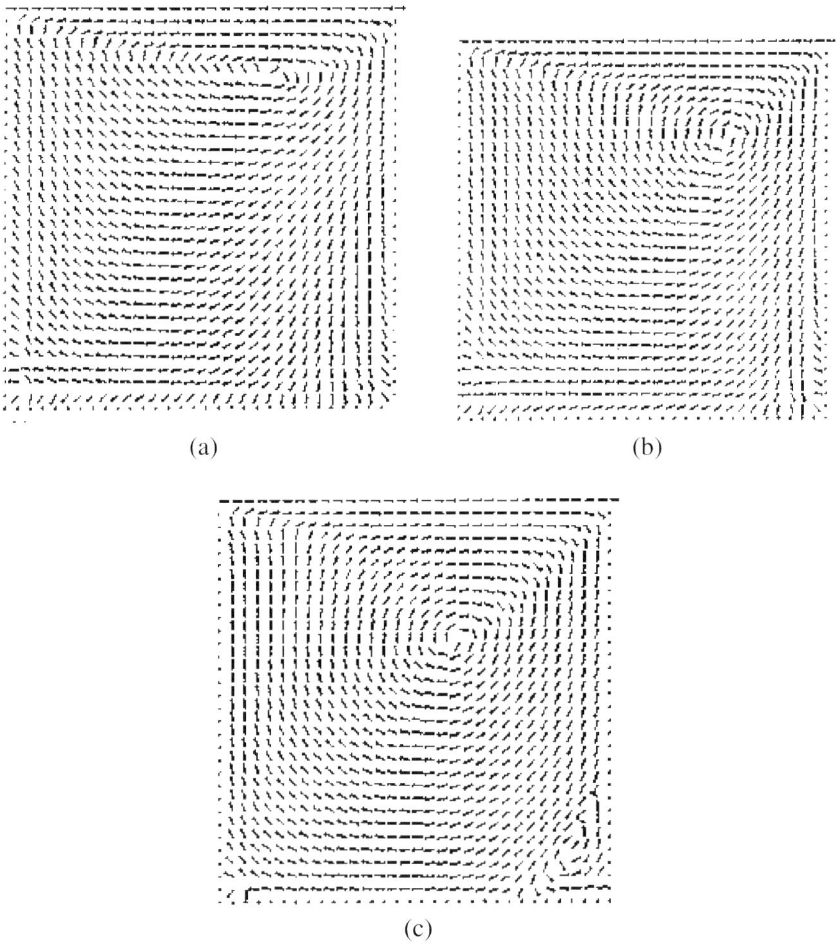

Fig. 7.3. A flow of gas in the space at $Re = 365.0$ and $Eu = 1.0$ for the moments of time: (a) $\hat{t} = 2.0$, (b) $\hat{t} = 5.0$, (c) $\hat{t} = 30.0$. The calculations are performed by the GHE.

Increasing the Reynolds number increases the difference between the flowfield patterns obtained from the generalized Euler equations and the Navier–Stokes equations.

Figure 7.2 illustrates the development of vortex flow with $Re = 6.24$, $Eu = 2.33$, $Kn = 0.13$, and the Mach number $M = 0.51$ for the moments of time $\hat{t} = 0.20, 0.50, 1.40$, respectively. The vortex that forms under these conditions corresponding to the intermediate range of the Knudsen numbers values evolves from the left corner of the space towards the center and stays above the middle of the space. The arrows in Figures 7.2–7.4 and subsequent figures indicate only the direction of the flow, and the arrow length does not depend on the magnitude of velocity vector.

Figures 7.3 and 7.4 correspond to the flow mode with $Re = 365$, $Eu = 1$, $Kn = 0.00343$, and $M = 0.775$. Figure 7.3 corresponds to calculations by the generalized

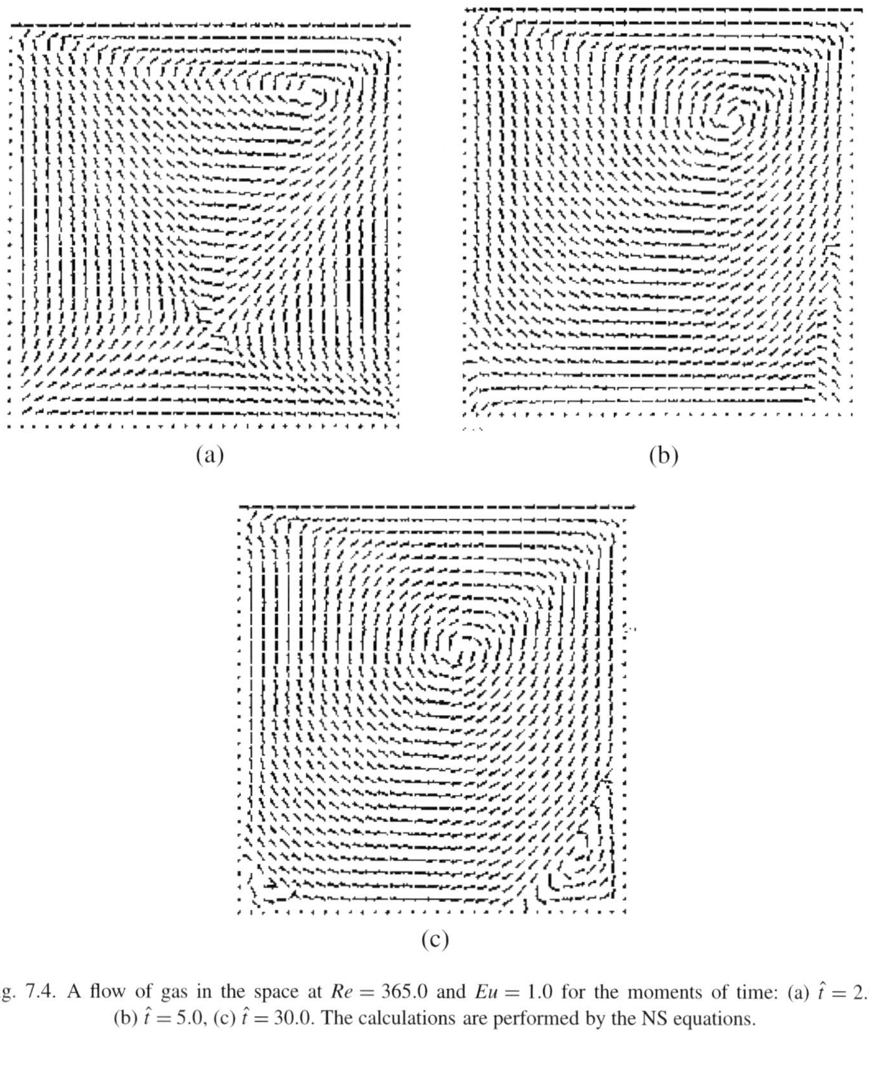

Fig. 7.4. A flow of gas in the space at $Re = 365.0$ and $Eu = 1.0$ for the moments of time: (a) $\hat{t} = 2.0$, (b) $\hat{t} = 5.0$, (c) $\hat{t} = 30.0$. The calculations are performed by the NS equations.

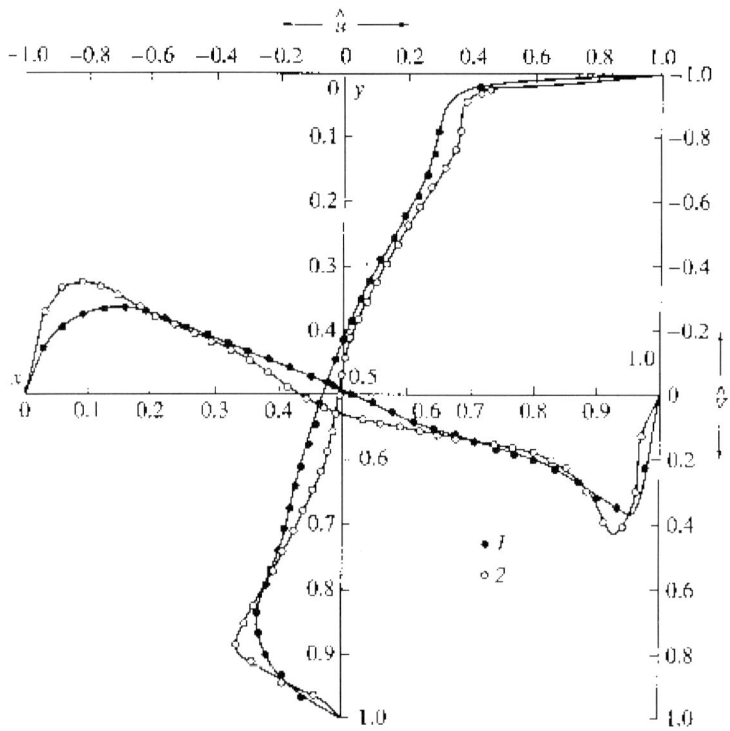

Fig. 7.5. The profiles of longitudinal and transverse velocity at $Re = 365.0$ and $Eu = 1.0$ in the quasi-steady mode. Curves 1 and 2 correspond to the GHE and NS equations, respectively.

hydrodynamic equations (GHE), and Figure 7.4 to calculations by the Navier–Stokes equations (NS) for the same moments of time, namely for $\hat{t} = 2.0, 5.0, 30.0$, which enables one to identify the differences in the behavior of flow during transition to the quasi-steady mode. In the essentially unsteady region, the patterns of flow differ considerably. However, for the moments of time $\hat{t} > 5.0$ a convergence of the qualitative patterns of flow begins. A "central" eddy and two bottom secondary eddies form, and, at $\hat{t} = 30.0$, the flow patterns become very similar from the qualitative standpoint. However, in the quasi-steady mode, the profiles of longitudinal and transverse velocities differ as well, as is illustrated by Figure 7.5, in which the curves 1 are derived by solving the GHE, and the curves 2 by solving NS equations.

As the Reynolds number increases, the flow pattern for the hydrodynamic GHE and NS models in the transition mode become ever more different.

Results for $Re = 3200$, $Eu = 1.0$, $Kn = 0.0003915$, and $M = 0.775$ at (dimensionless) instants of time $\hat{t} = 4.0, 9.5$, and 230.0 are shown in Figures 7.6 and 7.7 for the generalized Euler equations and the Navier–Stokes equations, respectively. Notice that the concept of a quasi-stationary regime becomes rather vague in this case.

Figure 7.8 shows the position of the center of the central vortex at large times. Points 1–11 were calculated from the generalized hydrodynamical equations for the di-

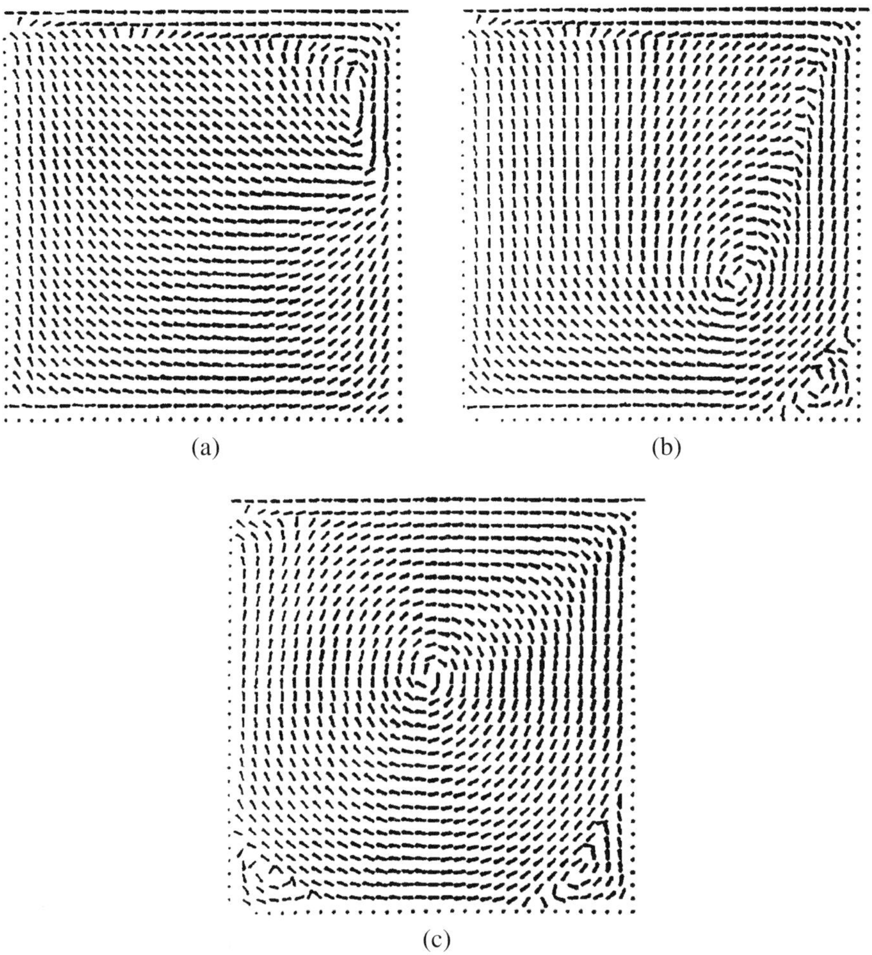

Fig. 7.6. Gas flow in a cavity at times: (a) $\hat{t} = 4.9$, (b) $\hat{t} = 9.5$, and (c) $\hat{t} = 230.0$. Calculations are made using the generalized Euler equations for $Re = 3200$, $Eu = 1.0$.

mensionless instants of time $\hat{t} = 201.0, 202.0, 203.5, 204.0, 205.0, 205.8, 207.0, 208.0, 209.0, 210.0$, and 211.5. It turns out that the vortex center performs a rotational motion.

The lid-driven flow in a rectangular cavity has been used extensively as a test bed for computational schemes. From experimental point of view this type of flow belongs to so-called "benchmark experiments" providing relevant date base for numerical modelers. Moreover, an experimental program was beginning as a result of large differences in data obtained by various schemes.

It is a well-known fact that two-dimensional Navier–Stokes calculations for an incompressible isothermal fluid (water) for the central cross section of a cavity correlate very poorly with experimental data (Kozeff and Street, 1984a, 1984b, 1984c) even if the length of the cavity is much larger than its width. Nor do three-dimensional Navier–

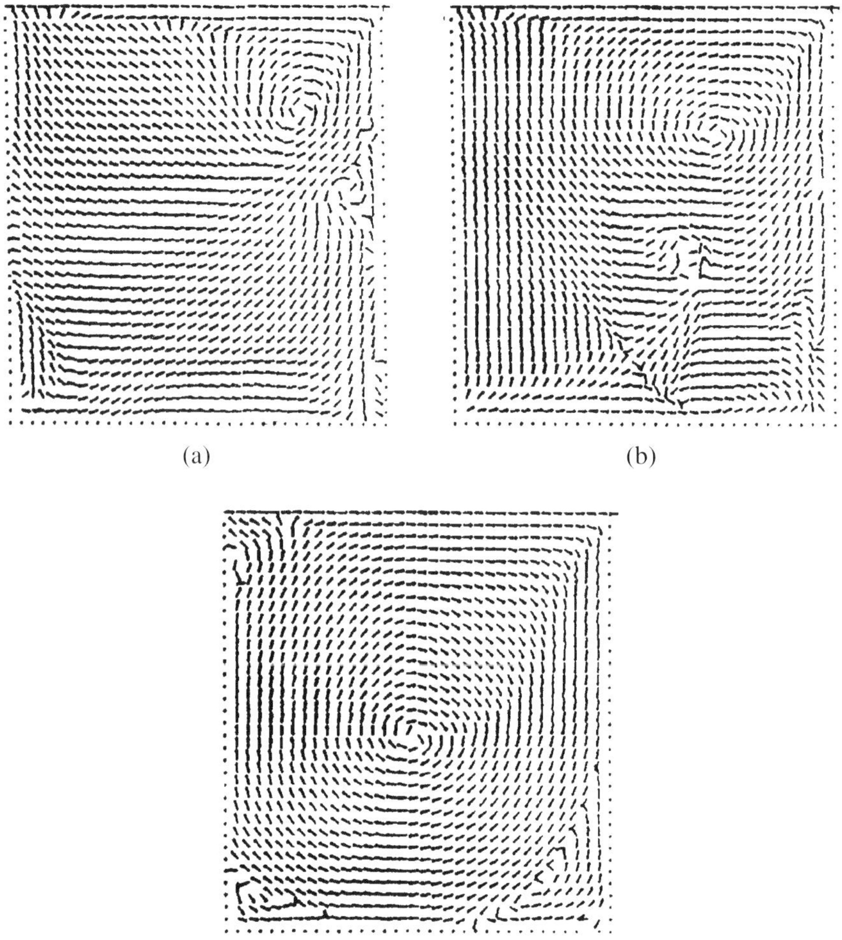

(a) (b) (c)

Fig. 7.7. Gas flow in a cavity at times: (a) $\hat{t} = 4.0$, (b) $\hat{t} = 9.5$, and (c) $\hat{t} = 230.0$. Calculations are made using the Navier–Stokes equations for $Re = 3200$, $Eu = 1.0$.

Stokes calculations improve the picture (Arnal et al., 1992; Verstappen and Veldman, 1996). It has been found that three-dimensional Navier–Stokes calculated results obtained on coarser meshes may agree better with experimental data than do formally more accurate solutions (Arnal et al., 1992).

Detailed experiments had been carried by J. Koseff and R. Street with lid-driven cavities (Kozeff and Street, 1984a, 1984b, 1984c). The central component of their experimental facility is a rectangular cavity of a width of 150 mm in the direction of lid motion, a maximum vertical depth of 925 mm and a span of 450 mm transverse to the direction of lid motion. The qualitative positions (found in experiments in the symmetry plane) of upper secondary eddy, primary eddy, upstream secondary eddy, and

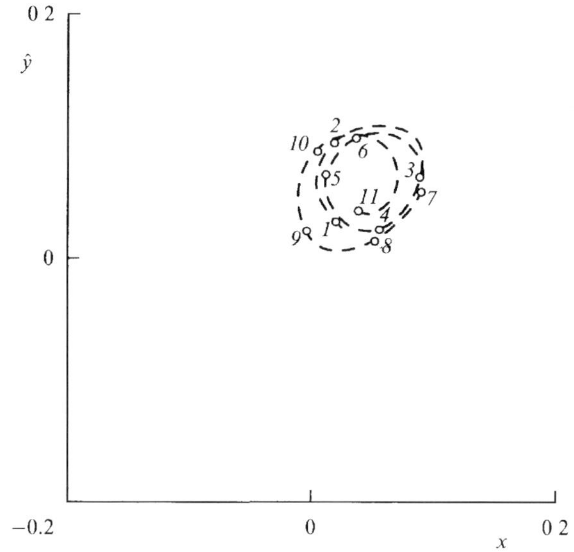

Fig. 7.8. Position of the center of the central vortex (relative to the center of the cavity) for instants of time $\hat{t} = 201.0, 202.0, 203.5, 204.0, 205.0, 205.8, 207.0, 208.0, 209.0, 210.0$, and 211.5 (points 1–11, respectively). Calculations were made using the generalized hydrodynamical equations for $Re = 3200$, $Eu = 1.0$.

downstream secondary eddy correspond to GHE calculations for rather high Reynolds numbers. In the Koseff and Street experiments the visualization method was used for water flows in the range of Reynolds numbers from 1000 to 10 000. In particular, they studied the following aspects of the flow in a square cavity: (1) the presence of turbulence in the flow; (2) the three-dimensionality of the flow and associated flow structures; (3) the size of the downstream secondary eddy as a function of Reynolds number; (4) the formation of Taylor-type instabilities during the first 30 seconds or so after the lid is started.

Velocity measurements were made for Reynolds numbers of 3200 and 10 000 on two vertical planes. These planes were the symmetry plane, a plane parallel to it, and one 17.5 mm from the end-wall (Kozeff and Street, 1984b).

Unfortunately, we have no experimental results on the situation we are considering – i.e., gas flow in a cavity – and as to the comparison of experimental and theoretical flow data for a gas and a liquid (even at the same Re), this requires great caution. Nevertheless, the calculated Re dependence of the ratio of the size of the bottom near-wall vortex D_3 (downstream secondary eddy) to the cavity width L (see Figure 7.9) is in general agreement with experiment (Kozeff and Street, 1984a, 1984b, 1984c). Interestingly, because of the growing oscillations of hydrodynamical quantities, it is found that even in the quasi-stationary regime fluctuations in the position of the vortex, D_3/L, increase in magnitude. As seen in Figure 7.9, the region of fluctuations is represented by an expanding band, which shows a transition to fully developed turbulence.

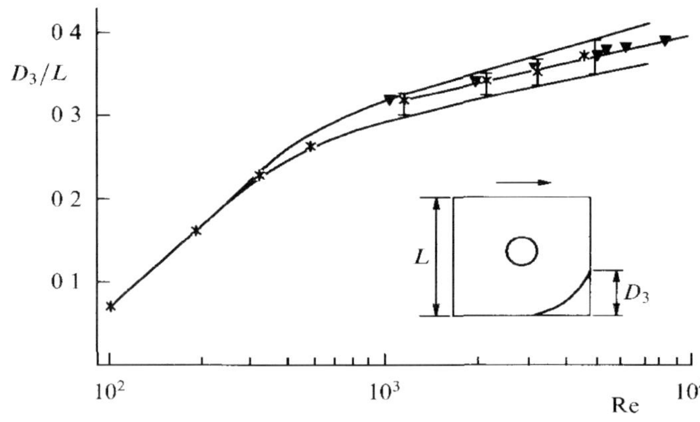

Fig. 7.9. Relative size D_3/L of the bottom vortex plotted versus Reynolds number Re for $Eu = 1.0$. Solid lines represent theoretical results.

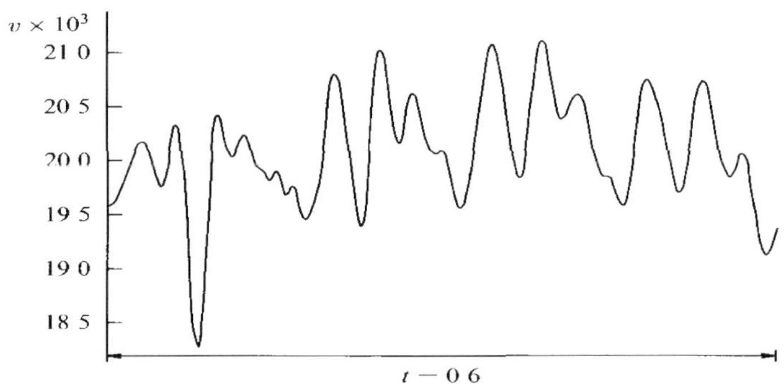

Fig. 7.10. Oscillations of the absolute value of velocity $\hat{v}_0$ at the point ($\hat{x} = 0.13$, $\hat{y} = 0.87$) over a period of $\hat{t} = 0.6$ in the near "quasi-stationary" flow regime for $Re = 3200$, $Eu = 1.0$. Zero time $\hat{t}_0 = 185.0$.

Figure 7.10 demonstrates oscillations in the absolute magnitude of the dimensionless velocity $\hat{v}_0$ at the point ($\hat{x} = 0.13$, $\hat{y} = 0.87$) over a dimensionless time period $\hat{t} = 0.6$ for $Re = 3200$. Note the irregular nature of the oscillations. Thus, already at $Re = 3200$ the flow starts to exhibit typical features of a turbulent regime.

Figures 7.11 and 7.12 contain comparison of horizontal and vertical velocity profiles for numerical and experimental results (2D and 3D variants for symmetry plane), $Re = 3200$, $Re = 10\,000$ correspondingly for the GHE and Navier–Stokes models and also for known k–ε turbulence model. All results obtained by the finite element method for GHE are modified for description of liquid media (Fedoseyev and Alexeev, 1998a, 1998b).

Details of numerical schemes applied for GHE solution are discussed below. Note that the use of the generalized hydrodynamical equations with viscous terms makes it possible to construct the extremely effective difference schemes, thus making these equations even more attractive.

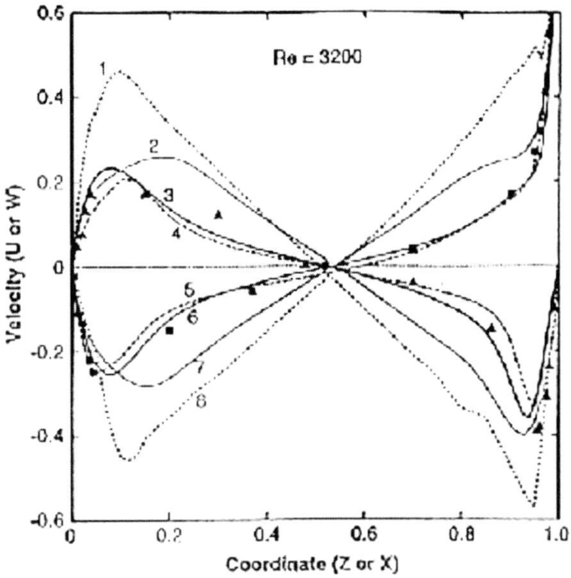

Fig. 7.11. Comparison of horizontal velocity profiles (1–4) for numerical (solid and dashed lines) and experimental (squares) results, and vertical velocity profiles (5–8) for numerical (solid and dashed lines) and experimental (triangles) results, $Re = 3200$: 1 – NS, 2 – k–ε model, 3 – GHE (2D), 4 – GHE (3D), 5 – GHE (3D), 6 – GHE (2D), 7 – k–ε model, 8 – NS.

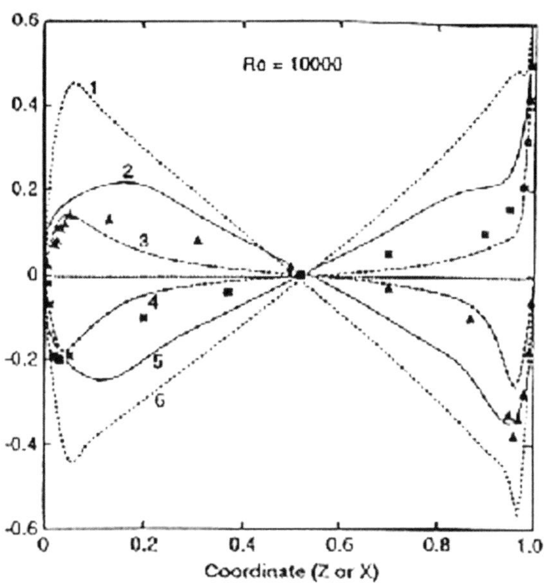

Fig. 7.12. Comparison of horizontal velocity profiles (1–3) for numerical (dashed lines) and experimental (squares) results, and vertical velocity profiles (4–6) for numerical (dashed lines) and experimental (triangles) results, $Re = 10^4$: 1 – NS, 2 – k–ε model, 3 – GHE (2D), 4 – GHE (2D), 5 – k–ε model, 6 – NS.

7.2. Application of the generalized hydrodynamic equations: to the investigation of gas flows in channels with a step

In this section generalized hydrodynamic equations, which include explicitly the Kolmogorov fluctuations of hydrodynamic quantities, are used for mathematical simulation of vortex gas flows. The calculations are performed in the range of Reynolds number values from 64 to 10 000 for two-dimensional unsteady nonisothermal flow of compressible gas in a channel with a step.

Turbulent flows of gases and liquids have been the subject of intensive study for well over a hundred years, which explains the large number of possible applications. Considering here backward-facing step flow as investigated above, the lid-driven flow in a rectangular cavity has been used extensively as a test bed for computational schemes. This type of flow also belongs to a set of benchmarks providing relevant date base for numerical modelers.

We will treat the problem in the following formulation. Let a steady flow of gas in a flat channel be initially accompanied by a momentary dip in the bottom that transforms the flow into an unsteady, generally speaking, turbulent, flow in a channel with a step. For the purposes of mathematical simulation, use is made of the geometry of a flat channel with a square step, as shown in Figure 7.13.

The flow on the left in the narrow channel zone exhibits the behavior of Poiseuille flow and has a parabolic velocity profile. We will demonstrate that generalized hydrodynamic equations, derived using the locally Maxwellian approximation (generalized Euler's equations, GEuE), lead to Poiseuille flow in standard assumptions: (a) the flow is one-dimensional and steady; (b) the density $\rho = const$; (c) the hydrodynamic velocity depends on y alone, $v_0 = v_{0x}(y)$; (d) the static pressure depends on x alone, $p = p(x)$.

We will first turn to the continuity equation, which is written as

$$\frac{\partial}{\partial \mathbf{r}} \rho \mathbf{v}_0 - \tau \left(\Delta p + \frac{\partial^2}{\partial \mathbf{r} \partial \mathbf{r}} : \rho \mathbf{v}_0 \mathbf{v}_0 \right) = 0. \tag{7.2.1}$$

In the foregoing assumptions, Eq. (7.2.1) is satisfied identically. Indeed, the first and third terms of the left-hand part of Eq. (7.2.1) tend to zero by virtue of the conditions (a)–(c), and the equation

$$\Delta p = 0 \tag{7.2.2}$$

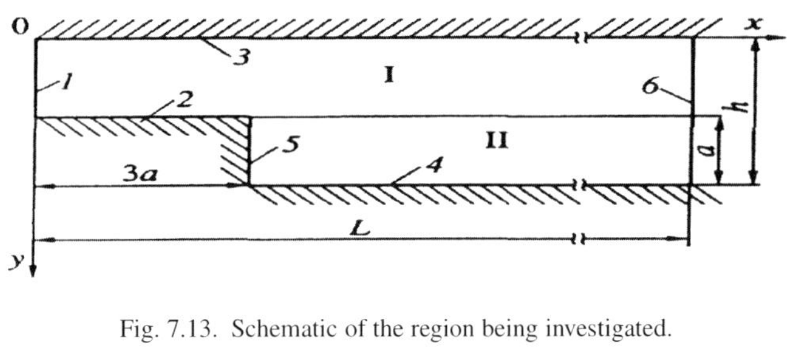

Fig. 7.13. Schematic of the region being investigated.

reduces in this case to the condition

$$\frac{\partial^2 p}{\partial x^2} = 0,$$

which leads to the equality

$$\frac{\partial p}{\partial x} = const, \tag{7.2.3}$$

which is one of the conditions of realization of Poiseuille flow.

We will now treat the generalized equation of motion in a projection onto the x-axis. In view of the conditions (a)–(c) and Eq. (7.2.3), we find from Eq. (7.1.2),

$$\frac{\partial p}{\partial x} = \tau^{(0)} p \frac{\partial^2 u}{\partial y^2}. \tag{7.2.4}$$

However, in view of relation $\tau = \Pi\mu/p$, we derive the known Poiseuille relation

$$\frac{\partial p}{\partial x} = \Pi\mu \frac{\partial^2 u}{\partial y^2}, \tag{7.2.5}$$

in which the factor Π allows for the model of interaction of gas particles. The integration of Eq. (7.2.5) simultaneously with (7.2.3) yields a parabolic velocity profile. Consequently, the GHE under the above-formulated conditions leads to a parabolic velocity profile which is adopted as the boundary condition for the equation of motion on the left-hand boundary of the calculation region.

We will introduce the characteristic quantities of flow that are used in what follows as scales in reducing the GEE to the dimensionless form: ρ_∞, p_∞, v_∞, and μ_∞ denote the density, static pressure, velocity, and dynamic viscosity at the channel inlet; the scale of length h is the channel width at the outlet; and the time scale has the form of h/v_∞. The similarity criteria, i.e., the Reynolds and Euler numbers, are written as $Re = \rho_\infty v_\infty h/\mu_\infty$ and $Eu = p_\infty/(\rho_\infty v_\infty^2)$. The force of gravity is neglected, so that there are two similarity criteria in the picture. The dimensionless quantities are further marked by the sign ^ above the respective symbol. We will conventionally divide the flow region into two, "top" and "bottom", parts that are shown in Figure 7.13 at I and II, respectively; this is necessary solely for convenience of description of the initial conditions and calculation results. We will further describe other geometric characteristics of the calculation region, shown in Figure 7.13, namely, L, the length of the region being investigated; a, the height of the backward-facing step (in the given concrete conditions, $a = h/2$); 1, the front section of the region being investigated (flow inlet section); 6, the rear section of the region being investigated (flow outlet section); 2, the step plane; 3, the top plane; 4, the bottom plane; and 5, the vertical wall of the step.

It is assumed that a steady laminar flow with the velocity profile in the form of a Poiseuille parabola exists in region I at the initial moment of time, while in region II

Table 7.1
Initial values of flow parameters in region I

Velocities and their derivatives		Pressure and density and their derivatives	
$\hat{u}(\hat{t}=0)=f(\hat{y})$,	$\left(\frac{\partial \hat{u}}{\partial \hat{t}}\right)_{\hat{t}=0}=0$	$\hat{p}(\hat{t}=0)=1$,	$\left(\frac{\partial \hat{p}}{\partial \hat{t}}\right)_{\hat{t}=0}=0$
$\hat{v}(\hat{t}=0)=0$,	$\left(\frac{\partial \hat{v}}{\partial \hat{t}}\right)_{\hat{t}=0}=0$	$\hat{\rho}(\hat{t}=0)=1$,	$\left(\frac{\partial \hat{\rho}}{\partial \hat{t}}\right)_{\hat{t}=0}=0$

Table 7.2
Initial values of flow parameters in region II

Velocities and their derivatives		Pressure and density and their derivatives	
$\hat{u}(\hat{t}=0)=0$,	$\left(\frac{\partial \hat{u}}{\partial \hat{t}}\right)_{\hat{t}=0}=0$	$\hat{p}(\hat{t}=0)=1$,	$\left(\frac{\partial \hat{p}}{\partial \hat{t}}\right)_{\hat{t}=0}=0$
$\hat{v}(\hat{t}=0)=0$,	$\left(\frac{\partial \hat{v}}{\partial \hat{t}}\right)_{\hat{t}=0}=0$	$\hat{\rho}(\hat{t}=0)=1$,	$\left(\frac{\partial \hat{\rho}}{\partial \hat{t}}\right)_{\hat{t}=0}=0$

the flow is absent (in other words, the flow velocity is zero). The results of further calculations readily transform to the model of flow of liquid (water) in a channel; in so doing, it is possible to omit the energy equation and assume the liquid to be incompressible.

The systems of dimensional hydrodynamic equations (7.1.1)–(7.1.4) and corresponding dimensionless equations (7.1.5)–(7.1.8) are applied in following two-dimensional flow calculations. The difference with Section 7.1 consists mainly in geometry of flow boundaries and boundary conditions. We begin with initial conditions contained in Tables 7.1 and 7.2.

In Table 7.1, $f(\hat{y})=k_1\hat{y}^2+k_2\hat{y}+k_3$ and the coefficients k_i $(i=1,\ldots,3)$ are found from the conditions

$$\begin{cases} f(0)=0, \\ \dfrac{1}{2}f(h-a)=1, \\ f(h-a)=0. \end{cases} \tag{7.2.6}$$

The boundary values of flow in the calculation region must be treated in application to each geometric element. Although the flow region is not too complex, for convenience in calculations we will write out the boundary conditions on each surface (Figure 7.13).

The boundary conditions specified in Table 7.3 call for further comment. They correspond to the conditions of no-slip on the wall and non-percolation, and the wall is assumed to be absolutely heat-resistant. For plane 1, the conditions are preassigned at a distance of $3a$ from the step. This distance may be varied with a view to eliminating the nonphysical reactive effect of flow behind the step on the initial flow in the narrow zone of the channel.

Table 7.3
Boundary conditions

For section 1	For plane 2	For plane 3
$x=0$	$y=h-a$	$y=0$
$\hat{\bar{u}}(\hat{y})=f(\hat{y})$	$\hat{u}(x)=0$	$\hat{u}(x)=0$
$\hat{\bar{v}}(\hat{y})=0$	$\hat{v}(x)=0$	$\hat{v}(x)=0$
$\frac{\partial^2 \hat{\bar{p}}}{\partial \hat{x}^2}=0$	$\frac{\partial \hat{p}}{\partial \hat{y}}=0$	$\frac{\partial \hat{p}}{\partial \hat{y}}=0$
$\frac{\partial \hat{\bar{\rho}}}{\partial x}=0$	$\frac{\partial \hat{\rho}}{\partial \hat{y}}=0$	$\frac{\partial \hat{\rho}}{\partial \hat{y}}=0$
For plane 4	**For plane 5**	**For section 6**
$y=h$	$x=3a$	$x=L$
$\hat{u}(x)=0$	$\hat{u}(y)=0$	$\frac{\partial \hat{\bar{u}}}{\partial \hat{x}}=0$
$\hat{v}(x)=0$	$\hat{v}(y)=0$	$\frac{\partial \hat{\bar{v}}}{\partial \hat{x}}=0$
$\frac{\partial \hat{p}}{\partial \hat{y}}=0$	$\frac{\partial \hat{p}}{\partial \hat{x}}=0$	$\frac{\partial^2 \hat{\bar{p}}}{\partial \hat{x}^2}=0$
$\frac{\partial \hat{\rho}}{\partial \hat{y}}=0$	$\frac{\partial \hat{\rho}}{\partial \hat{x}}=0$	$\frac{\partial \hat{\bar{\rho}}}{\partial x}=0$

At sections 1 and 6, the so-called "soft" boundary conditions are written relative to the averaged (rather than true) hydrodynamic quantities. Such conditions proved to be very effective in the case of numerical realization and eliminated the reactive effect upstream due to the numerical effect of cutting off the region.

The calculations (Alexeev and Mikhailov, 2001) were performed using an explicit difference scheme of the first order of accuracy with respect to time and of the second order of accuracy on the coordinates with constant dimensionless steps with respect to time and coordinates. The properties of generalized hydrodynamic equations are such that they enable one to realize stable numerical calculations in a wide range of Reynolds number values from units to tens of thousands. In calculations within this problem, the evolution of flow was studied for the values of Re from 64 to 10 000 and $Eu = 1.0$. Special attention was given to the values of Re of 1000 to 3200 as transition values between the laminar and turbulent modes of flow.

Table 7.4 gives the values of steps of the computational mesh and of the geometrical parameters involved in the calculations. Calculations with a smaller spatial step of the mesh failed to reveal substantial changes in the topology of flow in the treated range of Reynolds number values. The similarity criteria, i.e., the Knudsen and Mach numbers, may be represented in terms of Re and Eu using the average velocity of molecules for a Maxwellian distribution,

$$Kn = \Pi \sqrt{\frac{8}{\pi}} \frac{1}{Re\sqrt{Eu}}, \tag{7.2.7}$$

Table 7.4
Values of the parameters involved in the calculations

$Re = 64$	$Re = 365$	$Re = 10\,000$
$\Delta\hat{t} = 1.0 \times 10^{-4}$	$\Delta\hat{t} = 1.0 \times 10^{-3}$	$\Delta\hat{t} = 1.0 \times 10^{-4}$
$\Delta\hat{x} = \Delta\hat{y} = 0.0333$	$\Delta\hat{x} = \Delta\hat{y} = 0.0333$	$\Delta\hat{x} = \Delta\hat{y} = 0.0333$
$\hat{h} = 1.0$	$\hat{h} = 1.0$	$\hat{h} = 1.0$
$\hat{L} = 12.0$	$\hat{L} = 12.0$	$\hat{L} = 15.0$
$Re = 1000$	$Re = 3200$	
$\Delta\hat{t} = 1.0 \times 10^{-3}$	$\Delta\hat{t} = 1.0 \times 10^{-4}$	
$\Delta\hat{x} = \Delta\hat{y} = 0.0333$	$\Delta\hat{x} = \Delta\hat{y} = 0.0333$	
$\hat{h} = 1.0$	$\hat{h} = 1.0$	
$\hat{L} = 12.0$	$\hat{L} = 15.0$	

Table 7.5
Parameters of mathematical simulation

$Re = 64$,	$Kn = 1.995 \times 10^{-2}$,	$Eu = 1.0$,	$M = 0.7747$
$Re = 1000$,	$Kn = 1.277 \times 10^{-3}$,	$Eu = 1.0$,	$M = 0.7747$
$Re = 10\,000$,	$Kn = 1.277 \times 10^{-4}$,	$Eu = 1.0$,	$M = 0.7747$
$Re = 365$,	$Kn = 3.498 \times 10^{-3}$,	$Eu = 1.0$,	$M = 0.7747$
$Re = 3200$,	$Kn = 3.989 \times 10^{-3}$,	$Eu = 1.0$,	$M = 0.7747$

$$Kn = \Pi\sqrt{\frac{8\gamma}{\pi}\frac{M}{Re}}, \tag{7.2.8}$$

where γ is the heat capacity ratio, $\gamma = C_p/C_V$. The respective rescaling of parameters is given in Table 7.5.

Demonstrations of the calculation results using "snapshots" of the flow cannot fully describe its behavior in the calculation region. This is due to the fact that the parameters of a flow of liquid, even if it is laminar, do not remain unvaried in time at every point. We will treat in more detail one of the options of calculation of flow, namely, that with $Re = 1000$.

It follows from Figure 7.14 ($Re = 1000$ and $\hat{t} = 10$) that region V is some vortex formation that develops fairly actively with time. Such vortexes arise both during evolution of the region of return flow indicated in Figure 7.14 by the symbol P (a detailed view of return flow is given in Figure 7.15) and during stagnation of flow in the vicinity of the bottom wall. In viewing successively the entire pattern of flow in time, one can clearly see how independent and fairly intense vortexes (a detailed view of vortex V is given in Figure 7.16), swirling clockwise, originate and start moving along the flow to the outlet behind section 6 (see Figure 7.13). The transverse dimensions of vortexes V are not constant and amount to 0.20–0.30 of the flow region height h. The extended form of vortex V in Figure 7.14 does not correspond to the actual pattern; it is defined by the scale difference along the Ox and Oy axes. The evolution of bottom vortexes strongly resembles the soliton behavior, except for the fact that they are vortexes proper rather

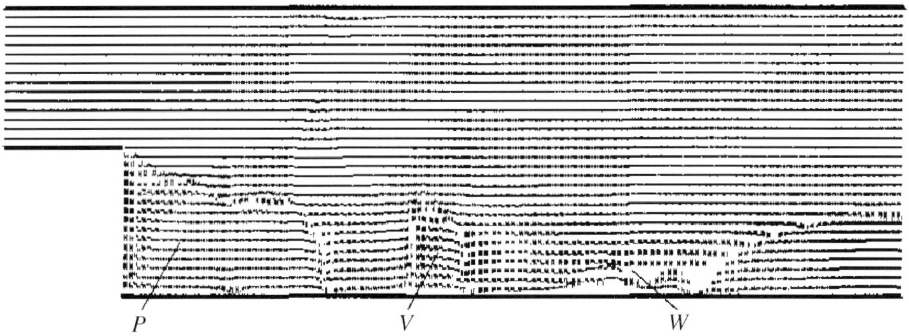

Fig. 7.14. A general view of the flow at $Re = 1000$ at the moment of time $\hat{t} = 10$ (for better illustration, the scale equality of the coordinates is not observed).

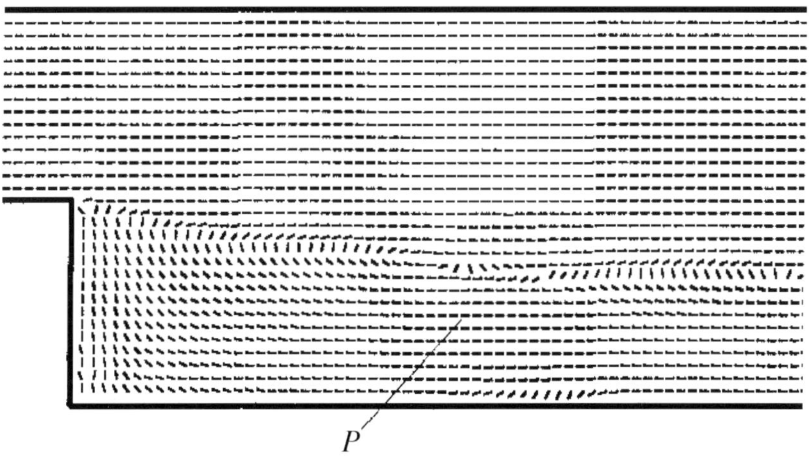

Fig. 7.15. A detailed view of return flow behind the step ($Re = 1000$, $\hat{t} = 10$).

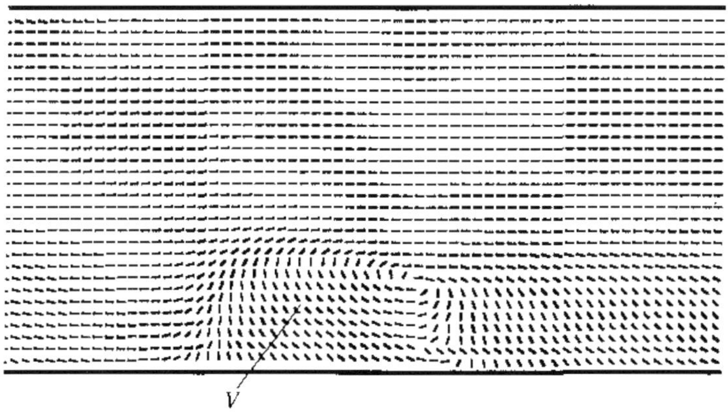

Fig. 7.16. A detailed view of vortex V for $Re = 1000$ and $\hat{t} = 10$.

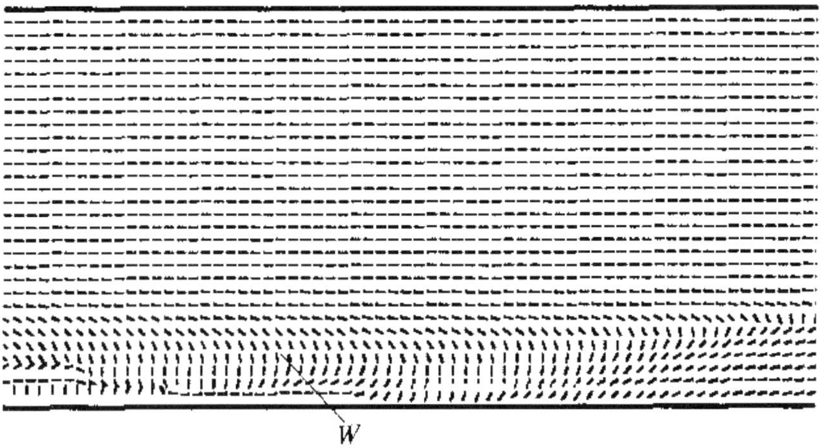

Fig. 7.17. A detailed view of vortex zones W for $Re = 1000$ and $\hat{t} = 10$.

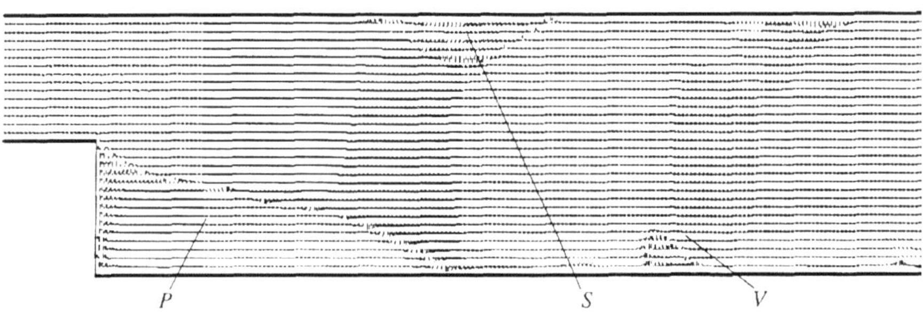

Fig. 7.18. A general view of flow at $Re = 1000$ at the moment of time $\hat{t} = 110$ (for better illustration, the scale equality of the coordinates is not observed).

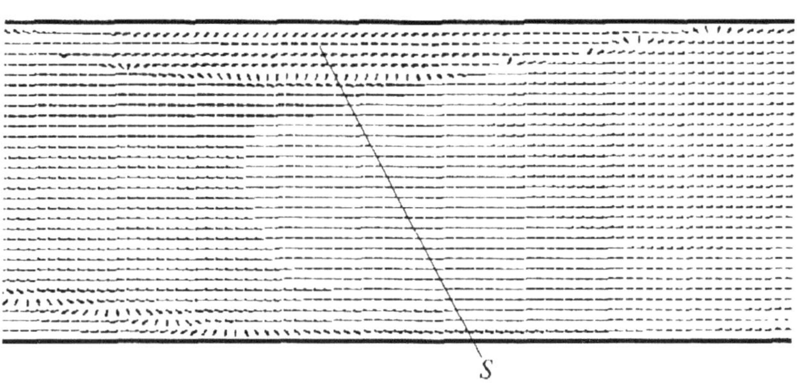

Fig. 7.19. A detailed view of wall vortex S at $Re = 1000$ and $\hat{t} = 110$.

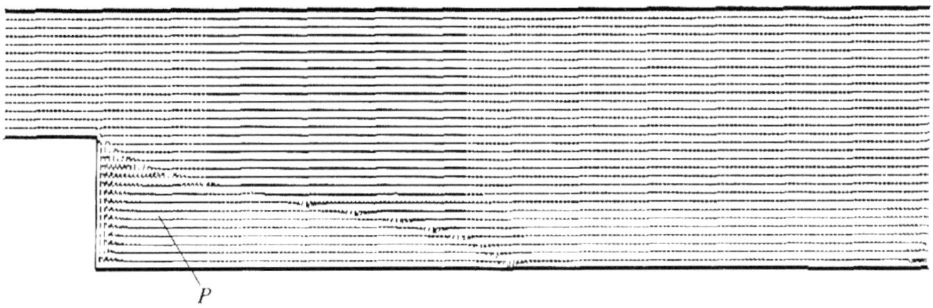

Fig. 7.20. A general view of flow at $Re = 1000$ at the moment of time $\hat{t} > 140$; the scale equality of the coordinates is not observed.

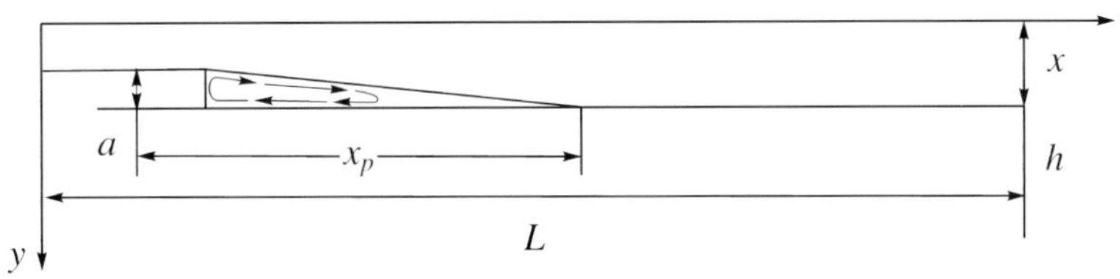

Fig. 7.21. A diagrammatic view of the point of flow attachment behind the step.

than single waves. In their motion, vortexes V "push out" the vortex formations shown in Figure 7.14 as W (a detailed view of W is given in Figure 7.17) from the calculation region to take their place and later move beyond the calculation region. Owing to the special boundary conditions at the flow section (Table 7.3, section 6), the vortex formations are not "reflected" but "pass" unobstructed to the non-calculation region behind the section. The motion of bottom vortexes downstream may lead to additional erosion of the channel bottom.

However, in passing the calculation region, vortex V, in addition to pushing out the vortex formations, introduces some disturbance of its own. This disturbance promotes the formation of wall vortexes on the top plane of the flow region (3 in Figure 7.13) that also move along the flow towards section 6 and exist until the moment the flow reaches the quasi-stationary state. It is well seen in Figure 7.18 ($Re = 1000$ and $\hat{t} = 110$), namely, region S (a more detailed view of region S is given in Figure 7.19), while vortexes of the V-type degenerate, become smaller, and finally disappear.

At $Re = 1000$, in contrast to higher values of Re, the flow reaches the quasi-stationary state (Figure 7.20) in which no further intense origination of vortexes is observed either on the bottom boundary of flow behind the step (4 in Figure 7.13) or on the top boundary (3 in Figure 7.13). One can clearly see the characteristic point of flow attachment behind the step.

For a more correct representation of the characteristic point of flow attachment to the bottom wall, we will treat a schematic, but properly scaled, Figure 7.21 in which, according to Figure 7.13, h is the height of the slot, a is the height of the step, L is the total length of the calculation region, and x_p is the position of the point of flow attachment.

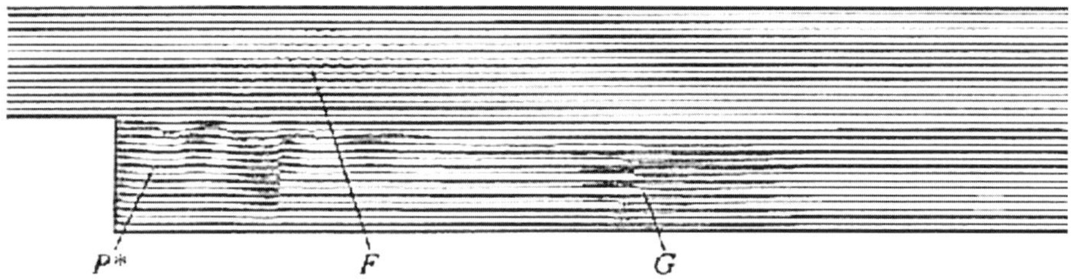

Fig. 7.22. A general view of flow at $Re = 3200$ at the moment of time $\hat{t} = 50$; the scale equality of the coordinates is not observed.

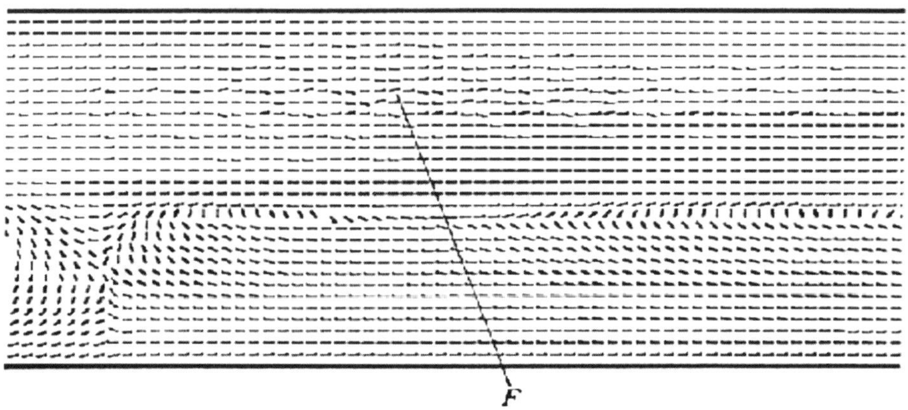

Fig. 7.23. A detailed view of turbulent "spot" in region F at $Re = 3200$ and $\hat{t} = 50$.

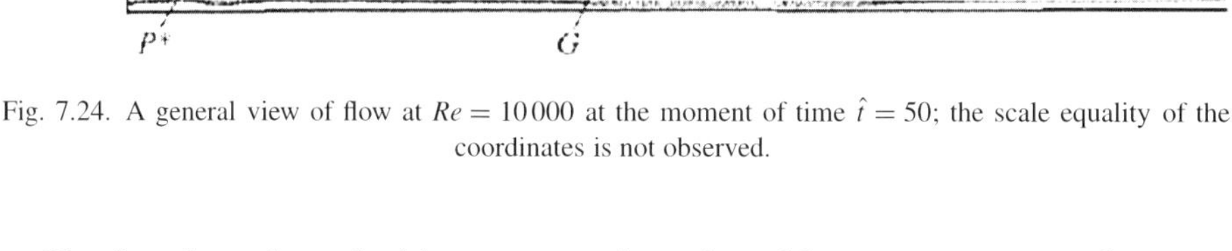

Fig. 7.24. A general view of flow at $Re = 10\,000$ at the moment of time $\hat{t} = 50$; the scale equality of the coordinates is not observed.

The flow in a channel with a step was investigated by numerous researches (see, e.g., Eaton, 1981). Moreover, this type of flow is included in the system of benchmark experiments or test experiments aimed at estimating the quality of numerical algorithms. Note that the flow topology and the position of the point of flow attachment are in good agreement with the experimental data. The position of x_p is usually located within five

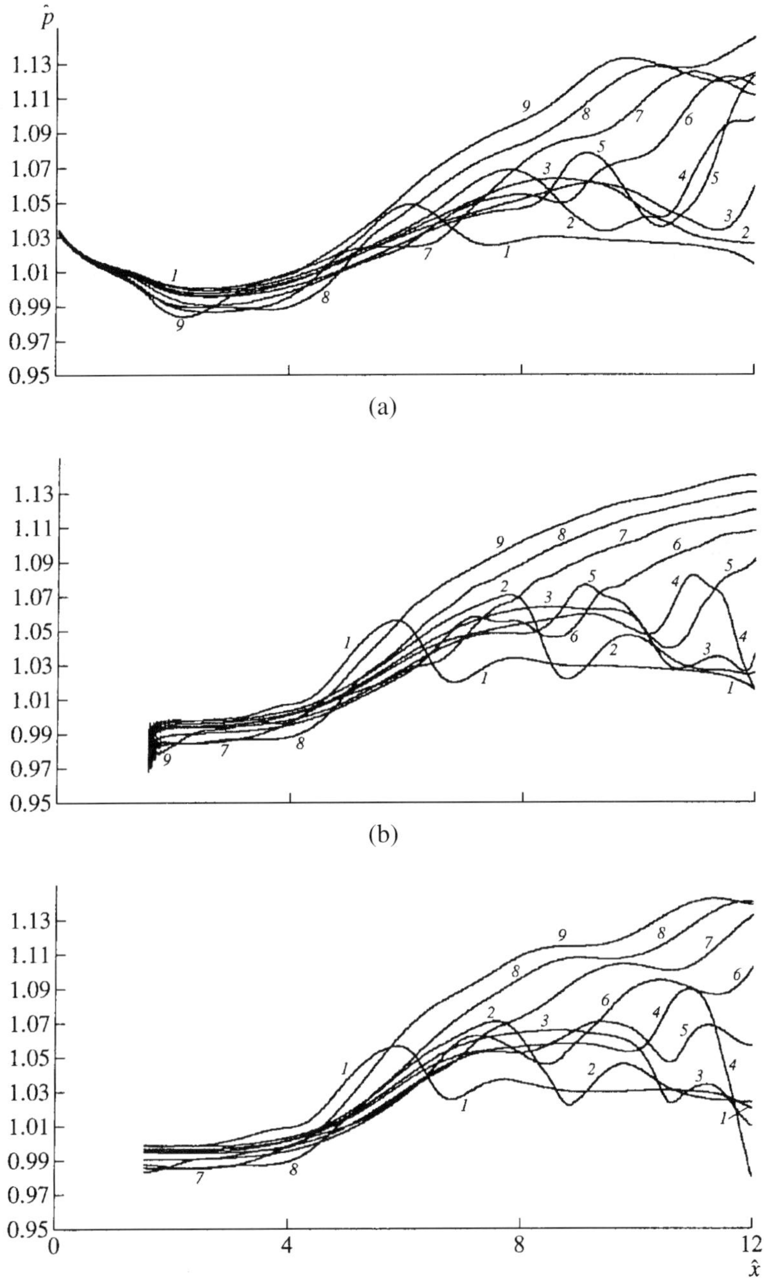

Fig. 7.25. The distribution of pressure at $Re = 1000$ along the lines $\hat{y} = const$ for different moments of time: (a) $\hat{y} = 0.1$, (b) $\hat{y} = 0.5$, (c) $\hat{y} = 0.9$; $1 - \hat{t} = 20$, $2 - \hat{t} = 30$, $3 - \hat{t} = 40$, $4 - \hat{t} = 50$, $5 - \hat{t} = 60$, $6 - \hat{t} = 70$, $7 - \hat{t} = 80$, $8 - \hat{t} = 90$, $9 - \hat{t} = 100$.

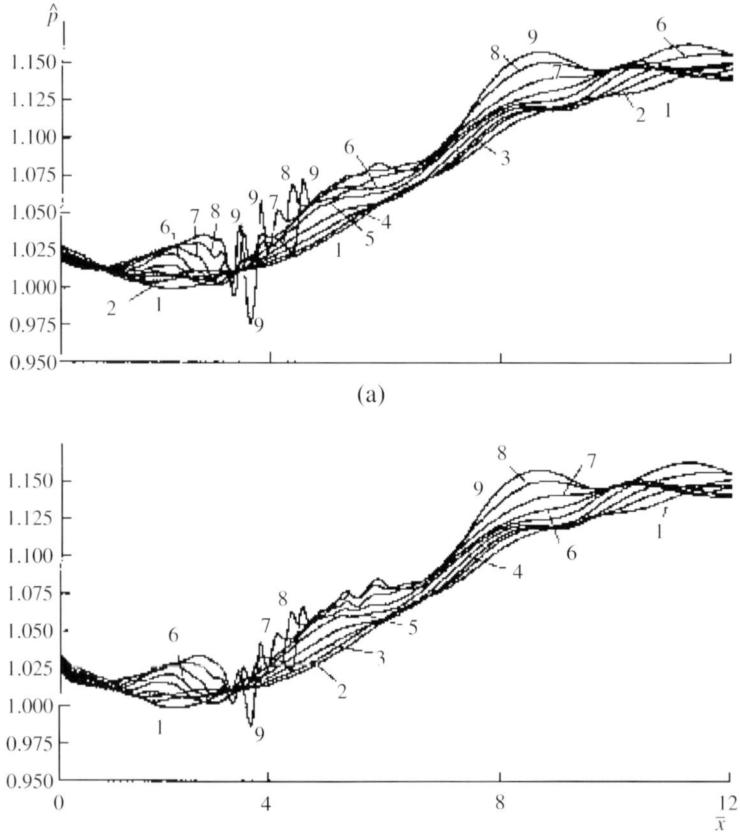

(a)

(b)

Fig. 7.26. The distribution of pressure at $Re = 3200$ along the lines $\hat{y} = const$ for different moments of time: (a) $\hat{y} = 0.1$, (b) $\hat{y} = 0.2$, (c) $\hat{y} = 0.5$, (d) $\hat{y} = 0.9$; $1 - \hat{t} = 10$, $2 - \hat{t} = 15$, $3 - \hat{t} = 20$, $4 - \hat{t} = 25$, $5 - \hat{t} = 30$, $6 - \hat{t} = 35$, $7 - \hat{t} = 40$, $8 - \hat{t} = 45$, $9 - \hat{t} = 50$.

to eight times the step height a, but the position of this point depends strongly on both the flow parameters and the singularities of the experimental setup.

We will point out yet another, not physical but numerical, effect that is observed in the case of insufficient length of calculation region, namely, the existence of a strong dependence of the value x_p on the relative length L/h of the calculation region. This dependence ceased to have an effect when ratio L/h in our calculations exceeded ten. Table 7.6 gives values of x_p for different values of Re, obtained in the calculations.

Given for illustration in Figures 7.22–7.24 are patterns of flows for other values of Re, in addition to the case $Re = 1000$ that has already been discussed in detail. Note the absence from Figure 7.22 of the clearly defined region of return flow behind the step. This region P^* is a turbulized region of return flow with an unstable structure. The region shown at G in Figure 7.22 is nothing but a turbulized structure moving (similarly

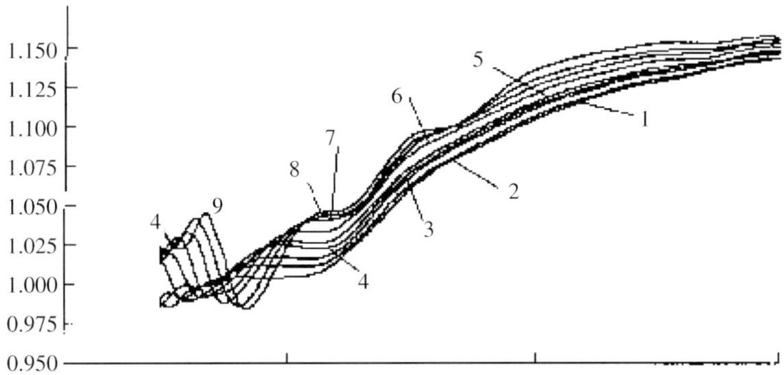

(c)

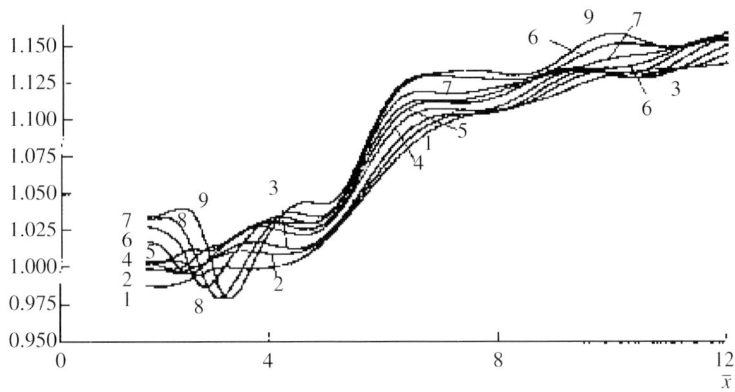

(d)

Fig. 7.26. (Continued).

Table 7.6
Values of the coordinate of the point of flow attachment (for $Re < 3200$, the values of x_p are given for the flow that reached the quasi-stationary state)

1	$Re = 64$	$x_p = 5$
2	$Re = 365$	$x_p = 7$
3	$Re = 1000$	$x_p = 10$
4	$Re = 3200$	$x_p = 5$–12 indeterminacy due to instability of the point of flow attachment
5	$Re = 10\,000$	x_p indeterminacy due to strong turbulization of the flow

to region V in Figure 7.14) along the flow. Note further region F, which is a "spot" of small-scale turbulence shown in more detail in Figure 7.23. Also of interest is the representation of the general view of the flow at $Re = 10\,000$. Figure 7.24 illustrates the respective flow. Also marked in Figure 7.24 are regions analogous with those of the flow shown in Figure 7.22; however, the absence of region F does not mean that this region does not arise with the given value of Reynolds number: the process of its origination and evolution occurs at subsequent moments of time. We do not illustrate the flows for $Re < 1000$. The flow topology in this case becomes simpler and less interesting. After a short time of relaxation, such flows repeat in general terms the flow pattern given in Figure 7.20.

Mathematical simulation enables one to obtain very extensive information about the hydrodynamic parameters of flow that may hardly be represented graphically to any degree of detail. Figures 7.25 and 7.26 give only the space and time evolution of pressure in a channel. We do not believe it necessary to provide the data of mathematical simulation, based on the Navier–Stokes equations. The appropriate comparison and comments may be found for another flow topology in Alexeev and Mikhailov (1999) and Fedoseyev and Alexeev (1998a, 1998b).

Note that the generalized hydrodynamic equations represent an effective tool for solving problems in hydrodynamics and gas dynamics.

7.3. Vortex and turbulent flow of viscous gas in channel with flat plate

We will treat the problem in the following formulation. Let an initially steady-state flow of gas in a flat channel be accompanied in some part of the channel by a momentary separation of flow that transforms the flow into an unsteady-state, generally speaking, turbulent flow. For the purposes of mathematical simulation, we use the geometry of a flat channel with a rectangular plate of finite length, as shown in Figure 7.27.

On the left (in a wide zone of the channel before the plate), the flow exhibits a Poiseuille pattern and has a parabolic velocity profile. It was demonstrated in the previous Section 7.2, that the generalized hydrodynamic equations – derived with the help of locally Maxwellian distribution function (generalized Euler equations, GEuE) – lead to

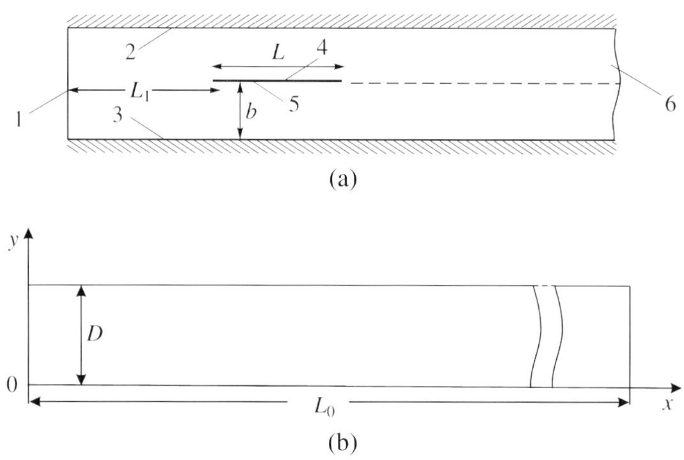

Fig. 7.27. (a) The flow scheme and (b) the system of coordinates, surfaces are shown on which boundary conditions are preassigned (the broken line corresponds to a flow past a semi-infinite plane).

Poiseuille flow with standard assumptions: (a) the flow is one-dimensional and steady-state; (b) the density $\rho = const$; (c) the hydrodynamic velocity depends on y alone, $v_0 = v_{0x}(y)$; (d) the static pressure depends on x alone, $p = p(x)$.

Generalized Euler equations (GEuE) under the conditions formulated above lead to a parabolic velocity profile, which is adopted as the boundary condition for the equation of motion on the left-hand boundary of the calculation region.

We will introduce the characteristic quantities of flow that are used in what follows as the scales in reducing the GEuEs to a dimensionless form: ρ_∞, p_∞, $v_{0\infty} = u_\infty$, and μ_∞ denote the density, static pressure, velocity, and dynamic viscosity at the channel inlet. The channel width at the inlet is used as the scale of length D, and the time scale has the form D/u_∞. The similarity criteria, i.e., the Reynolds and Euler numbers, are written as $Re = \rho_\infty u_\infty D/\mu_\infty$ and $Eu = p_\infty/(\rho_\infty v_\infty^2)$ The dimensionless quantities are further marked by the sign ˆ above the respective symbol.

The geometric characteristics of the calculation region are given in Figure 7.27, namely, D, the width of the region (channel) being investigated; L, the length of the plate in the channel; L_1, the distance from the front section to the beginning of the plate; b, the distance from the top boundary of the channel to the plate; and L_0, the length of the calculation region.

The boundary and initial conditions are preassigned on the following surfaces and sections of the flow region being investigated: 1, the front section of the region (flow inlet section); 2, the top plane; 3, the bottom plane; 4, the tip surface of the plate; 5, the bottom surface of the plate; 6, the outlet section of the region (flow outlet section).

Table 7.7 gives the initial conditions for the flow being simulated. These conditions correspond to the existence, at the initial moment of time, of a laminar flow with the velocity profile in the form of Poiseuille parabola. The plate is introduced into the flow at the time zero.

In Table 7.7, $f(\hat{y}) = k_1\hat{y}^2 + k_2\hat{y} + k_3$, and the coefficients k_i $(i = 1, \ldots, 3)$ are found from the conditions $f(0) = 0,\ f(D/2) = 1,\ f(D) = 0$.

The boundary values in the calculation region of flow must be treated in application to each geometric element. For this purpose, the boundary conditions written out on each surface (see Figure 7.27) are given in Table 7.8. The boundary conditions given in Table 7.8 call for some comment. They correspond to the conditions of no-slip on the wall and non-percolation, the wall is assumed to be absolutely non-heat-conducting, and the variation of density in the vicinity of a solid surface is taken to be minor. For the section 1, the conditions are preassigned at the distance L_1 from the plate tip. This distance may be varied with a view to eliminating the nonphysical reactive effect of flow behind the plate on the initial flow. At the section 6, the so-called “soft” boundary conditions are written relative to the averaged (rather than true) hydrodynamic quantities. Such conditions turned out to be very effective in the case of numerical realization and eliminated the upstream reactive effect due to the numerical effect of cutting off the region. The calculations were performed using an explicit difference scheme of the first order of accuracy with respect to time and of the second order of accuracy with respect to coordinates with constant dimensionless time and coordinate steps.

In calculations (Alexeev and Mikhailov, 2003) within the framework of this problem, the behavior of flow was studied for the values of Re from 1000 to 10 000 and $Eu = 1.0$.

Table 7.7
Initial values of flow parameters

Velocities and their derivatives	Pressure and density and their derivatives
$\hat{u}(\hat{t}=0)=0,\quad \left(\frac{\partial \hat{u}}{\partial \hat{t}}\right)_{\hat{t}=0}=0$	$\hat{p}(\hat{t}=0)=1,\quad \left(\frac{\partial \hat{p}}{\partial \hat{t}}\right)_{\hat{t}=0}=0$
$\hat{v}(\hat{t}=0)=0,\quad \left(\frac{\partial \hat{v}}{\partial \hat{t}}\right)_{\hat{t}=0}=0$	$\hat{\rho}(\hat{t}=0)=1,\quad \left(\frac{\partial \hat{\rho}}{\partial \hat{t}}\right)_{\hat{t}=0}=0$

Table 7.8
The boundary conditions for the calculated region of flow

For section 1	For surface 2	For surface 3
$\hat{\hat{u}}(0,y)=f(y)$	$\hat{u}(x,D)=0$	$\hat{u}(x,0)=0$
$\hat{\hat{v}}(0,y)=0$	$\hat{v}(x,D)=0$	$\hat{v}(x,0)=0$
$\left(\frac{\partial^2 \hat{\hat{p}}}{\partial \hat{x}^2}\right)_{\hat{x}=0}=0$	$\left(\frac{\partial \hat{p}}{\partial \hat{y}}\right)_{\hat{y}=1}=0$	$\left(\frac{\partial \hat{p}}{\partial \hat{y}}\right)_{\hat{y}=0}=0$
$\left(\frac{\partial^2 \hat{\hat{\rho}}}{\partial \hat{x}^2}\right)_{\hat{x}=0}=0$	$\left(\frac{\partial \hat{\rho}}{\partial \hat{y}}\right)_{\hat{y}=1}=0$	$\left(\frac{\partial \hat{\rho}}{\partial \hat{y}}\right)_{\hat{y}=0}=0$
For surface 4	**For surface 5**	**For section 6**
$\widehat{u}(x,b)=0$	$\widehat{u}(x,b)=0$	$\left(\frac{\partial \hat{\hat{u}}}{\partial \hat{x}}\right)_{x=L_0}=0$
$\widehat{v}(x,b)=0$	$\widehat{v}(x,b)=0$	$\left(\frac{\partial \hat{\hat{v}}}{\partial \hat{x}}\right)_{x=L_0}=0$
$\left(\frac{\partial \hat{p}}{\partial \hat{y}}\right)_{\hat{y}=\hat{b}+0}=0,$	$\left(\frac{\partial \hat{p}}{\partial \hat{y}}\right)_{\hat{y}=\hat{b}-0}=0,$	$\left(\frac{\partial^2 \hat{\hat{p}}}{\partial \hat{x}^2}\right)_{x=L_0}=0$
$\left(\frac{\partial \hat{\rho}}{\partial \hat{y}}\right)_{\hat{y}=\hat{b}+0}=0$	$\left(\frac{\partial \hat{\rho}}{\partial \hat{y}}\right)_{\hat{y}=\hat{b}-0}=0$	$\left(\frac{\partial^2 \hat{\hat{\rho}}}{\partial \hat{x}^2}\right)_{x=L_0}=0$

Special attention was given to the range of variation of *Re* values from 1000 to 3200, as the region of transition values between the laminar and turbulent modes of flow.

Table 7.9 gives the values of steps of the computational mesh and the geometric parameters of the problem for the employed values of *Re*. The similarity criteria (Knudsen and Mach numbers) may be calculated in terms of *Re* and *Eu* using the average velocity of molecules for a Maxwellian distribution (see (7.2.7), (7.2.8)). The respective rescaling of parameters is given in Table 7.9.

We will now discuss the results of calculation of flow past a plate of finite length. Relevant illustrations are given for the flow past a plate whose length is equal to the channel depth. The plate is located at the channel mid-depth. Also given for some typical cross sections of the channel are flow velocity profiles corresponding to different moments of dimensionless time. The cross sections selected for illustration are located

Table 7.9
Parameters involved in the calculations

$Re = 1000$	$Re = 3200$	$Re = 10\,000$
$\Delta\hat{t} = 1.0 \times 10^{-3}$ $\Delta\hat{y} = 0.0333$ $\hat{D} = 1.0,\ \hat{b} = 0.5$ $\hat{L} = 1.0,\ \hat{L}_1 = 2.5$	$\Delta\hat{t} = 1.0 \times 10^{-4}$ $\Delta\hat{y} = 0.0333$ $\hat{D} = 1.0,\ \hat{b} = 0.5$ $\hat{L} = 1.0,\ \hat{L}_1 = 2.5$	$\Delta\hat{t} = 1.0 \times 10^{-5}$ $\Delta\hat{x} = \Delta\hat{y} = 0.0333$ $\hat{D} = 1.0,\ \hat{b} = 0.5$ $\hat{L} = 1.0,\ \hat{L}_1 = 2.5$
1. $Re = 1000$ $Kn = 1.277 \times 10^{-3}$ $Eu = 1.0,\ M = 0.7747$	2. $Re = 3200$ $Kn = 3.989 \times 10^{-3}$ $Eu = 1.0,\ M = 0.7747$	3. $Re = 10\,000$ $Kn = 1.277 \times 10^{-4}$ $Eu = 1.0,\ M = 0.7747$

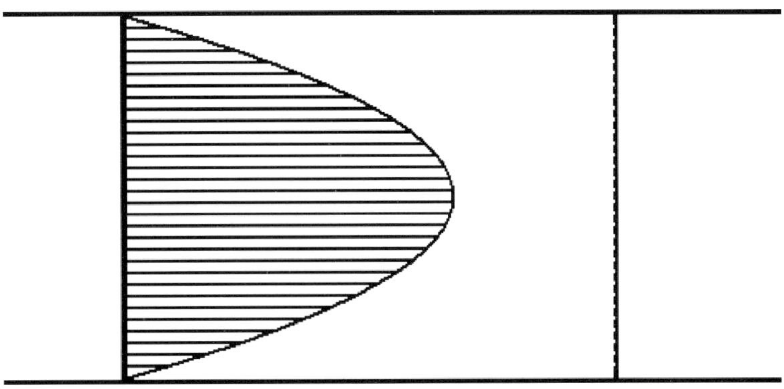

Fig. 7.28. Velocity profiles on the Ox (left) and Oy (right) axes at a distance $L_x = D/2$ to the left of the beginning of the plate; $Re = 1000$, $\hat{t} = 0.2$.

as follows: (a) before the plate at a distance equal to half its length, (b) immediately before the plate, (c) in the middle of the plate, (d) before the very end of the plate, and (e) behind the plate at a distance equal to half its length. The mathematical simulation helps to produce a large volume of information including illustrative information about the topology of flow. Therefore, we will first make a description of flow past a plate of finite length at $Re = 1000$ (Figures 7.28–7.42).

For the initial moment of time, a Poiseuille parabolic profile of flow is retained in section (a), the perturbation due to the plate in this section is still low, and the velocity component v is extremely small. A vertical velocity component emerges due to the deceleration of flow as a result of the effect of the friction forces, and a characteristic asymmetric profile of the velocity component u is formed. In the region where the flow comes off the plate, a marked vorticity of unsteady-state flow arises, which is reflected in the pulsations of the transverse velocity profile. Quite away from the plate downstream the flow, the longitudinal velocity profile becomes parabolic again, and the pulsations of the transverse velocity are small. By the moment of time $\hat{t} = 3.0$, the perturbations of the

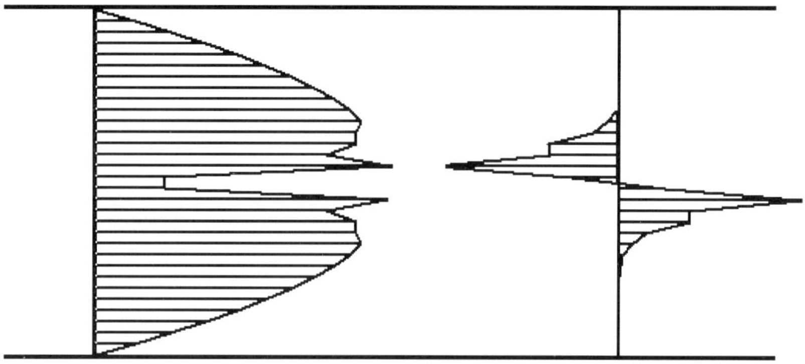

Fig. 7.29. Velocity profiles on the Ox (left) and Oy (right) axes at a distance $L_x = \Delta x$ to the left of the beginning of the plate; $Re = 1000$, $\hat{t} = 0.2$.

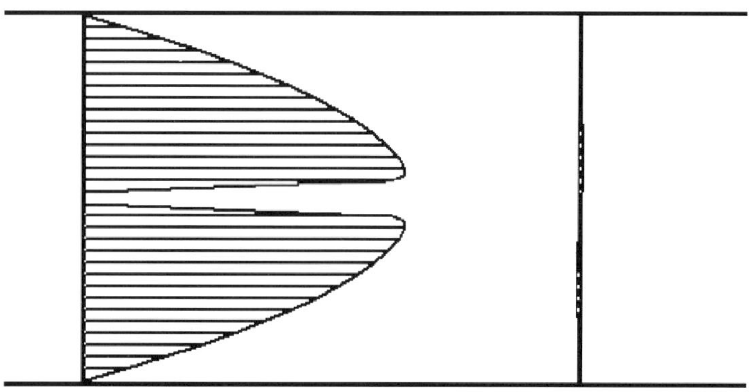

Fig. 7.30. Velocity profiles on the Ox (left) and Oy (right) axes at a distance $L_x = D/2$ to the right of the beginning of the plate (middle of the plate); $Re = 1000$, $\hat{t} = 0.2$.

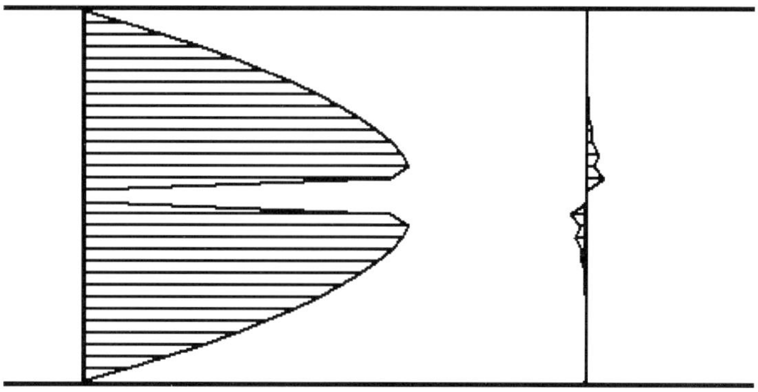

Fig. 7.31. Velocity profiles on the Ox (left) and Oy (right) axes at a distance $L_x = \Delta x$ to the left of the end of the plate; $Re = 1000$, $\hat{t} = 0.2$.

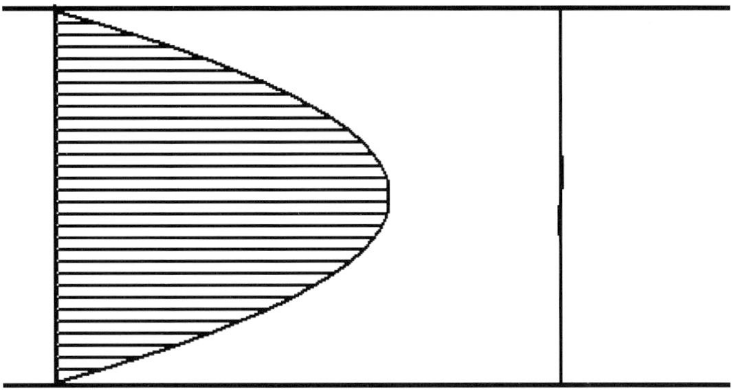

Fig. 7.32. Velocity profiles on the Ox (left) and Oy (right) axes at a distance $L_x = D/2$ to the right of the end of the plate; $Re = 1000$, $\hat{t} = 0.2$.

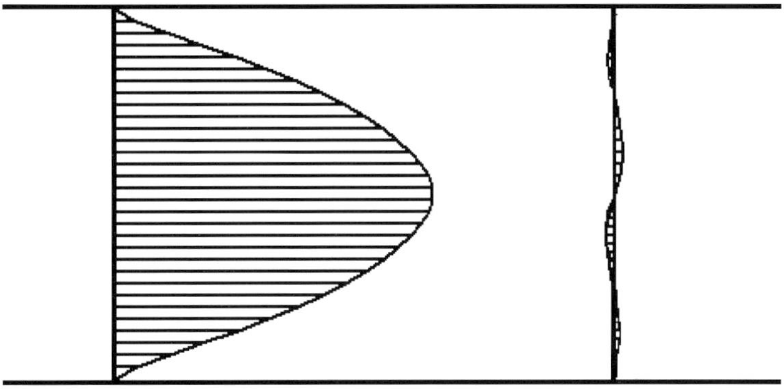

Fig. 7.33. Velocity profiles on the Ox (left) and Oy (right) axes at a distance $L_x = D/2$ to the left of the beginning of the plate; $Re = 1000$, $\hat{t} = 3.0$.

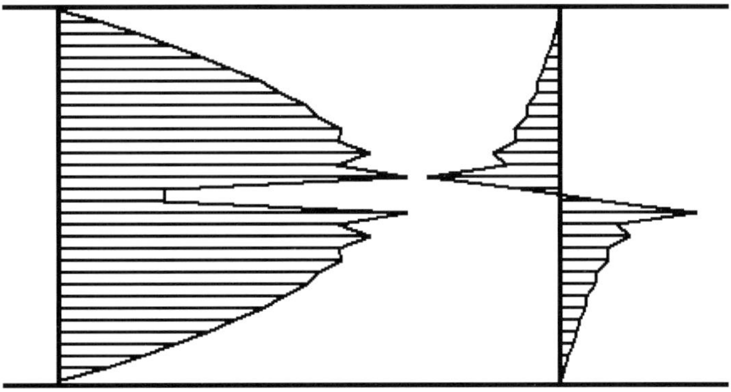

Fig. 7.34. Velocity profiles on the Ox (left) and Oy (right) axes at a distance $L_x = \Delta x$ to the left of the beginning of the plate; $Re = 1000$, $\hat{t} = 3.0$.

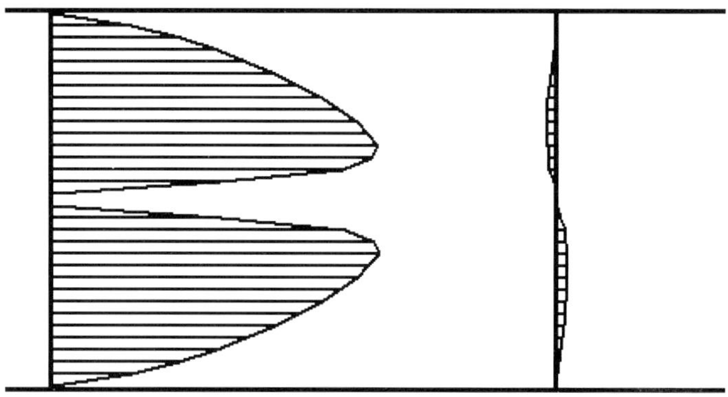

Fig. 7.35. Velocity profiles on the Ox (left) and Oy (right) axes at a distance $L_x = D/2$ to the right of the beginning of the plate (middle of the plate); $Re = 1000$, $\hat{t} = 3.0$.

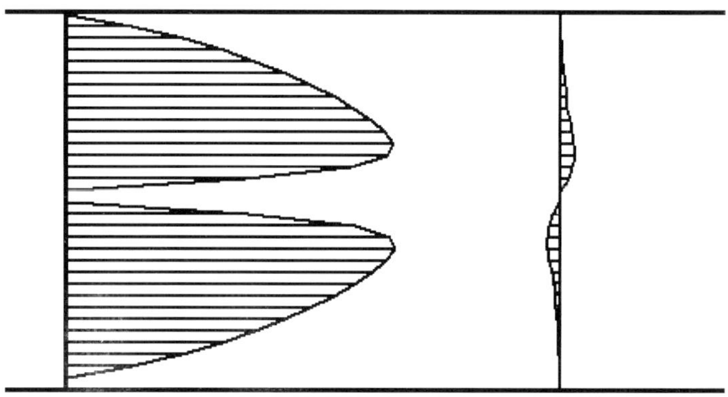

Fig. 7.36. Velocity profiles on the Ox (left) and Oy (right) axes at a distance $L_x = \Delta x$ to the left of the end of the plate; $Re = 1000$, $\hat{t} = 3.0$.

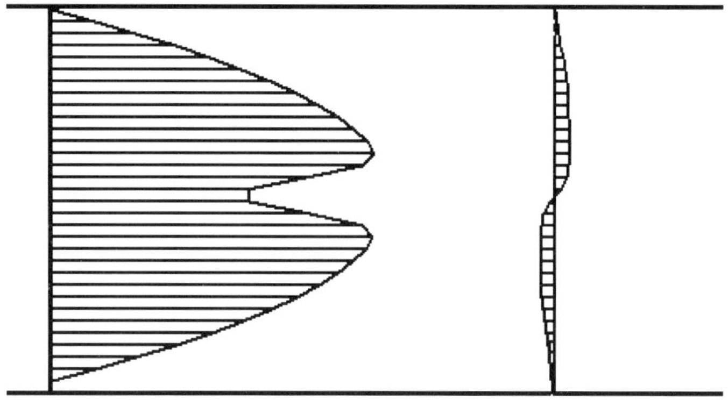

Fig. 7.37. Velocity profiles on the Ox (left) and Oy (right) axes at a distance $L_x = D/2$ to the right of the end of the plate; $Re = 1000$, $\hat{t} = 3.0$.

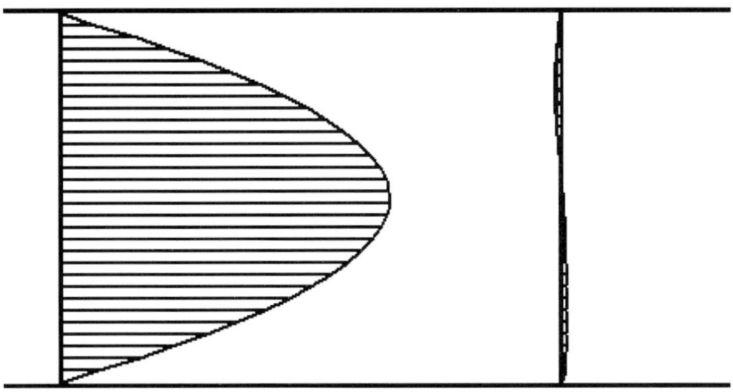

Fig. 7.38. Velocity profiles on the Ox (left) and Oy (right) axes at a distance $L_x = D/2$ to the left of the beginning of the plate; $Re = 1000$, $\hat{t} = 25.0$.

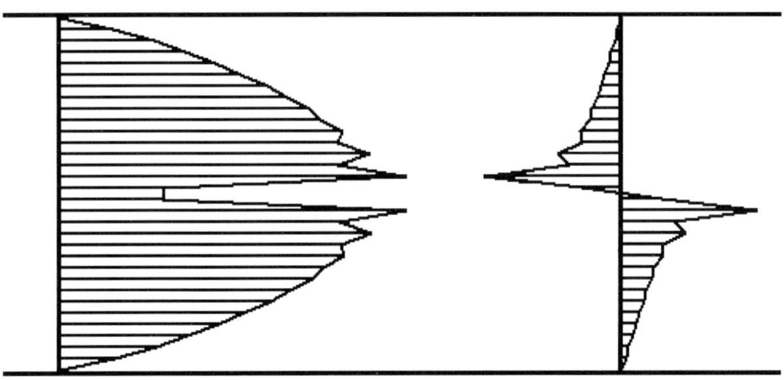

Fig. 7.39. Velocity profiles on the Ox (left) and Oy (right) axes at a distance $L_x = \Delta x$ to the left of the beginning of the plate; $Re = 1000$, $\hat{t} = 25.0$.

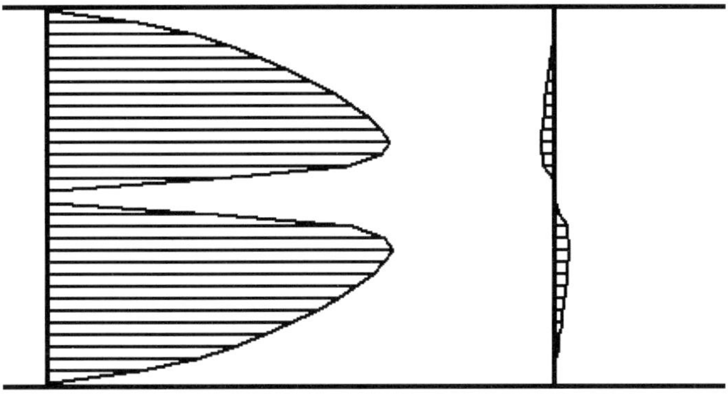

Fig. 7.40. Velocity profiles on the Ox (left) and Oy (right) axes at a distance $L_x = D/2$ to the right of the beginning of the plate (middle of the plate); $Re = 1000$, $\hat{t} = 25.0$.

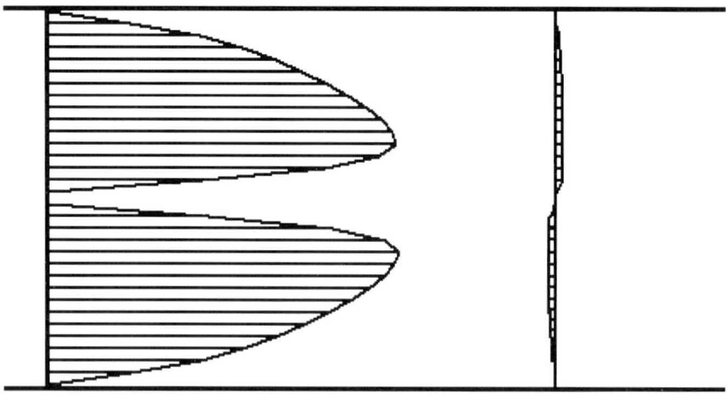

Fig. 7.41. Velocity profiles on the Ox (left) and Oy (right) axes at a distance $L_x = \Delta x$ to the left of the end of the plate; $Re = 1000$, $\hat{t} = 25.0$.

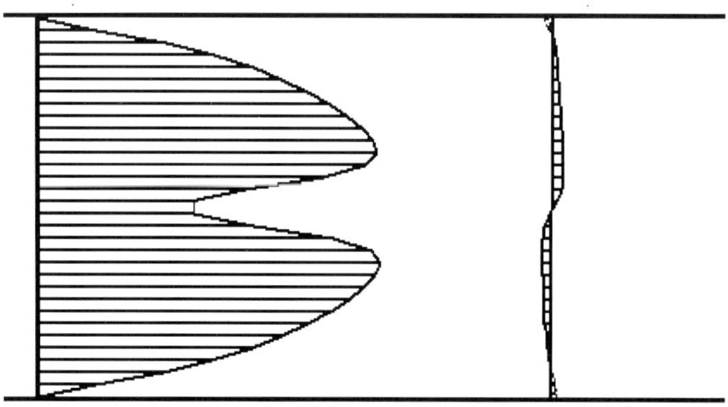

Fig. 7.42. Velocity profiles on the Ox (left) and Oy (right) axes at a distance $L_x = D/2$ to the right of the end of the plate; $Re = 1000$, $\hat{t} = 25.0$.

hydrodynamic characteristics propagate upstream, and a marked profile of transverse velocity arises in the cross section (a) as well.

We will now turn to the results of calculation of the flow past a plate at $Re = 3200$, which corresponds to the mode of transition from the laminar to turbulent flow. By the moment of time $\hat{t} = 2.0$, in addition to forming in the vicinity of the plate surface and before and behind the plate, vortexes begin to form at the top and bottom surfaces defining the channel width as well. Figures 7.43–7.48 give the topology of flow in the vicinity of plate, $Re = 3200$, $\hat{t} = 2.0$ and the profiles of components of flow velocity in the cross sections (a)–(e) for the moment of time $\hat{t} = 5.0$. The dash length is not proportional to the modulus of velocity and corresponds only to the velocity direction.

By this moment of time, the top and bottom wall vortexes weaken but become more extended; with time, an unsteady-state flow changes to a quasi-steady-state mode. In particular, the fluctuations of the longitudinal velocity profile at the middle of the plate

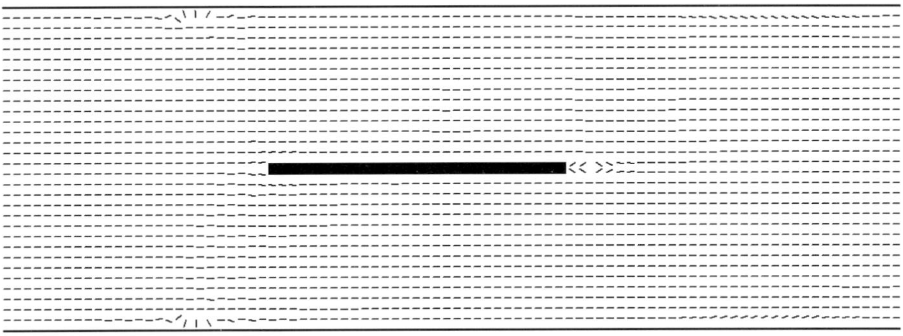

Fig. 7.43. The topology of flow in the vicinity of plate, $Re = 3200$, $\hat{t} = 2.0$. The dash length is not proportional to the modulus of velocity.

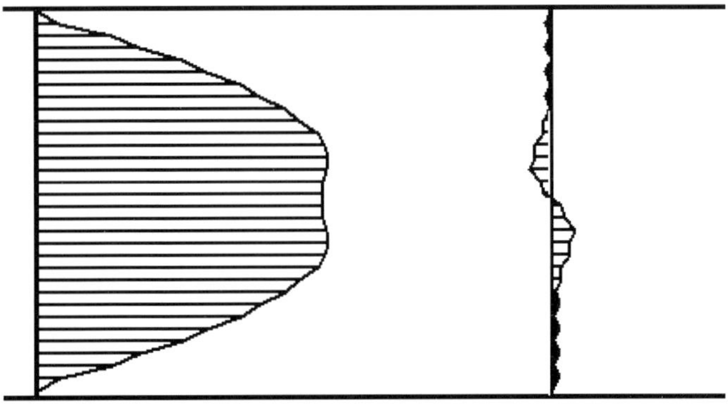

Fig. 7.44. Velocity profiles on the Ox (left) and Oy (right) axes at a distance $L_x = D/2$ to the left of the beginning of the plate; $Re = 3200$, $\hat{t} = 5.0$.

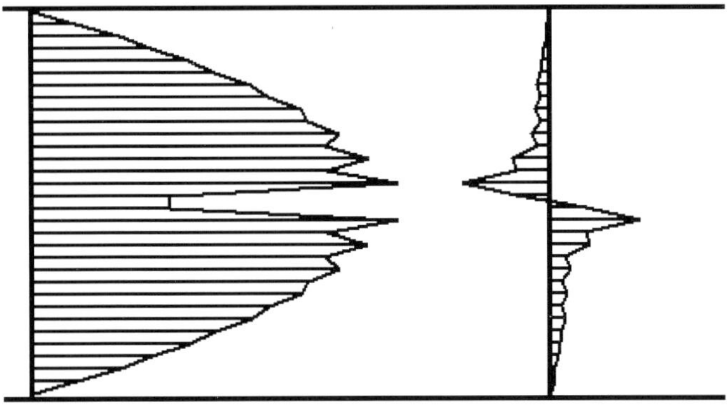

Fig. 7.45. Velocity profiles on the Ox (left) and Oy (right) axes at a distance $L_x = \Delta x$ to the left of the beginning of the plate; $Re = 3200$, $\hat{t} = 5.0$.

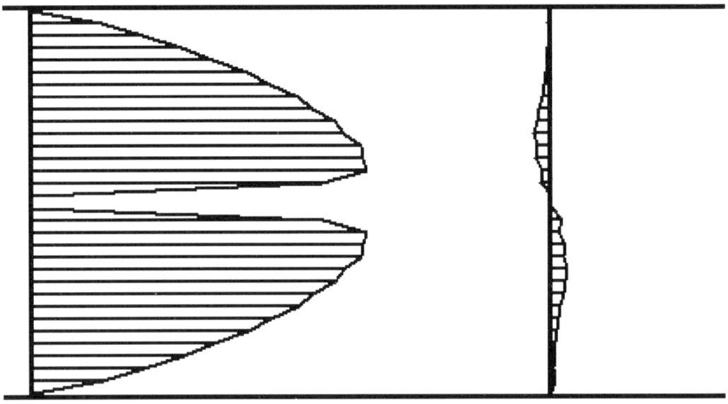

Fig. 7.46. Velocity profiles on the Ox (left) and Oy (right) axes at a distance $L_x = D/2$ to the right of the beginning of the plate; $Re = 3200$, $\hat{t} = 5.0$.

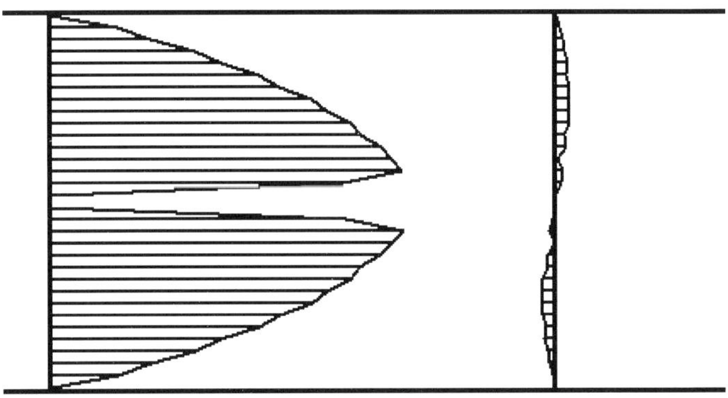

Fig. 7.47. Velocity profiles on the Ox (left) and Oy (right) axes at a distance $L_x = \Delta x$ to the left of the end of the plate; $Re = 3200$, $\hat{t} = 5.0$.

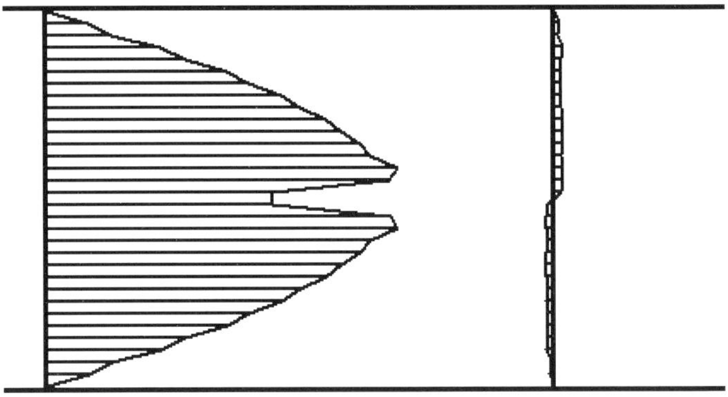

Fig. 7.48. Velocity profiles on the Ox (left) and Oy (right) axes at a distance $L_x = D/2$ to the right of the end of the plate; $Re = 3200$, $\hat{t} = 5.0$.

decrease somewhat, and in the cross section (e), the fluctuations of the transverse velocity of flow decrease.

Let, at the initial moment of time, a flow in a channel abruptly separate in two parallel flows. In this case, from the hydrodynamic standpoint, the problem reduces to a flow past a semi-infinite plate of finite thickness that suddenly appeared in the channel. The boundary conditions preassigned at the outlet of any of two channels formed are similar to the conditions at the section 6 (see Table 7.8).

Figures 7.49–7.52 illustrate, for a flow mode at $Re = 1000$ and $\hat{t} = 50.0$, the spatial evolution of the profiles of longitudinal and transverse velocities of flow in channels for sections located at distances L_x from the beginning of the semi-infinite plate equal to Δx (step of the space grid on the Ox-axis), $D/2$, D and $10D$, respectively. The effect of a semi-infinite plate on the upstream rearrangement of flow is by and large similar to the perturbation of flow due to a plate of finite length; therefore, it is not treated in what follows. Note that, at a distance of approximately D, the longitudinal velocity profiles

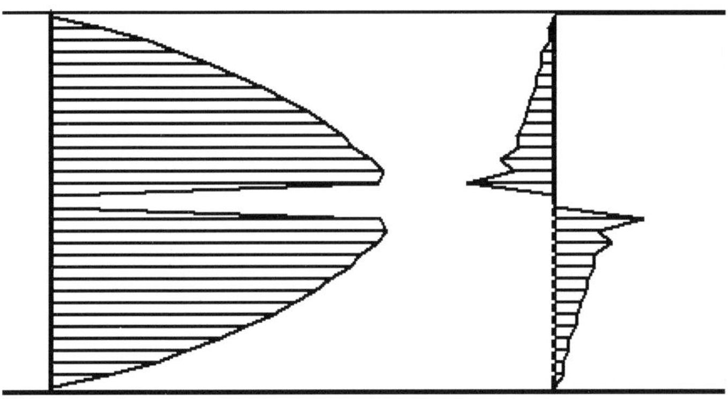

Fig. 7.49. Velocity profiles on the Ox (left) and Oy (right) axes at a distance $L_x = \Delta x$ to the right of the beginning of the plate; $Re = 1000$, $\hat{t} = 50.0$.

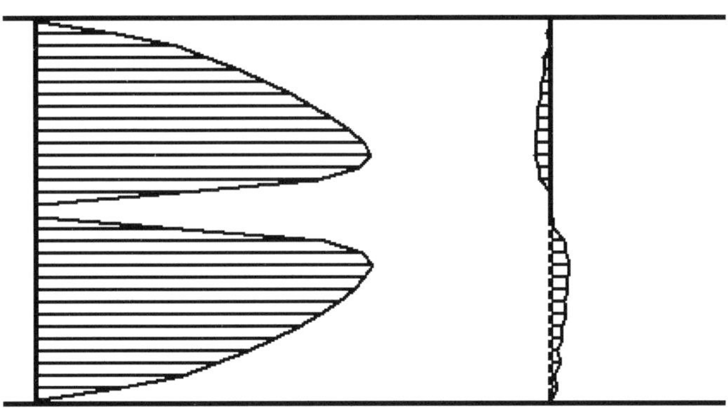

Fig. 7.50. Velocity profiles on the Ox (left) and Oy (right) axes at a distance $L_x = D/2$ to the right of the beginning of the plate; $Re = 1000$, $\hat{t} = 50.0$.

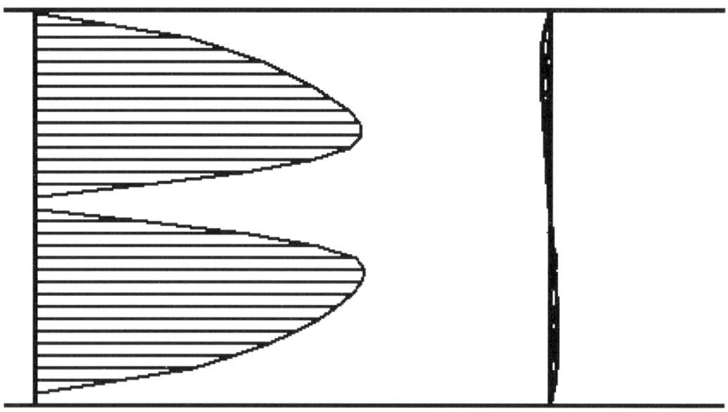

Fig. 7.51. Velocity profiles on the Ox (left) and Oy (right) axes at a distance $L_x = D$ to the right of the beginning of the plate; $Re = 1000$, $\hat{t} = 50.0$.

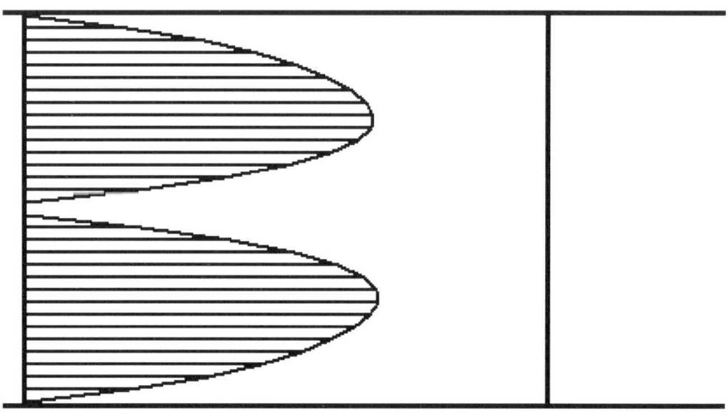

Fig. 7.52. Velocity profiles on the Ox (left) and Oy (right) axes at a distance $L_x = 10D$ to the right of the beginning of the plate; $Re = 1000$, $\hat{t} = 50.0$.

do not become parabolic: the maximum velocity is shifted from the center of channel to the plate. This shift is caused by the fact that, at the inlet to the wide part of the channel at the initial moment of time, a Poiseuille profile has a maximum of longitudinal velocity located on the level of the plate that suddenly appeared in the channel. By the moment of time $\hat{t} = 50.0$, the longitudinal velocity profiles in the section $10D$ assume a parabolic form, and the fluctuations of longitudinal velocity are very low.

For a better illustration of the processes occurring during a flow past flat bodies in a channel, we will consider the distribution of dimensionless pressure over the channel length. The graphs are plotted for two values of the Reynolds number over two segments parallel to the channel walls (see Figures 7.53–7.56). It is interesting to investigate the reaction of the generalized hydrodynamic equations to a variation of the boundary conditions. If the "soft" condition at the inlet to the calculation region $\mathrm{d}^2\hat{\bar{p}}/\mathrm{d}x^2 = 0$ is replaced by the "rigid" condition $\hat{p} = 1$, the pressure perturbations propagating upstream will not go beyond the limits of the calculation region. As a result, a system of

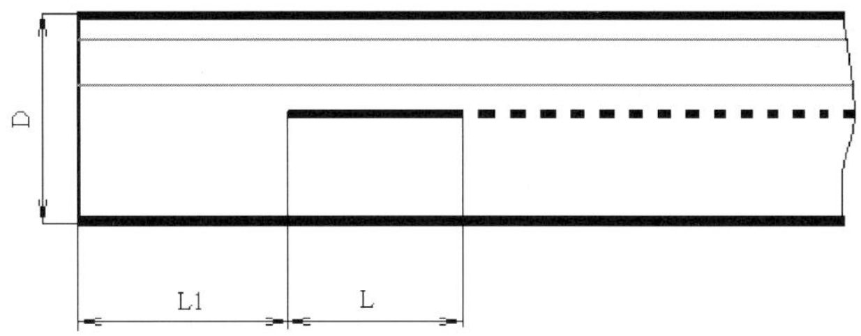

Fig. 7.53. A schematic representation of pressure chart plotting lines; each line is located at a distance of 0.15 from the nearest surface.

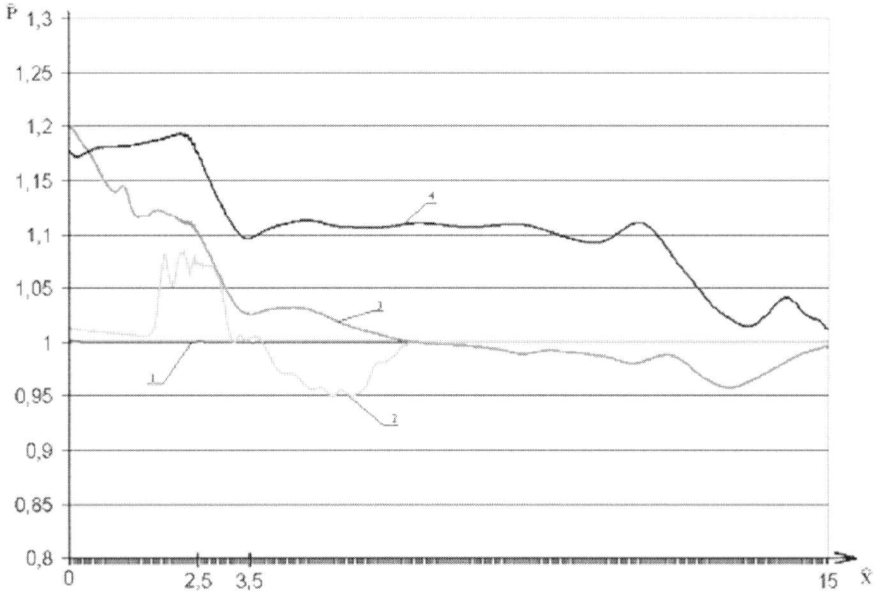

Fig. 7.54. The distribution of dimensionless pressure over the channel length on a line in the vicinity of the channel wall at $Re = 1000$ at different moments of time ($1 - \hat{t} = 0.2$; $2 - \hat{t} = 3.0$; $3 - \hat{t} = 25.0$; $4 - \hat{t} = 100.0$). The coordinates of the beginning and end of a plate placed in the channel are indicated.

standing pressure waves will be formed in the vicinity of the left-hand boundary of the calculation region; in so doing, the pattern of downstream variation of pressure changes little.

It is known that extensive experimental data exist on the drag coefficients in a flow past a plate, as well as prediction data based, as a rule, on the boundary layer theory. We will compare the results of mathematical simulation using the generalized hydrodynamic equations (GEE) to the Blasius theory.

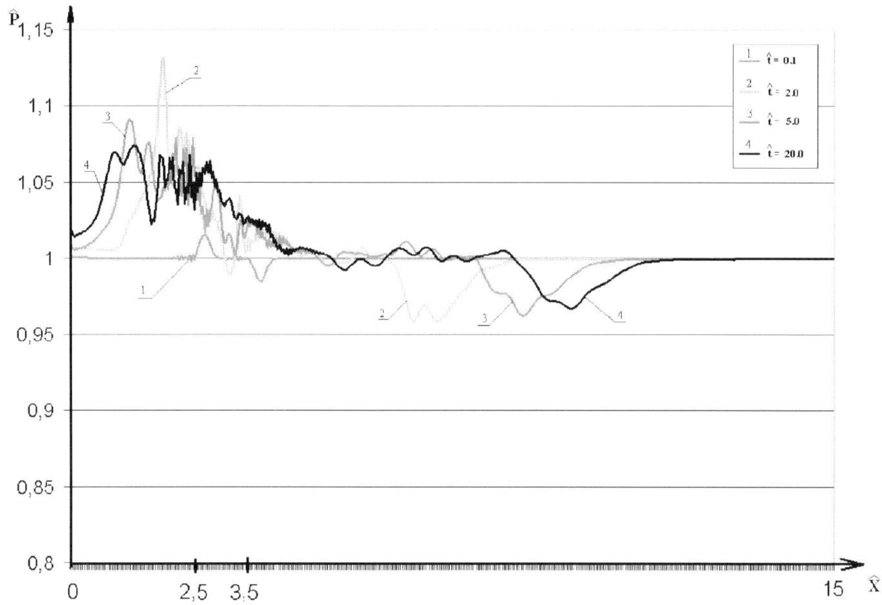

Fig. 7.55. The distribution of dimensionless pressure over the channel length on a line in the vicinity of the channel wall at $Re = 3200$ at different moments of time. The coordinates of the beginning and end of a plate placed in the channel are indicated.

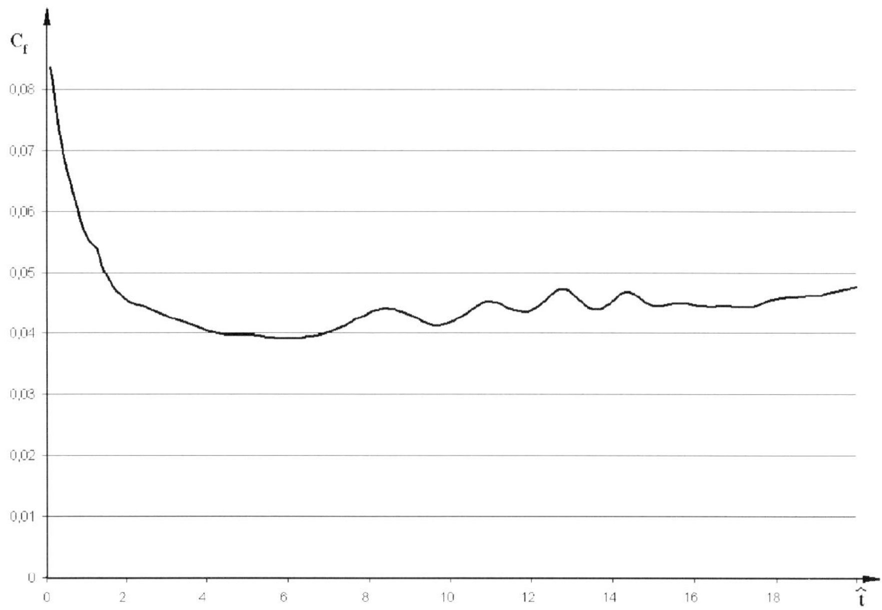

Fig. 7.56. The time variation of the coefficient of total drag for a plate of finite length at $Re = 1000$. The number of calculation points over the plate length is 300.

We will write the Blasius formula for the coefficient c_f of total drag of a plate of finite length L:

$$c_f = \frac{1.328}{\sqrt{Re_L}} \tag{7.3.1}$$

where Re_L is the Reynolds number calculated using the plate length L and the incident flow velocity. The Blasius formula (7.3.1) was derived within the laminar boundary layer (Schlichting, 1964). In the general case in a dimensionless form, the total drag coefficient may be represented as

$$c_f = \frac{2W}{\rho_\infty u_\infty^2 S/2}, \tag{7.3.2}$$

where W is drag of one side of the plate

$$W = h \int_{x=0}^{L} \tau_0 \, \mathrm{d}x. \tag{7.3.3}$$

In relations (7.3.2), (7.3.3) τ_0 is local drag coefficient of the plate corresponding to its surface w,

$$\tau_0 = \mu \left(\frac{\mathrm{d}u}{\mathrm{d}y} \right)_w, \tag{7.3.4}$$

S is an area of the plate streamed by flow, $S = 2Lh$, h is width of the plate.

As a result one obtains the coefficient of total drag

$$c_f = \frac{\int_{(L)} \mu (\mathrm{d}u/\mathrm{d}y)_w \, \mathrm{d}x}{\rho_\infty u_\infty^2 L/2}, \tag{7.3.5}$$

(the integral determined in Eq. (7.3.5) is calculated by the plate length), or in dimensionless form

$$c_f = \frac{2 \int_{(\hat{L})} \hat{\mu} (\mathrm{d}\hat{u}/\hat{y})_w \, \mathrm{d}\hat{x}}{Re}. \tag{7.3.6}$$

The values of c_f calculated by the foregoing formula using the grid functions for $Re = 1000$ are given in Table 7.10, and for $Re = 3200$ in Table 7.11; given in the same table for comparison are values found by formula (7.3.1). The Blasius formula corresponds to the steady-state mode of flow; therefore, the values of c_f are duplicated in the right-hand column for all moments of time.

It follows from Tables 7.10, 7.11 that, at the initial moments of time, the value of total drag coefficient significantly exceeds the value of c_f calculated by the Blasius formula.

Figure 7.56 shows the time evolution of the coefficient of total drag of a plate at $Re = 1000$. At the initial moments of time, the drag coefficient significantly exceeds the

Table 7.10
The coefficient of total drag of a plate, found by numerical calculation and by Blasius formula, $Re = 1000$

	c_f (numerical calculation)	c_f (Blasius formula)
$\hat{t} = 0.2$	0.07908	0.041995
$\hat{t} = 3.0$	0.04730	0.041995
$\hat{t} = 25.0$	0.04826	0.041995
$\hat{t} = 100.0$	0.04931	0.041995

Table 7.11
The coefficient of total drag of a plate, found by numerical calculation and by Blasius formula, $Re = 3200$

	c_f (numerical calculation)	c_f (Blasius formula)
$\hat{t} = 0.1$	0.03289	0.024246
$\hat{t} = 2.0$	0.02876	0.024246
$\hat{t} = 5.0$	0.02898	0.024246
$\hat{t} = 30.0$	0.02952	0.024246

plate resistance in a steady-state laminar flow; then, it decreases with time, experiences oscillation, and approaches the value calculated by the Blasius formula while remaining above it. Figure 7.57 corresponds to the variation of the local dimensionless friction coefficient over the plate length for different moments of time. The Blasius curve is given along with the results of mathematical simulation using the generalized hydrodynamic equations (GHE). Figures 7.58 and 7.59 are similar to Figures 7.56 and 7.57, but they are plotted for $Re = 3200$. The results of calculation of local coefficients of friction on a plate using the Blasius formula and the GHEs differ significantly. However, the total (integral) drag of the plate (which, in fact, represents the objective of experimental measurements) for these models in the quasi-steady-state mode in the investigated range of Reynolds number values differ little from one another (within 15%, see Tables 7.10 and 7.11). Therefore, the total drag is a "conservative" quantity.

Parameters reflecting the energy of turbulent pulsations are often used in the theory of turbulent flows (see, for example, Labusov and Lapin, 1996). In order to demonstrate the results of appropriate mathematical simulation, we will introduce the parameter ε related to the square of pulsating components of velocity,

$$\varepsilon = \sqrt{\frac{1}{3}\left[\left(\hat{u}_{\infty}^{\mathrm{fl}}\right)^2 + \left(\hat{v}_{\infty}^{\mathrm{fl}}\right)^2\right]} \times 100\%, \tag{7.3.7}$$

where

$$\hat{u}^{\mathrm{fl}} = \frac{\Pi \hat{\mu}}{\hat{p}} \frac{1}{Re\, Eu}\left[\frac{\partial \hat{v}_{0x}}{\partial \hat{t}} + \left(\hat{\mathbf{v}}_0 \cdot \frac{\partial}{\partial \hat{\mathbf{r}}}\right)\hat{v}_{0x} + \frac{Eu}{\hat{\rho}}\frac{\partial \hat{p}}{\partial \hat{x}}\right], \tag{7.3.8}$$

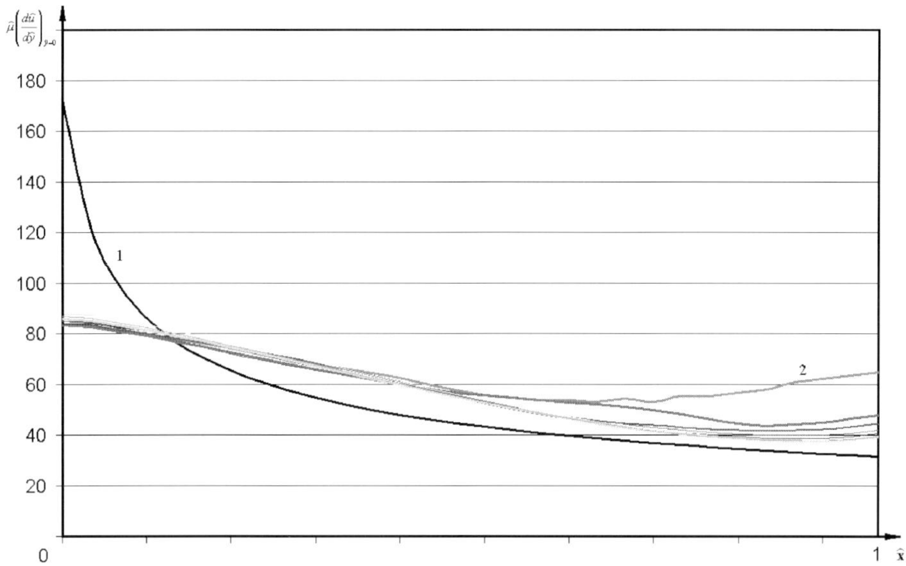

Fig. 7.57. The variation of the local drag coefficient for a plate of finite length. The dimensionless coordinates 0 and 1 indicate the beginning and the end of the plate, $Re = 1000$: 1 – calculation by the Blasius formula; 2 – numerical calculation, time $\hat{t} = 1.0$ (the remaining curves for the subsequent moments of time are fairly close and, therefore, not numbered).

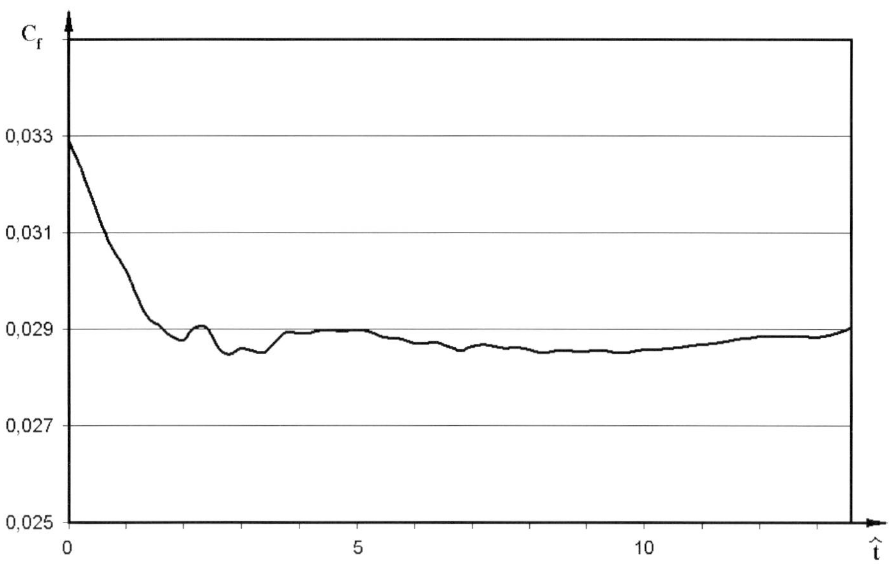

Fig. 7.58. The time variation of the coefficient of total drag for a plate of finite length at $Re = 3200$.

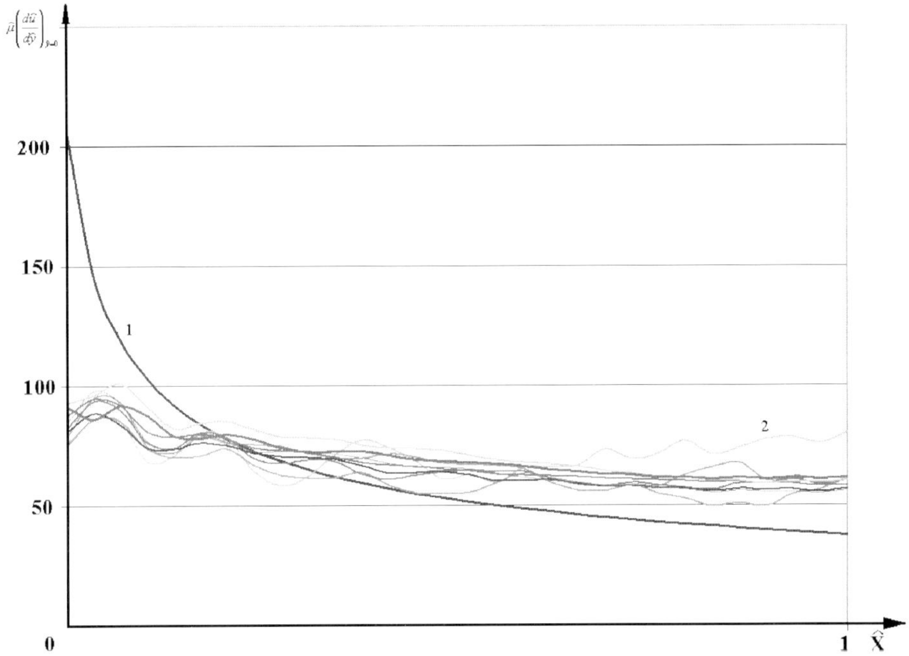

Fig. 7.59. The variation of the local drag coefficient for a plate of finite length. The dimensionless coordinates 0 and 1 indicate the beginning and the end of the plate, $Re = 3200$: 1 – calculation by the Blasius formula; 2 – numerical calculation, dimensionless time $\hat{t} = 1.0$ (the remaining curves for the subsequent moments of time are fairly close and, therefore, not numbered).

$$\hat{v}^{\mathrm{fl}} = \frac{\Pi\hat{\mu}}{\hat{p}}\frac{1}{Re\,Eu}\left[\frac{\partial\hat{v}_{0y}}{\partial\hat{t}} + \left(\hat{\mathbf{v}}_0\cdot\frac{\partial}{\partial\hat{\mathbf{r}}}\right)\hat{v}_{0y} + \frac{Eu}{\hat{\rho}}\frac{\partial\hat{p}}{\partial\hat{y}}\right]. \tag{7.3.9}$$

These pulsating components of velocity may, of course, be calculated both in the "external" (as indicated by the subscript ∞ with pulsating components of velocity) and in the "internal" flow.

Figure 7.60 gives, for different moments of time, the results of calculation of the parameter ε along the top longitudinal line indicated in Figure 7.53. While, at $Re = 1000$, a certain stabilization of the predicted parameter ε occurs with time, at $Re = 3200$ only a certain range of variation of energy parameters is observed.

Note that the "snapshots" of the flow, used to demonstrate the calculation results, cannot fully describe the flow behavior in the calculation region. This is due to the fact that the parameters of a flow of liquid, even if this flow is laminar, do not remain unvaried in time at every point. We will point out yet another, not physical but numerical, effect that is observed in the case of insufficient length of the calculation region. The calculation results are strongly dependent on the relative length of the calculation region. However, no such dependence is observed in the case of an adequate length of the calculation region of flow. The choice of the length of the calculation region is based on the comparison of calculation results for different lengths of the region.

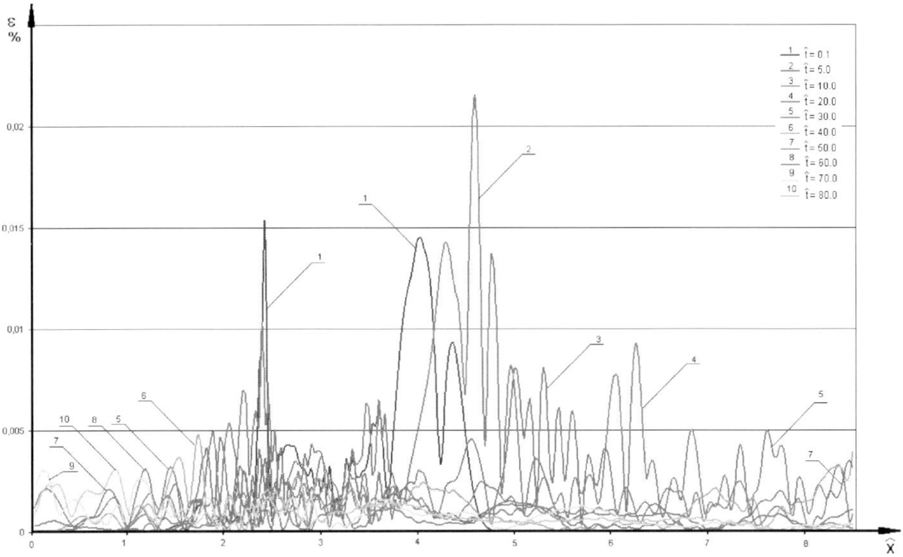

Fig. 7.60. The values of ε along the top line (Figure 7.53) in the flow for different time moments; $Re = 3200$.

It is interesting to notice that many recent works (see, for example, Tij and Santos, 1994; Malek, Baras and Garcia, 1997; Uribe and Garcia, 1999; Aoki, Takata and Nakanishi, 2002) have shown that the pressure and temperature profiles in plane Poiseuille flow exhibit a different qualitative behavior from the profiles obtained by Navier–Stokes equations. As comment to these results, Uribe and Garcia (1999) wrote that "Poiseuille flow is the first scenario in which the Navier–Stokes equations have been shown to be susceptible to significant improvement for a flow with relatively small gradients".

In more detail the problem formulation looks as follows. Stationary Poiseuille flow is confined between two rigid parallel plates. Two cases are considered. In the acceleration-driven case the boundary conditions are taken as periodic in the flow direction and external force – for example gravity force – is applied in this (x) direction. In the pressure-driven case the boundary conditions are set to create the pressure gradient but without external forces. The results of Navier–Stokes calculations are compared with data of Direct Simulation Monte Carlo (DSMC) (Bird, 1994; Alexander, Garcia and Alder, 1995).

The typical simulated flow is a hard sphere of gas for which $Kn = 0.1$, $M = 0.5$, $Eu = 0.5$ then $Re \sim 5$. In boundary conditions for Navier–Stokes equations (NS) the slip and jump corrections for fully accommodating surface were introduced (Uribe and Garcia, 1999).

Two main discrepancies were discovered between NS and DSMC modeling. For acceleration-driven case, in DSMC one observes a non-constant pressure profile and a temperature dip in the center of the channel (Uribe and Garcia, 1999). For NS and DSMC, pressure profiles in y direction have opposite curvatures. Moreover, the DSMC data indicate a reverse temperature jump at the wall – gas temperature near the wall is slightly greater than the wall temperature – despite that the global temperature gradient normal to walls is negative.

All these interesting effects need additional investigations. Comparison indicated results with data of modeling based on the generalized Euler equations lead to conclusion that mentioned effects can be observed but as a result of non-stationary non-one-dimensional calculations.

As an additional explanation let us consider the stationary one-dimensional acceleration-driven case for hard sphere of gas, with σ the corresponding particle's diameter. Let us suppose that

$$\rho = \rho(y), \quad u = u(y), \quad p = p(y), \quad T = T(y).$$

In this case the generalized Euler equations (2.7.49)–(2.7.51) lead to the system of hydrodynamic equations

$$\frac{\partial p}{\partial y} = 0, \tag{7.3.10}$$

$$T^2 \frac{\partial^2 u}{\partial y^2} + \frac{T}{2} \frac{\partial T}{\partial y} \frac{\partial u}{\partial y} + \frac{16 p \sigma^2}{5 k_B \Pi} g \sqrt{\frac{\pi m T}{k_B}} = 0, \tag{7.3.11}$$

$$\frac{\partial^2 T}{\partial y^2} + \frac{1}{2T} \left(\frac{\partial T}{\partial y} \right)^2 + \frac{2m}{5 k_B} \left(\frac{\partial u}{\partial y} \right)^2 + \frac{2m^2}{5 k_B^2 T} g^2 = 0. \tag{7.3.12}$$

For this case and in this approximation, Navier–Stokes equations is written as follows:

$$\frac{\partial p}{\partial y} = 0, \tag{7.3.13}$$

$$T^2 \frac{\partial^2 u}{\partial y^2} + \frac{T}{2} \frac{\partial T}{\partial y} \frac{\partial u}{\partial y} + \frac{16 p \sigma^2}{5 k_B} g \sqrt{\frac{\pi m T}{k_B}} = 0, \tag{7.3.14}$$

$$\frac{\partial^2 T}{\partial y^2} + \frac{1}{2T} \left(\frac{\partial T}{\partial y} \right)^2 + \frac{4m}{15 k_B} \left(\frac{\partial u}{\partial y} \right)^2 = 0. \tag{7.3.15}$$

The main difference between systems of Eqs. (7.3.10)–(7.3.12) and (7.3.13)–(7.3.15) consists, in appearance in the generalized Euler energy equation, of an additional term which contains – opposite to NS energy equation – gravitation in explicit form (more precisely – after transaction to the dimensionless form of energy equation – the ratio of gravitational energy to thermal energy, or in general case, the ratio of energy of external field to thermal energy).

Note in conclusion that the generalized hydrodynamic equations represent an effective tool for solving problems in gas dynamics.

CHAPTER 8

Generalized Boltzmann Physical Kinetics in Physics of Plasma and Liquids

We now proceed to apply the generalized Boltzmann kinetic theory to plasmas and liquids, where self-consistent forces with appropriately cut-off radius of their action are introduced to expand the capabilities of the GBE.

8.1. Extension of generalized Boltzmann physical kinetics for the transport processes description in plasma

Let us call that the dimensionless equation of the Bogolyubov–Born–Green–Kirkwood–Yvon (BBGKY) hierarchy for the s-particle distribution function f_s $(s = 1, \ldots, N)$ (N-number of particles in the system) has the form

$$\frac{\partial \hat{f}_s}{\partial \hat{t}_b} + \sum_{i=1}^{s} \hat{\mathbf{v}}_{ib} \cdot \frac{\partial \hat{f}_s}{\partial \hat{\mathbf{r}}_b} + \sum_{i,j=1}^{s} \widehat{\mathbf{F}}_{ij} \cdot \frac{\partial \hat{f}_s}{\partial \hat{\hat{v}}_{ib}} + \alpha \sum_{i=1}^{s} \widehat{\mathbf{F}}_i \cdot \frac{\partial \hat{f}_s}{\partial \hat{\mathbf{v}}_{ib}}$$

$$= -\varepsilon \frac{1}{N} \sum_{i=1}^{s} \sum_{j=s+1}^{N} \int \mathbf{F}_{ij} \cdot \frac{\partial}{\partial \hat{\mathbf{v}}_{ib}} \hat{f}_{s+1}\big(\hat{t}, \widehat{\Omega}_1, \ldots, \widehat{\Omega}_s, \widehat{\Omega}_j\big)\, \mathrm{d}\widehat{\Omega}_j, \qquad (8.1.1)$$

where $\hat{f}_s = f_s v_{0b}^{3s} n^{-s}$; v_{0b} is the characteristic collision velocity; n is the number density of particles; $\alpha = F_{0\lambda}/F_0$ is the ratio of the scales of the internal and external forces; $\mathrm{d}\Omega_j = \mathrm{d}\mathbf{r}_j\, \mathrm{d}\mathbf{v}_j$ is an elementary phase volume of the particle j whose position is determined by the radius-vector $\mathbf{r}_j$ and whose velocity is $\mathbf{v}_j$. We employ dot notation for a scalar product. A spatial variable is non-dimensionalized by introducing the interaction length r_b, and characteristic time scale is set by $r_b v_{0b}^{-1}$; ε corresponds to the number of particles which is contained in the interaction volume v_{int} and serves as a small parameter in the kinetic theory of rarefied neutral gases. There are actually at least three groups of scales to consider in a rarefied gas. Apart from r_b, v_{0b}, $t_{0b} = r_b/v_{0b}$, there exist "mean free path" λ-scales (the mean free path λ, the mean free-flight velocity $v_{0\lambda}$, and the characteristic time scale $\lambda/v_{0\lambda}$) and L-parameters corresponding to hydrodynamic flow parameters (the characteristic hydrodynamic dimension L, the hydrodynamic velocity v_{0L}, and the hydrodynamic time L/v_{0L}).

The fundamental aspect of plasma physics is the presence of multi-particle interaction. The choice of the characteristic scales which determine the evolution of a plasma volume and are used in the method of many scales below is discussed in Appendix 5. Let us introduce a small parameter $\varepsilon = nr_b^3 = \upsilon_{\text{int}}$ assuming that the interaction energy per particle is much less than the particle's kinetic energy. We also assume that the plasma is non-degenerate and employ the multi-scale approach. In the discussion to follow we shall concern ourselves with describing a physical system at the level of one-particle distribution function f_1 on the scales $r_b \equiv l,\ \lambda,\ L$ (l – the Landau length, λ – the mean free path of a probe particle between two "close" collisions, and L – hydrodynamic scale). Note that the mean free path of a plasma particle is introduced as

$$\lambda_n = \Lambda^{-1}\lambda, \tag{8.1.2}$$

with Λ being the Coulomb logarithm. The mean free path λ_n or the corresponding mean time between the collisions are involved in the definition of kinetic coefficients (Braginski, 1963). In the multi-scale method, $\hat{f}_s$ is expressed in the form of an asymptotic series

$$\hat{f}_s = \sum_{\upsilon=0}^{\infty} \hat{f}_s^{\nu}(\hat{t}_b, \hat{\mathbf{r}}_{ib}, \hat{\mathbf{v}}_{ib}; \hat{t}_\lambda, \hat{\mathbf{r}}_{i\lambda}, \hat{\mathbf{v}}_{i\lambda}; \hat{t}_L, \hat{\mathbf{r}}_{iL}, \hat{\mathbf{v}}_{iL})\varepsilon^{\upsilon}, \tag{8.1.3}$$

in which the functions $\hat{f}_s^{\nu}$ depend on all the three types of variables.

From the above BBGKY equation, taking the derivatives of the composite functions on the left-hand side and then equating the coefficients of ε^0 and ε^1, we find that

$$\frac{\partial \hat{f}_1^0}{\partial \hat{t}_b} + \hat{\mathbf{v}}_{ib} \cdot \frac{\partial \hat{f}_1^0}{\partial \hat{\mathbf{r}}_{1b}} + \alpha \widehat{\mathbf{F}}_1 \cdot \frac{\partial \hat{f}_1^0}{\partial \hat{\mathbf{v}}_{1b}} = 0, \tag{8.1.4}$$

$$\begin{aligned}
&\frac{\partial \hat{f}_1^1}{\partial \hat{t}_b} + \hat{\mathbf{v}}_{ib} \cdot \frac{\partial \hat{f}_1^1}{\partial \hat{\mathbf{r}}_{1b}} + \alpha \widehat{\mathbf{F}}_1 \cdot \frac{\partial \hat{f}_1^1}{\partial \hat{\mathbf{v}}_{1b}} + \varepsilon_2 \frac{\partial \hat{f}_1^0}{\partial \hat{t}_\lambda} + \hat{\mathbf{v}}_{ib} \cdot \frac{\partial \hat{f}_1^0}{\partial \hat{\mathbf{r}}_{1\lambda}} + \varepsilon_2 \widehat{\mathbf{F}}_1 \cdot \frac{\partial \hat{f}_1^0}{\partial \hat{\mathbf{v}}_{1\lambda}} \\
&\quad + \varepsilon_1\varepsilon_2\varepsilon_3 \frac{\partial \hat{f}_1^0}{\partial \hat{t}_L} + \varepsilon_1 \hat{\mathbf{v}}_{ib} \cdot \frac{\partial \hat{f}_1^0}{\partial \hat{\mathbf{r}}_{1L}} + \frac{\varepsilon_2}{\varepsilon_3} \widehat{\mathbf{F}}_1 \cdot \frac{\partial \hat{f}_1^0}{\partial \hat{\mathbf{v}}_{1L}} \\
&= -\sum_{\delta=1}^{\eta} \frac{N_\delta}{N} \int \widehat{\mathbf{F}}_{1,\, j\in N_\delta} \cdot \frac{\partial}{\partial \hat{\mathbf{v}}_{1b}} \hat{f}^0_{2,\, j\in N_\delta}\, \mathrm{d}\widehat{\Omega}_{j\in N_\delta},
\end{aligned} \tag{8.1.5}$$

where η is the number of components in the mixture, N_δ is the number of particles of the δth kind, $\varepsilon_1 = \lambda/L$ (the Knudsen number), $\varepsilon_2 = \upsilon_{0\lambda}/\upsilon_{0b}$, $\varepsilon_3 = \upsilon_{0L}/\upsilon_{0\lambda}$. The integration in Eq. (8.1.5) is performed on the r_b scale. Importantly, no restriction is placed on the value of the Knudsen number. Eq. (8.1.4) shows that the function $\hat{f}_1^0$ does not change along the phase trajectory on the r_b-scale – in other words, following the integration on the r_b-scale we have

$$\hat{f}_1^0 = \hat{f}_1^0(\hat{t}_\lambda, \hat{\mathbf{v}}_{1\lambda}, \hat{\mathbf{r}}_{1\lambda}; \hat{t}_L, \hat{\mathbf{v}}_{1L}, \hat{\mathbf{r}}_{1L}). \tag{8.1.6}$$

If the last function is known, $\hat{f}_1^1$ needs to be found from Eq. (8.1.5). This is possible if certain additional assumptions are posed on the function $\hat{f}_2^0$ on the right-hand, integral part of the expression (8.1.5). Thus we see that the system of equations contains linked terms. In real life, the dependence (8.1.6) is unknown in advance. Then Eq. (8.1.5) can serve to determine $\hat{f}_1^0$ on the λ- and L-scales, but in this case it becomes doubly linked with respect to both the lower index "2" and the upper index "1". As a result, the problem of breaking the equations arises.

Let us now write the analogue of Eq. (8.1.4) for the two-particle function $\hat{f}_2^0$ dependent on time and on the dynamic variables for the particles 1 and j:

$$\frac{\partial \hat{f}_2^0}{\partial \hat{t}_b} + \hat{\mathbf{v}}_{1b} \cdot \frac{\partial \hat{f}_2^0}{\partial \hat{\mathbf{r}}_{1b}} + \hat{\mathbf{v}}_{j\in N_\delta,b} \cdot \frac{\partial \hat{f}_2^0}{\partial \hat{\mathbf{r}}_{j\in N_\delta,b}} + \widehat{\mathbf{F}}_{1,\,j\in N_\delta} \cdot \frac{\partial \hat{f}_2^0}{\partial \hat{\mathbf{v}}_{1b}}$$
$$+ \widehat{\mathbf{F}}_{j\in N_\delta,\,1} \cdot \frac{\partial \hat{f}_2^0}{\partial \hat{\mathbf{v}}_{j\in N_\delta}} + \alpha \widehat{\mathbf{F}}_1 \cdot \frac{\partial \hat{f}_2^0}{\partial \hat{\mathbf{v}}_{1b}} + \alpha \widehat{\mathbf{F}}_{j\in N_\delta} \cdot \frac{\partial \hat{f}_2^0}{\partial \hat{\mathbf{v}}_{j\in N_\delta,\,1}} = 0. \tag{8.1.7}$$

Introducing the new variable $\hat{\mathbf{x}}_{1,\,j\in N_\delta} = \hat{\mathbf{r}}_{1,b} - \hat{\mathbf{r}}_{j\in N_\delta,\,b}$, we find from Eq. (8.1.7) that:

$$-\widehat{\mathbf{F}}_{1,\,j\in N_\delta} \cdot \frac{\partial \hat{f}_2^0}{\partial \hat{\mathbf{v}}_{1b}} = \frac{\partial \hat{f}_2^0}{\partial \hat{t}_b} + \hat{\mathbf{v}}_{1b} \cdot \frac{\partial \hat{f}_2^0}{\partial \hat{\mathbf{r}}_{1b}} + (\hat{\mathbf{v}}_{1b} - \hat{\mathbf{v}}_{j\in N_\delta,\,b}) \cdot \frac{\partial \hat{f}_2^0}{\partial \hat{\mathbf{x}}_{1,\,j\in N_\delta}}$$
$$+ \widehat{\mathbf{F}}_{j\in N_\delta,\,1} \cdot \frac{\partial \hat{f}_2^0}{\partial \hat{\mathbf{v}}_{j\in N_\delta,\,b}} + \alpha \widehat{\mathbf{F}}_1 \cdot \frac{\partial \hat{f}_2^0}{\partial \hat{\mathbf{v}}_{1b}} + \alpha \widehat{\mathbf{F}}_{j\in N_\delta} \cdot \frac{\partial \hat{f}_2^0}{\partial \hat{\mathbf{v}}_{j\in N_\delta,\,b}}. \tag{8.1.8}$$

Using the last equation, we obtain the representation for the integral in Eq. (8.1.5):

$$-\int \widehat{\mathbf{F}}_{1,\,j\in N_\delta} \cdot \frac{\partial}{\partial \hat{\mathbf{v}}_{1b}} \hat{f}_2^0(\hat{t}, \widehat{\Omega}_1, \widehat{\Omega}_{j\in N_\delta})\, \mathrm{d}\widehat{\Omega}_{j\in N_\delta}$$
$$= \int (\hat{\mathbf{v}}_{1b} - \hat{\mathbf{v}}_{j\in N_\delta,\,b}) \cdot \frac{\partial \hat{f}_2^0}{\partial \hat{\mathbf{x}}_{1,\,j\in N_\delta}} \mathrm{d}\widehat{\Omega}_{j\in N_\delta}$$
$$+ \int \left(\frac{\partial \hat{f}_2^0}{\partial \hat{t}_b} + \hat{\mathbf{v}}_{1b} \cdot \frac{\partial \hat{f}_2^0}{\partial \hat{\mathbf{r}}_{1b}} + \alpha \widehat{\mathbf{F}}_1 \cdot \frac{\partial \hat{f}_2^0}{\partial \hat{\mathbf{v}}_{1b}} + \alpha \widehat{\mathbf{F}}_{j\in N_\delta} \cdot \frac{\partial \hat{f}_2^0}{\partial \hat{\mathbf{v}}_{j\in N_\delta,\,b}} \right) \mathrm{d}\widehat{\Omega}_{j\in N_\delta}$$
$$+ \int \widehat{\mathbf{F}}_{j\in N_\delta,\,1} \cdot \frac{\partial \hat{f}_2^0}{\partial \hat{\mathbf{v}}_{j\in N_\delta,\,b}} \mathrm{d}\widehat{\Omega}_{j\in N_\delta}. \tag{8.1.9}$$

The last integral on the right-hand side of Eq. (8.1.9) can be written in the form

$$\int \widehat{\mathbf{F}}_{j\in N_\delta,\,1} \cdot \frac{\partial \hat{f}_2^0}{\partial \hat{\mathbf{v}}_{j\in N_\delta,\,b}} \mathrm{d}\widehat{\Omega}_{j\in N_\delta}$$
$$= \int \left[\int \frac{\partial}{\partial \hat{\mathbf{v}}_{j\in N_\delta,\,b}} \cdot \left(\widehat{\mathbf{F}}_{j\in N_\delta,\,1} \hat{f}_2^0 \right) \mathrm{d}\hat{\mathbf{v}}_{j\in N_\delta} \right] \mathrm{d}\hat{\mathbf{r}}_{j\in N_\delta}. \tag{8.1.10}$$

But the inner integral can be transformed by the Gauss theorem into an integral taken over an infinitely distant surface in the velocity space, which vanishes because $\hat{f}_2^0 \to 0$ for $\hat{v}_j \to \infty$. Let us now introduce two-particle correlation function $\widehat{W}_2(\hat{t}, \widehat{\Omega}_1, \widehat{\Omega}_{j\in N_\delta})$ (hereinafter f_j is the one-particle function corresponding to the particles N_j):

$$\hat{f}_2^0(\hat{t}, \widehat{\Omega}_1, \widehat{\Omega}_j) \\ = \hat{f}_1^0(\hat{t}, \widehat{\Omega}_1) \hat{f}_{j\in N_\delta}(\hat{t}, \widehat{\Omega}_{j\in N_\delta}) + \widehat{W}_2^0(\hat{t}, \widehat{\Omega}_1, \widehat{\Omega}_{j\in N_\delta}). \tag{8.1.11}$$

The next-to-last integral in Eq. (8.1.9) then becomes

$$\begin{aligned} &\int \left(\frac{\partial \hat{f}_2^0}{\partial \hat{t}_b} + \hat{\mathbf{v}}_{1b} \cdot \frac{\partial \hat{f}_2^0}{\partial \hat{\mathbf{r}}_{1b}} + \alpha \widehat{\mathbf{F}}_1 \cdot \frac{\partial \hat{f}_2^0}{\partial \hat{\mathbf{v}}_{1b}} + \alpha \widehat{\mathbf{F}}_{j\in N_\delta} \cdot \frac{\partial \hat{f}_2^0}{\partial \hat{\mathbf{v}}_{j\in N_\delta}} \right) \mathrm{d}\widehat{\Omega}_{j\in N_\delta} \\ &= \int \left[\hat{f}_{j\in N_\delta}^0 \left(\frac{\partial \hat{f}_1^0}{\partial \hat{t}_b} + \hat{\mathbf{v}}_{1b} \cdot \frac{\partial \hat{f}_1^0}{\partial \hat{\mathbf{r}}_{1b}} + \alpha \widehat{\mathbf{F}}_1 \cdot \frac{\partial \hat{f}_1^0}{\partial \hat{\mathbf{v}}_{1b}} \right) \right] \mathrm{d}\widehat{\Omega}_{j\in N_\delta} \\ &+ \int \hat{f}_1^0 \frac{\partial \hat{f}_{j\in N_\delta}^0}{\partial \hat{t}_b} \mathrm{d}\widehat{\Omega}_{j\in N_\delta} + \alpha \int \frac{\partial}{\partial \hat{\mathbf{v}}_{j\in N_\delta, b}} \cdot \left(\widehat{\mathbf{F}}_{j\in N_\delta} \hat{f}_1^0 \hat{f}_{j\in N_\delta}^0 \right) \mathrm{d}\widehat{\Omega}_{j\in N_\delta} \\ &+ \int \left(\frac{\partial \widehat{W}_2^0}{\partial \hat{t}_b} + \hat{\mathbf{v}}_{1b} \cdot \frac{\partial \widehat{W}_2^0}{\partial \hat{\mathbf{r}}_{1b}} + \alpha \widehat{\mathbf{F}}_1 \cdot \frac{\partial \widehat{W}_2^0}{\partial \hat{\mathbf{v}}_{1b}} + \alpha \widehat{\mathbf{F}}_{j\in N_\delta} \cdot \frac{\partial \widehat{W}_2^0}{\partial \hat{\mathbf{v}}_{j\in N_\delta}} \right) \mathrm{d}\widehat{\Omega}_{j\in N_\delta}. \end{aligned} \tag{8.1.12}$$

In expression (8.1.12), the first integral on the right is zero because of the relation (8.1.4), and the third integral is zero for the same reasons as in Eq. (8.1.10). The situation with the second and fourth integrals, however, requires a more detailed treatment. Consider first the integral

$$A = \int \hat{f}_1^0 \frac{\partial \hat{f}_{j\in N_\delta}^0}{\partial \hat{t}_b} \mathrm{d}\widehat{\Omega}_{j\in N_\delta}. \tag{8.1.13}$$

The dynamic variables determining the motion of the given trial particles 1 and j are correlated one with another in the collision of the particles, i.e., on the r_b-scale. In the center-of-mass system, the equations of motion for these particles are written as

$$\dot{\mathbf{v}}_{1b} = \mathbf{F}_{1j}, \qquad \dot{\mathbf{v}}_{jb} = \mathbf{F}_{j1}, \qquad \mathbf{p}_1 = -\mathbf{p}_j, \tag{8.1.14}$$

where an over dot denotes differentiation with respect to time, and $\mathbf{p}$ is the particle momentum.

Using Eqs. (8.1.14) and integrating by parts, we arrive at the relation

$$A \cong -\widehat{\mathbf{F}}_{1\delta}^a \cdot \frac{\partial \hat{f}_1}{\partial \hat{\mathbf{v}}_{1b}}, \tag{8.1.15}$$

where $\widehat{\mathbf{F}}^{a}_{1\delta}$ is the average force acting on particle 1 during its collision with particle j which has an arbitrary velocity and an arbitrary position on the r_b-scale (particles j belong to the species δ):

$$\widehat{\mathbf{F}}^{a}_{1\delta} = \int \hat{f}_j \widehat{\mathbf{F}}_{1j} \, \mathrm{d}\hat{\mathbf{v}}_{j\in N_\delta} \, \mathrm{d}\hat{\mathbf{r}}_{j\in N_\delta}. \tag{8.1.16}$$

Thus, the integral A vanishes provided that the self-consistent force of internal nature can be neglected, in particular, in comparison with the external force acting on particle 1. We next transform the integral A further by using the series (8.1.2), to obtain

$$A \cong -\widehat{\mathbf{F}}^{a}_{1\delta} \cdot \frac{\partial \hat{f}_1^0}{\partial \hat{\mathbf{v}}_{1b}} - \varepsilon \widehat{\mathbf{F}}^{a}_{1\delta} \cdot \frac{\partial \hat{f}_1^1}{\partial \hat{\mathbf{v}}_{1b}}. \tag{8.1.17}$$

The last term in Eq. (8.1.17) ensures, as we shall see below, that the generalized kinetic equation is written in a symmetrical form. Now consider the fourth – the last – integral on the right-hand side of Eq. (8.1.12). To do this, we write down an equation for the two-particle function f_2 of the Bogolyubov chain, in which, in this case, we do not separate the groups of particles belonging to a certain chemical component. The two-particle function f_2 corresponds to the dynamical variables of particles $N_1,\ N_2$ and is written in the form $f_2 = f_2(1,2)$.

Thus, one finds

$$\begin{aligned}
&\frac{\partial f_2}{\partial t} + \mathbf{v}_1 \cdot \frac{\partial f_2}{\partial \mathbf{r}_1} + \mathbf{v}_2 \cdot \frac{\partial f_2}{\partial \mathbf{r}_2} + \mathbf{F}_{12} \cdot \frac{\partial f_2}{\partial \mathbf{v}_1} + \mathbf{F}_{21} \cdot \frac{\partial f_2}{\partial \mathbf{v}_2} + \mathbf{F}_1 \cdot \frac{\partial f_2}{\partial \mathbf{v}_1} + \mathbf{F}_2 \cdot \frac{\partial f_2}{\partial \mathbf{v}_2} \\
&\quad = \int \Big\{ \mathbf{F}_{13} \cdot \frac{\partial}{\partial \mathbf{v}_1} \big[f_1(1) f_1(2) f_1(3) + f_1(1) W_2(2,3) + f_1(2) W_2(1,3) \\
&\qquad + f_1(3) W_2(1,2) \big] + \mathbf{F}_{23} \cdot \frac{\partial}{\partial \mathbf{v}_2} \big[f_1(1) f_1(2) f_1(3) + f_1(1) W_2(2,3) \\
&\qquad + f_1(2) W_2(1,3) + f_1(3) W_2(1,2) \big] \Big\} \, \mathrm{d}\Omega_3,
\end{aligned} \tag{8.1.18}$$

where the three-particle function f_3 has been approximated by using correlation functions as follows:

$$\begin{aligned}
f_3(\Omega_1, \Omega_2, \Omega_3, t) = {} & f_1(\Omega_1, t) f_1(\Omega_2, t) f_1(\Omega_3, t) + f_1(\Omega_1, t) W_2(\Omega_2, \Omega_3, t) \\
& + f_1(\Omega_2, t) W_2(\Omega_1, \Omega_3, t) + f_1(\Omega_3, t) W_2(\Omega_1, \Omega_2, t) \\
& + W_3(\Omega_1, \Omega_2, \Omega_3, t).
\end{aligned} \tag{8.1.19}$$

The effect of the correlation function $W_3(\Omega_1, \Omega_2, \Omega_3, t)$ is here neglected.

Eq. (8.1.18) written in the zeroth approximation in ε as an equation for finding f_2^0 is identical with Eq. (8.1.7) only when the zero-order correlation functions are zero, viz.

$$\begin{aligned}
&W_2^0(\Omega_2, \Omega_3, t) = 0, \qquad W_2^0(\Omega_1, \Omega_3, t) = 0, \\
&W_2^0(\Omega_1, \Omega_2, t) = 0, \qquad W_3^0(\Omega_1, \Omega_2, \Omega_3, t) = 0,
\end{aligned} \tag{8.1.20}$$

and when the interaction forces determining the effect of the third particle on the first and the second ones during their "close" collision are small, i.e., $F_{13} \approx 0,\ F_{23} \approx 0$. Hence, in the multi-scale approach, polarization terms on the right-hand side of Eq. (8.1.18) appear in the next, of small order $o(\varepsilon^2)$ approximation.

Thus, in the multi-scale approach, the last integral on the right-hand side of Eq. (8.1.12) vanishes because of the condition (8.1.20). The integral relation (8.1.9) can be written in the form

$$\begin{aligned}
&\int \widehat{\mathbf{F}}_{1,\,j\in N_\delta} \cdot \frac{\partial}{\partial \hat{\mathbf{v}}_{1b}} \hat{f}_2^0\big(\hat{t}, \widehat{\Omega}_1, \widehat{\Omega}_{j\in N_\delta}\big)\, \mathrm{d}\widehat{\Omega}_{j\in N_\delta} \\
&= \int (\hat{\mathbf{v}}_{1b} - \hat{\mathbf{v}}_{j\in N_\delta}) \cdot \frac{\partial \hat{f}_2^0}{\partial \hat{\mathbf{x}}_{1,\,j\in N_\delta}}\, \mathrm{d}\widehat{\Omega}_{j\in N_\delta} \\
&\quad - \widehat{\mathbf{F}}^a_{1\delta} \cdot \frac{\partial \hat{f}_1^0}{\partial \hat{\mathbf{v}}_{1b}} - \varepsilon \widehat{\mathbf{F}}^a_{1\delta} \cdot \frac{\partial \hat{f}_1^1}{\partial \hat{\mathbf{v}}_{1b}}.
\end{aligned} \tag{8.1.21}$$

We now introduce the cylindrical coordinate system $\hat{l},\ \hat{b},\ \varphi$ with the origin at point $\mathbf{r}_1$ and $\hat{l}$-axis parallel to the vector of the relative velocity of the colliding particles 1 and j. Then, in terms of $\hat{b}$ (dimensionless impact parameter) and φ (azimuth angle), the first term on the right-hand side of Eq. (8.1.21) is written as

$$\begin{aligned}
\hat{J}^{\,\mathrm{st},0} &= \sum_{\delta=1}^{\eta} \frac{N_\delta}{N} \int \hat{g}_{j\in N_\delta,\,1} \left[\int_{-\infty}^{+\infty} \frac{\partial \hat{f}_2^0}{\partial \hat{l}}\, \mathrm{d}\hat{l} \right] \hat{b}\, \mathrm{d}\hat{b}\, \mathrm{d}\varphi\, \mathrm{d}\hat{\mathbf{v}}_{j\in N_\delta,\,b} \\
&= \sum_{\delta=1}^{\eta} \frac{N_\delta}{N} \int \big[\hat{f}_2^0(+\infty) - \hat{f}_2^0(-\infty)\big] \hat{g}_{j\in N_\delta,\,1} \hat{b}\, \mathrm{d}\hat{b}\, \mathrm{d}\varphi\, \mathrm{d}\hat{\mathbf{v}}_{j\in N_\delta,\,b}.
\end{aligned} \tag{8.1.22}$$

The integration in Eq. (8.1.22) was performed on the r_b-scale, i.e., the distribution functions $\hat{f}_2^0(+\infty),\ \hat{f}_2^0(-\infty)$ are calculated for the velocities $\hat{\mathbf{v}}'_1,\ \hat{\mathbf{v}}'_{j\in N_\delta}$ and $\hat{\mathbf{v}}_1,\ \hat{\mathbf{v}}_{j\in N_\delta}$ in the situation where the particles are found outside of their region of interaction – in other words, before or after the collision (with primed velocities in the latter case). If before the collision the conditions of molecular chaos are fulfilled on the λ-scale, then the two-particle DFs can be expressed as a product of one-particle DFs. In this case $\hat{J}^{\,\mathrm{st},0}$ is the Boltzmann collision integral:

$$\hat{J}^{\,\mathrm{st},0} = \sum_{\delta=1}^{\eta} \frac{N_\delta}{N} \int \big[\hat{f}_1'^{\,0} \hat{f}'^{\,0}_{j\in N_\delta} - \hat{f}_1^{\,0} \hat{f}^{\,0}_{j\in N_\delta}\big] \hat{g}_{j\in N_\delta,\,1} \hat{b}\, \mathrm{d}\hat{b}\, \mathrm{d}\varphi\, \mathrm{d}\hat{\mathbf{v}}_{j\in N_\delta}. \tag{8.1.23}$$

Lenard (1960) and Balescu (1960) solved the equation for the correlation function W_2 under the assumptions of a weakened initial correlation, no time delay, and spatially uniform DF f_1. The corresponding collision integral (the Balescu–Lenard collision integral) incorporates the polarization of the plasma and allows elimination of the logarithmic divergence of the Boltzmann collision integral for a Coulomb plasma

(Lenard, 1960; Balescu, 1960, 1975; Klimontovich, 1975). If, however, the Boltzmann collision integral is still used in plasma description, a cut-off procedure involving Debye screening must be employed.

Using expressions (8.1.20), the kinetic equation (8.1.5) is in the form

$$\begin{aligned}
&\frac{\partial \hat{f}_1^1}{\partial \hat{t}_b} + \hat{\mathbf{v}}_{1b} \cdot \frac{\partial \hat{f}_1^1}{\partial \hat{\mathbf{r}}_{1b}} + \left(\alpha \widehat{\mathbf{F}}_1 + \varepsilon \widehat{\mathbf{F}}_1^a\right) \cdot \frac{\partial \hat{f}_1^1}{\partial \hat{\mathbf{v}}_{1b}} + \varepsilon_2 \frac{\partial \hat{f}_1^0}{\partial \hat{t}_\lambda} + \hat{\mathbf{v}}_{1b} \cdot \frac{\partial \hat{f}_1^0}{\partial \hat{\mathbf{r}}_{1\lambda}} \\
&\quad + \varepsilon_1 \varepsilon_2 \varepsilon_3 \frac{\partial \hat{f}_1^0}{\partial \hat{t}_L} + \varepsilon_1 \hat{\mathbf{v}}_{1b} \cdot \frac{\partial \hat{f}_1^0}{\partial \hat{\mathbf{r}}_{1L}} + \widehat{\mathbf{F}}_1^a \cdot \frac{\partial \hat{f}_1^0}{\partial \hat{\mathbf{v}}_{1b}} + \varepsilon_2 \widehat{\mathbf{F}}_1 \cdot \frac{\partial \hat{f}_1^0}{\partial \hat{\mathbf{v}}_{1\lambda}} \\
&\quad + \frac{\varepsilon_2}{\varepsilon_3} \widehat{\mathbf{F}}_1 \cdot \frac{\partial \hat{f}_1^0}{\partial \hat{\mathbf{v}}_{1L}} = \hat{J}^{\,\mathrm{st},0},
\end{aligned} \tag{8.1.24}$$

where

$$\widehat{\mathbf{F}}_1^a = \sum_{\delta=1}^{\eta} \frac{N_\delta}{N} \widehat{\mathbf{F}}_{1\delta}^a.$$

It should be emphasized that in its dimensional form the factor $\varepsilon \widehat{\mathbf{F}}_{1\delta}^a$ can be written in the form

$$\varepsilon \mathbf{F}_{1\delta}^a = \frac{1}{v_{0b}^2 / r_b} \int f_{j \in N_\delta} \mathbf{F}_{1,\, j \in N_\delta}\, \mathrm{d}\mathbf{v}_{j \in N_\delta}\, \mathrm{d}\mathbf{r}_{j \in N_\delta}, \tag{8.1.25}$$

if it is remembered that

$$\varepsilon = n r_b^3, \quad \hat{f} = f v_{0b}^3 n^{-1}, \quad \hat{v} = \frac{v}{v_{0b}}, \quad \hat{r} = \frac{r}{r_b}, \quad \widehat{F}_{1j} = \frac{F_{1j}}{v_{0b}^2 / r_b}.$$

The scale of the internal force F_{1j} corresponds to choosing the Landau length l as r_b. Let us now write Eq. (8.1.24) in the form (cf. Eq. (1.3.58))

$$\frac{D_1 \hat{f}_1^1}{D \hat{t}_b} + \frac{\mathrm{d}_1 \hat{f}_1^0}{\mathrm{d} \hat{t}_{b,\lambda,L}} = \hat{J}^{\,\mathrm{st},0}, \tag{8.1.26}$$

where we have introduced the notation

$$\frac{D_1 \hat{f}_1^1}{D \hat{t}_b} = \frac{\partial \hat{f}_1^1}{\partial \hat{t}_b} + \hat{\mathbf{v}}_{1b} \cdot \frac{\partial \hat{f}_1^1}{\partial \hat{\mathbf{r}}_{1b}} + \left(\alpha \widehat{\mathbf{F}}_1 + \varepsilon \widehat{\mathbf{F}}_1^a\right) \cdot \frac{\partial \hat{f}_1^1}{\partial \hat{\mathbf{v}}_{1b}}, \tag{8.1.27}$$

$$\begin{aligned}
\frac{\mathrm{d}_1 \hat{f}_1^0}{\mathrm{d} \hat{t}_{b,\lambda,L}} &= \varepsilon_2 \frac{\partial \hat{f}_1^0}{\partial \hat{t}_\lambda} + \hat{\mathbf{v}}_{1b} \cdot \frac{\partial \hat{f}_1^0}{\partial \hat{\mathbf{r}}_{1\lambda}} + \varepsilon_1 \varepsilon_2 \varepsilon_3 \frac{\partial \hat{f}_1^0}{\partial \hat{t}_L} + \varepsilon_1 \hat{\mathbf{v}}_{1b} \cdot \frac{\partial \hat{f}_1^0}{\partial \hat{\mathbf{r}}_{1L}} \\
&\quad + \widehat{\mathbf{F}}_1^a \cdot \frac{\partial \hat{f}_1^0}{\partial \hat{\mathbf{v}}_{1b}} + \varepsilon_2 \widehat{\mathbf{F}}_1 \cdot \frac{\partial \hat{f}_1^0}{\partial \hat{\mathbf{v}}_{1\lambda}} + \frac{\varepsilon_2}{\varepsilon_3} \widehat{\mathbf{F}}_1 \cdot \frac{\partial \hat{f}_1^0}{\partial \hat{\mathbf{v}}_{1L}}.
\end{aligned} \tag{8.1.28}$$

We now wish to use Eq. (8.1.26) for describing the evolution of the distribution function $\hat{f}_1^0$ – but the trouble is, this equation involves a single-order term $D_1\hat{f}_1^1/D_1\hat{t}_b$ linked with respect to upper index. So we are faced with the problem of how to approximate this term – a problem which is similar in a sense to that of approximating the two-particle distribution function in the collision integral. Using the series (8.1.3) allows an exact representation for the term of interest:

$$\frac{D_1\hat{f}_1^1}{D\hat{t}_b} = \frac{D}{D\hat{t}_b}\left[\frac{\partial \hat{f}_1}{\partial \varepsilon}\right]_{\varepsilon=0}. \tag{8.1.29}$$

The term $D_1\hat{f}_1^1/D\hat{t}_b$ describes how the distribution function varies over times of the order of the collision time, or equivalently on the r_b-scale. If this term is left out of account then, from the viewpoint of the derivation of the hierarchy of kinetic equations, this means that

(1) the distribution function does not vary on the r_b-scale (provided we also neglect the average internal force that gives rise to the second and third terms on the right in Eq. (8.1.21));
(2) the particles are point-like and structureless;
(3) changes in DF due to collisions take place instantaneously and are described by the source term $\hat{J}^{\,\mathrm{st},0}$.

In the field description, however, the DF f_1 on the interaction scale (r_b-scale) depends on ε through the dynamic variables $\mathbf{r}$, $\mathbf{v}$, t related by the laws of classical mechanics, thus allowing the approximation

$$\begin{aligned}
&\frac{D_1\hat{f}_1^0}{D\hat{t}_b}\left[\left(\frac{\partial \hat{f}_1}{\partial \varepsilon}\right)_{\varepsilon=0}\right] \\
&\quad\cong \frac{D_1}{D(-\hat{t}_b)}\left[\frac{\partial \hat{f}_1^0}{\partial(-\hat{t}_b)}\left(\frac{\partial(-\hat{t}_b)}{\partial\varepsilon}\right)_{\varepsilon=0} + \frac{\partial \hat{f}_1}{\partial \hat{\mathbf{r}}_{1b}}\cdot\frac{\partial \hat{\mathbf{r}}_b}{\partial(-\hat{t}_b)}\left(\frac{\partial(-\hat{t}_b)}{\partial\varepsilon}\right)_{\varepsilon=0}\right. \\
&\qquad\left. + \frac{\partial \hat{f}_1}{\partial \hat{\mathbf{v}}_{1b}}\cdot\frac{\partial \hat{\mathbf{v}}_b}{\partial(-\hat{t}_b)}\left(\frac{\partial(-\hat{t}_b)}{\partial\varepsilon}\right)_{\varepsilon=0}\right] = -\frac{D_1}{D\hat{t}_b}\left[\left(\frac{\partial \hat{t}_b}{\partial\varepsilon}\right)_{\varepsilon=0}\frac{D_1\hat{f}_1}{D\hat{t}_b}\right] \\
&\quad\cong -\frac{D_1}{D\hat{t}_b}\left[\left(\frac{\partial \hat{t}_b}{\partial\varepsilon}\right)_{\varepsilon=0}\frac{D_1\hat{f}_1^0}{D\hat{t}_b}\right].
\end{aligned} \tag{8.1.30}$$

In expression (8.1.30) we have introduced an approximation proceeded against the direction of time, which corresponds to the condition of there being no correlations for $t_0 \to -\infty$, with t_0 being a certain instant of time on the r_b-scale at which the particles start to interact.

In this way, Markov processes are separated out from all stochastic processes possible in the system.

For the particles of the chemical sort α ($\alpha = 1, \ldots, \eta$) we employ the following normalized DF:

$$f_\alpha = \frac{f_1 N_\alpha}{N}, \qquad \int f_\alpha \, \mathrm{d}\mathbf{v}_\alpha = n_\alpha, \qquad \int n_\alpha \, \mathrm{d}\mathbf{r} = N_\alpha. \tag{8.1.31}$$

In Eq. (8.1.31), f_1 is a one-particle DF. Returning to the expression (8.1.26) written in the dimensional form, we convolute the multi-scale substantial derivatives to obtain (Alekseev, 2003)

$$\frac{Df_\alpha}{Dt} - \frac{D}{Dt}\left[\tau_\alpha \frac{Df_\alpha}{Dt}\right] = \sum_{\beta=1}^{\eta} \int \left[f'_\alpha f'_\beta - f_\alpha f_\beta\right] g_{\beta\alpha} b \, \mathrm{d}b \, \mathrm{d}\varphi \, \mathrm{d}\mathbf{v}_\beta, \tag{8.1.32}$$

where

$$\frac{D}{Dt} = \frac{\partial}{\partial t} + \mathbf{v}_\alpha \cdot \frac{\partial}{\partial \mathbf{r}} + \mathbf{F}_\alpha^{\mathrm{sc}} \cdot \frac{\partial}{\partial \mathbf{v}_\alpha}, \qquad \mathbf{F}_\alpha^{\mathrm{sc}} = \mathbf{F}_\alpha + \mathbf{F}_\alpha^{a}. \tag{8.1.33}$$

Let us comment on Eq. (8.1.32).

(1) We consider that the particle numbered 1 in the multi-component mixture belongs to a component α, which is exactly what the subscript α on the symbol of the DF indicates. Note also that we dropped the superscript 0 from this symbol: carrying it no longer makes sense because all the equations hereinafter already contain only functions of zero order (in the sense of the series expansion in terms of the density parameter ε).

(2) The parameter τ_α is written in the form (see also (1.3.69))

$$\tau_{1 \in N_\alpha} = \frac{\varepsilon}{[\partial \varepsilon / \partial t]_{\varepsilon=0}}, \tag{8.1.34}$$

where ε is the number of particles of all kinds that find themselves in the interaction volume of an a particle by the instant of time t; introducing $\varepsilon^{\mathrm{eq}}$ (the "equilibrium" particle density in the close interaction volume), Eq. (8.1.34) is written in a typical relaxation form

$$\frac{\partial \varepsilon}{\partial t} = -\frac{\varepsilon(t) - \varepsilon^{\mathrm{eq}}}{\tau_\alpha}. \tag{8.1.35}$$

The denominator in Eq. (8.1.34) is interpreted as the number of particles that find themselves within the interaction volume of a certain particle belonging to the αth component per unit time; the derivative is calculated under the additional condition $\varepsilon = 0$.

Clearly, this number is equal to the number of collisions occurring in the interaction volume per unit time. Hence, τ_α is the mean time between collisions of a particle of the αth sort with particles of all other sorts. The procedure includes the action, during the collision of the particles, of the self-consistent force $\mathbf{F}^{\mathrm{sc}}$ being the sum of the external force and the force $\mathbf{F}^{a}$ of internal origin.

As the derivation of formula (8.1.34) suggests, τ_α is determined by close collisions occurring in the plasma.

By analogy with expression (8.1.2), we have

$$\tau_\alpha^n = \Lambda^{-1}\tau_\alpha, \tag{8.1.36}$$

where τ_α^n is the mean time between collisions.

In the hydrodynamic approximation, the time τ_α can be expressed in terms of the viscosity μ_α of the component α (Braginski, 1963; Trubnikov, 1963); for example, for ions one has

$$\tau_\alpha = \Lambda \Pi \mu_\alpha p_\alpha^{-1}. \tag{8.1.37}$$

Eq. (8.1.37) involves the coefficient Π, which is determined by the interaction model (for ions $\Pi = 1.04$ (Braginski, 1963; Trubnikov, 1963)) and the static pressure

$$p_\alpha = n_\alpha k_B T_\alpha. \tag{8.1.38}$$

The generalized Boltzmann equation is invariant under the Galileo transformation and has a correct free-molecular and Maxwellian asymptotic behavior.

We shall now write down the system of generalized hydrodynamic equations. These equations have been obtained previously for gaseous systems in an external field of forces. The distinguishing feature of the generalized Enskog equations we display below is the inclusion of the self-consistent forces $\mathbf{F}^{\mathrm{sc}}$ (see formulas (8.1.33)).

The continuity equation for the component α is given by

$$\begin{aligned}
&\frac{\partial}{\partial t}\left\{\rho_\alpha - \tau_\alpha\left[\frac{\partial \rho_\alpha}{\partial t} + \frac{\partial}{\partial \mathbf{r}}\cdot(\rho_\alpha \bar{\mathbf{v}}_\alpha)\right]\right\} \\
&\quad + \frac{\partial}{\partial \mathbf{r}}\cdot\left\{\rho_\alpha \bar{\mathbf{v}}_\alpha - \tau_\alpha\left[\frac{\partial}{\partial t}(\rho_\alpha \bar{\mathbf{v}}_\alpha) + \frac{\partial}{\partial \mathbf{r}}\cdot \rho_\alpha \overline{\mathbf{v}_\alpha \mathbf{v}_\alpha} - \rho_\alpha \mathbf{F}_\alpha^{(1)\mathrm{sc}}\right.\right. \\
&\quad \left.\left. - \frac{q_\alpha}{m_\alpha}\rho_\alpha \bar{\mathbf{v}}_\alpha \times \mathbf{B}^{\mathrm{sc}}\right]\right\} = R_\alpha,
\end{aligned} \tag{8.1.39}$$

the equation of motion is written as

$$\begin{aligned}
&\frac{\partial}{\partial t}\left\{\rho_\alpha \bar{\mathbf{v}}_\alpha - \tau_\alpha\left[\frac{\partial}{\partial t}(\rho_\alpha \bar{\mathbf{v}}_\alpha) + \frac{\partial}{\partial \mathbf{r}}\cdot \rho_\alpha \overline{\mathbf{v}_\alpha \mathbf{v}_\alpha} - \rho_\alpha \mathbf{F}_\alpha^{(1)\mathrm{sc}} - \frac{q_\alpha}{m_\alpha}\rho_\alpha \bar{\mathbf{v}}_\alpha \times \mathbf{B}^{\mathrm{sc}}\right]\right\} \\
&\quad - \mathbf{F}_\alpha^{(1)\mathrm{sc}}\left[\rho_\alpha - \tau_\alpha\left(\frac{\partial \rho_\alpha}{\partial t} + \frac{\partial}{\partial \mathbf{r}}\cdot(\rho_\alpha \bar{\mathbf{v}}_\alpha)\right)\right] - \frac{q_\alpha}{m_\alpha}\left\{\rho_\alpha \bar{\mathbf{v}}_\alpha - \tau_\alpha\left[\frac{\partial}{\partial t}(\rho_\alpha \bar{\mathbf{v}}_\alpha)\right.\right. \\
&\quad \left.\left. + \frac{\partial}{\partial \mathbf{r}}\cdot \rho_\alpha \overline{\mathbf{v}_\alpha \mathbf{v}_\alpha} - \rho_\alpha \mathbf{F}_\alpha^{(1)\mathrm{sc}} - \frac{q_\alpha}{m_\alpha}\rho_\alpha \bar{\mathbf{v}}_\alpha \times \mathbf{B}^{\mathrm{sc}}\right]\right\} \times \mathbf{B}^{\mathrm{sc}} + \frac{\partial}{\partial \mathbf{r}}\cdot\left\{\rho_\alpha \overline{\mathbf{v}_\alpha \mathbf{v}_\alpha}\right. \\
&\quad - \tau_\alpha\left[\frac{\partial}{\partial t}(\rho_\alpha \overline{\mathbf{v}_\alpha \mathbf{v}_\alpha}) + \frac{\partial}{\partial \mathbf{r}}\cdot \rho_\alpha \overline{(\mathbf{v}_\alpha \mathbf{v}_\alpha)\mathbf{v}_\alpha} - \rho_\alpha \mathbf{F}_\alpha^{(1)\mathrm{sc}}\bar{\mathbf{v}}_\alpha - \rho_\alpha \bar{\mathbf{v}}_\alpha \mathbf{F}_\alpha^{(1)\mathrm{sc}}\right. \\
&\quad \left.\left. - \frac{q_\alpha}{m_\alpha}\rho_\alpha \overline{(\mathbf{v}_\alpha \times \mathbf{B}^{\mathrm{sc}})\mathbf{v}_\alpha} - \frac{q_\alpha}{m_\alpha}\rho_\alpha \overline{\mathbf{v}_\alpha(\mathbf{v}_\alpha \times \mathbf{B}^{\mathrm{sc}})}\right]\right\} = \bar{\mathbf{I}}_{\alpha,\mathrm{mot}}
\end{aligned} \tag{8.1.40}$$

and the equation of energy has the form

$$\frac{\partial}{\partial t}\left\{\frac{\rho_\alpha \overline{v_\alpha^2}}{2}+\varepsilon_\alpha n_\alpha-\tau_\alpha\left[\frac{\partial}{\partial t}\left(\frac{\rho_\alpha \overline{v_\alpha^2}}{2}+\varepsilon_\alpha n_\alpha\right)+\frac{\partial}{\partial \mathbf{r}}\cdot\left(\frac{1}{2}\rho_\alpha \overline{v_\alpha^2 \mathbf{v}_\alpha}+\varepsilon_\alpha n_\alpha \bar{\mathbf{v}}_\alpha\right)\right.\right.$$
$$\left.\left.-\rho_\alpha \mathbf{F}_\alpha^{(1)\,\mathrm{sc}}\cdot\bar{\mathbf{v}}_\alpha\right]\right\}+\frac{\partial}{\partial \mathbf{r}}\cdot\left\{\frac{1}{2}\rho_\alpha \overline{v_\alpha^2 \mathbf{v}_\alpha}+\varepsilon_\alpha n_\alpha \bar{\mathbf{v}}_\alpha\right.$$
$$-\tau_\alpha\left[\frac{\partial}{\partial t}\left(\frac{1}{2}\rho_\alpha \overline{v_\alpha^2 \mathbf{v}_\alpha}+\varepsilon_\alpha n_\alpha \bar{\mathbf{v}}_\alpha\right)+\frac{\partial}{\partial \mathbf{r}}\cdot\left(\frac{1}{2}\rho_\alpha \overline{v_\alpha^2 \mathbf{v}_\alpha \mathbf{v}_\alpha}+\varepsilon_\alpha n_\alpha \overline{\mathbf{v}_\alpha \mathbf{v}_\alpha}\right)\right.$$
$$\left.\left.-\rho_\alpha \mathbf{F}_\alpha^{(1)\,\mathrm{sc}}\cdot\overline{\mathbf{v}_\alpha \mathbf{v}_\alpha}-\frac{1}{2}\rho_\alpha \overline{v_\alpha^2 \mathbf{F}_\alpha^{\mathrm{sc}}}-\varepsilon_\alpha n_\alpha \overline{\mathbf{F}_\alpha^{\mathrm{sc}}}\right]\right\}$$
$$-\left\{\rho_\alpha \mathbf{F}_\alpha^{(1)\,\mathrm{sc}}\cdot\bar{\mathbf{v}}_\alpha-\tau_\alpha\left[\mathbf{F}_\alpha^{(1)\,\mathrm{sc}}\cdot\left(\frac{\partial}{\partial t}(\rho_\alpha \bar{\mathbf{v}}_\alpha)+\frac{\partial}{\partial \mathbf{r}}\cdot\rho_\alpha \overline{\mathbf{v}_\alpha \mathbf{v}_\alpha}\right.\right.\right.$$
$$\left.\left.\left.-\rho_\alpha \mathbf{F}_\alpha^{(1)\,\mathrm{sc}}-q_\alpha n_\alpha \bar{\mathbf{v}}_\alpha\times\mathbf{B}^{\mathrm{sc}}\right)\right]\right\}=\bar{I}_{\alpha,\mathrm{en}}\quad(\alpha=1,\ldots,\eta),\tag{8.1.41}$$

where $\mathbf{F}_\alpha^{\mathrm{sc}}$ is the total self-consistent force acting on the unit mass of species of the αth kind, $\mathbf{F}_\alpha^{(1)\,\mathrm{sc}}$ is the component of the self-consistent force independent of the velocity of the charged particle, $\mathbf{B}^{\mathrm{sc}}$ is the magnetic induction, q_α the charge of the particle α, ε_α its internal energy, and ρ_α the density of component α; the bar indicates an average over the velocities.

Thus, the generalized Enskog hydrodynamic equations involve self-consistent forces due to the collective nature of plasma particle interactions. In the following sections we discuss the applicability of the above theory to plasma physics problems.

8.2. Dispersion equations of plasma in generalized Boltzmann theory

The generalized Boltzmann equation describes how the one-particle distribution function f_α $(\alpha=1,\ldots,\eta)$ in a η-component gas mixture changes over times of the order of the time between collisions, of the order of the hydrodynamic flow time, and, unlike the conventional Boltzmann equation, over a time of the order of the collision time. The GBE for a plasma medium has the form

$$\frac{Df_\alpha}{Dt}-\frac{D}{Dt}\left(\tau_\alpha\frac{Df_\alpha}{Dt}\right)=J_\alpha,\tag{8.2.1}$$

where

$$\frac{D}{Dt}=\frac{\partial}{\partial t}+\mathbf{v}_\alpha\cdot\frac{\partial}{\partial \mathbf{r}}+\mathbf{F}_\alpha\cdot\frac{\partial}{\partial \mathbf{v}_\alpha}\tag{8.2.2}$$

is the substantial (particle) derivative containing the self-consistent force $\mathbf{F}_\alpha$, J_α is the classical (Boltzmann) collision integral, and τ_α is the mean time between the close particle collisions. In the hydrodynamic regime τ_α can be expressed in terms of the

Coulomb logarithm Λ, viscosity μ_α, static pressure p_α, and the coefficient Π dependent on the particle interaction model (see Eq. (8.1.37)).

The generalized Boltzmann equation in general and that for plasma in particular have a fundamentally important feature that the additional GBE terms prove to be of the order of the Knudsen number. This does not mean that in the hydrodynamic limit (small Kn) these terms may be neglected: the Knudsen number in this case appears as a small parameter of the higher derivative in the GBE. Consequently, the additional GBE terms (as compared to the BE) are significant for any Kn, and the order of magnitude of the difference between the BE and GBE solutions is impossible to tell beforehand.

In this connection, it is of interest to apply the GBE model to obtain the dispersion relation for a plasma in the absence of a magnetic field. In doing so, we will make the same assumptions that were used in the BE-based derivation, namely:

(a) the integral collision term is neglected;
(b) the evolution of electrons and ions in a self-consistent electric field corresponds to a nonstationary one-dimensional model;
(c) the distribution functions for ions, f_i, and for electrons, f_e, deviate little from their equilibrium counterparts f_{0i} and f_{0e};
(d) a wave number k and a complex frequency ω are appropriate to the wave mode considered;
(e) the quadratic GBE terms determining the deviation from the equilibrium DF are neglected, and
(f) the self-consistent forces F_i and F_e are small.

Results of the calculations made under these assumptions are given in Appendix 6. Proceeding now to the dispersion relation, we lift one of these assumptions, the first, by introducing the Bhatnagar–Gross–Krook (BGK) collision term

$$J_\alpha = -\frac{f_\alpha - f_{0\alpha}}{\nu_\alpha^{-1}} \tag{8.2.3}$$

into the right-hand side of the GBE. Here, $f_{0\alpha}$ and $\nu_\alpha^{-1} = \tau_{\mathrm{rel},\alpha}$ are respectively a certain equilibrium distribution function and the relaxation time for species of the αth kind. Using Eqs. (A6.9) and (8.2.1), we arrive at the dispersion relation

$$\begin{aligned} 1 = -\frac{e^2}{\varepsilon_0 k}\Bigg\{ & \frac{1}{m_e}\int_{-\infty}^{+\infty} \frac{(\partial f_{0e}/\partial u)[\mathrm{i} - \tau_e(\omega - ku)]}{\mathrm{i}(\omega - ku) - \tau_e(\omega - ku)^2 - \nu_e}\,\mathrm{d}u \\ & + \frac{1}{m_i}\int_{-\infty}^{+\infty} \frac{(\partial f_{0i}/\partial u)[\mathrm{i} - \tau_i(\omega - ku)]}{\mathrm{i}(\omega - ku) - \tau_i(\omega - ku)^2 - \nu_i}\,\mathrm{d}u\Bigg\}. \end{aligned} \tag{8.2.4}$$

In the Boltzmann kinetic theory, the analogue of Eq. (8.2.4) is the equation (Artsimovich and Sagdeev, 1979)

$$1 = -\frac{e^2}{\varepsilon_0 k}\left\{\frac{1}{m_e}\int_{-\infty}^{+\infty}\frac{(\partial f_{0e}/\partial u)}{\omega - ku + \mathrm{i}\nu_e}\,\mathrm{d}u + \frac{1}{m_i}\int_{-\infty}^{+\infty}\frac{(\partial f_{0i}/\partial u)}{\omega - ku + \mathrm{i}\nu_i}\,\mathrm{d}u\right\}. \tag{8.2.5}$$

To solve Eq. (8.2.4), we take advantage of the additional conjectures. Let us assume that the ions are at rest and that both the temperature and average velocity of the electrons are zero. Then the electron distribution function can be expressed in terms of the delta function:

$$f_{0e}(u) = n_e \delta(u). \tag{8.2.6}$$

Upon integration by parts in Eq. (8.2.4), we arrive at the equation (the subscript "e" on ν_e and τ_e is dropped for brevity)

$$1 + \frac{e^2 n_e}{\varepsilon_0 m_e} \int_{-\infty}^{+\infty} \frac{\delta(u)\{[1 + \mathrm{i}\tau(\omega - ku)]^2 + \nu\tau\}}{[\mathrm{i}(\omega - ku) - \tau(\omega - ku)^2 - \nu]^2} \, \mathrm{d}u = 0. \tag{8.2.7}$$

In the special case of Boltzmann collisionless plasma, Eq. (8.2.7) leads to the classical formula

$$1 - \frac{e^2 n_e}{\varepsilon_0 m_e} \int_{-\infty}^{+\infty} \frac{\delta(u)}{(\omega - ku)^2} \, \mathrm{d}u = 0. \tag{8.2.8}$$

Using the properties of the delta function and performing the integration in Eq. (8.2.7), it is found that

$$\omega_e^2 = -\frac{(\nu\tau + \omega^2\tau^2 - \mathrm{i}\omega\tau)^2}{\tau^2(1 + \nu\tau - \omega^2\tau^2 + 2\mathrm{i}\omega\tau)}, \tag{8.2.9}$$

with $\omega_e = \sqrt{e^2 n_e / \varepsilon_0 m_e}$ being the plasma frequency.

Let us consider the limiting cases inherent in Eq. (8.2.9).

If $\nu \sim \omega' \sim |\omega''|$, $\omega'\tau \ll 1$, then

$$\omega_e^2 = \omega'^2 + 2\mathrm{i}\omega'(\omega'' + \nu) - (\omega'' + \nu)^2, \tag{8.2.10}$$

and separating the real and imaginary parts of relation (8.2.10) leads to the result

$$\omega' = \omega_e, \qquad \omega'' = -\nu. \tag{8.2.11}$$

Let us introduce now in consideration the generalized Maxwell equations (GME) (see, for example, (I.52))

$$\begin{gathered}
\frac{\partial}{\partial \mathbf{r}} \cdot \mathbf{B} = 0, \qquad \frac{\partial}{\partial \mathbf{r}} \cdot \mathbf{D} = \rho^a, \\
\frac{\partial}{\partial \mathbf{r}} \times \mathbf{E} = -\frac{\partial \mathbf{B}}{\partial t}, \qquad \frac{\partial}{\partial \mathbf{r}} \times \mathbf{H} = \mathbf{j}^a + \frac{\partial \mathbf{D}}{\partial t},
\end{gathered} \tag{8.2.12}$$

where

$$\rho^a = \rho - \rho^{\mathrm{fl}}, \qquad \mathbf{j}^a = \mathbf{j} - \mathbf{j}^{\mathrm{fl}}. \tag{8.2.13}$$

Fluctuations ρ^{fl}, $\mathbf{j}^{\text{fl}}$ – which were calculated in the frame of GBE – can be taken from Table 3.2:

$$\rho^{\text{fl}} = \sum_{\alpha} \tau_{\alpha} \left\{ \frac{\partial \rho_{\alpha}}{\partial t} + \frac{\partial}{\partial \mathbf{r}} \cdot (\rho_{\alpha} \bar{\mathbf{v}}_{\alpha}) \right\}, \tag{8.2.14}$$

$$\mathbf{j}^{\text{fl}} = \sum_{\alpha} \tau_{\alpha} \left\{ \frac{\partial}{\partial t} (q_{\alpha} n_{\alpha} \bar{\mathbf{v}}_{\alpha}) + \frac{\partial}{\partial \mathbf{r}} \cdot (q_{\alpha} n_{\alpha} \overline{\mathbf{v}_{\alpha} \mathbf{v}_{\alpha}}) - q_{\alpha} n_{\alpha} \overline{\mathbf{F}}_{\alpha} \right\}. \tag{8.2.15}$$

For example, for one-dimensional case (see also Appendix 6) we have

$$\delta n_e^{\text{fl}} = \tau_e \left(\frac{\partial \delta n_e}{\partial t} + \bar{u}_e \frac{\partial \delta n_e}{\partial x} \right), \tag{8.2.16}$$

$$\delta n_i^{\text{fl}} = \tau_e \left(\frac{\partial \delta n_i}{\partial t} + \bar{u}_i \frac{\partial \delta n_i}{\partial x} \right), \tag{8.2.17}$$

where $\bar{u}_e$, $\bar{u}_i$ are hydrodynamic velocities of electrons and ions.

Eqs. (8.2.12)–(8.2.17) lead to modification of Eq. (8.2.4):

$$\begin{aligned} 1 = -\frac{e^2}{\varepsilon_0 k} \Bigg\{ & \frac{1}{m_e} \int_{-\infty}^{+\infty} \frac{(\partial f_{0e}/\partial u)[\mathrm{i} - \tau_e(\omega - ku)]}{\mathrm{i}(\omega - ku) - \tau_e(\omega - ku)^2 - \nu_e} \, \mathrm{d}u \big[1 + \mathrm{i}\tau_e(\omega - k\bar{u}_e)\big] \\ & + \frac{1}{m_i} \int_{-\infty}^{+\infty} \frac{(\partial f_{0i}/\partial u)[\mathrm{i} - \tau_i(\omega - ku)]}{\mathrm{i}(\omega - ku) - \tau_i(\omega - ku)^2 - \nu_i} \, \mathrm{d}u \\ & \times \big[1 + \mathrm{i}\tau_i(\omega - k\bar{u}_i)\big] \Bigg\}. \end{aligned} \tag{8.2.18}$$

Assuming again that the ions are at rest and that both the temperature and average velocity of the electrons are zero, we obtain

$$\omega_e^2 = -\frac{(\nu + \omega^2 \tau - \mathrm{i}\omega)^2}{(1 + \nu\tau - \omega^2\tau^2 + 2\mathrm{i}\omega\tau)(1 + \mathrm{i}\omega\tau)}. \tag{8.2.19}$$

For considered above limit case $\nu \sim \omega' \sim |\omega''|$, $\omega'\tau \ll 1$ the dispersion equation (8.2.19) has the same solution as (8.2.9) (see (8.2.11)).

Let us show now that GME can lead to reasonable solutions of dispersion equation when in classical case the corresponding dispersion equation has no physical sense.

One obtains for the limit case

$$\omega'\tau \ll 1, \qquad |\omega''|\tau \gg 1, \qquad \nu \sim \omega' \tag{8.2.20}$$

from Eq. (8.2.19):

$$\omega_e^2 = -\frac{(\tau\omega''^2 - 2\mathrm{i}\tau\omega'\omega'' - \nu)^2}{\tau^2(\tau\omega''^2 - 2\mathrm{i}\tau\omega'\omega'' + \nu)(\mathrm{i}\omega' - \omega'')} \tag{8.2.21}$$

and therefore

$$\omega' = 0, \qquad \omega'' = \tau\omega_e^2. \tag{8.2.22}$$

Then conditions (8.2.20) – and by the way conditions

$$\omega'\tau \ll 1, \qquad |\omega''|\tau \gg 1, \qquad \nu \sim \omega'' \tag{8.2.23}$$

– lead to erasing of instability. Interesting to notice that for the mentioned conditions (8.2.20), (8.2.23) dispersion equation (8.2.9) has no solutions.

Now let the ions be at rest, and the electron component have the following velocity distribution:

$$f_{0e}(u) = n_e\delta(u - u_1). \tag{8.2.24}$$

The generalized dispersion equation found with taking into account GME has the form

$$1 + \frac{e^2 n_e}{\varepsilon_0 m_e}\left[1 + \mathrm{i}\tau(\omega - \bar{u}k)\right] \times \int_{-\infty}^{+\infty} \frac{\delta(u - u_1)\{[1 + \mathrm{i}\tau(\omega - ku)]^2 + \nu\tau\}}{[\mathrm{i}(\omega - ku) - \tau(\omega - ku)^2 - \nu]^2}\,\mathrm{d}u = 0 \tag{8.2.25}$$

and after integration in (8.2.25) one obtains

$$1 + \frac{e^2 n_e}{\varepsilon_0 m_e} A \frac{[1 + \mathrm{i}\tau(\omega - ku_1)]^2 + \nu\tau}{[\mathrm{i}(\omega - ku_1) - \tau(\omega - ku_1)^2 - \nu]^2} = 0, \tag{8.2.26}$$

where

$$A = 1 - \tau\omega'' + \mathrm{i}\tau(\omega' - \bar{u}k). \tag{8.2.27}$$

Dispersion equation written with the help of classical Maxwell equations has the form

$$1 + \frac{e^2 n_e}{\varepsilon_0 m_e} \frac{1 + \mathrm{i}\tau(\omega - ku_1)]^2 + \nu\tau}{[\mathrm{i}(\omega - ku_1) - \tau(\omega - ku_1)^2 - \nu]^2} = 0. \tag{8.2.28}$$

It can be shown that (8.2.26) has physical solutions when dispersion equation (8.2.28) has no solutions at all. With this aim we consider the case when the velocity u_1 of electron beam is equal to hydrodynamic velocity $\bar{u}$ and electrons are trapped by the wave of electrical field:

$$u_1 = \bar{u}, \qquad \omega' = k\bar{u}. \tag{8.2.29}$$

In this case it follows from Eqs. (8.2.26) and (8.2.28):

$$1 + \omega_e^2 \tau^2 \frac{x(x^2 + \nu\tau)}{[x(1-x) + \nu\tau]^2} = 0, \tag{8.2.30}$$

$$1 + \omega_e^2 \tau^2 \frac{x^2 + \nu\tau}{[x(1-x) + \nu\tau]^2} = 0, \tag{8.2.31}$$

where

$$\omega_e^2 = \frac{e^2 n_e}{\varepsilon_0 m_e}, \qquad x = 1 - \omega''\tau. \tag{8.2.32}$$

Because x is a real number, (8.2.31) has no solutions in physical sense, but Eq. (8.2.30) – by the negative values of x – has solutions which correspond of developing of instability in the system. In particular, if $|x| \ll 1$, $|x| \ll \nu\tau$, then it follows from (8.2.30)

$$x = -\frac{1}{\omega_e \tau} \frac{\nu}{\omega_e} \tag{8.2.33}$$

or

$$\omega''\tau = 1 + \frac{1}{\omega_e \tau} \frac{\nu}{\omega_e}. \tag{8.2.34}$$

This solution has a clear physical meaning.

Now let the ions be at rest, and the electron component have a Maxwellian velocity distribution:

$$f_{0e} = n_e \left(\frac{m_e}{2\pi k_B T} \right)^{3/2} \mathrm{e}^{-m_e u^2 / 2k_B T}, \tag{8.2.35}$$

where k_B is the Boltzmann constant. Then Eq. (8.2.4) becomes

$$1 + \frac{e^2 n_e}{\varepsilon_0 k m_e} \left(\frac{m_e}{2\pi k_B T} \right)^{1/2} \times \int_{-\infty}^{+\infty} \frac{[\mathrm{i} - \tau(\omega - ku)](\partial/\partial u)\, \mathrm{e}^{-m_e u^2/2k_B T}}{\mathrm{i}(\omega - ku) - \tau(\omega - ku)^2 - \nu} \, \mathrm{d}u = 0, \tag{8.2.36}$$

where we have reintroduced the notation ($u \equiv V_x$) for the velocity of the one-dimensional, unsteady wave motion.

From the above equation one derives the expression

$$1+\frac{1}{r_D^2k^2}\left[1-\sqrt{\frac{m_e}{2\pi k_B T}}\right.\\ \left.\times\int_{-\infty}^{+\infty}\frac{\{[\mathrm{i}-\tau(\omega-ku)]\omega-\nu\}\,\mathrm{e}^{-m_e u^2/2k_B T}}{\mathrm{i}(\omega-ku)-\tau(\omega-ku)^2-\nu}\,\mathrm{d}u\right]=0, \quad (8.2.37)$$

where $r_D=\sqrt{\varepsilon_0 k_B T/n_e e^2}$ is the Debye–Hueckel radius.

Introducing now the dimensionless variables

$$\begin{aligned} \hat{u}&=u\sqrt{\frac{m_e}{2k_B T}}, \qquad &\widehat{\omega}&=\omega\frac{1}{k}\sqrt{\frac{m_e}{2k_B T}},\\ \hat{\nu}&=\nu\frac{1}{k}\sqrt{\frac{m_e}{2k_B T}}, \qquad &\hat{\tau}&=\tau k\sqrt{\frac{2k_B T}{m_e}} \end{aligned} \quad (8.2.38)$$

we can rewrite Eq. (8.2.37) in the form

$$1+\frac{1}{r_D^2k^2}\left[1-\sqrt{\frac{1}{\pi}}\int_{-\infty}^{+\infty}\frac{\{[\mathrm{i}-\hat{\tau}(\widehat{\omega}-\hat{u})]\widehat{\omega}-\hat{\nu}\}\,\mathrm{e}^{-\hat{u}^2}}{\mathrm{i}(\widehat{\omega}-\hat{u})-\hat{\tau}(\widehat{\omega}-\hat{u})^2-\hat{\nu}}\,\mathrm{d}\hat{u}\right]=0. \quad (8.2.39)$$

Now consider a situation in which the denominator of the complex integrand in Eq. (8.2.39) becomes zero. The quadratic equation

$$\hat{\tau}y^2-\mathrm{i}y+\hat{\nu}=0, \quad y=\widehat{\omega}-\hat{u} \quad (8.2.40)$$

has the roots

$$\begin{aligned} y_1&=\frac{\mathrm{i}}{2\hat{\tau}}\left(1+\sqrt{1+4\hat{\tau}\hat{\nu}}\right),\\ y_2&=\frac{\mathrm{i}}{2\hat{\tau}}\left(1-\sqrt{1+4\hat{\tau}\hat{\nu}}\right). \end{aligned} \quad (8.2.41)$$

Hence, Eq. (8.2.39) can be rewritten as

$$1+\frac{1}{r_D^2k^2}\left[1+\frac{1}{\hat{\tau}\sqrt{\pi}}\int_{-\infty}^{+\infty}\frac{\{[\mathrm{i}+\hat{\tau}(\hat{u}-\widehat{\omega})]\widehat{\omega}-\hat{\nu}\}\,\mathrm{e}^{-\hat{u}^2}}{(\hat{u}-\hat{u}_1)(\hat{u}-\hat{u}_2)}\,\mathrm{d}u\right]=0, \quad (8.2.42)$$

where

$$\hat{u}_1=\widehat{\omega}-y_1, \qquad \hat{u}_2=\widehat{\omega}-y_2. \quad (8.2.43)$$

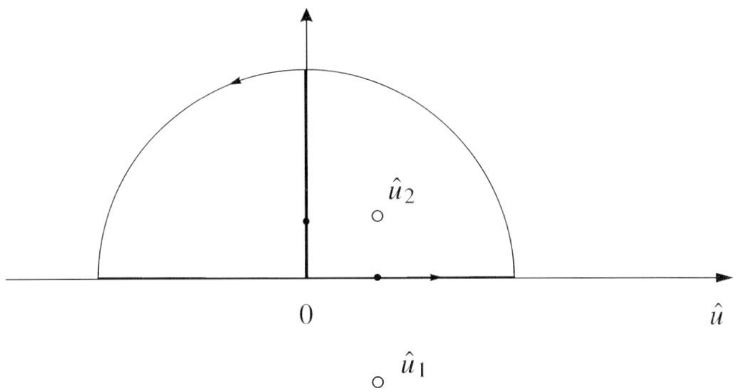

Fig. 8.1. Integration contour for evaluating complex integrals in Eq. (8.2.24). The two open circles are the possible positions of the poles $\hat{u}_1$ and $\hat{u}_2$ (see Eqs. (8.2.26) and (8.2.27)) involved in the solution of the dispersion equation in the regime of damped plasma oscillations.

Let us transform Eq. (8.2.42) to the following:

$$1+\frac{1}{r_D^2k^2}\left\{1+\frac{1}{\sqrt{\pi}}\left[\left(\frac{\mathrm{i}\hat{\nu}+0.5\widehat{\omega}}{\sqrt{1+4\hat{\tau}\hat{\nu}}}-0.5\widehat{\omega}\right)\int_{-\infty}^{+\infty}\frac{\mathrm{e}^{-\hat{u}^2}}{\hat{u}_1-\hat{u}}\,\mathrm{d}\hat{u}\right.\right.$$
$$\left.\left.-\left(\frac{\mathrm{i}\hat{\nu}+0.5\widehat{\omega}}{\sqrt{1+4\hat{\tau}\hat{\nu}}}+0.5\widehat{\omega}\right)\int_{-\infty}^{+\infty}\frac{\mathrm{e}^{-\hat{u}^2}}{\hat{u}_2-\hat{u}}\,\mathrm{d}\hat{u}\right]\right\}=0. \tag{8.2.44}$$

The last equation contains improper Cauchy type integrals, which can be evaluated using the theory of residues. Let us first analyze the conditions under which plasma waves can be damped. This requires, first, that (see Eq. (A6.4)) the imaginary part of the complex frequency fulfil the condition

$$\omega''<0 \tag{8.2.45}$$

and, second, that the poles involved in the calculation of the integrals in Eq. (8.2.24) lie in the upper half-plane (see Figure 8.1, in which the integration contour is shown).

Since

$$\hat{u}_1=\widehat{\omega}'+\mathrm{i}\left(\widehat{\omega}''-\frac{1+\sqrt{1+4\hat{\tau}\hat{\nu}}}{2\hat{\tau}}\right), \tag{8.2.46}$$

$$\hat{u}_2=\widehat{\omega}'+\mathrm{i}\left(\widehat{\omega}''-\frac{1-\sqrt{1+4\hat{\tau}\hat{\nu}}}{2\hat{\tau}}\right), \tag{8.2.47}$$

the second condition is fulfilled for the integral

$$I_2=\int_{-\infty}^{+\infty}\frac{\mathrm{e}^{-\hat{u}^2}}{\hat{u}-\hat{u}_2}\,\mathrm{d}\hat{u}$$

if the inequality

$$\widehat{\omega}'' + \frac{\sqrt{1+4\hat{\tau}\hat{\mathrm{v}}}-1}{2\hat{\tau}} > 0 \tag{8.2.48}$$

is satisfied. A similar condition for the integral

$$I_1 = \int_{-\infty}^{+\infty} \frac{\mathrm{e}^{-\hat{u}^2}}{\hat{u}-\hat{u}_1}\,\mathrm{d}\hat{u} \tag{8.2.49}$$

cannot be satisfied. For this integral, the poles lie in the lower half-plane, and hence

$$I_2 = 2\pi\mathrm{i}\,\mathrm{res}(\hat{u}=\hat{u}_2), \tag{8.2.50}$$

$$I_1 = 0. \tag{8.2.51}$$

Then Eq. (8.2.44) produces the dispersion relation, which admits a damped plasma wave solution:

$$\mathrm{e}^{\hat{u}_2^2}\frac{1+r_D^2k^2}{2\sqrt{\pi}} = \frac{\hat{\mathrm{v}}}{\sqrt{1+4\hat{\tau}\hat{\mathrm{v}}}} - \frac{\mathrm{i}\widehat{\omega}}{2}\left(1+\frac{1}{\sqrt{1+4\hat{\tau}\hat{\mathrm{v}}}}\right), \tag{8.2.52}$$

where

$$\begin{aligned}\hat{u}_2^2 &= \left[\widehat{\omega}' + \mathrm{i}\left(\widehat{\omega}'' + \frac{\sqrt{1+4\hat{\tau}\hat{\mathrm{v}}}-1}{2\hat{\tau}}\right)\right]^2 = \widehat{\omega}'^2 - \widehat{\omega}''^2 - \widehat{\omega}''\frac{\sqrt{1+4\hat{\tau}\hat{\mathrm{v}}}-1}{\hat{\tau}} \\ &\quad - \frac{1+2\hat{\tau}\hat{\mathrm{v}}-\sqrt{1+4\hat{\tau}\hat{\mathrm{v}}}}{2\hat{\tau}^2} + \mathrm{i}\left(2\widehat{\omega}'' + \frac{\sqrt{1+4\hat{\tau}\hat{\mathrm{v}}}-1}{\hat{\tau}}\right)\widehat{\omega}'.\end{aligned} \tag{8.2.53}$$

The relaxation time τ_{rel} can be estimated in terms of the mean time τ between close collisions and the Coulomb logarithm (Trubnikov, 1963):

$$\tau_{\mathrm{rel}} = \tau\Lambda^{-1}. \tag{8.2.54}$$

We can then write

$$\tau\mathrm{v} = \Lambda, \qquad \hat{\tau}\hat{\mathrm{v}} = \Lambda. \tag{8.2.55}$$

Now, in Eq. (8.2.52) we write down the complex part of the exponential in the trigonometrical form:

$$\begin{aligned}&\exp\left[-\mathrm{i}\widehat{\omega}'\left(2\widehat{\omega}'' + \frac{\sqrt{1+4\hat{\tau}\hat{\mathrm{v}}}-1}{\hat{\tau}}\right)\right] \\ &= \cos\left[\widehat{\omega}'\left(2\widehat{\omega}'' + \frac{\sqrt{1+4\hat{\tau}\hat{\mathrm{v}}}-1}{\hat{\tau}}\right)\right] - \mathrm{i}\sin\left[\widehat{\omega}'\left(2\widehat{\omega}'' + \frac{\sqrt{1+4\hat{\tau}\hat{\mathrm{v}}}-1}{\hat{\tau}}\right)\right]\end{aligned}$$

and then separate the real and imaginary parts. For the real part we have

$$\frac{1+r_D^2k^2}{2\sqrt{\pi}}\exp\left\{\widehat{\omega}'^2-\widehat{\omega}''^2-\widehat{\omega}''\hat{\mathrm{v}}\frac{\sqrt{1+4\Lambda}-1}{\Lambda}-\hat{\mathrm{v}}^2\frac{1+2\Lambda-\sqrt{1+4\Lambda}}{2\Lambda^2}\right\}$$
$$=\left[\frac{\hat{\mathrm{v}}}{\sqrt{1+4\Lambda}}+0.5\widehat{\omega}''+\frac{0.5\widehat{\omega}''}{\sqrt{1+4\Lambda}}\right]\cos\left[\widehat{\omega}'\left(2\widehat{\omega}''+\hat{\mathrm{v}}\frac{\sqrt{1+4\Lambda}-1}{\Lambda}\right)\right]$$
$$-0.5\widehat{\omega}'\left[1+\frac{1}{\sqrt{1+4\Lambda}}\right]\sin\left[\widehat{\omega}'\left(2\widehat{\omega}''+\hat{\mathrm{v}}\frac{\sqrt{1+4\Lambda}-1}{\Lambda}\right)\right]. \tag{8.2.56}$$

Similarly, for the imaginary part we find

$$0.5\widehat{\omega}'\left[1+\frac{1}{\sqrt{1+4\Lambda}}\right]\cos\left[\widehat{\omega}'\left(2\widehat{\omega}'+\hat{\mathrm{v}}\frac{\sqrt{1+4\Lambda}-1}{\Lambda}\right)\right]$$
$$+\left[\frac{\hat{\mathrm{v}}}{\sqrt{1+4\Lambda}}+0.5\widehat{\omega}''+\frac{0.5\widehat{\omega}''}{\sqrt{1+4\Lambda}}\right]$$
$$\times\sin\left[\widehat{\omega}'\left(2\widehat{\omega}''+\hat{\mathrm{v}}\frac{\sqrt{1+4\Lambda}-1}{\Lambda}\right)\right]=0. \tag{8.2.57}$$

The system of complicated transcendent equations (8.2.56), (8.2.57) can generally be solved only by computer. If, however, the Coulomb logarithm Λ is large enough for terms of order $\Lambda^{-1/2}$ to be negligible, then a common calculator will do. The system of Eqs. (8.2.56), (8.2.57) in this case simplifies to

$$\frac{1+r_D^2k^2}{\sqrt{\pi}}\mathrm{e}^{\widehat{\omega}'^2-\widehat{\omega}''^2}=\widehat{\omega}''\cos(2\widehat{\omega}'\widehat{\omega}'')-\widehat{\omega}'\sin(2\widehat{\omega}'\widehat{\omega}'')=0, \tag{8.2.58}$$

$$\widehat{\omega}'\cos(2\widehat{\omega}'\widehat{\omega}'')+\widehat{\omega}''\sin(2\widehat{\omega}'\widehat{\omega}'')=0. \tag{8.2.59}$$

Let us introduce the notation

$$\alpha=2\widehat{\omega}'\widehat{\omega}'',\qquad \beta=1+r_D^2k^2 \tag{8.2.60}$$

and note that

$$\widehat{\omega}'^2=-\frac{1}{2}\alpha\,\mathrm{tg}\,\alpha,\qquad \widehat{\omega}''^2=-\frac{1}{2}\alpha\,\mathrm{ctg}\,\alpha,\qquad \widehat{\omega}'^2-\widehat{\omega}''^2=\alpha\,\mathrm{ctg}\,2\alpha.$$

Then from Eqs. (8.2.58), (8.2.59) one finds

$$-\sin 2\alpha\,\mathrm{e}^{2\alpha\,\mathrm{ctg}\,2\alpha}=\frac{\pi}{\beta^2}\alpha.$$

Now, if one introduces the variable $\gamma=-2\alpha=-4\widehat{\omega}'\widehat{\omega}''$, the problem reduces to the transcendent equation

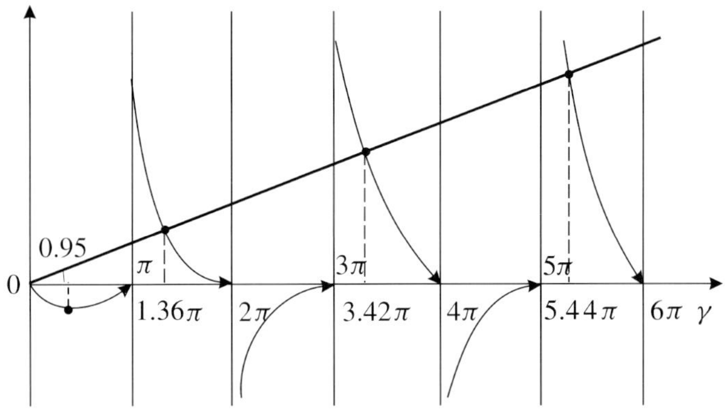

Fig. 8.2. Graphical solution of Eq. (8.2.61), producing a discrete spectrum, examine the asymptotic behavior of the dispersion relations.

$$-\mathrm{e}^{\gamma\,\mathrm{ctg}\,\gamma}\sin\gamma=\frac{\pi}{2\beta^2}\gamma, \tag{8.2.61}$$

which can be solved graphically or iteratively (Alekseev, 2000a).

The graphical solution for $\beta^2=1$ (corresponding to $r_D^2k^2\ll 1$) is illustrated in Figure 8.2, which shows that in this case a discrete plasma oscillation spectrum appears even for an unbounded medium. The first seven values of γ are $\gamma_1=1.361\pi$, $\gamma_2=3.418\pi$, $\gamma_3=5.439\pi$, $\gamma_4=7.451\pi$, $\gamma_5=9.460\pi$, $\gamma_6=11.465\pi$, $\gamma_7=13.469\pi$.

In the asymptotic limit of large, odd positive integers n ($n\geqslant 301$), we have

$$\gamma_n=\left(n+\frac{1}{2}\right)\pi. \tag{8.2.62}$$

The dimensionless frequencies $\widehat{\omega}'_n$, $\widehat{\omega}''_n$ are calculated using the formulas

$$\widehat{\omega}'_n=\frac{1}{2}\sqrt{-\gamma_n\,\mathrm{tg}\,\frac{\gamma_n}{2}}, \tag{8.2.63}$$

$$\widehat{\omega}''_n=-\frac{1}{2}\sqrt{-\gamma_n\,\mathrm{ctg}\,\frac{\gamma_n}{2}}. \tag{8.2.64}$$

Their asymptotic values are given by

$$\widehat{\omega}'_n=\frac{\sqrt{\gamma_n}}{2}=\frac{1}{2}\sqrt{\pi\left(n+\frac{1}{2}\right)}, \tag{8.2.65}$$

$$\widehat{\omega}''_n=-\frac{\sqrt{\gamma_n}}{2}=-\frac{1}{2}\sqrt{\pi\left(n+\frac{1}{2}\right)}. \tag{8.2.66}$$

For example, $\widehat{\omega}'_1=1.296$, $\widehat{\omega}''_1=-0.824$, $\widehat{\omega}'_2=1.866$, $\widehat{\omega}''_2=-1.438$ and $\widehat{\omega}'_3=2.275$, $\widehat{\omega}''_3=-1.877$.

In the classical Boltzmann kinetic theory, the search for damped wave modes in collisionless plasma leads to necessity of taking complex integrals of the type (A6.12), whose integrands have no poles in the upper half-plane (the "upper" being specified by the choice of the solution in the form (A6.4)). This problem is overcome by an artifice, the Landau rule for making a detour from below around a pole located on the real axis. We will show how the dispersion relation (8.2.42) produces results corresponding to "classical" damping in collisional and collisionless plasmas. To do this, let us

$$\begin{aligned}
&\frac{1+r_D^2k^2}{2\sqrt{\pi}}\exp\left\{\widehat{\omega}'^2-\widehat{\omega}''^2-\widehat{\omega}''\hat{\nu}\frac{\sqrt{1+4\hat{\tau}\hat{\nu}}-1}{\hat{\tau}\hat{\nu}}-\hat{\nu}^2\frac{1+2\hat{\tau}\hat{\nu}-\sqrt{1+4\hat{\tau}\hat{\nu}}}{2(\hat{\tau}\hat{\nu})^2}\right\}\\
&=\left[\frac{\hat{\nu}}{\sqrt{1+4\hat{\tau}\hat{\nu}}}+0.5\widehat{\omega}''\left(1+\frac{1}{\sqrt{1+4\hat{\tau}\hat{\nu}}}\right)\right]\\
&\quad\times\cos\left[\widehat{\omega}'\left(2\widehat{\omega}''+\hat{\nu}\frac{\sqrt{1+4\hat{\tau}\hat{\nu}}-1}{\hat{\tau}\hat{\nu}}\right)\right]\\
&\quad-0.5\widehat{\omega}'\left[1+\frac{1}{\sqrt{1+4\hat{\tau}\hat{\nu}}}\right]\sin\left[\widehat{\omega}'\left(2\widehat{\omega}''+\hat{\nu}\frac{\sqrt{1+4\hat{\tau}\hat{\nu}}-1}{\hat{\tau}\hat{\nu}}\right)\right],
\end{aligned}\tag{8.2.67}$$

$$\begin{aligned}
&0.5\widehat{\omega}'\left[1+\frac{1}{\sqrt{1+4\hat{\tau}\hat{\nu}}}\right]\cos\left[\widehat{\omega}'\left(2\widehat{\omega}''+\hat{\nu}\frac{\sqrt{1+4\hat{\tau}\hat{\nu}}-1}{\hat{\tau}\hat{\nu}}\right)\right]\\
&\quad+\left[\frac{\hat{\nu}}{\sqrt{1+4\hat{\tau}\hat{\nu}}}+0.5\widehat{\omega}''\left(1+\frac{1}{\sqrt{1+4\hat{\tau}\hat{\nu}}}\right)\right]\\
&\quad\times\sin\left[\widehat{\omega}'\left(2\widehat{\omega}''+\hat{\nu}\frac{\sqrt{1+4\hat{\tau}\hat{\nu}}-1}{\hat{\tau}\hat{\nu}}\right)\right]=0.
\end{aligned}\tag{8.2.68}$$

The passage to the case of classical collisions is achieved by proceeding to the limit $\tau\to 0$ at a fixed frequency ν. The indeterminate forms involved in the calculation are evaluated by expanding the corresponding terms in a power series of a small parameter $\hat{\tau}\hat{\nu}$ and retaining the first two terms in the expansion in Eqs. (8.2.67), (8.2.68) (in the last term in the curly brackets in Eq. (8.2.67), the quadratic term is also retained, though). We follow this procedure to give

$$\begin{aligned}
&\frac{1+r_D^2k^2}{2\sqrt{\pi}}\exp\left\{\widehat{\omega}'^2-(\widehat{\omega}''+\nu)^2\right\}\\
&=[\widehat{\omega}''+\hat{\upsilon}]\cos\left[2\widehat{\omega}'(\widehat{\omega}''+\hat{\nu})\right]-\widehat{\omega}'\sin\left[2\widehat{\omega}'(\widehat{\omega}''+\hat{\upsilon})\right],
\end{aligned}\tag{8.2.69}$$

$$\widehat{\omega}'\cos\left[2\widehat{\omega}'(\widehat{\omega}''+\hat{\nu})\right]+[\widehat{\omega}''+\hat{\nu}]\sin\left[2\widehat{\omega}'(\widehat{\omega}''+\hat{\nu})\right]=0.\tag{8.2.70}$$

Eqs. (8.2.69) and (8.2.70) can be brought to the same form as the system of Eqs. (8.2.58), (8.2.59):

$$\frac{1+r_D^2k^2}{2\sqrt{\pi}}e^{\widehat{\omega}'^2-\widehat{\omega}''^2}=\widehat{\omega}''\cos(2\widehat{\omega}'\widehat{\omega}'')-\widehat{\omega}'\sin(2\widehat{\omega}'\widehat{\omega}''),\tag{8.2.71}$$

$$\widehat{\omega}' \cos(2\widehat{\omega}'\widehat{\omega}'') + \widehat{\omega}'' \sin(2\widehat{\omega}'\widehat{\omega}'') = 0, \tag{8.2.72}$$

by replacing ω'' with the variable $\omega_1'' = \omega'' + \nu$. It should also be noted that, in the asymptotics we are considering, an additional factor 0.5 appears on the left-hand side of Eq. (8.2.71). Eqs. (8.2.71) and (8.2.72) are then solved in exactly the same manner to give (for large n):

$$\widehat{\omega}_n' = \frac{1}{2}\sqrt{\pi\left(n+\frac{1}{2}\right)}, \qquad \widehat{\omega}_n'' = -\frac{1}{2}\sqrt{\pi\left(n+\frac{1}{2}\right)} - \hat{\nu}. \tag{8.2.73}$$

The relevant pole here lies in the upper half-plane (see Figure 8.1) inside the integration contour and has the ordinate $\omega'' + v$.

We are now in a position to determine the frequency spectrum corresponding to the classical collisionless damping regime (Landau damping). For this we proceed to the limit $\nu \to 0$ in (8.2.53). Using the notation introduced in Eq. (8.2.60), it then follows from Eqs. (8.2.51) and (8.2.52) that

$$-\sin 2\alpha\, e^{2\alpha \operatorname{ctg} 2\alpha} = \frac{4\pi}{\beta^2}\alpha. \tag{8.2.74}$$

If we introduce the variable $\gamma = -2\alpha = -4\widehat{\omega}'\widehat{\omega}''$, then the problem reduces to the transcendent equation

$$-e^{\gamma \operatorname{ctg} \gamma} \sin\gamma = \frac{2\pi}{\beta^2}\gamma \tag{8.2.75}$$

which can be solved graphically or iteratively. The graphical solution for the $\beta^2 = 1$, corresponding to the long-wave length limit $r_D^2 k^2 \ll 1$, is illustrated in Figure 8.3. This

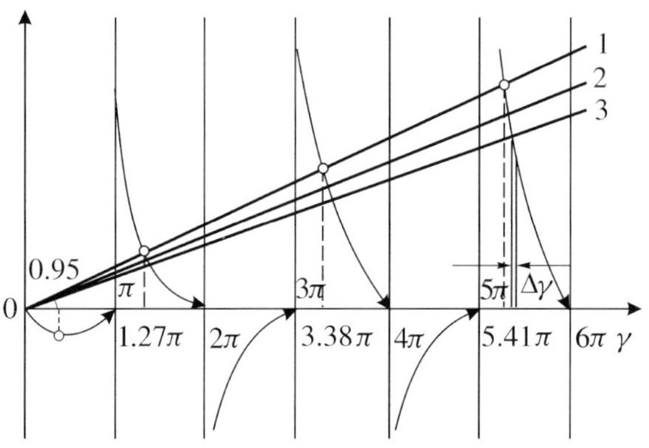

Fig. 8.3. Graphical solution of the dispersion equation (8.2.75).

solution shows that in the case of interest a discrete oscillation spectrum appears even for an unbounded medium. The first seven values of γ are

$$\gamma_1 = 1.271\pi, \quad \gamma_2 = 3.379\pi, \quad \gamma_3 = 5.410\pi, \quad \gamma_4 = 7.432\pi,$$
$$\gamma_5 = 9.444\pi, \quad \gamma_6 = 11.452\pi, \quad \gamma_7 = 13.458\pi.$$

In the asymptotic limit of large, odd positive integers n ($n > 301$) we have

$$\gamma_n = \left(n + \frac{1}{2}\right)\pi. \tag{8.2.76}$$

The dimensionless frequencies $\widehat{\omega}'_n$, $\widehat{\omega}''_n$ are calculated using (8.2.43), (8.2.44), and their dimensional forms are represented as

$$\omega'_n = k\sqrt{-\frac{k_B T}{2m_e}\gamma_n \operatorname{tg}\frac{\gamma_n}{2}}, \qquad \omega''_n = -k\sqrt{-\frac{k_B T}{2m_e}\gamma_n \operatorname{ctg}\frac{\gamma_n}{2}}. \tag{8.2.77}$$

The asymptotic values of the frequencies are

$$\begin{aligned} \widehat{\omega}'_n &= \frac{\sqrt{\gamma_n}}{2} = \frac{1}{2}\sqrt{\pi\left(n + \frac{1}{2}\right)}, \\ \widehat{\omega}''_n &= -\frac{\sqrt{\gamma_n}}{2} = -\frac{1}{2}\sqrt{\pi\left(n + \frac{1}{2}\right)}. \end{aligned} \tag{8.2.78}$$

It is also worthwhile to list the first seven pairs of dimensionless frequencies (Alekseev, 2001):

$$\begin{aligned} &\widehat{\omega}'_1 = 1.484, \quad \widehat{\omega}''_1 = -0.673, \quad \widehat{\omega}'_2 = 1.979, \quad \widehat{\omega}''_2 = -1.341, \\ &\widehat{\omega}'_3 = 2.379, \quad \widehat{\omega}''_3 = -1.786, \quad \widehat{\omega}'_4 = 2.691, \quad \widehat{\omega}''_4 = -2.169, \\ &\widehat{\omega}'_5 = 2.975, \quad \widehat{\omega}''_5 = -2.493, \quad \widehat{\omega}'_6 = 3.235, \quad \widehat{\omega}''_6 = -2.780, \\ &\widehat{\omega}'_7 = 3.473, \quad \widehat{\omega}''_7 = -3.043. \end{aligned} \tag{8.2.79}$$

Let us investigate the short wave range approximation and consider the dispersion equation (8.2.75) which can be written in the form

$$-e^{\gamma \operatorname{ctg}\gamma} \sin\gamma = \xi\gamma, \tag{8.2.80}$$

where

$$\xi = \frac{2\pi}{\beta^2} = \frac{2\pi}{(1 + r_D^2 k^2)^2}. \tag{8.2.81}$$

In short wave approximation, the parameter ξ is very small $\xi \sim \lambda^4$ (here λ is the wave length) and, as before, we obtain the discrete character of solutions γ_n of Eq. (8.2.80). From the first glance all solutions will be contained in intervals

$$\frac{3}{2}\pi + 2\pi(n-1) < \gamma_n < 2\pi n, \quad n = 1, 2, \ldots.$$

But asymptotic behavior of the function $-e^{\gamma \operatorname{ctg} \gamma} \sin \gamma$ is rather complicated and it is better to use a direct numerical procedure of solution for Eq. (8.2.80) even by small ξ.

Let us consider the concrete examples of calculations. Suppose that $\xi = 0.001$, it corresponds to $r_D k = 5.62$. Even for so small value of coefficient ξ the difference of γ_1 from asymptotic value $\frac{3}{2}\pi$ is large, $\gamma_1 = 1.74\pi$. We introduce ε_n as a measure of difference between γ_n and its asymptotic value

$$\varepsilon_n = \gamma_n - 2\pi(n-1) - \frac{3}{2}\pi, \quad n = 1, 2, 3, \ldots. \tag{8.2.82}$$

Tables 8.1 and 8.2 contain γ_n, $\widehat{\omega}'_n$, $\widehat{\omega}''_n$, ε_n, $n = 1, 2, 3, \ldots, 7$, for $\xi = 0.001$ and $\xi = 0.0001$ correspondingly.

Now one computes the asymptotic solutions of Eq. (8.2.80). The solutions γ_n can be written in the form

$$\gamma_n = \varepsilon_n + 2\pi(n-1) + \frac{3}{2}\pi. \tag{8.2.83}$$

We investigate the $\varepsilon(n)$ dependence for large n, considering $\varepsilon(n)$ as a continuous function. Using relations

$$\sin\gamma = -\cos\varepsilon, \qquad \cos\gamma = \sin\varepsilon, \tag{8.2.84}$$

we rewrite (8.2.80)

$$\cos\varepsilon\, e^{-[(3/2)\pi+\varepsilon+2\pi(n-1)]\operatorname{tg}\varepsilon} = \xi\left[\frac{3}{2}\pi + \varepsilon + 2\pi(n-1)\right]. \tag{8.2.85}$$

After differentiating (8.2.85), it turns out

Table 8.1
$\xi = 0.001$

n	γ_n	$\widehat{\omega}'_n$	$\widehat{\omega}''_n$	ε_n
1	1.739π	0.770	−1.773	0.239π
2	3.611π	1.409	−2.012	0.111π
3	5.572π	1.866	−2.345	0.072π
4	7.556π	2.230	−2.661	0.056π
5	9.540π	2.570	−2.915	0.040π
6	11.529π	2.875	−3.150	0.029π
7	13.521π	3.153	−3.368	0.021π

Table 8.2
$\xi = 0.0001$

n	γ_n	$\widehat{\omega}'_n$	$\widehat{\omega}''_n$	ε_n
1	1.771π	0.723	−1.923	0.271π
2	3.675π	1.271	−2.270	0.175π
3	5.611π	1.757	−2.588	0.111π
4	7.575π	2.166	−2.747	0.075π
5	9.561π	2.488	−3.018	0.061π
6	11.553π	2.771	−3.275	0.053π
7	13.548π	3.024	−3.518	0.048π

$$\frac{d\varepsilon}{dn} = -\frac{2\pi\{\xi + \sin\varepsilon\, e^{-[(3/2)\pi+\varepsilon+2\pi(n-1)]\operatorname{tg}\varepsilon}\}}{\{2\sin\varepsilon + [(3/2)\pi + \varepsilon + 2\pi(n-1)](1/\cos\varepsilon)\}\, e^{-[(3/2)\pi+\varepsilon+2\pi(n-1)]\operatorname{tg}\varepsilon} + \xi}. \tag{8.2.86}$$

Eq. (8.2.86) is simplified after using the relations (8.2.80), (8.2.83), (8.2.84):

$$\frac{d\varepsilon}{dn} = -\frac{2\pi \sin\gamma(\sin\gamma - \gamma\cos\gamma)}{\sin^2\gamma - (2\sin\gamma\cos\gamma - \gamma)\gamma}. \tag{8.2.87}$$

For small ε and large n, Eq. (8.2.87) is written

$$n\frac{d\varepsilon}{dn} + \varepsilon = -\frac{1}{2\pi n} \tag{8.2.88}$$

and after integration,

$$\varepsilon = -\frac{\ln n}{4\pi n}. \tag{8.2.89}$$

Strictly speaking – taking into account the general solution for homogeneous equation – the indicated solution is of a more complicated form

$$\varepsilon = \left(C - \frac{\ln n}{4\pi}\right)\frac{1}{n}, \tag{8.2.90}$$

where C is constant which can be estimated from the condition: $\varepsilon_1 = C$ by $n = 1$.

Because $\gamma_1 = \frac{3}{2}\pi + \varepsilon_1$, and $\gamma_1 = 1.74\pi$ by $\xi = 0.001$, then $\varepsilon_1 = 0.24\pi$ and $C = 0.24\pi \approx 0.75$. Of course it is a rough estimation, the asymptotic theory being valid only for large n. Nevertheless, this estimation reflects two facts: (1) positive ε can be included in asymptotic formula for ε; (2) for large n the solution of mentioned above equation is not significant. It turns out therefore that for large n the difference ε becomes negative and solution γ is placed out of interval $(\frac{3}{2}\pi + 2\pi(n-1), 2\pi n)$. The interval $\frac{3}{2}\pi + 2\pi(n-1) < \gamma_n < 2\pi n$ contains γ_n if n are not too large. Notice also that $\varepsilon \to 0$

by $n \to \infty$. The integral number n can be estimated when the difference ε becomes negative. It follows from (8.2.85) that by $\varepsilon = 0$, this number n is defined by entire part of value

$$n = \left[\frac{1}{4} + \frac{1}{2\pi\xi}\right]. \tag{8.2.91}$$

Let us estimate also the extreme of function $\varepsilon(n)$ by the large n. It follows from (8.2.87) that the extreme condition has the form

$$\xi + \sin\varepsilon\, e^{-[(3/2)\pi+\varepsilon+2\pi(n-1)]\,\mathrm{tg}\,\varepsilon} = 0. \tag{8.2.92}$$

The asymptotic solution of this equation does not depend on ξ and looks as

$$\varepsilon = -\frac{1}{2\pi n}. \tag{8.2.93}$$

In conclusion we notice that Eq. (8.2.80) is valid – in the frame of formulated assumptions – in an entire interval of ξ definition, $0 < \xi < 2\pi$.

8.3. Generalized dispersion relations for plasma: theory and experiment

In this section, theoretical results are discussed in the context of Looney and Brown's experiments (Looney and Brown, 1954) on the detection of plasma waves excited by an electron beam. The experimental setup of Looney and Brown (see Figure 8.4) consists of a bulb about 10 cm in diameter, in which mercury plasma at a pressure as low as 3×10^{-3} mm Hg was created by electric discharge between two cathodes C and an anode ring A. An electron beam produced in a lateral tap was accelerated by a voltage of several hundred volts and then introduced into the plasma through a hole of diameter 1 mm in the bulb wall. In the region between the accelerating anode A and the anode disk D, with a separation of 1.5 cm, an ion cloud formed. The beam electrons excited oscillations in the region AD.

The oscillations were registered by a movable probe attached to the detector. The results of the experiment are presented in Figure 8.5, reproduced from the Looney and Brown's paper. Because the density of the electron beam is proportional to the discharge current, Looney and Brown presented their results as the dependences of the oscillation frequency squared on the electron number density n_e. The inset on the left shows the electric field distribution over the region AD, and the dashed straight line corresponds to the dispersion relation

$$\omega = \omega_{pe}, \tag{8.3.1}$$

where ω_{pe} is the Langmuir frequency of plasma electron oscillations. Eq. (8.3.1) follows from the one-dimensional hydrodynamic equation of motion without considering

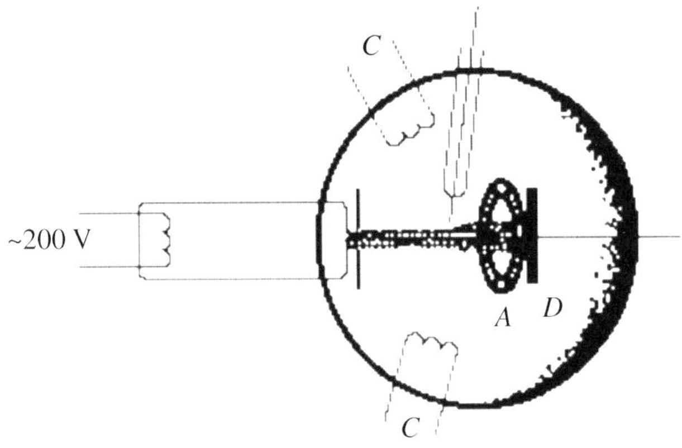

Fig. 8.4. Schematic of Looney and Brown's experiment on the excitation of plasma oscillations.

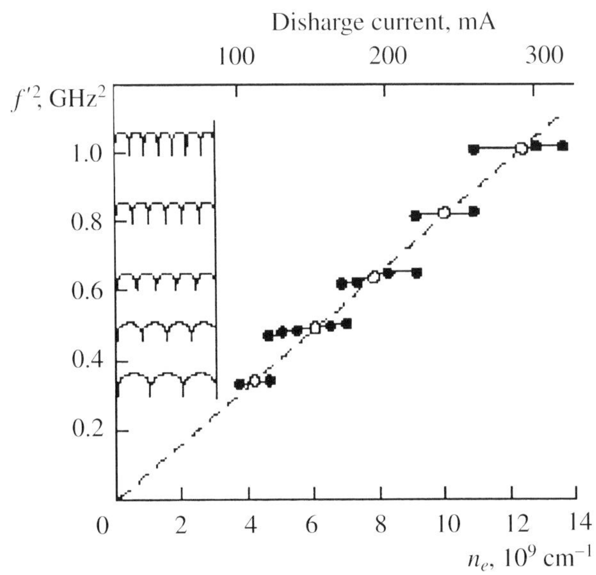

Fig. 8.5. Square of the observed frequency versus plasma electron number density, as measured in the Looney–Brown experiment. The inset shows the oscillations observed in the anode gap AD.

convective terms and the pressure gradient. In a more general form including the electron pressure, this equation becomes

$$\omega^2 - \omega_{pe}^2 = \frac{3}{2}k^2 v_T^2, \tag{8.3.2}$$

where $v_T^2 = 2k_B T/m_e$. The term on the right-hand side of Eq. (8.3.2) was dropped in constructing the dashed line in Figure 8.5 in order to achieve a better agreement between the theoretical and experimental results (Looney and Brown, 1954; Chen, 1984).

Looney and Brown noted the fundamental disagreement between the dispersion relations (8.3.1) and (8.3.2), which lead to a continuous oscillation spectrum, and the experimental data displaying a discrete oscillation spectrum. Furthermore, as the electron density increases, one can see from Figure 8.5 that the curve grows stepwise, with discontinuities and a slight increase in the slope of the steps within the confines of the plateau, i.e., the oscillation spectrum one observes is in fact of a band type. Band spectral structures were also seen in later experiments on the damping of electron waves in collisionless plasmas (see, for instance, Franklin, Hamberger and Smith, 1972). Neither Eqs. (8.3.1), (8.3.2) nor qualitative considerations based upon the theory of standing waves can explain these experiments. We proceed now to the interpretation of the experimental data based on the generalized dispersion relations (8.2.71), (8.2.72). We note from the beginning that a study on the level of dispersion relations is inadequate to give a complete picture of processes occurring in the system under study here. Therefore, the calculation of $\omega(\lambda)$ only reflects major qualitative and quantitative features of the system – provided we know the wavelengths of the observed waves and the hydrodynamic parameters, primarily the concentrations of the components and the ion and electron temperatures.

A. The dispersion relation (8.2.75) produces a discrete spectrum of solutions γ_n and, hence, of $\omega_n'(\lambda_n, T)$, $\omega_n''(\lambda_n, T)$. The discrete frequency spectrum is observed in experiment. To proceed to quantitative estimates, however, it is necessary to first estimate the beam temperature. From the requirement that the theoretical value of the square of the linear frequency

$$f_1'^2 = \widehat{\omega}_1'^2 \frac{2k_B T}{\lambda_1^2 m_e} \tag{8.3.3}$$

be equal to its experimental value ($f_1'^2 = 3 \times 10^{17}$ Hz^2), we have $T \approx 40$ eV. While the temperature was not measured directly in the experiment, there is indirect experimental evidence to support this value. Based on the parameters of the experiment, the wavelength $\lambda_1 \approx 1$ cm and $r_D^2 k^2 \approx 0.2$ for the lower frequency level. Consequently, this experiment fails to satisfy the conditions

$$r_D^2 k^2 \ll 1, \qquad \lambda \gg r_D,$$

which formally underlie the Landau-damping solution of the classical dispersion relation (Klimontovich, 1964). The solutions of Eq. (8.2.75) presented above has been found for the limiting case

$$\beta^2 = 1. \tag{8.3.4}$$

Note, however, that the region where the straight line $2\pi\gamma$ intersects the curve $\Phi(\gamma) = -e^{\gamma \operatorname{ctg} \gamma} \sin\gamma$ is that of the steep rise of the function $\Phi(\gamma)$ (see Figure 8.3). Consequently, varying the slope of this line – in the region of existence of solutions of Eq. (8.2.75) – has little effect on the solution γ_n. To estimate

$$f_n'^2 = \widehat{\omega}_n'^2 \frac{2k_B T}{\lambda_n^2 m_e} \tag{8.3.5}$$

one should take the calculated values of $\widehat{\omega}_n'^2$ and the experimental values of λ_n and T. The electron temperatures in experiments yielding the frequencies f_n' can be estimated from the simplest form of the beam equilibrium condition

$$p_e = p_{pl}, \tag{8.3.6}$$

where p_e is the pressure produced by the beam electrons, and p_{pl} is that of the mercury plasma. As a result, the quantitative agreement between the theory and experiment is quite reasonable (to within 20–50%) for the second to the fifth of the observed levels $f_n'^2$.

B. Under the condition (8.3.4), the straight line l in Figure 8.3 has the maximum slope possible. The straight line 2 corresponds to a certain nonzero value of $r_D^2 k^2$ and is drawn for illustrative purposes. Now suppose that the concentration n_e of the beam electrons begins to increase, whereas other plasma parameters remain, to a first approximation, unchanged. Increasing n_e reduces (see Eq. (A5.5)) the Debye–Hueckel radius r_D and increases the slope of the straight line 2 which now takes position 3. The straight line approaches position 1. Instead of a certain discrete set of γ_n we will have a set of possible intervals $\Delta\gamma_n$, and hence of intervals $\Delta\omega_n'$, $\Delta\omega_n''$, giving rise to the plateau regions of the experimentally revealed values of $\omega_n'^2$.

C. It is easily verified by direct calculation that the function

$$F(\gamma_n) = \frac{\gamma_n}{4}\,\mathrm{tg}\left(-\frac{\gamma_n}{2}\right) \tag{8.3.7}$$

increases with decreasing γ_n. Hence, within the confines of a plateau the square of the frequency $\omega_n'^2$ will grow slightly with concentration of the beam electrons:

$$\omega_{n,\gamma-\Delta\gamma}'^2 = \omega_{n,\gamma}'^2 + \mathrm{o}\big(|\Delta\gamma_n|\big). \tag{8.3.8}$$

This effect is also observed in experiment.

D. The square of the oscillation frequency of plasma waves, $\omega_n'^2$, is proportional to the wave energy. Hence, the energy of plasma waves is quantized, and as n grows we have the asymptotic expression

$$\widehat{\omega}_n'^2 = \frac{\pi}{4}\left(n + \frac{1}{2}\right) \tag{8.3.9}$$

and the squares of possible dimensionless frequencies become equally spaced:

$$\widehat{\omega}_{n+1}'^2 - \widehat{\omega}_n'^2 = \frac{\pi}{4}. \tag{8.3.10}$$

E. Let us see how the motion of ions affects the solution of the dispersion equation. The velocity distribution of electrons is taken to be Maxwellian; the thermal motion of ions is neglected. The ion distribution function is then written as

$$f_i = n_i\delta(\mathbf{v}_i). \tag{8.3.11}$$

The generalized dispersion relation in the collisionless limit becomes

$$1+\frac{1}{r_D^2k^2}\left[1+2\pi\mathrm{i}\widehat{\omega}\mathrm{e}^{-\widehat{\omega}'^2+\widehat{\omega}''^2-2\mathrm{i}\widehat{\omega}'\widehat{\omega}''}\right]=\frac{\omega_{pi}^2}{\omega^2}, \tag{8.3.12}$$

where ω_{pi} is the Langmuir ion frequency. Introducing the parameter

$$\varepsilon=\frac{m_e n_i}{2m_i n_e}, \tag{8.3.13}$$

Eq. (8.3.12) is written as

$$1+r_D^2k^2+2\pi\mathrm{i}\widehat{\omega}\mathrm{e}^{-\widehat{\omega}'^2+\widehat{\omega}''^2-2\mathrm{i}\widehat{\omega}'\widehat{\omega}''}=\frac{\varepsilon}{\widehat{\omega}^2}. \tag{8.3.14}$$

Under the conditions of the experiment being discussed, ε is a small quantity. In the general case, however – if $n_i \succ\succ n_e$ and if the parameter ε is not too small – it may happen that the ion motion must be accounted for. Eq. (8.3.14) can be solved perturbatively by expanding the frequencies in power series:

$$\widehat{\omega}'=\sum_{k=0}^{\infty}\varepsilon^k\widehat{\omega}'^{(k)}, \tag{8.3.15}$$

$$\widehat{\omega}''=\sum_{k=0}^{\infty}\varepsilon^k\widehat{\omega}''^{(k)}. \tag{8.3.16}$$

Then for $k=0$, in the first approximation, we arrive at a special case of Eqs. (8.2.69), (8.2.70) as a result of separating the real and imaginary parts of Eq. (8.3.14).

In the second approximation ($k=1$), we have

$$\widehat{\omega}'^{(1)}\left[\cos\frac{\gamma_0}{2}-\frac{\gamma_0}{2}\operatorname{ctg}\frac{\gamma_0}{2}\right]+\widehat{\omega}''^{(1)}\left[\sin\frac{\gamma_0}{2}+\frac{\gamma_0}{2}\right]$$
$$=\frac{\sin^2\gamma_0}{\gamma_0\sqrt{\pi}}\mathrm{e}^{(\gamma_0/2)\operatorname{ctg}\gamma_0}-\widehat{\omega}'^{(0)}, \tag{8.3.17}$$

$$\widehat{\omega}'^{(1)}\left[\frac{\beta}{\sqrt{\pi}}\widehat{\omega}'^{(0)}\mathrm{e}^{(\gamma_0/2)\operatorname{ctg}\gamma_0}-\frac{\gamma_0}{2}-\sin\frac{\gamma_0}{2}\right]$$
$$-\widehat{\omega}''^{(1)}\left[\frac{\beta}{\sqrt{\pi}}\widehat{\omega}''^{(0)}\mathrm{e}^{(\gamma_0/2)\operatorname{ctg}\gamma_0}+\cos\frac{\gamma_0}{2}+\frac{\gamma_0}{2}\operatorname{tg}\frac{\gamma_0}{2}\right]$$
$$=\frac{\sin 2\gamma_0}{2\gamma_0\sqrt{\pi}}\mathrm{e}^{(\gamma_0/2)\operatorname{ctg}\gamma_0}+\widehat{\omega}''^{(0)}, \tag{8.3.18}$$

where γ_0 is the first-order approximation to the solution of Eq. (8.3.14). We can estimate the magnitude of the effect by giving the second-order results for the coefficients in the expansions of (8.3.15), (8.3.16):

$$\widehat{\omega}'^{(1)} = 3.05, \qquad \widehat{\omega}''^{(1)} = -0.87.$$

The complete solution of the boundary value problem concerning the spatial and temporal evolution of plasma in the Looney–Brown apparatus can only be obtained by solving the GBE. Still, the dispersion relation we have considered reflects correctly the essential features of the experimental results.

We now proceed to compare the theoretical results with those of the computer experiment. Extensive simulations of the attenuation of Langmuir waves in plasma have been performed at the SB RAS Institute of Nuclear Physics (Novosibirsk) in the 1970s and 1980s (see, for example, Buchel'nikova and Matochkin, 1977a, 1977b, 1979, 1980). Of interest to us here is the formulation of the problem close to the classical Landau formulation (Landau, 1941a, 1946a, 1946b). The problem involves a one-dimensional plasma system subject to periodic boundary conditions. The velocity distribution of plasma electrons is taken to be Maxwellian, and ions are assumed to be at rest ($m_i/m_e = 10^4$) and distributed uniformly over the length of the system. It is also assumed that at some initial point in time the system is subjected to small electron velocity and electron density perturbations of the form

$$\begin{aligned} \frac{\delta n}{n_0} &= \frac{k_0 E_0}{4\pi e n_0} \sin(\omega_0 t - k_0 x), \\ \delta v &= \frac{\omega_0 E_0}{4\pi e n_0} \sin(\omega_0 t - k_0 x), \end{aligned} \tag{8.3.19}$$

corresponding to the linear wave

$$E(x,t) = E_0 \sin(\omega_0 t - k_0 x),$$

where $\omega_0^2 = \omega_{pe}^2 + \frac{3}{2}k_0^2 v_T^2$, $k_0 = 2\pi/\lambda_0$. The quantities E_0, φ_0, λ_0, ω_0, and k_0 are the initial values of the field amplitude, potential, wavelength, frequency, and wave number, respectively. The numerical integration is performed using the particles-in-cells method. The number of particles is not large (in Buchel'nikova and Matochkin, 1977a, 1977b, 1979, 1980, the authors usually put $N = 10^4$, with about 10^2 particles per cell). To reduce the initial noise level, the easy-start method is used, in which neither the coordinate nor velocity distribution functions of the electrons change from one cell to another. In this case, it was noted in Buchel'nikova and Matochkin (1979) that the noise level is determined by computation errors but increases for the computation scheme chosen; the noise level increases with increasing E_0 and with decreasing λ_0. The computation only makes sense until the noise level remains small compared to the harmonics of the effect under study that arise in the calculation.

The calculations in Buchel'nikova and Matochkin (1977a, 1977b, 1979, 1980) were performed over a wide range of initial wave parameters. The time dependence of the field strength is quite complicated, but the initial stage always corresponds to the wave damping regime in which an increase in the amplitude and the phase velocity v_ϕ in the range $e\varphi_0/(k_B T) \succ 1$ (and the corresponding decrease in the parameter $k_0 r_D$) dramatically increases the damping decrement compared to the Landau value (Landau, 1946b).

In theory and mathematical experiments the question can be put about the number of trapped electrons by travelling wave in electric field. But the mathematical expectation of the number of electrons moving with the phase velocity is zero. Then it is possible to speak only about a definite velocity interval, which contains phase velocity of wave, and about electrons (or probe particles in mathematical experiments) belonging to this interval.

In mentioned mathematical experiments they use also another possibility of calculation of the trapped particles. The mathematical experiments were realized with the initial number of particles in cell, which was not more than 100, and maximum initial velocity which was less than $2.15 v_T$, where v_T is the thermal electron velocity. In experiments (see also Table 8.3) the phase velocity was changed in the range of $2.46 v_T$ to $22.4 v_T$. Then it is possible to calculate the number of trapped probe particles the velocity of which – in the process of evolution of the physical system – becomes more than phase velocity v_ϕ. In the beginning all high energetic electrons in the tail of the Maxwell distribution function are cutting off. But the number of these "lost" electrons (estimation of authors of experiments) is not more than $\sim$1.6%.

From Table 8.3 it follows that the number of trapped electrons is not more than 20%. On the macroscopic level of the system description we can speak only about average (hydrodynamic) velocity $\bar{u}$ of electrons and about difference between $\bar{u}$ and v_ϕ.

Table 8.3 summarizes the results of the numerical simulation series I-1 to I-8 (Buchel'nikova and Matochkin, 1979, 1980). In the leftmost column of the table, the following definitions need some clarification: $\widehat{E}_0 = E_0 e \tau_{0p}^2/(m r_D)$ is the dimensionless, normalized wave amplitude; τ_{0p} is the period of plasma oscillations; $\Delta N/N$ is the fraction of trapped electrons (in %); γ is the damping decrement determined in the simulation experiment; $\gamma_A = -\omega_1''$ is the damping decrement found as the asymptotics of the solution to the generalized Boltzmann equation and calculated from the first decrement involved in the discrete spectrum of solutions. Table 8.3 also presents the damping decrement $\gamma_L = -\omega_L''$ calculated from a modified Landau formula (Clemmow and Dougherty, 1990). There is strong disagreement between the decrements γ, γ_A and γ_L in the long wavelength limit $k_0 r_D \prec\prec 1$.

This disagreement is easy to explain from the computational point of view. Let us write the Landau formula in its classic form

$$\frac{\gamma}{\omega_{pe}} = \sqrt{\frac{\pi}{8}} \frac{1}{k_0^3 r_D^3} \mathrm{e}^{-1/(k_0^2 r_D^2)}. \tag{8.3.20}$$

For $k_0 r_D \prec\prec 1$, the damping decrement calculated by Eq. (8.3.20) becomes very small, whereas its simulation counterpart does not differ much from the plasma frequency. Applying the GBE asymptotics to the solution of classical problem of Landau damping makes it possible, even in Landau's linear formulation, to obtain quite a satisfactory agreement with both physical and mathematical experiments.

Table 8.3
Comparison of analytical predictions with numerical simulation results

N	I-1	I-2	I-3	I-4	I-5	I-6	I-7	I-8
$v_\phi/\sqrt{\frac{k_BT}{m}}$	2.46	2.95	4.2	6.9	9.4	11.2	16	22.4
λ_0/r_D	11	15	24	42	58	70	100	140
k_0r_D	0.57	0.42	0.26	0.15	0.11	0.09	0.063	0.045
$(k_0r_D)^2$	3.3×10^{-1}	1.7×10^{-1}	6.8×10^{-2}	2.2×10^{-2}	1.2×10^{-2}	8×10^{-3}	3.9×10^{-3}	2×10^{-3}
$\widehat{E}_0$	1–60	11–60	26–70	70–170	119–250	170–250	240–450	333–591
$\sqrt{\frac{e\varphi_0}{m}}/\sqrt{\frac{k_BT}{m}}$	0.2–1.6	0.8–1.9	1.6–2.6	3.5–5.4	5.3–7.6	6.9–8.4	9.8–13.5	13.7–18.3
$\sqrt{\frac{e\varphi_0}{m}}/v_\phi$	8×10^{-2}–0.8	0.28–0.65	0.38–0.62	0.5–0.78	0.56–0.81	0.61–0.75	0.61–0.84	0.61–0.82
$e\varphi_0/k_BT$	4×10^{-2}–2.7	0.7–3.6	2.5–6.8	11.9–28.5	28–58.4	48–70.5	96.7–181	188–334
$\Delta N/N$, %	0–20	1–13	1–7	1–11	1–12	2–7.5	1–13	0.5–10
γ_L/ω_{pe}	0.32	0.17	3×10^{-3}	10^{-8}	4×10^{-17}	2×10^{-25}	6×10^{-53}	2×10^{-105}
γ/ω_{pe}	0.32–1	0.17–0.96	0.03–0.4	0.03–0.65	0.03–0.8	0.04–0.2	0.03–0.8	0.02–0.3
γ_A/ω_{pe}	0.522	0.4	0.247	0.143	0.105	0.0857	0.06	0.0428
γ/γ_L	1.0–3.4	1.0–5.6	5–60	$\sim 10^6$–10^8	$\sim 10^{15}$–10^{16}	$\sim 10^{23}$–10^{24}	$\sim 10^{50}$–10^{52}	$\sim 10^{102}$ –10^{104}

For investigation of plasma parameter perturbations Landau used, in the long wave approximation, the expansion (see also Klimontovich, 1964) of the function $1/(\omega - ku)$ in a complex power series

$$\frac{1}{\omega - ku} = \frac{1}{\omega}\left[1 + \frac{ku}{\omega} + \left(\frac{ku}{\omega}\right)^2 + \left(\frac{ku}{\omega}\right)^3 + \cdots\right]. \tag{8.3.21}$$

Only two first terms were used by Landau in the following calculations. The question about convergence of series (8.3.21) by all possible values of u, ω was not considered. The results of direct mathematical simulation can be considered as evidence of divergence of series (8.3.21) in domain of definition of u, ω.

Now let the ions be at rest, and the electron component have a Maxwellian velocity distribution. As a result, the dispersion equation – with taking into account the generalized Maxwell equation – can be written as

$$1 + \frac{A}{r_D^2 k^2}\left[1 - \sqrt{\frac{m_e}{2\pi k_B T}} \times \int_{-\infty}^{+\infty} \frac{\{[\mathrm{i} - \tau(\omega - ku)]\omega - \nu\}\,\mathrm{e}^{-m_e u^2/2k_B T}}{\mathrm{i}(\omega - ku) - \tau(\omega - ku)^2 - \nu}\,\mathrm{d}u\right] = 0, \tag{8.3.22}$$

where

$$A = 1 - \hat{\tau}\widehat{\omega}'' + \mathrm{i}\hat{\tau}\left(\widehat{\omega}' - \hat{\hat{u}}\right). \tag{8.3.23}$$

After integrating in (8.3.22) we have

$$\frac{A + r_D^2 k^2}{2\sqrt{\pi}} = A\,\mathrm{e}^{-w_2^2}\left[\frac{\hat{\nu}}{\sqrt{1 + 4\hat{\tau}\hat{\nu}}} - 0.5\widehat{\omega}\mathrm{i}\left(1 + \frac{1}{\sqrt{1 + 4\hat{\tau}\hat{\nu}}}\right)\right], \tag{8.3.24}$$

or

$$\begin{aligned} &\frac{A + r_D^2 k^2}{2\sqrt{\pi}} \exp\left\{\widehat{\omega}'^2 - \widehat{\omega}''^2 - \widehat{\omega}'\frac{\sqrt{1 + 4\hat{\tau}\hat{\nu}} - 1}{\hat{\tau}} - \frac{1 + 2\hat{\tau}\hat{\nu} - \sqrt{1 + 4\hat{\tau}\hat{\nu}}}{2\hat{\tau}^2}\right\} \\ &= A\left[\frac{\hat{\nu}}{\sqrt{1 + 4\hat{\tau}\hat{\nu}}} - \mathrm{i}\frac{1}{2}(\widehat{\omega}' + \mathrm{i}\widehat{\omega}'')\left(1 + \frac{1}{\sqrt{1 + 4\hat{\tau}\hat{\nu}}}\right)\right] \\ &\times \left[\cos\left(2\widehat{\omega}'' + \frac{\sqrt{1 + 4\hat{\tau}\hat{\nu}} - 1}{\hat{\tau}}\right)\widehat{\omega}' \right. \\ &\left. - \mathrm{i}\sin\left(2\widehat{\omega}'' + \frac{\sqrt{1 + 4\hat{\tau}\hat{\nu}} - 1}{\hat{\tau}}\right)\widehat{\omega}'\right]. \end{aligned} \tag{8.3.25}$$

The separation of real and imaginary parts of the dispersion equation (8.3.25) leads to the following relations. A real part:

$$\begin{aligned}
&\frac{1+r_D^2k^2-\hat{\tau}\widehat{\omega}''}{2\sqrt{\pi}}\\
&\quad\times\exp\left\{\widehat{\omega}'^2-\widehat{\omega}''^2-\widehat{\omega}''\frac{\sqrt{1+4\hat{\tau}\hat{\mathrm{v}}}-1}{\hat{\tau}}-\frac{1+2\hat{\tau}\hat{\mathrm{v}}-\sqrt{1+4\hat{\tau}\hat{\mathrm{v}}}}{2\hat{\tau}^2}\right\}\\
&=(1-\hat{\tau}\widehat{\omega}'')\left\{\left[\frac{\hat{\mathrm{v}}}{\sqrt{1+4\hat{\tau}\hat{\mathrm{v}}}}+0.5\widehat{\omega}''+0.5\widehat{\omega}''\frac{1}{\sqrt{1+4\hat{\tau}\hat{\mathrm{v}}}}\right]\right.\\
&\quad\times\cos\left[\widehat{\omega}'\left(2\widehat{\omega}''+\frac{\sqrt{1+4\hat{\tau}\hat{\mathrm{v}}}-1}{\hat{\tau}}\right)\right]\\
&\quad\left.-0.5\widehat{\omega}'\left(1+\frac{1}{\sqrt{1+4\hat{\tau}\hat{\mathrm{v}}}}\right)\sin\left[\widehat{\omega}'\left(2\widehat{\omega}''+\frac{\sqrt{1+4\hat{\tau}\hat{\mathrm{v}}}-1}{\hat{\tau}}\right)\right]\right\}\\
&\quad-\hat{\tau}(\widehat{\omega}'-\hat{\bar{u}})\left\{-0.5\widehat{\omega}'\left(1+\frac{1}{\sqrt{1+4\hat{\tau}\hat{\mathrm{v}}}}\right)\right.\\
&\quad\times\cos\left[\left(2\widehat{\omega}''+\frac{\sqrt{1+4\hat{\tau}\hat{\mathrm{v}}}-1}{\hat{\tau}}\right)\widehat{\omega}'\right]\\
&\quad-\left[\frac{\hat{\mathrm{v}}}{\sqrt{1+4\hat{\tau}\hat{\mathrm{v}}}}+0.5\widehat{\omega}''+\frac{0.5\widehat{\omega}''}{\sqrt{1+4\hat{\tau}\hat{\mathrm{v}}}}\right]\\
&\quad\left.\times\sin\left[\left(2\widehat{\omega}''+\frac{\sqrt{1+4\hat{\tau}\hat{\mathrm{v}}}-1}{\hat{\tau}}\right)\widehat{\omega}'\right]\right\}.
\end{aligned}\tag{8.3.26}$$

An imaginary part:

$$\begin{aligned}
&\frac{1}{2\sqrt{\pi}}(\widehat{\omega}'-\hat{\bar{u}})\\
&\quad\times\exp\left\{\widehat{\omega}'^2-\widehat{\omega}''^2-\widehat{\omega}''\frac{\sqrt{1+4\hat{\tau}\hat{\mathrm{v}}}-1}{\hat{\tau}}-\frac{1+2\hat{\tau}\hat{\mathrm{v}}-\sqrt{1+4\hat{\tau}\hat{\mathrm{v}}}}{2\hat{\tau}^2}\right\}\\
&=\hat{\tau}(\widehat{\omega}'-\hat{\bar{u}})\left\{\left[\frac{\hat{\mathrm{v}}}{\sqrt{1+4\hat{\tau}\hat{\mathrm{v}}}}+0.5\widehat{\omega}''+0.5\widehat{\omega}''\frac{1}{\sqrt{1+4\hat{\tau}\hat{\mathrm{v}}}}\right]\right.\\
&\quad\times\cos\left[\widehat{\omega}'\left(2\widehat{\omega}''+\frac{\sqrt{1+4\hat{\tau}\hat{\mathrm{v}}}-1}{\hat{\tau}}\right)\right]\\
&\quad\left.-0.5\widehat{\omega}'\left(1+\frac{1}{\sqrt{1+4\hat{\tau}\hat{\mathrm{v}}}}\right)\sin\left[\widehat{\omega}'\left(2\widehat{\omega}''+\frac{\sqrt{1+4\hat{\tau}\hat{\mathrm{v}}}-1}{\hat{\tau}}\right)\right]\right\}\\
&\quad+(1-\hat{\tau}\widehat{\omega}'')\left\{-0.5\widehat{\omega}'\left(1+\frac{1}{\sqrt{1+4\hat{\tau}\hat{\mathrm{v}}}}\right)\right.
\end{aligned}$$

$$\times \cos\left[\widehat{\omega}'\left(2\widehat{\omega}'' + \frac{\sqrt{1+4\hat{\tau}\hat{\nu}}-1}{\hat{\tau}}\right)\right]$$

$$-\left[\frac{\hat{\nu}}{\sqrt{1+4\hat{\tau}\hat{\nu}}} + 0.5\widehat{\omega}'' + \frac{0.5\widehat{\omega}''}{\sqrt{1+4\hat{\tau}\hat{\nu}}}\right]$$

$$\times \sin\left[\widehat{\omega}'\left(2\widehat{\omega}'' + \frac{\sqrt{1+4\hat{\tau}\hat{\nu}}-1}{\hat{\tau}}\right)\right]\Bigg\}. \tag{8.3.27}$$

The system of Eqs. (8.3.26), (8.3.27) differs from the system of Eqs. (8.2.67), (8.2.68), obtained with the help of classical Maxwell equations. But for electrons trapped by the wave ($\widehat{\omega}' = \hat{\hat{u}}$), Eq. (8.3.27) coincides with (8.2.68), and (8.3.26) differs from (8.2.67) by multiplier

$$B = \frac{1 + r_D^2 k^2 - \hat{\tau}\widehat{\omega}''}{1 + r_D^2 k^2}. \tag{8.3.28}$$

In this case the method of successive approximation can be realized with the use of (8.2.79) for the coefficient B calculation in the first approximation. Then this correction does not change the discrete character of solutions but leads to other numerical values of these solutions.

In conclusion we state that application of the generalized Maxwell equations can lead not only to correction of solutions but, as it was demonstrated, to physical solutions when classical dispersion equations have no solutions at all.

In the following consideration we intend to apply the main ideas of the generalized Boltzmann physical kinetics to kinetic and hydrodynamic theory of liquids. This theory is developed with due regard for the variation of the distribution function on a characteristic scale of the order of the time of vibration of particles in cells communicated with a free volume of liquid.

8.4. To the kinetic and hydrodynamic theory of liquids

As is known, the classical Boltzmann equation (BE), describing the processes of transfer in gases, is valid only on characteristic scales related to the hydrodynamic time of flow and the mean time between collisions. As it was established, the inclusion of the third possible scale, namely, the particle collision time, leads to the emergence of additional terms in the BE, which, generally speaking, are of the same order as the other terms in the BE.

A fundamental difference between a liquid and a rarefied gas consists in the many-particle interaction of their component particles. Nevertheless, it turns out that the use of fundamental concepts of the generalized Boltzmann kinetic theory (GBKT) and of the many-scale method leads to important results in the kinetic and hydrodynamic theory of liquids.

Let us turn to experimental data in the theory of liquids and theoretical models. In the theory of liquids, some or other microscopic models are usually employed. A liquid is investigated either as a non-ideal gas with many-particle interaction or from the standpoint of the theory of a crystal in which the long-range order is lost. The singularities of these models are described in Rezibois and De Lener (1980) and Fizicheskaya Entsiklopediya (1990), and we will only refer to two models that are often used in the theory of liquid, namely:

(1) The cell model, in which a liquid is treated as a deformed crystal with molecules localized in the vicinity of the "points" of deformed lattice.
(2) The hole theory in which it is assumed that the transition of molecules from one state to another is realized owing to the vacancies available in the deformed lattice.

Starting in 1924, Ya.I. Frenkel has been systematically developing the theory of liquid state as a generalization of the theory of real crystals involving the concepts of mobile holes, as well as the "concepts of thermal motion as alternation of small vibrations about some positions of equilibrium with abrupt variation of these positions" (Frenkel', 1945). The latter statement is equivalent to introducing the activation energy (or loosening energy using Frenkel's terminology) in the liquid theory. In principle, Frenkel's model reflects the experimental fact of existence of the short-range order in liquid. As a result, the particle motion assumes the behavior of "irregular vibrations with the mean frequency τ_0^{-1} close to that of the vibration frequency of particles in crystals, and with the amplitude defined by the size of the "free volume" offered to the given particle by its neighbors. The free lifetime of molecule in the temporary position of equilibrium between two activated jumps is related by

$$\tau = \tau_0 \,\mathrm{e}^{W/kT}, \tag{8.4.1}$$

where W is the activation energy" (Fizicheskaya Entsiklopediya, 1990). Relation (8.4.1) reflects the physically transparent fact that, with the activation energy (loosening energy, according to the preferable terminology of Frenkel) equal to zero, the particle loses its contact with neighbors during the characteristic vibration time τ_0, while with $W \to \infty$ the molecule does not change its environment at all.

We will discuss Frenkel's model in view of the experimental data of Hildebrand (Alekseev, 1997; Hildebrand, 1977). Hildebrand gives the experimental data on the viscosity of liquid in the form

$$\mu = \mu_a + \mu_b, \tag{8.4.2}$$

where μ_a is the viscosity that includes only the collective effects in the liquid, and μ_b is the viscosity that includes the "individuality" of the particles. This differentiation calls for additional explanation. For this purpose, consider Figure 8.6 which gives, by way of example, the viscosity μ, and the components μ_a and μ_b for C_3H_8 ($T = 410.9$ K). Also given in Figure 8.6 is the critical volume V_c, as well as the volume V_t; when this latter volume is reached (on the side of smaller volumes), the individuality of interacting particles starts showing up.

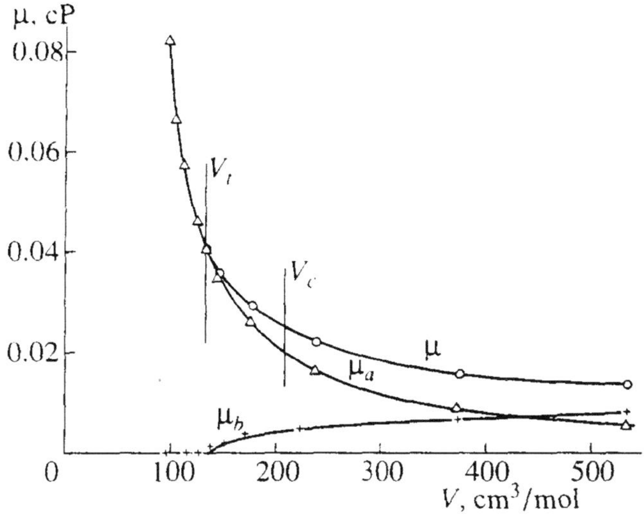

Fig. 8.6. The viscosity μ and viscosity components μ_a and μ_b, for C_3H_8 vs. volume V.

If μ_0 is the viscosity of rarefied gas,

$$\mu_0 = \frac{2(mk_BT)^{1/2}}{3\pi^{3/2}\sigma^2}, \tag{8.4.3}$$

where σ is the diameter of molecule and m is its mass, then

$$\mu_b = \mu_0\left(1 - \frac{V_t}{V}\right), \tag{8.4.4}$$

with $\mu_b = 0$ at $V < V_t$. Therefore, with rarefaction, a term emerges in the total viscosity, which is calculated by the kinetic theory of rarefied gas within the coefficient $1 - V_t/V$. At $V < V_t$, the last "traces" of pair interaction fully disappear, and the viscosity is defined by purely collective effects. The values of volume V_t are found experimentally and tabulated.

As a test of validity of the relation

$$\frac{\mu_b}{\mu_a} = 1 - \frac{V_t}{V}, \tag{8.4.5}$$

Figure 8.7 gives the data of Hildebrand on $(\mu - \mu_a)/\mu_0$ function of V. In so doing, the points in the plot for Ar correspond to the temperature of 323 and 373 K, those for CH_4 to 311 and 341 K, and those for C_3H_8 to 411 and 511 K. The construction of the plot of μ_b/μ_0 as a function of $1 - V_t/V$ leads to a universal dependence for different temperatures. The "collective" viscosity μ_a according to Hildebrand is approximated by the formula

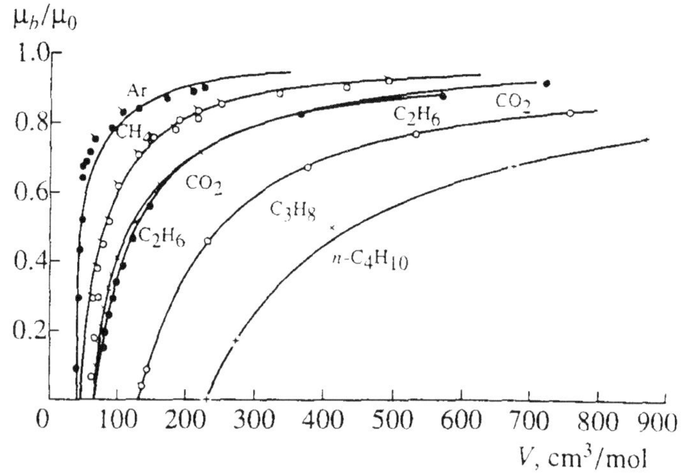

Fig. 8.7. The data of Hildebrand on $(\mu - \mu_a)/\mu_0$ function of V.

Table 8.4
Eigenvolume V_0 and coefficient B for different molecules

Molecule	V_0 (cm^3/mol)	B (cP^{-1})
C_3H_8	61	18.6
C_5H_{12}	92	18.5
C_6H_{14}	111	18.0
C_7H_{16}	130	18.8
C_8H_{18}	147	17.1
C_9H_{20}	165	16.5
$C_{10}H_{22}$	183	15.2
$C_{12}H_{26}$	217	14.1
CCl_4	88.3	17.4
P_4	68.6	16.3
CS_2	50.5	14.0
$CHCl_3$	70	14.0
Hg	14.1	12.6

$$\mu_a = \frac{V_0}{B(V - V_0)}, \tag{8.4.6}$$

where V_0 is the eigenvolume of molecules. Table 8.4 contains the data of Hildebrand, according to which it follows that the coefficient B is both universal and little dependent on the sort of matter.

The values of B and V_0 are determined by the experimental data from the plot of dependence of the fluidity $1/\mu$ on the volume V in the liquid state region, when the correlations associated with pair interaction are insignificant. The values of V_t are also determined by the experimental data from Figure 8.7. Table 8.5 gives the values of these quantities according to Hildebrand's experiments and literature data.

Table 8.5
Parameters of the liquid state

Molecule	V_0 (cm^3/mol)	V_t	V_c	B (cP^{-1})
CH_4	32.0	40	99.4	35.6
C_2H_6	44.6	64	148	25.8
C_3H_8	61.5	130	205	22.0
n-C_4H_{10}	78.5	230	255	21.8
CO_2	28.7	60	94	15.7

The viscosity in liquid and in the transition state may be calculated by the formula

$$\mu = \mu_0 \frac{V - V_t}{V} + \frac{1}{B} \frac{V_0}{V - V_0}. \tag{8.4.7}$$

The viscosity μ_0, calculated by formulas of rarefied gas dynamics, may be expressed in terms of the mean time τ_r between collisions,

$$\tau_r p = \Pi \mu_0, \tag{8.4.8}$$

where Π is the parameter associated with the model of particle interaction; in the first approximation, for the hard-sphere model, $\Pi = 0.8$. From Eqs. (8.4.7) – after multiplication of the both sides of this equation by Π/p – and (8.4.8), it follows

$$\Pi \frac{\mu}{p} = \tau_r \frac{V - V_t}{V} + \frac{\Pi}{B} \frac{V_0}{V - V_0} \frac{1}{p}, \tag{8.4.9}$$

where p is the static pressure.

The time

$$\tau_r^* = \tau_r \frac{V - V_t}{V} \tag{8.4.10}$$

may be treated as the mean time between particle collisions in the transition mode. With $V < V_t$, the "sample" particle ceases to be predominantly influenced by one of the particles and senses the effect of a self-consistent force alone, in this case, $\tau_r^* = 0$.

We will now introduce the time

$$\tau_{\text{osc}} = \frac{\Pi}{B} \frac{V_0}{V - V_0} \frac{1}{p}. \tag{8.4.11}$$

Relation (8.4.9) may be represented in terms of Frenkel's model if one uses the equation of state in the form of Dieterici's equation which is "the best equation of state, depending on two constants" (Hirschfelder, Curtiss and Bird, 1954),

$$p(V - b) = RT\,\mathrm{e}^{-a/(RTV)}. \tag{8.4.12}$$

The constant a is related to the measure of "cohesive forces" of molecules; and the constant b is proportional to the volume of molecules and taken equal to V_0. The first Dieterici equation describes well the region of the critical state of a matter. For example, the critical coefficient s, defined by the thermodynamic parameters of a matter, p_c, V_c, and T_c in the critical state, $s = RT_c/(p_c V_c) = 3.695$, which correlates with the experimental data better than $s = 2.67$ according to van der Waals. With high values of V, Dieterici's equation transforms to van der Waals' equation. It follows from (8.4.10) and (8.4.11) that

$$\tau_{\text{osc}} = \tau_0 \, \mathrm{e}^{W/(RT)}, \tag{8.4.13}$$

where

$$\tau_0 = \frac{\Pi V_0}{BRT}, \tag{8.4.14}$$

$$W = \frac{a}{V}. \tag{8.4.15}$$

But formula (8.4.13) is Frenkel's formula for the time of residence of molecule in the cell, τ_0 is the particle vibration period, and W is the activation energy (or "loosening energy", according to Frenkel).

As found by Hevesy (Frenkel', 1945, p. 104), the activation energy W decreases as the temperature rises. This effect is readily taken into account by the second Dieterici equation,

$$p(V - V_0) = \mathrm{e}^{-a/(RT^{\gamma} V)}, \tag{8.4.16}$$

which contains the temperature correction T^{γ}, $\gamma = 1.27$. In this case,

$$W = \frac{a}{VT^{0.27}}. \tag{8.4.17}$$

Further detailing of relations (8.4.14) and (8.4.15) is possible within Dieterici's equation. In view of

$$a = 2RT_c V_c, \qquad V_0 = \frac{1}{2} V_c, \tag{8.4.18}$$

we find that, for the first Dieterici equation,

$$\tau_0 = \frac{\Pi V_c}{2BRT}, \qquad W = \frac{2RT_c V_c}{V}. \tag{8.4.19}$$

Needless to say, formulas of the type of (8.4.19) can be expected to coincide with the experimental data only by the order of magnitude. The prospects for rigorous calculation of the above-identified parameters are associated with further development of the trend in statistical physics, based on the calculation of thermodynamic parameters "from the first principles" (see, for example, Krieger, Lukin and Semenov, 1986; Semenov, 1984), and with the methods of molecular dynamics. For example, for mercury at $T = 300$ K, Eq. (8.4.19) yields $\tau_0 = 1.4 \times 10^{-13}$ s. Note that the typical value of the particle vibration period in the cell, given by Frenkel' (1945, p. 182) is 10^{-13} s.

Therefore, the data of Hildebrand may in fact be treated from the standpoint of Frenkel's theory. Note only that formula (8.4.13) was written by Frenkel for the liquid state (when the pair interaction effects are fully suppressed).

In what follows, we will *only* need Frenkel's model to introduce the *scales* of quantities in the kinetic theory of liquids.

We will now write (8.4.9) as

$$\Pi \frac{\mu}{p} = \tau_r^* + \tau_0 \, \mathrm{e}^{W/(RT)}, \tag{8.4.20}$$

and introduce the small parameter $\varepsilon = nr_{\text{cell}}^3$, where r_{cell} is the characteristic radius of the cell. The parameter ε may be represented as

$$\varepsilon = nr_{\text{cell}}^3 = \frac{N}{V}\frac{V - V_0}{N} = \frac{V - V_0}{V}, \tag{8.4.21}$$

where N is the number of particles in the system, and V is the molar volume. We will introduce Hildebrand's parameter $k = (V - V_0)/V_0$ and note that $\varepsilon = k/(k+1)$. Tables 8.6 and 8.7 contain the experimentally obtained values of ε for some liquids, as well as the values of viscosity of liquids and the pressure dependence of the parameter ε.

For liquid HeII, the eigenvolume of an atom is equal about 9 Å, and on average 45.8 Å^3 of "empty" space corresponds to every atom.

It follows from Tables 8.6 and 8.7 that the quantity ε may be treated as an adequate small parameter in constructing the theory.

It becomes obvious now that, in constructing the kinetic theory of liquid, one must take into account the following three groups of scales:

Table 8.6
Parameter ε and the viscosity of some liquids

Liquid	T (°C)	μ (cP)	ε
CCl_4	20	0.97	0.0557
	40	0.74	0.0723
C_6H_6	20	0.650	0.0741
	40	0.492	0.1007
CS_2	20	0.366	0.1554
	40	0.349	0.1625
$(C_4F_9)_3N$	20		0.0148
	42.2		0.0393
C_5ONF_{11}	20		0.1007
	42.2		0.1197

Table 8.7
The pressure dependence of some parameters for n-decanes

p (atm)	V (cm^3)	$V - V_0$	ε
13.6	232.2	4.821	0.0208
27.2	231.0	4.687	0.0203
54.4	229.9	4.597	0.0200
68.0	227.9	4.396	0.0193
204.0	220.4	3.631	0.0165
340.0	214.8	3.081	0.0143
408.0	212.3	2.829	0.0133

(I) the cell scale corresponding to the vibrations of molecule in the "blocked" state, namely, $v_{0\,\mathrm{cell}}$ – the scale of velocity of vibratory motion of particle in the "blocked" state, r_{cell} – the scale of cell size ($r_{\mathrm{cell}} = V_{\mathrm{cell}}^{1/3}$, where V_{cell} is the cell volume), and τ_0 – the period of particle vibration in the cell. We will refer to this group of scales as the "τ_0-scale";

(II) the scale of the cell state associated with the characteristic time of particle residence in the cell, or the "τ-scale", namely, $v_{0\tau}$ – the scale of particle velocity in the τ-scale (generally speaking, this scale of velocity does not coincide with $v_{0\,\mathrm{cell}}$, because a high-energy particle is capable of overcoming the energy barrier associated with the activation energy), r_τ – the scale of length in the τ-scale (the characteristic distance covered by a particle during the time of residence in the cell), and τ – the residence time of particle in the cell (for example, under normal conditions for water, $\tau \sim 10^2\tau_0$); and

(III) the hydrodynamic scale that needs no additional explanations, namely, v_L – the hydrodynamic velocity of flow, L – the hydrodynamic length, and τ_L – the hydrodynamic time.

In what follows, these groups of scales will be used to derive the kinetic equation for liquid.

We will use the Bogolyubov chain equation for the s-particle distribution function f_s,

$$\frac{\partial f_s}{\partial t} + \sum_{i=1}^{s} \mathbf{v}_i \cdot \frac{\partial f_s}{\partial \mathbf{r}_i} + \sum_{i=1}^{s} \mathbf{F}_i \cdot \frac{\partial f_s}{\partial \mathbf{v}_i} + \sum_{i,j=1}^{s} \mathbf{F}_{ij} \cdot \frac{\partial f_s}{\partial \mathbf{v}_i}$$
$$= -\frac{1}{N} \sum_{i=1}^{s} \sum_{j=s+1}^{N} \int \mathbf{F}_{ij} \cdot \frac{\partial}{\partial \mathbf{v}_i} f_{s+1}\, \mathrm{d}\Omega_{s+1}, \tag{8.4.22}$$

written in standard notation, where, in particular, $\mathbf{F}_i$ and $\mathbf{F}_{ij}$ are external and internal forces, respectively. We will make Eq. (8.4.22) dimensionless on the τ_0-scale, using the following scales: $n^s v_{0\,\mathrm{cell}}^{-3s}$ for f_s, and $v_{0\,\mathrm{cell}}^2/r_{\mathrm{cell}}$, F_0^{ex} for the internal and external forces, respectively. Then, Eq. (8.4.22) assumes the form

$$\frac{\partial \hat{f}_s}{\partial \hat{t}_{\text{cell}}} + \sum_{i=1}^{s} \hat{\mathbf{v}}_{i\,\text{cell}} \cdot \frac{\partial \hat{f}_s}{\partial \hat{\mathbf{r}}_{i\,\text{cell}}} + \alpha \sum_{i=1}^{s} \widehat{\mathbf{F}}_i \cdot \frac{\partial \hat{f}_s}{\partial \hat{\mathbf{v}}_{i\,\text{cell}}} + \sum_{i,j=1}^{s} \widehat{\mathbf{F}}_{ij} \cdot \frac{\partial \hat{f}_s}{\partial \hat{\mathbf{v}}_{i\,\text{cell}}}$$

$$= -\varepsilon \frac{1}{N} \sum_{i=1}^{s} \sum_{j=s+1}^{N} \widehat{\mathbf{F}}_{ij} \cdot \frac{\partial}{\partial \hat{\mathbf{v}}_{i\,\text{cell}}} \hat{f}_{s+1}\, \mathrm{d}\widehat{\Omega}_{s+1}, \tag{8.4.23}$$

where $\alpha = F_0^{\text{ex}}/F_{0\,\text{cell}}$. We will follow the many-scale method and represent the dimensionless s-particle distribution function as a function of three groups of scales and ε,

$$\hat{f}_s = \hat{f}_s(\hat{t}_{\text{cell}}, \hat{\mathbf{r}}_{i\,\text{cell}}, \hat{\mathbf{v}}_{i\,\text{cell}}; \hat{t}_\tau, \hat{\mathbf{r}}_{i\tau}, \hat{\mathbf{v}}_{i\tau}; \hat{t}_L, \hat{\mathbf{r}}_{iL}, \hat{\mathbf{v}}_{iL}; \varepsilon), \tag{8.4.24}$$

and introduce the expansion

$$\hat{f}_s = \sum_{\upsilon=0}^{\infty} \hat{f}_s^{\,\upsilon}(\hat{t}_{\text{cell}}, \hat{\mathbf{r}}_{i\,\text{cell}}, \hat{\mathbf{v}}_{i\,\text{cell}}; \hat{t}_\tau, \hat{\mathbf{r}}_{i\tau}, \hat{\mathbf{v}}_{i\tau}; \hat{t}_L, \hat{\mathbf{r}}_{iL}, \hat{\mathbf{v}}_{iL};)\varepsilon^{\upsilon}. \tag{8.4.25}$$

The derivatives in the left-hand part of Eq. (8.4.23) are calculated by the rule of differentiation of composite function,

$$\frac{\mathrm{d}\hat{f}_s}{\mathrm{d}\hat{t}_{\text{cell}}} = \frac{\partial \hat{f}_s}{\partial \hat{t}_{\text{cell}}} + \varepsilon\varepsilon_2 \frac{\partial \hat{f}_s}{\partial \hat{t}_\tau} + \varepsilon\varepsilon_1\varepsilon_2 \frac{\partial \hat{f}_s}{\partial \hat{t}_L}, \tag{8.4.26}$$

$$\frac{\mathrm{d}\hat{f}_s}{\mathrm{d}\hat{\mathbf{r}}_{i\,\text{cell}}} = \frac{\partial \hat{f}_s}{\partial \hat{\mathbf{r}}_{i\,\text{cell}}} + \varepsilon\varepsilon_2\varepsilon_3 \frac{\partial \hat{f}_s}{\partial \hat{\mathbf{r}}_{i\tau}} + \varepsilon\varepsilon_1\varepsilon_2\varepsilon_3\varepsilon_4 \frac{\partial \hat{f}_s}{\partial \hat{\mathbf{r}}_{iL}}, \tag{8.4.27}$$

$$\frac{\mathrm{d}\hat{f}_s}{\mathrm{d}\hat{\mathbf{v}}_{i\,\text{cell}}} = \frac{\partial \hat{f}_s}{\partial \hat{\mathbf{v}}_{i\,\text{cell}}} + \varepsilon\varepsilon_2 \frac{F_{0\,\text{cell}}}{F_{0\tau}} \frac{\partial \hat{f}_s}{\partial \hat{\mathbf{v}}_{i\tau}} + \varepsilon\varepsilon_2\varepsilon_4 \frac{F_{0\,\text{cell}}}{F_{0\tau}} \frac{\partial \hat{f}_s}{\partial \hat{\mathbf{v}}_{iL}}, \tag{8.4.28}$$

using the notation

$$\varepsilon_1 = \frac{\tau}{\tau_L}, \quad \varepsilon_2 = \frac{pV}{RT}, \quad \varepsilon_3 = \frac{v_{0\,\text{cell}}}{v_{0\tau}},$$

$$\varepsilon_4 = \frac{v_{0\tau}}{v_{0L}}, \quad F_{0\tau} = \frac{v_{0\tau}}{\tau}, \quad F_{0\,\text{cell}} = \frac{v_{0\,\text{cell}}^2}{r_{\text{cell}}};$$

the symbol ˆ indicates dimensionless quantities. One can introduce the analog of Knudsen number,

$$Kn_{\text{fl}} = \frac{\tau v_{0\tau}}{\tau_L v_{0L}} = \varepsilon_1\varepsilon_4. \tag{8.4.29}$$

No restrictions are introduced on the above-identified parameters.

We will now substitute series (8.4.25) into approximation derivatives (8.4.26)–(8.4.28), and the derived expressions – into Eq. (8.4.23). Besides, in the right-hand part of (8.4.23), we also use the notation $\hat{f}_{s+1}$ in the form of the above-mentioned series. We will equate the expressions for the same values of ε.

For ε^0, we find

$$\frac{\partial \hat{f}_s^0}{\partial \hat{t}_{\text{cell}}} + \sum_{i=1}^{s} \hat{\mathbf{v}}_{i\,\text{cell}} \cdot \frac{\partial \hat{f}_s^0}{\partial \hat{\mathbf{r}}_{i\,\text{cell}}} + \sum_{i,j=1}^{s} \widehat{\mathbf{F}}_{ij} \cdot \frac{\partial \hat{f}_s^0}{\partial \hat{\mathbf{v}}_{i\,\text{cell}}} + \alpha \sum_{i=1}^{s} \widehat{\mathbf{F}}_i \cdot \frac{\partial \hat{f}_s^0}{\partial \hat{\mathbf{v}}_{i\,\text{cell}}} = 0. \tag{8.4.30}$$

For ε^1, we derive

$$\begin{aligned}
&\frac{\partial \hat{f}_s^1}{\partial \hat{t}_{\text{cell}}} + \sum_{i=1}^{s} \hat{\mathbf{v}}_{i\,\text{cell}} \cdot \frac{\partial \hat{f}_s^1}{\partial \hat{\mathbf{r}}_{i\,\text{cell}}} + \sum_{i,j=1}^{s} \widehat{\mathbf{F}}_{ij} \cdot \frac{\partial \hat{f}_s^1}{\partial \hat{\mathbf{v}}_{i\,\text{cell}}} + \alpha \sum_{i=1}^{s} \widehat{\mathbf{F}}_i \cdot \frac{\partial \hat{f}_s^1}{\partial \hat{\mathbf{v}}_{i\,\text{cell}}} + \varepsilon_2 \frac{\partial \hat{f}_s^0}{\partial \hat{t}_\tau} \\
&\quad + \varepsilon_2 \varepsilon_3 \sum_{i=1}^{s} \hat{\mathbf{v}}_{i\,\text{cell}} \cdot \frac{\partial \hat{f}_s^0}{\partial \hat{\mathbf{r}}_{i\tau}} + \varepsilon_2 \frac{F_{0\,\text{cell}}}{F_{0\tau}} \sum_{i,j=1}^{s} \widehat{\mathbf{F}}_{ij} \cdot \frac{\partial \hat{f}_s^0}{\partial \hat{\mathbf{v}}_{i\tau}} \\
&\quad + \alpha \varepsilon_2 \frac{F_{0\,\text{cell}}}{F_{0\tau}} \sum_{i=1}^{s} \widehat{\mathbf{F}}_i \cdot \frac{\partial \hat{f}_s^0}{\partial \hat{\mathbf{v}}_{i\tau}} + \varepsilon_1 \varepsilon_2 \frac{\partial \hat{f}_s^0}{\partial \hat{t}_L} + \varepsilon_1 \varepsilon_2 \varepsilon_3 \varepsilon_4 \sum_{i=1}^{s} \hat{\mathbf{v}}_{i\,\text{cell}} \cdot \frac{\partial \hat{f}_s^0}{\partial \hat{\mathbf{r}}_{iL}} \\
&\quad + \varepsilon_2 \varepsilon_4 \frac{F_{0\,\text{cell}}}{F_{0\tau}} \sum_{i,j=1}^{s} \widehat{\mathbf{F}}_{ij} \cdot \frac{\partial \hat{f}_s^0}{\partial \hat{\mathbf{v}}_{iL}} + \alpha \varepsilon_2 \varepsilon_4 \frac{F_{0\,\text{cell}}}{F_{0\tau}} \sum_{i=1}^{s} \widehat{\mathbf{F}}_i \cdot \frac{\partial \hat{f}_s^0}{\partial \hat{\mathbf{v}}_{iL}} \\
&= -\frac{1}{N} \sum_{i=1}^{s} \sum_{j=s+1}^{N} \widehat{\mathbf{F}}_{ij} \cdot \frac{\partial}{\partial \hat{\mathbf{v}}_{i\,\text{cell}}} \hat{f}_{s+1}^0 \, \mathrm{d}\widehat{\Omega}_j.
\end{aligned} \tag{8.4.31}$$

For $s = 1$, with due regard for the indistinguishability of particles and for the condition of $s \ll N$, Eq. (8.4.31) yields

$$\begin{aligned}
&\frac{\partial \hat{f}_1^1}{\partial \hat{t}_{\text{cell}}} + \hat{\mathbf{v}}_{1\text{cell}} \cdot \frac{\partial \hat{f}_1^1}{\partial \hat{\mathbf{r}}_{1\,\text{cell}}} + \alpha \widehat{\mathbf{F}}_1 \cdot \frac{\partial \hat{f}_1^1}{\partial \hat{\mathbf{v}}_{1\,\text{cell}}} + \varepsilon_2 \frac{\partial \hat{f}_1^0}{\partial \hat{t}_\tau} + \varepsilon_2 \varepsilon_3 \hat{\mathbf{v}}_{1\,\text{cell}} \cdot \frac{\partial \hat{f}_1^0}{\partial \hat{\mathbf{r}}_{1\tau}} \\
&\quad + \alpha \varepsilon_2 \frac{F_{0\,\text{cell}}}{F_{0\tau}} \widehat{\mathbf{F}}_1 \cdot \frac{\partial \hat{f}_1^0}{\partial \hat{\mathbf{v}}_{1\tau}} + \varepsilon_1 \varepsilon_2 \frac{\partial \hat{f}_1^0}{\partial \hat{t}_L} + \varepsilon_1 \varepsilon_2 \varepsilon_3 \varepsilon_4 \hat{\mathbf{v}}_{1\,\text{cell}} \cdot \frac{\partial \hat{f}_1^0}{\partial \hat{\mathbf{r}}_{1L}} \\
&\quad + \alpha \varepsilon_2 \varepsilon_4 \frac{F_{0\,\text{cell}}}{F_{0\tau}} \widehat{\mathbf{F}}_1 \cdot \frac{\partial \hat{f}_1^0}{\partial \hat{\mathbf{v}}_{1L}} = -\int \widehat{\mathbf{F}}_{12} \cdot \frac{\partial}{\partial \hat{\mathbf{v}}_{1\,\text{cell}}} \hat{f}_2^0 \, \mathrm{d}\widehat{\Omega}_2.
\end{aligned} \tag{8.4.32}$$

We will now write Eq. (8.4.23) for $s = 2$ and use the derived relation to represent the integral in the right part of (8.4.32) in the form

$$-\int \widehat{\mathbf{F}}_{12} \cdot \frac{\partial}{\partial \hat{\mathbf{v}}_{1\,\text{cell}}} \hat{f}_2^0 \,\mathrm{d}\widehat{\Omega}_2 = \int (\hat{\mathbf{v}}_{1\,\text{cell}} - \hat{\mathbf{v}}_{2\,\text{cell}}) \cdot \frac{\partial \hat{f}_2^0}{\partial \hat{\mathbf{x}}_{12}} \,\mathrm{d}\widehat{\Omega}_2$$
$$+ \int \left[\frac{\partial \hat{f}_2^0}{\partial \hat{t}_{\text{cell}}} + \hat{\mathbf{v}}_{1\,\text{cell}} \cdot \frac{\partial \hat{f}_2^0}{\partial \hat{\mathbf{r}}_{1\,\text{cell}}} + \alpha \widehat{\mathbf{F}}_1 \cdot \frac{\partial \hat{f}_2^0}{\partial \hat{\mathbf{v}}_{1\,\text{cell}}} + \alpha \widehat{\mathbf{F}}_2 \cdot \frac{\partial \hat{f}_2^0}{\partial \hat{\mathbf{v}}_{2\,\text{cell}}} \right] \mathrm{d}\widehat{\Omega}_2$$
$$+ \int \widehat{\mathbf{F}}_{21} \cdot \frac{\partial \hat{f}_2}{\partial \hat{\mathbf{v}}_{2\,\text{cell}}} \,\mathrm{d}\widehat{\Omega}_2, \tag{8.4.33}$$

where $\hat{\mathbf{x}}_{12} = \hat{\mathbf{r}}_{1\,\text{cell}} - \hat{\mathbf{r}}_{2\,\text{cell}}$ defines distance between certain liquid particles designated by the numbers 1 and 2. The last integral term in (8.4.33) vanishes, because it may be transformed to integrals over the surface that is infinitely removed in the velocity space, where the distribution function goes to zero. Using Eq. (8.4.30) written for $s = 1$ (see also (8.1.9), (8.1.12)), as a result the collision integral assumes the form

$$-\int \widehat{\mathbf{F}}_{12} \cdot \frac{\partial}{\partial \hat{\mathbf{v}}_{1\,\text{cell}}} \hat{f}_2^0 \,\mathrm{d}\widehat{\Omega}_2 = A + B + C, \tag{8.4.34}$$

where

$$A = \int \hat{f}_{1,1}^0 \frac{\partial \hat{f}_{1,2}^0}{\partial \hat{t}_{\text{cell}}} \,\mathrm{d}\widehat{\Omega}_2, \tag{8.4.35}$$

$$B = \int (\hat{\mathbf{v}}_{1\,\text{cell}} - \hat{\mathbf{v}}_{2\,\text{cell}}) \cdot \frac{\partial \hat{f}_2^0}{\partial \hat{\mathbf{x}}_{12}} \,\mathrm{d}\widehat{\Omega}_2, \tag{8.4.36}$$

$$C = \int \left[\frac{D W_2^0}{D \hat{t}_{\text{cell}}} \right] \mathrm{d}\widehat{\Omega}_2, \tag{8.4.37}$$

where the correlation function W_2^0 is introduced

$$f_2^0(t, \mathbf{r}_1, \mathbf{v}_1, \mathbf{r}_2, \mathbf{v}_2)$$
$$= f_1^0(t, \mathbf{r}_1, \mathbf{v}_1) f_1^0(t, \mathbf{r}_2, \mathbf{v}_2) + W_2^0(t, \mathbf{r}_1, \mathbf{v}_1, \mathbf{r}_2, \mathbf{v}_2). \tag{8.4.38}$$

As it was shown in Section 8.1, integral A leads to appearance of self-consistent forces in substantive derivatives, integral B may be transformed to the Boltzmann form. Integral C is equal to zero as containing the correlation functions of *zeroth* order in ε.

We return to dimensional variables to derive the kinetic equation that is valid for all three scales in a lower approximation in s

$$\frac{D f_{1,1}^1}{Dt} + \frac{D f_{1,1}^0}{Dt} = \int \left[f_{1,1}^{01} f_{1,2}^{01} - f_{1,1}^0 f_{1,2}^0 \right] g b \,\mathrm{d}b \,\mathrm{d}\varphi \,\mathrm{d}\hat{\mathbf{v}}_2. \tag{8.4.39}$$

Here, the prime indicates the parameters of particles after interaction; integration is performed on the τ_0-scale. Eq. (8.4.39) remains linked by the superscript in the first term of the left-hand part of (8.4.39), where the notation

$$\frac{D}{Dt} \equiv \frac{\partial}{\partial t} + \mathbf{v}_1 \cdot \frac{\partial}{\partial \mathbf{r}_1} + \mathbf{F}_1^{sc} \cdot \frac{\partial}{\partial \mathbf{v}_1} \tag{8.4.40}$$

is introduced, $\mathbf{F}_1^{sc}$ is the effective force acting on the particle on the cell scale.

We will treat

$$\frac{Df_{1,1}^{1}}{Dt} = \frac{D}{Dt}\left(\frac{\partial f_{1,1}}{\partial \varepsilon}\right)_{\varepsilon=0}, \tag{8.4.41}$$

introduce the assumption of the ε-dependence of the one-particle distribution function on the cell scale in terms of the dynamic variables $\mathbf{r},\ \mathbf{v},\ t$, and use the irreversibility of time. Then,

$$\frac{Df_{1,1}^{1}}{Dt} = -\frac{D}{Dt}\left[\frac{1}{(\partial\varepsilon/\partial t)_{\varepsilon=0}}\frac{Df_{1,1}^{0}}{Dt}\right]. \tag{8.4.42}$$

Consider now the relaxation form $(\partial\varepsilon/\partial t)_{\varepsilon=0}$. Physically, this means that it is necessary to find out how many particles will enter the cell per unit time under additional condition of $\varepsilon = 0$. Relaxation time can be estimated as mean time τ^* of particle residence in the cell, then

$$\left(\frac{\partial\varepsilon}{\partial t}\right)_{\varepsilon=0} = \tau^{*-1}. \tag{8.4.43}$$

From (8.4.20), there follows the representation of the time of particle residence in the cell

$$\tau^* = \tau_0\,\mathrm{e}^{W/RT} + \tau_r^* \tag{8.4.44}$$

and the possibility of expressing τ^* in terms of hydro-dynamic quantities,

$$\tau^* = \Pi\frac{\mu}{p}. \tag{8.4.45}$$

The second term in the right-hand part of (8.4.44) includes the correction of the time of particle residence in the cell for pair interaction, when the system in the liquid state approaches the gaseous state. Namely,

$$\tau^* = \begin{cases} \tau_0\,\mathrm{e}^{W/(RT)} + \tau_r^* & (V \geqslant V_t), \\ \tau_0\,\mathrm{e}^{W/(RT)} & (V \leqslant V_t), \end{cases} \qquad \tau_r^* = \tau_r\frac{V - V_t}{V}. \tag{8.4.46}$$

As a result, the kinetic equation for liquid assumes the form (in what follows, the superscript ^{0} is omitted)

$$\frac{Df_{1,1}}{Dt} - \frac{D}{Dt}\left(\tau^* \frac{Df_{1,1}}{Dt}\right) = \int \left[f'_{1,1} f'_{1,2} - f_{1,1} f_{1,2}\right] gb \, \mathrm{d}b \, \mathrm{d}\varphi \, \mathrm{d}\mathbf{v}_2. \qquad (8.4.47)$$

The kinetic equation (8.4.47) transforms asymptotically to the generalized Boltzmann equation for rarefied gases. It should be noticed that corrections of local Boltzmann collision integral – related to correlation functions – begin in the following approximation in ε. In this sense situation is analogous to developed plasma kinetic theory where corresponding collision integral incorporates effects of plasma polarization in following approximations.

Further derivation of generalized hydrodynamic Enskog equations may be performed analogously of the previous considerations. In so doing, the parameter τ^* defined by relation (8.4.46) is introduced into the equations.

For a one-component non-reacting medium in the absence of magnetic field, generalized hydrodynamic equations take the following form:

– continuity equation,

$$\frac{\partial}{\partial t}\left\{\rho - \tau^*\left[\frac{\partial \rho}{\partial t} + \frac{\partial}{\partial \mathbf{r}} \cdot (\rho \bar{\mathbf{v}})\right]\right\} + \frac{\partial}{\partial \mathbf{r}} \cdot \left\{\rho \bar{\mathbf{v}} - \tau^*\left[\frac{\partial}{\partial t}(\rho \bar{\mathbf{v}}) + \frac{\partial}{\partial \mathbf{r}} \cdot (\rho \overline{\mathbf{v}\mathbf{v}}) - \rho \mathbf{F}^{(1)}\right]\right\} = 0, \qquad (8.4.48)$$

– equation of motion,

$$\frac{\partial}{\partial t}\left\{\rho \bar{\mathbf{v}} - \tau^*\left[\frac{\partial}{\partial t}(\rho \bar{\mathbf{v}}) + \frac{\partial}{\partial \mathbf{r}} \cdot \rho \overline{\mathbf{v}\mathbf{v}} - \rho \mathbf{F}^{(1)}\right]\right\} - \mathbf{F}^{(1)}\left[\rho - \tau^*\left(\frac{\partial \rho}{\partial t} + \frac{\partial}{\partial \mathbf{r}} \cdot \rho \bar{\mathbf{v}}\right)\right] + \frac{\partial}{\partial \mathbf{r}} \cdot \left\{\rho \overline{\mathbf{v}\mathbf{v}} - \tau^*\left[\frac{\partial}{\partial t}(\rho \overline{\mathbf{v}\mathbf{v}}) + \frac{\partial}{\partial \mathbf{r}} \cdot \rho \overline{(\mathbf{v}\mathbf{v})\mathbf{v}} - \rho \mathbf{F}^{(1)} \bar{\mathbf{v}} - \rho \bar{\mathbf{v}} \mathbf{F}^{(1)}\right]\right\} = 0, \qquad (8.4.49)$$

– energy equation,

$$\frac{\partial}{\partial t}\left\{\frac{\rho \overline{v^2}}{2} - \tau^*\left[\frac{\partial}{\partial t}\frac{\rho \overline{v^2}}{2} + \frac{\partial}{\partial \mathbf{r}} \cdot \frac{1}{2}\rho \overline{v^2 \mathbf{v}} - \mathbf{F}^{(1)} \cdot \rho \bar{\mathbf{v}}\right]\right\} + \frac{\partial}{\partial \mathbf{r}} \cdot \left\{\frac{1}{2}\rho \overline{v^2 \mathbf{v}} - \tau^*\left[\frac{\partial}{\partial t}\frac{1}{2}\rho \overline{v^2 \mathbf{v}} + \frac{\partial}{\partial \mathbf{r}} \cdot \frac{1}{2}\rho \overline{v^2 \mathbf{v}\mathbf{v}} - \rho \mathbf{F}^{(1)} \cdot \overline{\mathbf{v}\mathbf{v}} - \frac{1}{2}\rho \overline{v^2} \mathbf{F}^{(1)}\right]\right\} - \left\{\rho \mathbf{F}^{(1)} \cdot \bar{\mathbf{v}} - \tau^*\left[\mathbf{F}^{(1)} \cdot \left(\frac{\partial}{\partial t}(\rho \bar{\mathbf{v}}) + \frac{\partial}{\partial \mathbf{r}} \cdot \rho \overline{\mathbf{v}\mathbf{v}} - \rho \mathbf{F}^{(1)}\right)\right]\right\} = 0, \qquad (8.4.50)$$

where ρ is the density of liquid, the bar over expressions depending on the own velocity of particles is indicative of averaging over the velocity using the one-particle distribution function.

Note in conclusion that τ^* transforms to the value of τ_r for rarefied gas, this providing for a "through" description of the liquid–gas system using generalized hydrodynamic equations (8.4.48)–(8.4.50).

Write down now GHE using notations of averaged values, for example

$$\rho^a = \rho - \tau^* \left[\frac{\partial \rho}{\partial t} + \frac{\partial}{\partial \mathbf{r}} \cdot (\rho \bar{\mathbf{v}}) \right]. \tag{8.4.51}$$

Then density ρ^a corresponds to mass contained *inside* the control unit volume filled with the particles of *finite* sizes. The other velocity moments have analogous physical sense. As a result, we have for the mentioned case:

- continuity equation

$$\frac{\partial \rho^a}{\partial t} + \frac{\partial}{\partial \mathbf{r}} \cdot (\rho \bar{\mathbf{v}})^a = 0, \tag{8.4.52}$$

- momentum equation

$$\frac{\partial}{\partial t}(\rho \bar{\mathbf{v}})^a - \left(\rho \mathbf{F}^{(1)}\right)^a + \frac{\partial}{\partial \mathbf{r}} \cdot (\rho \overline{\mathbf{v}\mathbf{v}})^a = 0, \tag{8.4.53}$$

- energy equation

$$\frac{\partial}{\partial t}\left(\rho \overline{v^2}\right)^a + \frac{\partial}{\partial \mathbf{r}} \cdot \left(\rho \overline{v^2 \mathbf{v}}\right)^a - 2\left(\rho \mathbf{F}^{(1)} \cdot \bar{\mathbf{v}}\right)^a = 0. \tag{8.4.54}$$

Systems of generalized hydrodynamic equations written in form (8.4.48)–(8.4.50) and (8.4.52)–(8.4.54) can be considered from the position of theory of super-fluidity developed by Landau (1941b, 1947). In phenomenological theory of superfluidity, hydrodynamic parameters – in particular density of mass flux and density of quantum liquid – are represented as a sum of normal and superfluid components:

$$\mathbf{j} = \mathbf{j}_s + \mathbf{j}_n, \qquad \rho = \rho_s + \rho_n. \tag{8.4.55}$$

Obviously, the structure of GHE coincides with Landau equations if to correlate now the superfluid component (s-component) with averaged values in GHE (for example, $\mathbf{j}^a$ and ρ^a) and normal component (n-component) with fluctuation parameters in GHE (for example $\mathbf{j}^{\mathrm{fl}}$ and ρ^{fl}). In this case Eqs. (8.4.52)–(8.4.54) have the character of Euler equations. In general case, the separation of normal and super-fluid components of liquid is not possible, the flow of liquid helium is said to consider as non-viscous flow in viscous media.

As it was indicated above, the generalized Boltzmann equation in particular and the generalized Boltzmann kinetic theory on the whole are really "working" theoretical instrument for solving applied problems. Certainly, tremendous quantity of applied problems can be considered. But first of all the following problems can be indicated:

(1) Calculations of turbulent flows (with taking into account the explicit form of Kolmogorov micro turbulent fluctuations) from the first principles of physics.
(2) Calculations of the flows by the intermediate Knudsen numbers and as the consequence the "through" calculations of gas dynamics flows including the structure of shock waves.
(3) Calculations of flows in plasma devices and gas discharge devices where the particle collisions are significant.
(4) Investigation of transport processes in ionosphere.
(5) Investigation of transport processes in semiconductors.

To conclude, the results presented here are only a part of what the generalized kinetic theory has produced during about two decades of its development – the years, one is safe to say, which have showed it to be a highly effective tool for solving many physical problems in areas where the classical theory runs into difficulties.

Appendices

Appendix 1. Derivation of energy equation for invariant $E_\alpha = m_\alpha V_\alpha^2/2 + \varepsilon_\alpha$

Generalized hydrodynamic Enskog energy equation for invariant $E_\alpha = m_\alpha V_\alpha^2/2 + \varepsilon_\alpha$ is written as

$$\int E_\alpha \left\{ \frac{\partial f_\alpha}{\partial t}\left[1 - \frac{\partial \tau_\alpha}{\partial t} - \mathbf{v}_\alpha \cdot \frac{\partial \tau_\alpha}{\partial \mathbf{r}}\right] + \mathbf{v}_\alpha \cdot \frac{\partial f_\alpha}{\partial \mathbf{r}}\left[1 - \frac{\partial \tau_\alpha}{\partial t} - \mathbf{v}_\alpha \cdot \frac{\partial \tau_\alpha}{\partial \mathbf{r}}\right] \right.$$
$$+ \mathbf{F}_\alpha \cdot \frac{\partial f_\alpha}{\partial \mathbf{v}_\alpha}\left[1 - \frac{\partial \tau_\alpha}{\partial t} - \mathbf{v}_\alpha \cdot \frac{\partial \tau_\alpha}{\partial \mathbf{r}}\right]$$
$$- \tau_\alpha \left[\frac{\partial^2 f_\alpha}{\partial t^2} + 2\frac{\partial^2 f_\alpha}{\partial \mathbf{r}\partial t} \cdot \mathbf{v}_\alpha + \frac{\partial^2 f_\alpha}{\partial \mathbf{r}\partial \mathbf{r}} : \mathbf{v}_\alpha \mathbf{v}_\alpha + 2\frac{\partial^2 f_\alpha}{\partial \mathbf{v}_\alpha \partial t} \cdot \mathbf{F}_\alpha \right.$$
$$+ \frac{\partial \mathbf{F}_\alpha}{\partial t} \cdot \frac{\partial f_\alpha}{\partial \mathbf{v}_\alpha} + \mathbf{F}_\alpha \cdot \frac{\partial f_\alpha}{\partial \mathbf{r}} + \frac{q_\alpha}{m_\alpha}\frac{\partial f_\alpha}{\partial \mathbf{v}_\alpha}\left(\mathbf{F}_\alpha^{(1)} \times \mathbf{B}\right)$$
$$+ \frac{\partial f_\alpha}{\partial \mathbf{v}_\alpha} \cdot \left[\mathbf{B}(\mathbf{v}_\alpha \cdot \mathbf{B}) - B^2 \mathbf{v}_\alpha\right]\left(\frac{q_\alpha}{m_\alpha}\right)^2 + \frac{\partial f_\alpha}{\partial \mathbf{v}_\alpha}\mathbf{v}_\alpha : \frac{\partial}{\partial \mathbf{r}}\mathbf{F}_\alpha + \frac{\partial^2 f_\alpha}{\partial \mathbf{v}_\alpha \partial \mathbf{v}_\alpha} : \mathbf{F}_\alpha \mathbf{F}_\alpha$$
$$\left.\left. + 2\frac{\partial^2 f_\alpha}{\partial \mathbf{v}_\alpha \partial \mathbf{r}} : \mathbf{v}_\alpha \mathbf{F}_\alpha \right]\right\} \mathrm{d}\mathbf{v}_\alpha = J_\alpha^{\mathrm{en}}. \tag{A1.1}$$

Transformation of the terms on the left-hand side of Eq. (A1.1) is realized in a similar manner as it was indicated in Section 2.7. Then in transformations that follow we present as reference source the results of integral calculations which deserve further comments.

(1) $$\int E_\alpha \frac{\partial f_\alpha}{\partial t}\,\mathrm{d}\mathbf{v}_\alpha = \frac{\partial}{\partial t}\left(n_\alpha \overline{E}_\alpha\right) + \mathbf{j}_\alpha \cdot \frac{\partial \mathbf{v}_0}{\partial t}. \tag{A1.2}$$

Internal energy $\widetilde{U}_\alpha$ of particles of species α (per mass unit) is defined as

$$n_\alpha \overline{E}_\alpha = \rho_\alpha \widetilde{U}_\alpha. \tag{A1.3}$$

Then

$$n_\alpha \overline{E}_\alpha = \frac{3}{2} n_\alpha k_B T_\alpha + \varepsilon_\alpha n_\alpha = \frac{3}{2} p_\alpha + \varepsilon_\alpha n_\alpha = \rho_\alpha \widetilde{U}_\alpha. \tag{A1.4}$$

For gas mixture we have

$$\rho\widetilde{U} = \frac{3}{2}p + \sum_{\alpha} \varepsilon_\alpha n_\alpha. \tag{A1.5}$$

(2)
$$\int E_\alpha \mathbf{v}_\alpha \cdot \frac{\partial f_\alpha}{\partial \mathbf{r}} \mathrm{d}\mathbf{v}_\alpha = \frac{\partial}{\partial \mathbf{r}} \cdot (\mathbf{v}_0 \rho_\alpha \widetilde{U}_\alpha) + \frac{\partial}{\partial \mathbf{r}} \cdot \mathbf{q}_\alpha^{\mathrm{T}} + \mathbf{j}_\alpha \cdot \left(\mathbf{v}_0 \cdot \frac{\partial}{\partial \mathbf{r}}\right)\mathbf{v}_0 - \rho_\alpha \overline{\mathbf{V}_\alpha \cdot \left(\mathbf{V}_\alpha \cdot \frac{\partial}{\partial \mathbf{r}}\right)\mathbf{V}_\alpha}. \tag{A1.6}$$

Let us recall that flux densities of heat and mass can be written as

$$\mathbf{q}_\alpha^{\mathrm{T}} = \frac{1}{2}\rho_\alpha \overline{V_\alpha^2 \mathbf{V}_\alpha} + \varepsilon_\alpha n_\alpha \overline{\mathbf{V}}_\alpha$$

$$\mathbf{j}_\alpha = m_\alpha n_\alpha \overline{\mathbf{V}}_\alpha.$$

(3)
$$\int E_\alpha \mathbf{F}_\alpha \cdot \frac{\partial f_\alpha}{\partial \mathbf{v}_\alpha} \mathrm{d}\mathbf{v}_\alpha = -\rho_\alpha \overline{\mathbf{F}_\alpha \cdot \mathbf{V}_\alpha}. \tag{A1.7}$$

(4)
$$\int E_\alpha \mathbf{v}_\alpha \cdot \frac{\partial \tau_\alpha}{\partial \mathbf{r}} \frac{\partial f_\alpha}{\partial t} \mathrm{d}\mathbf{v}_\alpha = \mathbf{v}_0 \cdot \frac{\partial \tau_\alpha}{\partial \mathbf{r}} \frac{\partial}{\partial t}(\rho_\alpha \widetilde{U}_\alpha) + \frac{\partial \tau_\alpha}{\partial \mathbf{r}} \cdot \frac{\partial \mathbf{q}_\alpha^{\mathrm{T}}}{\partial t} + \frac{\partial \tau_\alpha}{\partial \mathbf{r}} \cdot \mathbf{v}_0 \left(\mathbf{j}_\alpha \cdot \frac{\partial \mathbf{v}_0}{\partial t}\right) + \frac{\partial \tau_\alpha}{\partial \mathbf{r}} \cdot \left(\overset{\leftrightarrow}{P}_\alpha \cdot \frac{\partial \mathbf{v}_0}{\partial t}\right), \tag{A1.8}$$

where pressure tensor $\overset{\leftrightarrow}{P}_\alpha$ is

$$\overset{\leftrightarrow}{P}_\alpha = \rho_\alpha \overline{\mathbf{V}_\alpha \mathbf{V}_\alpha}.$$

(5)
$$\begin{aligned} \int E_\alpha \mathbf{v}_\alpha \cdot \frac{\partial f_\alpha}{\partial \mathbf{r}} \left(\mathbf{v}_\alpha \cdot \frac{\partial \tau_\alpha}{\partial \mathbf{r}}\right) \mathrm{d}\mathbf{v}_\alpha &= \frac{\partial \tau_\alpha}{\partial \mathbf{r}} \cdot \left[\frac{\partial}{\partial \mathbf{r}} \cdot (\mathbf{v}_0 \mathbf{v}_0 \rho_\alpha \widetilde{U}_\alpha)\right] + \frac{\partial \tau_\alpha}{\partial \mathbf{r}} \cdot \left[\frac{\partial}{\partial \mathbf{r}} \cdot \mathbf{v}_0 \mathbf{q}_\alpha^{\mathrm{T}}\right] \\ &+ \frac{\partial \tau_\alpha}{\partial \mathbf{r}} \cdot \left[\frac{\partial}{\partial \mathbf{r}} \cdot \mathbf{q}_\alpha^{\mathrm{T}} \mathbf{v}_0\right] + \frac{\partial \tau_\alpha}{\partial \mathbf{r}} \cdot \left[\frac{\partial}{\partial \mathbf{r}} \cdot n_\alpha \overline{E_\alpha \mathbf{V}_\alpha \mathbf{V}_\alpha}\right] \\ &+ \frac{\partial \tau_\alpha}{\partial \mathbf{r}} \cdot \left[\rho_\alpha \overline{\mathbf{V}_\alpha \cdot \left(\mathbf{v}_0 \mathbf{V}_\alpha \cdot \frac{\partial}{\partial \mathbf{r}}\right)} \mathbf{v}_0\right] + \frac{\partial \tau_\alpha}{\partial \mathbf{r}} \cdot \left[\rho_\alpha \overline{\mathbf{V}_\alpha \cdot \left(\mathbf{V}_\alpha \mathbf{v}_0 \cdot \frac{\partial}{\partial \mathbf{r}}\right)} \mathbf{v}_0\right] \\ &+ \frac{\partial \tau_\alpha}{\partial \mathbf{r}} \cdot \left[\rho_\alpha \overline{\mathbf{V}_\alpha \cdot (\mathbf{V}_\alpha \mathbf{V}_\alpha \cdot} \frac{\partial}{\partial \mathbf{r}}) \mathbf{v}_0\right] + \frac{\partial \tau_\alpha}{\partial \mathbf{r}} \cdot \left[\mathbf{v}_0 \mathbf{v}_0 \cdot \left(\mathbf{j}_\alpha \cdot \frac{\partial}{\partial \mathbf{r}}\right) \mathbf{v}_0\right] \\ &= \frac{\partial \tau_\alpha}{\partial \mathbf{r}} \frac{\partial}{\partial \mathbf{r}} : n_\alpha \overline{E_\alpha \mathbf{v}_\alpha \mathbf{v}_\alpha} + \rho_\alpha \overline{\left(\frac{\partial \tau_\alpha}{\partial \mathbf{r}} \cdot \mathbf{v}_\alpha\right) \mathbf{V}_\alpha} \cdot \left(\mathbf{v}_\alpha \cdot \frac{\partial}{\partial \mathbf{r}}\right) \mathbf{v}_0. \end{aligned} \tag{A1.9}$$

(6) $$\int E_\alpha\left(\mathbf{F}_\alpha\cdot\frac{\partial f_\alpha}{\partial\mathbf{v}_\alpha}\right)\mathbf{v}_\alpha\cdot\frac{\partial\tau_\alpha}{\partial\mathbf{r}}\,d\mathbf{v}_\alpha=-\frac{\partial\tau_\alpha}{\partial\mathbf{r}}\cdot\left(n_\alpha\overline{E_\alpha\mathbf{F}_\alpha}\right)$$
$$-\left(\frac{\partial\tau_\alpha}{\partial\mathbf{r}}\cdot\mathbf{v}_0\right)n_\alpha\overline{\mathbf{V}_\alpha\cdot\mathbf{F}_\alpha}-n_\alpha\overline{\left(\frac{\partial\tau_\alpha}{\partial\mathbf{r}}\cdot\mathbf{V}_\alpha\right)\mathbf{V}_\alpha\cdot\mathbf{F}_\alpha}. \tag{A1.10}$$

(7) $$\int E_\alpha\frac{\partial f_\alpha}{\partial t}\mathbf{v}_\alpha\cdot\frac{\partial\tau_\alpha}{\partial\mathbf{r}}\,d\mathbf{v}_\alpha$$
$$=\frac{\partial\mathbf{q}_\alpha^{\mathrm{T}}}{\partial t}\cdot\frac{\partial\tau_\alpha}{\partial\mathbf{r}}+\frac{\partial\tau_\alpha}{\partial\mathbf{r}}\cdot\frac{\partial}{\partial t}\left(\mathbf{v}_0\rho_\alpha\widetilde{U}_\alpha\right)+\rho_\alpha\overline{\mathbf{v}_\alpha\mathbf{V}_\alpha}:\frac{\partial\mathbf{v}_0}{\partial t}\frac{\partial\tau_\alpha}{\partial\mathbf{r}}. \tag{A1.11}$$

(8) $$\int E_\alpha\frac{\partial^2 f_\alpha}{\partial t^2}\,d\mathbf{v}_\alpha$$
$$=\frac{\partial^2}{\partial t^2}\left(\rho_\alpha\widetilde{U}_\alpha\right)+\rho_\alpha\left(\frac{\partial\mathbf{v}_0}{\partial t}\right)^2-\mathbf{j}_\alpha\cdot\frac{\partial^2\mathbf{v}_0}{\partial t^2}+2\frac{\partial\mathbf{j}_\alpha}{\partial t}\cdot\frac{\partial\mathbf{v}_0}{\partial t}. \tag{A1.12}$$

(9) $$\int E_\alpha\frac{\partial^2 f_\alpha}{\partial\mathbf{r}\partial t}\cdot\mathbf{v}_\alpha\,d\mathbf{v}_\alpha$$
$$=\frac{\partial^2}{\partial\mathbf{r}\partial t}\cdot\left[\mathbf{q}_\alpha^{\mathrm{T}}+\mathbf{v}_0 n_\alpha\varepsilon_\alpha+\mathbf{v}_0\cdot\overleftrightarrow{P}_\alpha+\frac{3}{2}p_\alpha\mathbf{v}_0+\frac{\rho_\alpha v_0^2}{2}\mathbf{v}_0\right]$$
$$-\mathbf{v}_0\cdot\left[\frac{\partial^2}{\partial\mathbf{r}\partial t}\cdot\left(\overleftrightarrow{P}_\alpha+\rho_\alpha\mathbf{v}_0\mathbf{v}_0\right)\right]+\frac{1}{2}v_0^2\frac{\partial^2}{\partial\mathbf{r}\partial t}\cdot\rho_\alpha\mathbf{v}_0. \tag{A1.13}$$

(10) $$\int E_\alpha\frac{\partial^2 f_\alpha}{\partial\mathbf{r}\partial\mathbf{r}}:\mathbf{v}_\alpha\mathbf{v}_\alpha\,d\mathbf{v}_\alpha$$
$$=\frac{1}{2}\frac{\partial^2}{\partial\mathbf{r}\partial\mathbf{r}}:\rho_\alpha\overline{v_\alpha^2\mathbf{v}_\alpha\mathbf{v}_\alpha}-\left[\left(\frac{\partial^2}{\partial\mathbf{r}\partial\mathbf{r}}:\rho_\alpha\overline{\mathbf{v}_\alpha\mathbf{v}_\alpha\right)\mathbf{v}_\alpha}\right]\cdot\mathbf{v}_0$$
$$+\frac{1}{2}v_0^2\frac{\partial^2}{\partial\mathbf{r}\partial\mathbf{r}}:\rho_\alpha\overline{\mathbf{v}_\alpha\mathbf{v}_\alpha}+\frac{\varepsilon_\alpha}{m_\alpha}\frac{\partial^2}{\partial\mathbf{r}\partial\mathbf{r}}:\rho_\alpha\overline{\mathbf{v}_\alpha\mathbf{v}_\alpha}$$
$$=\frac{1}{2}\frac{\partial^2}{\partial\mathbf{r}\partial\mathbf{r}}:v_0^2\overleftrightarrow{P}_\alpha+\frac{\partial^2}{\partial\mathbf{r}\partial\mathbf{r}}:\left\{\rho_\alpha(\mathbf{V}_\alpha\mathbf{V}_\alpha)(\mathbf{v}_0\cdot\mathbf{V}_\alpha)\right\}$$
$$+\frac{1}{2}\frac{\partial^2}{\partial\mathbf{r}\partial\mathbf{r}}:\rho_\alpha\overline{(\mathbf{V}_\alpha\mathbf{V}_\alpha)V_\alpha^2}+2\frac{\partial^2}{\partial\mathbf{r}\partial\mathbf{r}}:\left\{\rho_\alpha\overline{(\mathbf{V}_\alpha\mathbf{v}_0)\mathbf{V}_\alpha}\cdot\mathbf{v}_0\right\}$$
$$+\frac{\partial^2}{\partial\mathbf{r}\partial\mathbf{r}}:\left\{\rho_\alpha\overline{(\mathbf{V}_\alpha\mathbf{v}_0)V_\alpha^2}\right\}+\frac{1}{2}\frac{\partial^2}{\partial\mathbf{r}\partial\mathbf{r}}:\left\{\rho_\alpha(\mathbf{v}_0\mathbf{v}_0)v_0^2\right\}$$
$$+\frac{3}{2}\frac{\partial^2}{\partial\mathbf{r}\partial\mathbf{r}}:\left\{p_\alpha(\mathbf{v}_0\mathbf{v}_0)\right\}-\mathbf{v}_0\cdot\frac{\partial^2}{\partial\mathbf{r}\partial\mathbf{r}}:\left\{\left(\overleftrightarrow{P}_\alpha+\rho_\alpha\mathbf{v}_0\mathbf{v}_0\right)\mathbf{v}_0\right\}$$
$$-2\mathbf{v}_0\cdot\frac{\partial^2}{\partial\mathbf{r}\partial\mathbf{r}}:\left\{\rho_\alpha\overline{(\mathbf{V}_\alpha\mathbf{v}_0)\mathbf{V}_\alpha}\right\}-\mathbf{v}_0\cdot\frac{\partial^2}{\partial\mathbf{r}\partial\mathbf{r}}:\left\{\rho_\alpha\overline{(\mathbf{V}_\alpha\mathbf{V}_\alpha)\mathbf{V}_\alpha}\right\}$$
$$+\frac{1}{2}v_0^2\frac{\partial^2}{\partial\mathbf{r}\partial\mathbf{r}}:\rho_\alpha\mathbf{v}_0\mathbf{v}_0+\frac{\partial^2}{\partial\mathbf{r}\partial\mathbf{r}}:\left\{\varepsilon_\alpha\left[\overleftrightarrow{P}_\alpha+2\mathbf{v}_0\mathbf{j}_\alpha+\rho_\alpha\mathbf{v}_0\mathbf{v}_0\right]\right\}$$

$$+\frac{1}{2}v_0^2\frac{\partial^2}{\partial\mathbf{r}\partial\mathbf{r}}:\overleftrightarrow{P}_\alpha. \tag{A1.14}$$

(11)
$$\int E_\alpha\frac{\partial^2 f_\alpha}{\partial\mathbf{v}_\alpha\partial t}\cdot\mathbf{F}_\alpha\,\mathrm{d}\mathbf{v}_\alpha$$
$$=-\frac{\partial}{\partial t}\left(\mathbf{F}_\alpha^{(1)}\cdot\mathbf{j}_\alpha\right)-q_\alpha\frac{\partial}{\partial t}\left[n_\alpha\overline{(\mathbf{v}_\alpha\times\mathbf{B})\cdot\mathbf{V}_\alpha}\,\right]$$
$$+\rho_\alpha\overline{\frac{\partial}{\partial t}\left(\mathbf{F}_\alpha^{(1)}\cdot\mathbf{V}_\alpha\right)}+q_\alpha n_\alpha\overline{\frac{\partial}{\partial t}\left[(\mathbf{v}_\alpha\times\mathbf{B})\cdot\mathbf{V}_\alpha\right]}. \tag{A1.15}$$

(12)
$$\int E_\alpha\frac{\partial\mathbf{F}_\alpha}{\partial t}\cdot\frac{\partial f_\alpha}{\partial\mathbf{v}_\alpha}\,\mathrm{d}\mathbf{v}_\alpha=-\left[\frac{q_\alpha}{m_\alpha}\left(\mathbf{v}_0\times\frac{\partial\mathbf{B}}{\partial t}\right)+\frac{\partial\mathbf{F}_\alpha^{(1)}}{\partial t}\right]\cdot\mathbf{j}_\alpha.$$

(13)
$$\int E_\alpha\mathbf{F}_\alpha\cdot\frac{\partial f_\alpha}{\partial\mathbf{r}}\,\mathrm{d}\mathbf{v}_\alpha=\left(\mathbf{F}_\alpha^{(1)}\cdot\frac{\partial}{\partial\mathbf{r}}\right)\left(\rho_\alpha\widetilde{U}_\alpha\right)+\rho_\alpha\bar{\mathbf{v}}_\alpha\mathbf{F}_\alpha^{(1)}:\frac{\partial}{\partial\mathbf{r}}\mathbf{v}_0$$
$$-\frac{1}{2}\rho_\alpha\left(\mathbf{F}_\alpha^{(1)}\cdot\frac{\partial}{\partial\mathbf{r}}\right)v_0^2+\frac{q_\alpha}{m_\alpha}\frac{\partial}{\partial\mathbf{r}}\cdot\left[\rho_\alpha\tilde{U}_\alpha\mathbf{v}_0\times\mathbf{B}\right]+\frac{q_\alpha}{m_\alpha}\frac{\partial}{\partial\mathbf{r}}\cdot\left[\mathbf{q}_\alpha^{\mathrm{T}}\times\mathbf{B}\right]$$
$$+\frac{1}{2}q_\alpha n_\alpha\overline{v_\alpha^2\mathbf{v}_\alpha}\cdot\operatorname{rot}\mathbf{B}-q_\alpha n_\alpha\overline{\mathbf{v}_\alpha\cdot\operatorname{rot}\left[(\mathbf{v}_\alpha\cdot\mathbf{v}_0)\right]\mathbf{B}}$$
$$+\frac{1}{2}q_\alpha n_\alpha\bar{\mathbf{v}}_\alpha\cdot\operatorname{rot}\left(v_0^2\mathbf{B}\right)+\frac{q_\alpha}{m_\alpha}\varepsilon_\alpha n_\alpha\bar{\mathbf{v}}_\alpha\cdot\operatorname{rot}\mathbf{B}. \tag{A1.16}$$

(14)
$$\int E_\alpha\frac{\partial f_\alpha}{\partial\mathbf{v}_\alpha}\cdot\left(\mathbf{F}_\alpha^{(1)}\times\mathbf{B}\right)\mathrm{d}\mathbf{v}_\alpha=-\left(\mathbf{F}_\alpha^{(1)}\times\mathbf{B}\right)\cdot\mathbf{j}_\alpha. \tag{A1.17}$$

(15)
$$\int E_\alpha\frac{\partial f_\alpha}{\partial\mathbf{v}_\alpha}\cdot\mathbf{v}_\alpha\,\mathrm{d}\mathbf{v}_\alpha=-\left[3\rho_\alpha\tilde{U}_\alpha+\rho_\alpha\overline{V_\alpha^2}+\mathbf{v}_0\cdot\mathbf{j}_\alpha\right]. \tag{A1.18}$$

(16)
$$\int E_\alpha\frac{\partial f_\alpha}{\partial\mathbf{v}_\alpha}\mathbf{v}_\alpha:\frac{\partial}{\partial\mathbf{r}}\mathbf{F}_\alpha\,\mathrm{d}\mathbf{v}_\alpha$$
$$=-n_\alpha\overline{\mathbf{V}_\alpha\mathbf{v}_\alpha}:\frac{\partial}{\partial\mathbf{r}}\mathbf{F}_\alpha-\rho_\alpha\tilde{U}_\alpha\frac{\partial}{\partial\mathbf{r}}\cdot\mathbf{F}_\alpha^{(1)}+n_\alpha\frac{q_\alpha}{m_\alpha}\overline{E_\alpha\mathbf{v}_\alpha}\cdot\operatorname{rot}\mathbf{B}. \tag{A1.19}$$

(17)
$$\int E_\alpha\frac{\partial^2 f_\alpha}{\partial\mathbf{v}_\alpha\partial\mathbf{v}_\alpha}:\mathbf{F}_\alpha\mathbf{F}_\alpha\,\mathrm{d}\mathbf{v}_\alpha$$
$$=\rho_\alpha F_\alpha^{(1)2}+\left(\frac{q_\alpha}{m_\alpha}\right)^2\rho_\alpha\left\{\left[B^2\left(v_0^2-2\widetilde{U}_\alpha-\overline{V_\alpha^2}\right)\right]+\overline{(\mathbf{V}_\alpha\cdot\mathbf{B})^2}-(\mathbf{v}_0\cdot\mathbf{B})^2\right.$$
$$\left.+2(\mathbf{B}\cdot\mathbf{v}_0)\left(\mathbf{B}\cdot\overline{\mathbf{V}}_\alpha\right)\right\}. \tag{A1.20}$$

(18)
$$\int E_\alpha\frac{\partial^2 f_\alpha}{\partial\mathbf{v}_\alpha\partial\mathbf{r}}:\mathbf{v}_\alpha\mathbf{F}_\alpha\,\mathrm{d}\mathbf{v}_\alpha$$
$$=-\mathbf{F}_\alpha^{(1)}\cdot\left[\frac{\partial}{\partial\mathbf{r}}\cdot\overleftrightarrow{P}_\alpha+\frac{\partial}{\partial\mathbf{r}}\cdot\mathbf{v}_0\mathbf{j}_\alpha+\left(\mathbf{j}_\alpha\cdot\frac{\partial}{\partial\mathbf{r}}\right)\mathbf{v}_0+\rho_\alpha\left(\mathbf{v}_0\cdot\frac{\partial}{\partial\mathbf{r}}\right)\mathbf{v}_0\right]$$

$$
\begin{aligned}
&-\left(\mathbf{F}_{\alpha}^{(1)}\cdot\frac{\partial}{\partial\mathbf{r}}\right)\rho_{\alpha}\widetilde{U}_{\alpha}-\left(\mathbf{F}_{\alpha}^{(1)}\cdot\frac{\partial}{\partial\mathbf{r}}\mathbf{v}_{0}\right)\cdot\mathbf{j}_{\alpha}\\
&+\frac{\partial}{\partial\mathbf{r}}\cdot\left[\mathbf{v}_{0}\mathbf{v}_{0}\cdot(q_{\alpha}n_{\alpha}\bar{\mathbf{v}}_{\alpha}\times\mathbf{B})\right]+q_{\alpha}\frac{\partial}{\partial\mathbf{r}}\cdot\left[n_{\alpha}\mathbf{v}_{0}\cdot\overline{(\mathbf{v}_{\alpha}\times\mathbf{B})\mathbf{V}_{\alpha}}\right]\\
&+\overline{\left(\mathbf{v}_{0}\cdot q_{\alpha}n_{\alpha}\frac{\partial}{\partial\mathbf{r}}\right)\left[(\mathbf{v}_{0}\times\mathbf{B})\cdot\mathbf{V}_{\alpha}\right]}+\frac{q_{\alpha}}{m_{\alpha}}\overset{\leftrightarrow}{P}_{\alpha}:\frac{\partial}{\partial\mathbf{r}}(\mathbf{v}_{0}\times\mathbf{B})\\
&+(\mathbf{v}_{0}\times\mathbf{B})\cdot q_{\alpha}n_{\alpha}\overline{\left(\mathbf{V}_{\alpha}\cdot\frac{\partial}{\partial\mathbf{r}}\right)\mathbf{V}_{\alpha}}-\frac{\partial}{\partial\mathbf{r}}\cdot\left[(\mathbf{v}_{0}\times\mathbf{B})\frac{q_{\alpha}}{m_{\alpha}}\rho_{\alpha}\widetilde{U}_{\alpha}\right]\\
&-\frac{q_{\alpha}}{m_{\alpha}}\frac{\partial}{\partial\mathbf{r}}\cdot\left(\mathbf{q}_{\alpha}^{\mathrm{T}}\times\mathbf{B}\right)+\frac{3}{2}k_{B}\frac{q_{\alpha}}{m_{\alpha}}\left[(\mathbf{v}_{0}\times\mathbf{B})\cdot\frac{\partial}{\partial\mathbf{r}}\right]T_{\alpha}\\
&+\frac{1}{2}q_{\alpha}n_{\alpha}\overline{\left[(\mathbf{V}_{\alpha}\times\mathbf{B})\cdot\frac{\partial}{\partial\mathbf{r}}\right]V_{\alpha}^{2}}-\frac{q_{\alpha}}{m_{\alpha}}\rho_{\alpha}\widetilde{U}_{\alpha}\mathbf{v}_{0}\cdot\operatorname{rot}\mathbf{B}\\
&-\frac{q_{\alpha}}{m_{\alpha}}\mathbf{q}_{\alpha}^{\mathrm{T}}\cdot\operatorname{rot}\mathbf{B}.
\end{aligned}
\tag{A1.21}
$$

Using relations (A1.2)–(A1.21) for transformation of energy equation (A1.1), the following form of energy equation can be found for invariant E_{α}:

$$
\begin{aligned}
&\left[\frac{\partial}{\partial t}\left(\rho_{\alpha}\widetilde{U}_{\alpha}\right)+\mathbf{j}_{\alpha}\cdot\frac{\partial\mathbf{v}_{0}}{\partial t}+\frac{\partial}{\partial\mathbf{r}}\cdot\left(\mathbf{v}_{0}\rho_{\alpha}\widetilde{U}_{\alpha}\right)+\frac{\partial}{\partial\mathbf{r}}\cdot\mathbf{q}_{\alpha}^{\mathrm{T}}+\mathbf{j}_{\alpha}\cdot\left(\mathbf{v}_{0}\cdot\frac{\partial}{\partial\mathbf{r}}\right)\mathbf{v}_{0}\right.\\
&\left.-\rho_{\alpha}\overline{\mathbf{F}_{\alpha}\mathbf{V}_{\alpha}}-\rho_{\alpha}\overline{\mathbf{V}_{\alpha}\cdot\left(\mathbf{V}_{\alpha}\cdot\frac{\partial}{\partial\mathbf{r}}\right)\mathbf{V}_{\alpha}}\right]\left(1-\frac{\partial\tau_{\alpha}}{\partial t}\right)-\frac{\partial\mathbf{q}_{\alpha}^{\mathrm{T}}}{\partial t}\cdot\frac{\partial\tau_{\alpha}}{\partial\mathbf{r}}\\
&-\frac{\partial\tau_{\alpha}}{\partial\mathbf{r}}\cdot\frac{\partial}{\partial t}\left(\mathbf{v}_{0}\rho_{\alpha}\widetilde{U}_{\alpha}\right)-\rho_{\alpha}\overline{\mathbf{v}_{\alpha}\mathbf{V}_{\alpha}}:\frac{\partial\mathbf{v}_{0}}{\partial t}\frac{\partial\tau_{\alpha}}{\partial\mathbf{r}}-\frac{\partial\tau_{\alpha}}{\partial\mathbf{r}}\frac{\partial}{\partial\mathbf{r}}:n\alpha\overline{E_{\alpha}\mathbf{v}_{\alpha}\mathbf{v}_{\alpha}}\\
&-\rho_{\alpha}\overline{\left(\frac{\partial\tau_{\alpha}}{\partial\mathbf{r}}\cdot\mathbf{v}_{\alpha}\right)\mathbf{V}_{\alpha}}\cdot\left(\mathbf{v}_{\alpha}\cdot\frac{\partial}{\partial\mathbf{r}}\right)\mathbf{v}_{0}+\frac{\partial\tau_{\alpha}}{\partial\mathbf{r}}\left(n_{\alpha}\overline{E_{\alpha}\mathbf{F}_{\alpha}}\right)\\
&+\left(\frac{\partial\tau_{\alpha}}{\partial\mathbf{r}}\cdot\mathbf{v}_{0}\right)n_{\alpha}\overline{\mathbf{F}_{\alpha}\cdot\mathbf{V}_{\alpha}}+n_{\alpha}\overline{\left(\frac{\partial\tau_{\alpha}}{\partial\mathbf{r}}\cdot\mathbf{V}_{\alpha}\right)\mathbf{V}_{\alpha}\cdot\mathbf{F}_{\alpha}}-\tau_{\alpha}\left\{\rho_{\alpha}\bar{\mathbf{v}}_{\alpha}\mathbf{F}_{\alpha}^{(1)}:\frac{\partial}{\partial\mathbf{r}}\mathbf{v}_{0}\right.\\
&-\frac{1}{2}\rho_{\alpha}\left(\mathbf{F}_{\alpha}^{(1)}\cdot\frac{\partial}{\partial\mathbf{r}}\right)v_{0}^{2}+\frac{1}{2}q_{\alpha}n_{\alpha}\overline{v_{\alpha}^{2}\mathbf{v}_{\alpha}}\cdot\operatorname{rot}\mathbf{B}-q_{\alpha}n_{\alpha}\overline{\mathbf{v}_{\alpha}\cdot\operatorname{rot}\left[(\mathbf{v}_{\alpha}\cdot\mathbf{v}_{0})\mathbf{B}\right]}\\
&+\frac{1}{2}q_{\alpha}n_{\alpha}\mathbf{v}_{\alpha}\cdot\operatorname{rot}\left(v_{0}^{2}\mathbf{B}\right)+\frac{q_{\alpha}}{m_{\alpha}}\varepsilon_{\alpha}n_{\alpha}\bar{\mathbf{v}}_{\alpha}\cdot\operatorname{rot}\mathbf{B}-\left[\frac{q_{\alpha}}{m_{\alpha}}\left(\mathbf{v}_{0}\times\frac{\partial\mathbf{B}}{\partial t}\right)\right.\\
&\left.+\frac{\partial\mathbf{F}_{\alpha}^{(1)}}{\partial t}\right]\cdot\mathbf{j}_{\alpha}-\frac{q_{\alpha}}{m_{\alpha}}\left(\mathbf{F}_{\alpha}^{(1)}\times\mathbf{B}\right)\cdot\mathbf{j}_{\alpha}+\left(\frac{q_{\alpha}}{m_{\alpha}}\right)^{2}B^{2}\mathbf{v}_{0}\cdot\mathbf{j}_{\alpha}-n_{\alpha}\overline{\mathbf{V}_{\alpha}\mathbf{v}_{\alpha}}:\frac{\partial}{\partial\mathbf{r}}\mathbf{F}_{\alpha}\\
&-\rho_{\alpha}\widetilde{U}_{\alpha}\frac{\partial}{\partial\mathbf{r}}\cdot\mathbf{F}_{\alpha}^{(1)}+\frac{\partial^{2}}{\partial t^{2}}\left(\rho_{\alpha}\widetilde{U}_{\alpha}\right)+\rho_{\alpha}\left(\frac{\partial\mathbf{v}_{0}}{\partial t}\right)^{2}-\mathbf{j}_{\alpha}\cdot\frac{\partial^{2}\mathbf{v}_{0}}{\partial t^{2}}
\end{aligned}
$$

$$
\begin{aligned}
&+2\frac{\partial \mathbf{j}_\alpha}{\partial t}\cdot\frac{\partial \mathbf{v}_0}{\partial t}+\frac{1}{2}\frac{\partial^2}{\partial \mathbf{r}\partial \mathbf{r}}:\rho\overline{v_\alpha^2\mathbf{v}_\alpha\mathbf{v}_\alpha}\\
&-\left[\left(\frac{\partial^2}{\partial \mathbf{r}\partial \mathbf{r}}:\rho_\alpha\overline{\mathbf{v}_\alpha\mathbf{v}_\alpha}\right)\mathbf{v}_\alpha\right]\cdot\mathbf{v}_0+\frac{1}{2}v_0^2\frac{\partial^2}{\partial \mathbf{r}\partial \mathbf{r}}:\rho_\alpha\overline{\mathbf{v}_\alpha\mathbf{v}_\alpha}\\
&+\frac{\varepsilon_\alpha}{m_\alpha}\frac{\partial^2}{\partial \mathbf{r}\partial \mathbf{r}}:\rho_\alpha\overline{\mathbf{v}_\alpha\mathbf{v}_\alpha}+\rho_\alpha F_\alpha^{(1)2}+\rho_\alpha\left(\frac{q_\alpha}{m_\alpha}\right)^2\left[B^2v_0^2-(\mathbf{v}_0\cdot\mathbf{B})^2\right.\\
&\left.+(\mathbf{v}_0\cdot\mathbf{B})(\overline{\mathbf{V}}_\alpha\cdot\mathbf{B})\right]-2\mathbf{F}_\alpha^{(1)}\cdot\left[\frac{\partial}{\partial \mathbf{r}}\cdot\overleftrightarrow{P}_\alpha+\frac{\partial}{\partial \mathbf{r}}\cdot\mathbf{v}_0\mathbf{j}_\alpha+\left(\mathbf{j}_\alpha\cdot\frac{\partial}{\partial \mathbf{r}}\right)\mathbf{v}_0\right.\\
&\left.+\rho_\alpha\left(\mathbf{v}_0\cdot\frac{\partial}{\partial \mathbf{r}}\right)\mathbf{v}_0\right]-\left(\mathbf{F}_\alpha^{(1)}\cdot\frac{\partial}{\partial \mathbf{r}}\right)\rho_\alpha\widetilde{U}_\alpha-2\left(\mathbf{F}_\alpha^{(1)}\cdot\frac{\partial}{\partial \mathbf{r}}\mathbf{v}_0\right)\cdot\mathbf{j}_\alpha\\
&+2\frac{\partial}{\partial \mathbf{r}}\cdot\left[\mathbf{v}_0\mathbf{v}_0\cdot(q_\alpha n_\alpha\bar{\mathbf{v}}_\alpha\times\mathbf{B})\right]+2q_\alpha\frac{\partial}{\partial \mathbf{r}}\cdot\left[n_\alpha\mathbf{v}_0\overline{(\mathbf{v}_\alpha\times\mathbf{B})\mathbf{V}_\alpha}\right]\\
&+2\overline{\left(\mathbf{v}_0\cdot q_\alpha n_\alpha\frac{\partial}{\partial \mathbf{r}}\right)\left[(\mathbf{v}_0\times\mathbf{B})\cdot\mathbf{V}_\alpha\right]}+2\frac{q_\alpha}{m_\alpha}\overleftrightarrow{P}_\alpha:\frac{\partial}{\partial \mathbf{r}}(\mathbf{v}_0\times\mathbf{B})\\
&+2(\mathbf{v}_0\times\mathbf{B})\cdot q_\alpha n_\alpha\overline{\left(\mathbf{V}_\alpha\cdot\frac{\partial}{\partial \mathbf{r}}\right)\mathbf{V}_\alpha}-\frac{q_\alpha}{m_\alpha}\frac{\partial}{\partial \mathbf{r}}\cdot\left[(\mathbf{v}_0\times\mathbf{B})\rho_\alpha\widetilde{U}_\alpha\right]\\
&-\frac{q_\alpha}{m_\alpha}\frac{\partial}{\partial \mathbf{r}}\cdot\left(\mathbf{q}_\alpha^{\mathrm{T}}\times\mathbf{B}\right)+\frac{q_\alpha}{m_\alpha}\left[(\mathbf{v}_0\times\mathbf{B})\cdot\frac{\partial}{\partial \mathbf{r}}\right]\left(\rho_\alpha\overline{V_\alpha^2}\right)\\
&+\frac{q_\alpha}{m_\alpha}\rho_\alpha\overline{\left[(\mathbf{V}_\alpha\times\mathbf{B})\cdot\frac{\partial}{\partial \mathbf{r}}\right]V_\alpha^2}-\mathbf{v}_0\cdot\operatorname{rot}\mathbf{B}\frac{q_\alpha}{m_\alpha}\rho_\alpha\widetilde{U}_\alpha-\frac{q_\alpha}{m_\alpha}\mathbf{q}_\alpha^{\mathrm{T}}\cdot\operatorname{rot}\mathbf{B}\\
&-2\frac{q_\alpha}{m_\alpha}\frac{\partial}{\partial t}\left[\rho_\alpha\overline{(\mathbf{v}_\alpha\times\mathbf{B})\cdot\mathbf{V}_\alpha}\right]+2\frac{q_\alpha}{m_\alpha}\rho_\alpha\overline{\frac{\partial}{\partial t}\left[(\mathbf{v}_\alpha\times\mathbf{B})\cdot\mathbf{V}_\alpha\right]}\\
&-2\frac{\partial}{\partial t}\left(\mathbf{F}_\alpha^{(1)}\cdot\mathbf{j}_\alpha\right)+2\rho_\alpha\overline{\frac{\partial}{\partial t}\left(\mathbf{F}_\alpha^{(1)}\cdot\mathbf{V}_\alpha\right)}+\frac{\partial^2}{\partial \mathbf{r}\partial t}\cdot\rho_\alpha\overline{\mathbf{v}_\alpha v_\alpha^2}\\
&-2\mathbf{v}_0\cdot\left[\frac{\partial^2}{\partial \mathbf{r}\partial t}\cdot\rho_\alpha\overline{\mathbf{v}_\alpha\mathbf{v}_\alpha}\right]+v_0^2\frac{\partial^2}{\partial \mathbf{r}\partial t}\cdot\rho_\alpha\bar{\mathbf{v}}_\alpha\\
&\left.+2\frac{\partial^2}{\partial \mathbf{r}\partial t}\cdot(\varepsilon_\alpha n_\alpha\bar{\mathbf{v}}_\alpha)\right\}=J_\alpha^{\mathrm{en},V}.
\end{aligned}
\tag{A1.22}
$$

In the generalized Euler approximation (for local Maxwellian distribution function), the following relations are valid:

(19)
$$\rho_\alpha\overline{\mathbf{V}_\alpha\cdot\left(\mathbf{V}_\alpha\cdot\frac{\partial}{\partial \mathbf{r}}\right)\mathbf{V}_\alpha}=-\frac{\rho_\alpha\overline{V_\alpha^2}}{3}\frac{\partial}{\partial \mathbf{r}}\cdot\mathbf{v}_0. \tag{A1.23}$$

(20)
$$\frac{\partial\tau_\alpha^{(0)}}{\partial \mathbf{r}}\frac{\partial}{\partial \mathbf{r}}:n_\alpha\overline{E_\alpha\mathbf{v}_\alpha\mathbf{v}_\alpha}=\frac{\partial\tau_\alpha^{(0)}}{\partial \mathbf{r}}\frac{\partial}{\partial \mathbf{r}}:\left(\rho_\alpha\widetilde{U}_\alpha\mathbf{v}_0\mathbf{v}_0\right)$$

$$+\frac{\partial\tau_\alpha^{(0)}}{\partial\mathbf{r}}\frac{\partial}{\partial\mathbf{r}}:\left(\frac{5}{2}k_BT_\alpha\frac{p_\alpha}{m_\alpha}\overset{\leftrightarrow}{I}\right)+\frac{\partial\tau_\alpha^{(0)}}{\partial\mathbf{r}}\frac{\partial}{\partial\mathbf{r}}:\frac{\varepsilon_\alpha}{m_\alpha}p_\alpha\overset{\leftrightarrow}{I},\tag{A1.24}$$

where

$$p_\alpha=n_\alpha k_BT_\alpha.$$

(21)

$$\rho_\alpha\overline{\left(\frac{\partial\tau_\alpha^{(0)}}{\partial\mathbf{r}}\cdot\mathbf{v}_\alpha\right)\mathbf{V}_\alpha\cdot\left(\mathbf{v}_\alpha\cdot\frac{\partial}{\partial\mathbf{r}}\right)}\mathbf{v}_0$$

$$=\frac{\partial\tau_\alpha^{(0)}}{\partial\mathbf{r}}\cdot\mathbf{v}_0\left(\frac{\partial}{\partial\mathbf{r}}\cdot\mathbf{v}_0\right)p_\alpha+p_\alpha\frac{\partial\tau_\alpha^{(0)}}{\partial\mathbf{r}}\mathbf{v}_0:\frac{\partial}{\partial\mathbf{r}}\mathbf{v}_0.\tag{A1.25}$$

(22)

$$\frac{\partial^2}{\partial\mathbf{r}\partial\mathbf{r}}:\rho_\alpha\overline{V_\alpha^2\mathbf{V}_\alpha\mathbf{V}_\alpha}=5k_B\Delta\left(T_\alpha\frac{p_\alpha}{m_\alpha}\right),\tag{A1.26}$$

where operator $\Delta\equiv\partial^2/\partial x^2+\partial^2/\partial y^2+\partial^2/\partial z^2$ is Laplacian.

(23)

$$\frac{\partial^2}{\partial\mathbf{r}\partial\mathbf{r}}:\rho_\alpha\overline{(\mathbf{V}_\alpha\mathbf{v}_0)\mathbf{V}_\alpha\cdot\mathbf{v}_0}=\frac{\partial^2}{\partial\mathbf{r}\partial\mathbf{r}}:p_\alpha\mathbf{v}_0\mathbf{v}_0.\tag{A1.27}$$

(24)

$$\frac{\partial^2}{\partial\mathbf{r}\partial\mathbf{r}}:\rho_\alpha\overline{\mathbf{v}_\alpha\mathbf{v}_\alpha}=\frac{\partial^2}{\partial\mathbf{r}\partial\mathbf{r}}:\rho_\alpha\overline{\mathbf{v}_0\mathbf{v}_0}+\frac{\partial^2}{\partial\mathbf{r}\partial\mathbf{r}}:p_\alpha\overset{\leftrightarrow}{I}.\tag{A1.28}$$

(25)

$$\left[\overline{\left(\frac{\partial^2}{\partial\mathbf{r}\partial\mathbf{r}}:\rho_\alpha\mathbf{v}_0\mathbf{V}_\alpha\right)\mathbf{V}_\alpha}\right]\cdot\mathbf{v}_0=\frac{1}{3}\left(\frac{\partial^2}{\partial\mathbf{r}\partial\mathbf{r}}\cdot\rho_\alpha\overline{V_\alpha^2}\mathbf{v}_0\right)\cdot\mathbf{v}_0.\tag{A1.29}$$

(26)

$$\frac{\partial^2}{\partial\mathbf{r}\partial t}\cdot\left(\rho_\alpha\overline{\mathbf{v}_\alpha v_\alpha^2}\right)$$

$$=\frac{\partial^2}{\partial\mathbf{r}\partial t}\cdot\rho_\alpha\mathbf{v}_0v_0^2+\frac{\partial^2}{\partial\mathbf{r}\partial t}\cdot\rho_\alpha\mathbf{v}_0\overline{V_\alpha^2}+2\frac{\partial^2}{\partial\mathbf{r}\partial t}\cdot\left(\rho_\alpha\overline{\mathbf{V}_\alpha(\mathbf{v}_0\cdot\mathbf{V}_\alpha)}\right).\tag{A1.30}$$

As a result, the generalized Euler energy equation (for non-reacting mixture of gases ($\varepsilon_\alpha=0$) in the absence of external forces) is written as

$$\left[\frac{\partial}{\partial t}\left(\rho_\alpha\widetilde{U}_\alpha\right)+\frac{1}{3}\rho_\alpha\overline{V_\alpha^2}\frac{\partial}{\partial\mathbf{r}}\cdot\mathbf{v}_0+\frac{\partial}{\partial\mathbf{r}}\cdot\left(\mathbf{v}_0\rho_\alpha\widetilde{U}_\alpha\right)\right]\left(1-\frac{\partial\tau_\alpha^{(0)}}{\partial t}\right)$$

$$-\frac{\partial\tau_\alpha^{(0)}}{\partial\mathbf{r}}\cdot\frac{\partial}{\partial t}\left(\mathbf{v}_0\rho_\alpha\widetilde{U}_\alpha\right)-p_\alpha\overset{\leftrightarrow}{I}:\frac{\partial\mathbf{v}_0}{\partial t}\frac{\partial\tau_\alpha^{(0)}}{\partial\mathbf{r}}-\frac{\partial\tau_\alpha^{(0)}}{\partial\mathbf{r}}\frac{\partial}{\partial\mathbf{r}}:\left(\rho_\alpha\widetilde{U}_\alpha\mathbf{v}_0\mathbf{v}_0\right)$$

$$-\frac{\partial\tau_\alpha^{(0)}}{\partial\mathbf{r}}\frac{\partial}{\partial\mathbf{r}}:\left(\frac{5}{2}k_BT_\alpha\frac{p_\alpha}{m_\alpha}\overset{\leftrightarrow}{I}\right)-\frac{\partial\tau_\alpha^{(0)}}{\partial\mathbf{r}}\cdot\mathbf{v}_0p_\alpha\cdot\left(\frac{\partial}{\partial\mathbf{r}}\cdot\mathbf{v}_0\right)$$

$$-p_\alpha\frac{\partial\tau_\alpha^{(0)}}{\partial\mathbf{r}}\mathbf{v}_0:\frac{\partial}{\partial\mathbf{r}}\mathbf{v}_0-\tau_\alpha^{(0)}\left\{\frac{\partial^2}{\partial t^2}\left(\rho_\alpha\widetilde{U}_\alpha\right)+\rho_\alpha\left(\frac{\partial\mathbf{v}_0}{\partial t}\right)^2\right.$$

$$
\begin{aligned}
&+\frac{5}{2}k_B\Delta\left(T_\alpha\frac{p_\alpha}{m_\alpha}\right)+\frac{1}{2}\frac{\partial^2}{\partial\mathbf{r}\partial\mathbf{r}}:\rho_\alpha v_0^2\mathbf{v}_0\mathbf{v}_0+\frac{1}{2}\frac{\partial^2}{\partial\mathbf{r}\partial\mathbf{r}}:\left(\rho_\alpha\overline{V_\alpha^2\mathbf{v}_0\mathbf{v}_0}\right)\\
&+\frac{1}{2}\frac{\partial^2}{\partial\mathbf{r}\partial\mathbf{r}}:\left(v_0^2p_\alpha\overleftrightarrow{I}\right)+2\frac{\partial^2}{\partial\mathbf{r}\partial\mathbf{r}}:p_\alpha\mathbf{v}_0\mathbf{v}_0-\left[\left(\frac{\partial^2}{\partial\mathbf{r}\partial\mathbf{r}}:\rho_\alpha\mathbf{v}_0\mathbf{v}_0\right)\mathbf{v}_0\right]\cdot\mathbf{v}_0\\
&-\left[\left(\frac{\partial^2}{\partial\mathbf{r}\partial\mathbf{r}}:p_\alpha\overleftrightarrow{I}\right)\mathbf{v}_0\right]\cdot\mathbf{v}_0-\frac{2}{3}\left[\frac{\partial^2}{\partial\mathbf{r}\partial\mathbf{r}}\cdot\rho_\alpha\overline{V_\alpha^2\mathbf{v}_0}\right]\cdot\mathbf{v}_0+\frac{1}{2}\frac{\partial^2}{\partial\mathbf{r}\partial\mathbf{r}}:\rho_\alpha\mathbf{v}_0\mathbf{v}_0\\
&+\frac{1}{2}v_0^2\frac{\partial^2}{\partial\mathbf{r}\partial\mathbf{r}}:p_\alpha\overleftrightarrow{I}+\frac{\partial^2}{\partial\mathbf{r}\partial t}\cdot\rho_\alpha\mathbf{v}_0v_0^2+\frac{5}{3}\frac{\partial^2}{\partial\mathbf{r}\partial t}\cdot\rho_\alpha\mathbf{v}_0\overline{V_\alpha^2}\\
&-2\mathbf{v}_0\cdot\left[\frac{\partial^2}{\partial\mathbf{r}\partial t}\cdot\rho_\alpha\mathbf{v}_0\mathbf{v}_0\right]-2\mathbf{v}_0\left(\frac{\partial^2}{\partial\mathbf{r}\partial t}\cdot p_\alpha\overleftrightarrow{I}\right)\\
&\left.+v_0^2\frac{\partial^2}{\partial\mathbf{r}\partial t}\cdot\rho_\alpha\mathbf{v}_0\right\}=J_\alpha^{\mathrm{en},V}.
\end{aligned}
\tag{A1.31}
$$

Eq. (A1.31) can be simplified for single-species one-dimensional case

$$
\begin{aligned}
&\left[\frac{\partial}{\partial t}(\rho\widetilde{U})+p\frac{\partial v_0}{\partial x}+\frac{\partial}{\partial x}(\rho v_0\widetilde{U})\right]\left(1-\frac{\partial\tau^{(0)}}{\partial t}\right)\\
&-\left[\frac{\partial}{\partial t}(v_0\rho\widetilde{U})+p\frac{\partial v_0}{\partial t}+\frac{\partial}{\partial x}(v_0^2\rho\widetilde{U})+\frac{5}{2}\frac{\partial}{\partial x}\left(k_BT\frac{p}{m}\right)\right.\\
&\left.+2pv_0\frac{\partial v_0}{\partial x}\right]\frac{\partial\tau^{(0)}}{\partial x}-\tau^{(0)}\left\{\frac{\partial^2}{\partial t^2}(\rho\widetilde{U})+\rho\left(\frac{\partial v_0}{\partial t}\right)^2+\frac{5}{2}k_B\frac{\partial^2}{\partial x^2}\left(\frac{pT}{m}\right)\right.\\
&+\frac{1}{2}\frac{\partial^2}{\partial x^2}(\rho v_0^4)+4\frac{\partial^2}{\partial x^2}(pv_0^2)-v_0\frac{\partial^2}{\partial x^2}(\rho v_0^3)-3v_0\frac{\partial^2}{\partial x^2}(pv_0)\\
&+\frac{1}{2}v_0^2\frac{\partial^2}{\partial x^2}(\rho v_0^2)+\frac{1}{2}v_0^2\frac{\partial^2 p}{\partial x^2}+\frac{\partial^2}{\partial x\partial t}(\rho v_0^3)+5\frac{\partial^2}{\partial x\partial t}(pv_0)\\
&\left.-2v_0\frac{\partial^2}{\partial x\partial t}(\rho v_0^2)-2v_0\frac{\partial^2 p}{\partial x\partial t}+v_0^2\frac{\partial^2}{\partial x\partial t}(\rho v_0)\right\}=0,
\end{aligned}
\tag{A1.32}
$$

in this case $\rho\widetilde{U}=\frac{3}{2}p$.

Energy equation for invariant $m_\alpha v_\alpha^2/2+\varepsilon_\alpha$ can be obtained from the energy equation for invariant E_α as linear combination of hydrodynamic transport equations. Particularly we obtain energy of Eq. (2.7.54) after summation of Eq. (A1.32) (multiplied by two), one-dimensional motion of Eq. (2.7.53) (multiplied term-by-term by hydrodynamic velocity v_0), and continuity equation (2.7.52), multiplied by v_0^2. Then the question about the application of one or another form of equation is the question of easiness of the concrete calculation.

Appendix 2. Three-diagonal method of Gauss elimination technics for the differential third-order equation

Let us consider a highly effective numerical method for solution of boundary problems described by ordinary differential equation of the third order. This method is known in the West as three-diagonal method of Gauss elimination, and in the East as "progonka" ("sweep"). "Progonka" was applied in the theory of the gas dynamic boundary layer with chemical reactions as early as in the beginning of sixties (see, for example, Alexeev, 1967, 1982) and introduced by the author the iterative procedure of suppression of arising oscillations is now universally adopted element of numerical solution of differential equations.

The aim of this appendix is to show how to construct "progonka" method for differential equation

$$b_1 y''' + b_2 y'' + b_3 y' + b_4 y + b_5 = 0. \tag{A2.1}$$

It is assumed that coefficients b_i $(i = 1, \ldots, 5)$ are functions only of the independent variable x; in the following we intend to discuss how to avoid this restriction.

As an example, the boundary conditions are chosen in the form

$$\begin{aligned} &x = a\colon\ y = \alpha,\ y' = \beta, \\ &x = b\colon\ y = \gamma. \end{aligned} \tag{A2.2}$$

The second-order scheme is introduced using a uniform partition of the interval $[a, b]$ with the mesh width h $(x_0 = a, \ldots, x_n = b)$, $x_k = x_0 + kh$, $k = 0, \ldots, n$, and

$$y'_k = \frac{y_{k+1} - y_{k-1}}{2h}, \tag{A2.3}$$

$$y''_k = \frac{y_{k+1} - 2y_k + y_{k-1}}{h^2}, \tag{A2.4}$$

$$y'''_k = \frac{y_{k+2} - 2y_{k+1} + 2y_{k-1} - y_{k-2}}{2h^3}. \tag{A2.5}$$

Using (A2.1)–(A2.5), we find

$$\begin{aligned} &b_{1k}\frac{y_{k+2} - 2y_{k+1} + 2y_{k-1} - y_{k-2}}{2h^3} + b_{2k}\frac{y_{k+1} - 2y_k + y_{k-1}}{h^2} \\ &\quad + b_{3k}\frac{y_{k+1} - y_{k-1}}{2h} + b_{4k} y_k + b_{5k} = 0, \end{aligned} \tag{A2.6}$$

and for $k = 2, \ldots, n - 2$,

$$y_k = y_{k-2}\frac{-b_{1k}}{4hb_{2k} - 2h^3 b_{4k}} + y_{k-1}\frac{2b_{1k} + 2hb_{2k} - b_{3k}h^2}{4hb_{2k} - 2h^3 b_{4k}}$$

$$+ y_{k+1}\frac{-2b_{1k}+2hb_{2k}-b_{3k}h^2}{4hb_{2k}-2h^3b_{4k}} + y_{k+2}\frac{b_{1k}}{4hb_{2k}-2h^3b_{4k}}$$
$$+ \frac{2h^3b_{5k}}{4hb_{2k}-2h^3b_{4k}}. \tag{A2.7}$$

Newton formula leads to the approximation function

$$y(x_m - th) = y(x_m) - \frac{t}{1!}\Delta y_m + \frac{t(t-1)}{2!}\Delta^2 y_m - \frac{t(t-1)(t-2)}{3!}\Delta^3 y_m$$
$$+ \frac{t(t-1)(t-2)(t-3)}{4!}\Delta^4 y_m + R_5 \tag{A2.8}$$

in the point $x_m - th$, where t is an integer or a fractional number and finite differences can be calculated as

$$\begin{aligned}
&\Delta y_m = y_m - y_{m-1},\\
&\Delta y_{m-1} = y_{m-1} - y_{m-2},\\
&\Delta y_{m-2} = y_{m-2} - y_{m-3},\\
&\Delta y_{m-4} = y_{m-4} - y_{m-5},\\
&\Delta^2 y_m - \Delta y_m - \Delta y_{m-1} = y_m - 2y_{m-1} + y_{m-2},\\
&\Delta^2 y_{m-1} = \Delta y_{m-1} - \Delta y_{m-2} = y_{m-1} - 2y_{m-2} + y_{m-3},\\
&\Delta^3 y_m = \Delta^2 y_m - \Delta^2 y_{m-1} = y_m - 3y_{m-1} + 3y_{m-2} - y_{m-3},\\
&\Delta^3 y_{m-1} = \Delta^2 y_{m-1} - \Delta^2 y_{m-2} = y_{m-1} - 3y_{m-2} + 3y_{m-3} - y_{m-4},\\
&\Delta^4 y_m = \Delta^3 y_m - \Delta^3 y_{m-1} = y_m - 4y_{m-1} + 6y_{m-2} - 4y_{m-3} + y_{m-4}.
\end{aligned} \tag{A2.9}$$

Let us introduce the variable $x = x_m - th$, then $t = (x_m - x)/h$, and $t'_x = -h^{-1}$. After differentiating (A2.8) with respect to x, we find

$$\begin{aligned}
&y'_x = -\Delta y t'_x + \frac{2t-1}{2!}\Delta^2 y t'_x,\\
&y''_x = \Delta^2 y (t'_x)^2 - \frac{6t-6}{3!}\Delta^3 y (t'_x)^2 + \frac{12t^2-36t+22}{4!}\Delta^4 y (t'_x)^2,\\
&y'''_x = -\Delta^3 y (t'_x)^3 + \frac{24t-36}{4!}\Delta^4 y (t'_x)^3,
\end{aligned} \tag{A2.10}$$

or for the point x_{m-1},

$$y'_x(x_{m-1}) = \frac{y_m - y_{m-2}}{2h}, \tag{A2.11}$$
$$y''_x(x_{m-1}) = \frac{1}{h^2}\left(\Delta^2 y\right)_m - \frac{1}{12h^2}\left(\Delta^4 y\right)_m$$
$$= \frac{1}{h^2}\Big[y_m - 2y_{m-1} + y_{m-2} - \frac{1}{12}y_m + \frac{1}{3}y_{m-1} - \frac{1}{2}y_{m-2}$$

$$+\frac{1}{3}y_{m-3}-\frac{1}{12}y_{m-4}\Bigg]$$

$$=\frac{1}{h^2}\left[\frac{11}{12}y_m-\frac{5}{3}y_{m-1}+\frac{1}{2}y_{m-2}+\frac{1}{3}y_{m-3}-\frac{1}{12}y_{m-4}\right], \quad \text{(A2.12)}$$

$$y_x'''(x_{m-1})=\frac{1}{h^3}\left[\frac{3}{2}y_m-5y_{m-1}+6y_{m-2}-3y_{m-3}+\frac{1}{2}y_{m-4}\right]. \quad \text{(A2.13)}$$

After substitution of (A2.11)–(A2.13) into (A2.1) we have:

$$\begin{aligned} &y_m\left[18b_{1,m-1}+11hb_{2,m-1}+6h^2b_{3,m-1}\right]\\ &+y_{m-1}\left[-60b_{1,m-1}-20hb_{2,m-1}+12h^3b_{4,m-1}\right]\\ &+y_{m-2}\left[72b_{1,m-1}+6hb_{2,m-1}-6h^2b_{3,m-1}\right]\\ &+y_{m-3}[-36b_{1,m-1}+4hb_{2,m-1}]\\ &+y_{m-4}[6b_{1,m-1}-hb_{2,m-1}]+12h^3b_{5,m-1}=0. \end{aligned} \quad \text{(A2.14)}$$

Three point difference equation is written as

$$y_k=a_1^{(k)}y_{k+1}+a_2^{(k)}y_{k+2}+a_3^{(k)}, \quad \text{(A2.15)}$$

where $a_1^{(k)}$, $a_2^{(k)}$, $a_3^{(k)}$ are ("progonka") coefficients which should be calculated in every point k. On the left boundary ($k=0$) we have $y_0=\alpha$, then

$$\begin{aligned} &y_0=a_1^{(0)}y_1+a_2^{(0)}y_2+a_3^{(0)},\\ &a_1^{(0)}=0, \quad a_2^{(0)}=0, \quad a_3^{(0)}=\alpha. \end{aligned} \quad \text{(A2.16)}$$

For $k=1$ ($m=2$), from Eqs. (A2.9), (A2.10) it is found

$$y_0'=\frac{-y_2+4y_1-3y_0}{2h}, \quad \text{(A2.17)}$$

or

$$y_1=\frac{1}{4}y_2+\frac{1}{4}(2h\beta+3\alpha).$$

Then,

$$a_1^{(1)}=\frac{1}{4}, \qquad a_2^{(1)}=0, \qquad a_3^{(1)}=\frac{1}{4}(3\alpha+2h\beta). \quad \text{(A2.18)}$$

Let us find the values y_{k+1} and y_{k-2}, using the corresponding "progonka" coefficients and y_{k+1}, y_{k+2}. After substitution of expressions y_{k-1} and y_{k-2}, we find using (A2.7) and (A2.15)

$$y_{k-1} = a_1^{(k-1)} a_1^{(k)} y_{k+1} + a_1^{(k-1)} a_2^{(k)} y_{k+2} + a_1^{(k-1)} a_3^{(k)} + a_2^{(k-1)} y_{k+1} + a_3^{(k-1)}, \tag{A2.19}$$

$$y_{k-2} = y_{k+1}\left(a_1^{(k-2)} a_1^{(k-1)} a_1^{(k)} + a_1^{(k-2)} a_2^{(k-1)} + a_2^{(k-2)} a_1^{(k)}\right) + y_{k+2}\left(a_1^{(k-2)} a_1^{(k-1)} a_2^{(k)} + a_2^{(k-2)} a_2^{(k)}\right) + a_1^{(k-2)} a_3^{(k-1)} + a_2^{(k-2)} a_3^{(k)} + a_3^{(k-2)} + a_1^{(k-2)} a_1^{(k-1)} a_3^{(k)}, \tag{A2.20}$$

$$\begin{aligned} y_k = y_{k+1}\{&\left[-b_{1k}\left(a_1^{(k-2)} a_1^{(k-1)} a_1^{(k)} + a_1^{(k-2)} a_2^{(k-1)} + a_2^{(k-2)} a_1^{(k)}\right)\right] \\ &\times \left(4hb_{2k} - 2h^3 b_{4k}\right)^{-1} \\ &+ \left[\left(a_1^{(k-1)} a_1^{(k)} + a_2^{(k-1)}\right)\left(2b_{1k} + 2hb_{2k} - b_{3k}h^2\right)\right. \\ &+ \left.\left(-2b_{1k} + 2hb_{2k} + b_{3k}h^2\right)\right]\left(4hb_{2k} - 2h^3 b_{4k}\right)^{-1}\} \\ &+ y_{k+2}\left[-b_{1k}\left(a_1^{(k-2)} a_1^{(k-1)} a_2^{(k)} + a_2^{(k-2)} a_2^{(k)}\right)\right. \\ &+ \left. a_1^{(k-1)} a_2^{(k)}\left(2b_{1k} + 2hb_{2k} - b_{3k}h^2\right) + b_{1k}\right]\left(4hb_{2k} - 2h^3 b_{4k}\right)^{-1} \\ &+ \left[2h^3 b_{5k} - b_{1k}\left(a_1^{(k-2)} a_1^{(k-1)} a_3^{(k)} + a_1^{(k-2)} a_3^{(k-1)} + a_2^{(k-2)} a_3^{(k)}\right.\right. \\ &+ \left.\left. a_3^{(k-2)}\right)\right]\left(4hb_{2k} - 2h^3 b_{4k}\right)^{-1} + \left[\left(a_1^{(k-1)} a_3^{(k)} + a_3^{(k-1)}\right)\right. \\ &\times \left.\left(2b_{1k} + 2hb_{2k} - b_{3k}h^2\right)\right]\left(4hb_{2k} - 2h^3 b_{4k}\right)^{-1}. \end{aligned} \tag{A2.21}$$

But relation (A2.21) is an identity and as a result the following are valid:

$$a_1^{(k)} = \frac{-M_k a_2^{(k-1)} - 2b_{1k} + 2hb_{2k} + b_{3k}h^2}{M_k a_1^{(k-1)} + b_{1k} a_2^{(k-2)} + 4hb_{2k} - 2h^3 b_{4k}}, \tag{A2.22}$$

$$a_2^{(k)} = \frac{b_{1k}}{M_k a_1^{(k-1)} + b_{1k} a_2^{(k-2)} + 4hb_{2k} - 2h^3 b_{4k}}, \tag{A2.23}$$

$$a_3^{(k)} = \frac{-M_k a_3^{(k-1)} - b_{1k} a_3^{(k-2)} + 2h^3 b_{5k}}{M_k a_1^{(k-1)} + b_{1k} a_2^{(k-2)} + 4hb_{2k} - 2h^3 b_{4k}}, \tag{A2.24}$$

where

$$M_k = -2b_{1k} - 2hb_{2k} + h^2 b_{3k} + b_{1k} a_1^{(k-2)}.$$

On the right side of the interval the system of two equations for definition of y_{n-1}, y_{n-2} can be written. Really, the first equation of this system is

$$y_{n-2} = a_1^{(n-2)} y_{n-1} + a_2^{(n-2)} y_n + a_3^{(n-2)}, \tag{A2.25}$$

and $y_n = \gamma$.

But from (A2.14), it follows

$$\begin{aligned} & y_n\left[18b_{1,n-1} + 11hb_{2,n-1} + 6h^2 b_{3,n-1}\right] \\ & + y_{n-1}\left[-60b_{1,n-1} - 20hb_{2,n-1} + 12h^3 b_{4,n-1}\right] \\ & + y_{n-2}\left[72b_{1,n-1} + 6hb_{2,n-1} - 6h^2 b_{3,n-1}\right] \\ & + y_{n-3}[-36b_{1,n-1} + 4hb_{2,n-1}] + y_{n-4}[6b_{1,n-1} - hb_{2,n-1}] \\ & + 12h^3 b_{5,n-1} = 0, \end{aligned} \tag{A2.26}$$

and

$$y_{n-3} = a_1^{(n-3)} y_{n-2} + a_2^{(n-3)} y_{n-1} + a_3^{(n-3)}, \tag{A2.27}$$

$$y_{n-4} = a_1^{(n-4)} y_{n-3} + a_2^{(n-4)} y_{n-2} + a_3^{(n-4)}. \tag{A2.28}$$

Substitution of (A2.27), (A2.28) into (A2.26) leads to the second equation we need:

$$Ay_{n-2} = By_{n-1} + C, \tag{A2.29}$$

$$\begin{aligned} A = {} & 72b_{1,n-1} + 6hb_{2,n-1} - 6h^2 b_{3,n-1} - 36b_{1,n-1}a_1^{(n-3)} + 4hb_{2,n-1}a_1^{(n-3)} \\ & + \left(a_1^{(n-4)} a_1^{(n-3)} + a_2^{(n-4)}\right)(6b_{1,n-1} - hb_{2,n-1}), \end{aligned} \tag{A2.30}$$

$$\begin{aligned} B = {} & 60b_{1,n-1} + 20hb_{2,n-1} - 12h^3 b_{4,n-1} + a_2^{(n-3)}(36b_{1,n-1} - 4hb_{2,n-1}) \\ & - a_1^{(n-4)} a_2^{(n-3)}(6b_{1,n-1} - hb_{2,n-1}), \end{aligned} \tag{A2.31}$$

$$\begin{aligned} C = {} & -\gamma\left(18b_{1,n-1} + 11hb_{2,n-1} + 6h^2 b_{3,n-1}\right) \\ & + a_3^{(n-3)}(36b_{1,n-1} - 4hb_{2,n-1}) \\ & - \left(a_1^{(n-4)} a_3^{(n-3)} + a_3^{(n-4)}\right)(6b_{1,n-1} - hb_{2,n-1}) - 12h^3 b_{5,n-1}. \end{aligned} \tag{A2.32}$$

Practical application of the "progonka" method can be realized as follows:

(a) From relations (A2.16), (A2.18), (A2.22)–(A2.24) the Gauss ("progonka") coefficients should be found for $0 \leqslant k \leqslant n$.
(b) From the system of two Eqs. (A2.25), (A2.29), we find y_{n-1}, y_{n-2}.
(c) From the equation (backward "progonka")

$$y_{k-2} = a_1^{(k-2)} y_{k-1} + a_2^{(k-2)} y_k + a_3^{(k-2)}, \tag{A2.33}$$

we find the rest of discrete functions for $3 \leqslant k \leqslant n-1$.

For non-linear equations the formulated procedure can be realized after linearization of corresponding differential equation and introduction of iterative calculation. But as it was written in Alexeev (1967): "The formulated procedure led to increasing oscillations of numerical solutions. The way to avoid this non-stability consists in very simple but effective consideration, i.e., introduction of suppressive coefficients δ ("Daempfer coefficients," in the Eastern scientific literature). For construction of $(n+1)$-iteration, the outlet numerical massif of nth iteration $\varphi_i^{(n)}$ was used in corrected form $\varphi_i'^{(n)}$, which was connected by the

$$\varphi_i'^{(n)} = \varphi_i'^{(n-1)} + \delta\big(\varphi_i^{(n)} - \varphi_i'^{(n-1)}\big). \tag{A2.34}$$

The coefficient δ was selected by way of experiment during the process of numerical calculations. In many cases is sufficient to use $\delta \approx 0.1$."

Appendix 3. Some integral calculations in the generalized Navier–Stokes approximation

Some integrals should be calculated in the generalized Navier–Stokes approximation connected with the distribution function

$$f_\alpha = f_\alpha^{(0)} + f_\alpha^{(1)},$$

where

$$f_\alpha^{(1)} = f_\alpha^{(0)}\left\{-\mathbf{A}_\alpha \cdot \frac{\partial \ln T}{\partial \mathbf{r}} - \overleftrightarrow{\mathbf{B}}_\alpha : \frac{\partial}{\partial \mathbf{r}}\mathbf{v}_0 + n\sum_j \mathbf{C}_\alpha^{(j)} \cdot \mathbf{d}_j\right\}.$$

Here the results of calculations are presented:

(1)
$$\frac{\partial^2}{\partial\mathbf{r}\partial\mathbf{r}} : \sum_\alpha \rho_\alpha\overline{(\mathbf{V}_\alpha\mathbf{v}_0)\mathbf{V}_\alpha} = \frac{\partial}{\partial\mathbf{r}}\left[\frac{\partial}{\partial\mathbf{r}}\cdot(p\mathbf{v}_0)\right] - 2\frac{\partial^2}{\partial\mathbf{r}\partial\mathbf{r}} : \mathbf{v}_0\mu\overleftrightarrow{S}, \tag{A3.1}$$

where in accordance with the definition, $\partial^2/\partial\mathbf{r}\partial\mathbf{r} : \mathbf{v}_0\mu\overleftrightarrow{S}$ is vector with components,

$$\left[\frac{\partial^2}{\partial\mathbf{r}\partial\mathbf{r}} : \mathbf{v}_0\mu\overleftrightarrow{S}\right]_k = \sum_{i,j=1}^{3}\frac{\partial^2}{\partial r_i\partial r_j}\left\{\mu v_{0j}\left[\frac{1}{2}\left(\frac{\partial v_{0k}}{\partial r_i} + \frac{\partial v_{0i}}{\partial r_k}\right) - \frac{1}{3}\delta_{ki}\frac{\partial}{\partial\mathbf{r}}\cdot\mathbf{v}_0\right]\right\},$$

$$k = 1, 2, 3. \tag{A3.2}$$

(2)
$$\frac{\partial^2}{\partial\mathbf{r}\partial\mathbf{r}} : \sum_\alpha \rho_\alpha\overline{(\mathbf{V}_\alpha\mathbf{V}_\alpha)\mathbf{V}_\alpha} = -2\frac{\partial^2}{\partial\mathbf{r}\partial\mathbf{r}}\cdot\left(K\frac{\partial T}{\partial\mathbf{r}}\right) - \Delta\left(K\frac{\partial T}{\partial\mathbf{r}}\right)$$
$$+2\frac{\partial^2}{\partial\mathbf{r}\partial\mathbf{r}}\cdot\left[\frac{pn}{\rho}\sum_{\alpha,\beta} m_\beta\big(D_{\alpha\beta} - D_{\alpha\beta}^1\big)\mathbf{d}_\beta\right]$$
$$+\Delta\left(\frac{pn}{\rho}\sum_{\alpha,\beta} m_\beta\big(D_{\alpha\beta} - D_{\alpha\beta}^1\big)\mathbf{d}_\beta\right), \tag{A3.3}$$

where Δ is Laplacian, $K = \sum_\alpha K_\alpha$. See also (5.3.22), (5.3.27), (5.3.28).

(3)
$$\frac{\partial^2}{\partial\mathbf{r}\partial\mathbf{r}} : \sum_\alpha \rho_\alpha \overline{(\mathbf{V}_\alpha\mathbf{V}_\alpha)V_\alpha^2}$$
$$= 5k_B\Delta\left(T\sum_\alpha \frac{p_\alpha}{m_\alpha}\right) - \frac{112}{9}\frac{\partial^2}{\partial\mathbf{r}\partial\mathbf{r}} : \left[\overset{\leftrightarrow}{S}k_B T\sum_\alpha \frac{1}{m_\alpha}(\mu_\alpha - \mu_\alpha^1)\right]$$
$$- \frac{280}{27}\frac{\partial^2}{\partial\mathbf{r}\partial\mathbf{r}} : \left[\overset{\leftrightarrow}{D}k_B T\sum_\alpha \frac{1}{m_\alpha}(\mu_\alpha - \mu_\alpha^1)\right], \tag{A3.4}$$

where $\overset{\leftrightarrow}{D}$ is a tensor with components

$$D_{ij} = \delta_{ij}\frac{\partial v_{0i}}{\partial r_j},$$

and viscosity and "the first" viscosity are defined by the relations

$$\mu_\alpha = \frac{1}{2}k_B n_\alpha T b_{\alpha 0}, \tag{A3.5}$$

$$\mu_\alpha^1 = \frac{1}{2}k_B n_\alpha T b_{\alpha 1}, \tag{A3.6}$$

(4)
$$\frac{\partial^2}{\partial\mathbf{r}\partial\mathbf{r}} : \sum_\alpha \rho_\alpha \overline{(\mathbf{V}_\alpha\mathbf{v}_0)\mathbf{V}_\alpha\cdot\mathbf{v}_0} = \frac{\partial^2}{\partial\mathbf{r}\partial\mathbf{r}} : p\mathbf{v}_0\mathbf{v}_0 - 2\frac{\partial^2}{\partial\mathbf{r}\partial\mathbf{r}} : \mu\left(\mathbf{v}_0\mathbf{v}_0\cdot\overset{\leftrightarrow}{S}\right). \tag{A3.7}$$

(5)
$$\frac{\partial^2}{\partial\mathbf{r}\partial\mathbf{r}} : \sum_\alpha \rho_\alpha \overline{(\mathbf{V}_\alpha\mathbf{v}_0)V_\alpha^2}$$
$$= -5\frac{\partial^2}{\partial\mathbf{r}\partial\mathbf{r}} : K\frac{\partial T}{\partial\mathbf{r}}\mathbf{v}_0 + 5\frac{\partial^2}{\partial\mathbf{r}\partial\mathbf{r}} : \frac{p}{\rho}\sum_{\alpha,\beta} m_\beta\left(D_{\alpha\beta} - D_{\alpha\beta}^1\right)\mathbf{d}_\beta\mathbf{v}_0. \tag{A3.8}$$

(6)
$$\frac{\partial^2}{\partial\mathbf{r}\partial\mathbf{r}} : \sum_\alpha \rho_\alpha \overline{(\mathbf{V}_\alpha\mathbf{V}_\alpha)(\mathbf{V}_\alpha\cdot\mathbf{v}_0)} = -2\frac{\partial^2}{\partial\mathbf{r}\partial\mathbf{r}} : K\mathbf{v}_0\frac{\partial T}{\partial\mathbf{r}}$$
$$-\Delta\left(K\mathbf{v}_0\cdot\frac{\partial T}{\partial\mathbf{r}}\right) - 2\frac{\partial^2}{\partial\mathbf{r}\partial\mathbf{r}} : \frac{pn}{\rho}\mathbf{v}_0\sum_{\alpha,\beta} m_\alpha m_\beta\left(D_{\alpha\beta} - D_{\alpha\beta}^1\right)\mathbf{d}_\beta$$
$$-\Delta\left[\frac{pn}{\rho}\mathbf{v}_0\cdot\sum_{\alpha,\beta} m_\alpha m_\beta\left(D_{\alpha\beta} - D_{\alpha\beta}^1\right)\mathbf{d}_\beta\right]. \tag{A3.9}$$

Appendix 4. Three-diagonal method of Gauss elimination technique for the differential second-order equation

Let us consider the solution of the ordinary differential equation of the second-order of the three-diagonal method of Gauss elimination technics,

$$c_1 y'' + c_2 y' + c_3 y + c_4 = 0. \tag{A4.1}$$

The boundary conditions are chosen in the form

$$x = a, \qquad y = \alpha, \tag{A4.2}$$

$$x = b, \qquad y = \gamma. \tag{A4.3}$$

The second-order scheme $O(h^2)$ is introduced using the following approximation in the node k:

$$y'_k = \frac{y_{k+1} - y_{k-1}}{2h} + \frac{h^2}{6} f'''(\xi), \tag{A4.4}$$

$$y''_k = \frac{y_{k-1} - 2y_k + y_{k-1}}{h^2} + \frac{h^2}{12} f^{(IV)}(\xi) \tag{A4.5}$$

for the uniform partition of the interval $[a, b]$.

$$a = x_0 < x_1 < \cdots < x_{n-1} < x_n = b.$$

Therefore

$$x_k - x_{k-1} = \frac{b-a}{n}. \tag{A4.6}$$

Using (A4.4), (A4.5), we find the finite-difference approximation for Eq. (A4.1)

$$c_{1k} \frac{y_{k+1} - 2y_k + y_{k-1}}{h^3} + c_{2k} \frac{y_{k+1} - y_{k-1}}{2h} + c_{3k} y + c_{4k} = 0. \tag{A4.7}$$

Eqs. (A4.7) – together with boundary conditions $y_0 = \alpha$, $y_n = \gamma$ – constitutes a coupled system of linear algebraic equations of $(n + 1)$-order for obtaining unknown values of y_k.

Eq. (A4.7) in the point x_1 is written as:

$$c_{11} \frac{y_2 - 2y_1 + y_0}{h^2} + c_{21} \frac{y_2 - y_0}{2h} + c_{31} y_1 + c_{41} = 0. \tag{A4.8}$$

Because y_0 is a known value, Eq. (A4.8) can be transformed as:

$$y_1 = a_1^{(1)} y_2 + a_2^{(1)}, \tag{A4.9}$$

where a_1^1, a_2^1 are known coefficients. The analogue relations can be written for every step of calculations. Therefore

$$y_k = a_1^{(k)} y_{k+1} + a_2^{(k)}. \tag{A4.10}$$

Coefficients $a_1^{(k)}$, $a_2^{(k)}$ are Gauss coefficients in the three-diagonal method of elimination technique for the differential second-order equation ("progonka" coefficients, in the Eastern scientific literature).

It is convenient to formulate recurrence formulae. With this aim first of all let us find y_{k-1} as a function of y_{k+1}

$$\begin{cases} y_{k-1} = a_1^{(k-1)} y_k + a_2^{(k-1)}, \\ y_{k-1} = a_1^{(k-1)} y_{k+1} a_1^{(k)} + a_1^{(k-1)} a_2^{(k)} + a_2^{(k-1)}. \end{cases} \tag{A4.11}$$

After substitution of y_{k-1}, y_k from (A4.10), (A4.11) into (A4.8), we have

$$\begin{aligned} & c_{1k} y_{k+1} - 2c_{1k} a_1^{(k)} y_{k+1} - 2c_{1k} a_2^{(k)} + c_{1k} a_1^{(k-1)} y_{k+1} a_1^{(k)} + c_{1k} a_1^{(k-1)} a_2^{(k)} \\ & + c_{1k} a_2^{(k-1)} + c_{2k} \frac{h}{2} y_{k+1} - c_{2k} \frac{h}{2} a_1^{(k-1)} y_{k+1} a_1^{(k)} - c_{2k} \frac{h}{2} a_1^{(k-1)} a_2^{(k)} \\ & - c_{2k} \frac{h}{2} a_2^{(k-1)} + c_{3k} h^2 a_1^{(k)} y_{k+1} + c_{3k} h^2 a_2^{(k)} + h^2 c_{4k} = 0 \end{aligned} \tag{A4.12}$$

or

$$\begin{aligned} & y_{k+1} \left(c_{1k} - 2c_{1k} a_1^{(k)} + c_{1k} a_1^{(k-1)} a_1^{(k)} + c_{2k} \frac{h}{2} - c_{2k} \frac{h}{2} a_1^{(k-1)} a_1^{(k)} + c_{3k} h^2 a_1^{(k)} \right) \\ & + \left(-2c_{1k} a_2^{(k)} + c_{1k} a_1^{(k-1)} a_2^{(k)} + c_{1k} a_2^{(k-1)} - c_{2k} \frac{h}{2} a_1^{(k-1)} a_2^{(k)} \right. \\ & \left. - c_{2k} \frac{h}{2} a_2^{(k-1)} + c_{3k} h^2 a_2^{(k)} + h^2 b_{4k} \right) = 0. \end{aligned} \tag{A4.13}$$

But relation (A4.13) is an identity which is valid for every y_{k+1}. Then:

$$a_1^{(k)} = \frac{c_{1k} + c_{2k} h/2}{a_1^{(k-1)} [-c_{1k} + c_{2k} h/2] + 2c_{1k} - c_{3k} h^2}, \tag{A4.14}$$

$$a_2^{(k)} = \frac{-a_2^{(k-1)} (-c_{1k} + c_{2k} h/2) + c_{4k} h^2}{a_1^{(k-1)} (-c_{1k} + c_{2k} h/2) + 2c_{1k} - c_{3k} h^2}. \tag{A4.15}$$

It is not difficult to find the first pair of coefficients $a_1^{(0)}$, $a_2^{(0)}$. Really, if boundary condition ($k = 0$) has the form $y_0 = \alpha$, then from

$$y_0 = a_1^{(0)} y_1 + a_2^{(0)}$$

we have

$$a_1^{(0)} = 0, \qquad a_2^{(0)} = \alpha. \tag{A4.16}$$

Appendix 5. Characteristic scales in plasma physics

The fundamental feature of plasma physics is the existence of a multiparticle interaction in the system under study. Consequently, care must be exercised when choosing typical scales for describing the evolution of a plasma volume. The Landau length l over which the characteristic kinetic energy $k_B T$ of thermal motion equals the potential energy of interaction between charges e is determined by the relation

$$l = \frac{e^2}{4\pi\varepsilon_0 k_B T} = \frac{1.67 \times 10^{-5}}{T} \text{ m}, \tag{A5.1}$$

where ε_0 is the electrical permittivity of vacuum, and k_B is the Boltzmann constant.

Binary collisions for which impact parameters are less than or equal to the Landau length are said to be "close". It is useful to introduce the ratio of the Landau length to the average distance $n^{-1/3}$ between plasma particles:

$$\beta = \ln^{1/3} = 1.67 \times 10\ {}^{5} n^{1/3} T\ {}^{1}, \tag{A5.2}$$

where n is the number density of particles, in m^{-3}. While the interaction parameter β is usually small in laboratory plasma, the solar corona and the solar atmosphere, in the ionosphere and interstellar gas but for free electrons in a metal it can reach a value of $\approx 10^2$. The cross section σ_b for close collisions is determined by the relation

$$\sigma_b = \pi l^2, \tag{A5.3}$$

and the mean free path of a probe particle between binary close collisions is

$$\lambda = \frac{1}{\pi n l^2} = \frac{1}{\pi n^{1/3}\beta^2} = 1.1 \times 10^9 \frac{T^2}{n} \text{ m}. \tag{A5.4}$$

The pair interaction between particles in a plasma effectively extends to the distance determined by the Debye–Hueckel radius r_D defined as follows:

$$r_D = \sqrt{\frac{\varepsilon_0 k_B T}{ne^2}} = \frac{1}{n^{1/3}\sqrt{4\pi\beta}} = 0.69 \times 10^2 \sqrt{\frac{T}{n}} \text{ m}. \tag{A5.5}$$

It can be argued that collective plasma properties disappear in systems less than r_D in size. The following relationship between the characteristic plasma lengths should be noted:

$$l : n^{-1/3} : r_D : \lambda = \beta : 1 : \frac{1}{2\sqrt{\pi\beta}} : \frac{1}{\pi\beta^2}. \tag{A5.6}$$

Eq. (A5.6) should be complemented by the hydrodynamic scale L being the characteristic size of the system; usually L is much larger than λ.

The above list of characteristic plasma scales is not exhaustive, though for processes in rapidly alternating fields, when the distance a particle travels over a period of the field oscillation is less than the range of the forces involved. Additional scales may appear in the problem.

Appendix 6. Dispersion relations in the generalized Boltzmann kinetic theory neglecting the integral collision term

We are concerned with developing (within the GBE framework) the dispersion relation for plasma in the absence of a magnetic field. We make the same assumptions used in developing this relation within the BE model, namely:

(a) the integral collision term is neglected;
(b) the evolution of electrons and ions in a self-consistent electric field corresponds to a one-dimensional, unsteady model;
(c) distribution functions for ions f_i and electrons f_e deviate only slightly from their respective equilibrium values f_{0i} and f_{0e}:

$$f_i = f_{0i}(u) + \delta f_i(x, u, t), \tag{A6.1}$$

$$f_e = f_{0e}(u) + \delta f_e(x, u, t); \tag{A6.2}$$

(d) we consider a wave mode corresponding to a certain wave number k and a complex frequency ω, so that the solution of the GBE can be written in the form

$$\delta f_i = \langle \delta f_i \rangle \, \mathrm{e}^{\mathrm{i}(kx-\omega t)}, \tag{A6.3}$$

$$\delta f_e = \langle \delta f_e \rangle \, \mathrm{e}^{\mathrm{i}(kx-\omega t)}; \tag{A6.4}$$

(e) the quadratic terms in the GBE, determining the deviation from the equilibrium DFs, are neglected, and
(f) the self-consistent forces F_i and F_e are small:

$$F_i = -\frac{e}{m_i}\frac{\partial \varphi}{\partial x}, \tag{A6.5}$$

$$F_e = -\frac{e}{m_e}\frac{\partial \varphi}{\partial x}, \tag{A6.6}$$

where e is the absolute electron charge, m_i are the ion masses, m_e the electron mass, and finally

$$\varphi = \langle \varphi \rangle \, \mathrm{e}^{\mathrm{i}(kx-\omega t)}. \tag{A6.7}$$

Under these assumptions, the GBE is written as follows (we seek the solution for the ion plasma component, to be specific):

$$\frac{\partial f_i}{\partial t} + u\frac{\partial f_i}{\partial x} + F_i\frac{\partial f_i}{\partial u} - \tau_i\left\{\frac{\partial^2 f_i}{\partial t^2} + 2u\frac{\partial^2 f_i}{\partial t\partial x} + u^2\frac{\partial^2 f_i}{\partial x^2} + 2F_i\frac{\partial^2 f_i}{\partial t\partial u}\right.$$
$$\left. + \frac{\partial F_i}{\partial t}\frac{\partial f_i}{\partial u} + F_i\frac{\partial f_i}{\partial x} + u\frac{\partial F_i}{\partial x}\frac{\partial f_i}{\partial u} + F_i^2\frac{\partial^2 f_i}{\partial u^2} + 2uF_i\frac{\partial^2 f_i}{\partial u\partial x}\right\} = 0. \quad \text{(A6.8)}$$

Using the assumptions listed above, we find the relations

$$\begin{aligned}
&\frac{\partial f_i}{\partial t} = -\mathrm{i}\omega\delta f_i, \qquad u\frac{\partial f_i}{\partial x} = \mathrm{i}ku\delta f_i,\\
&F_i\frac{\partial f_i}{\partial u} = -\frac{e}{m_i}\frac{\partial\varphi}{\partial x}\frac{\partial f_{0i}}{\partial u}, \qquad \frac{\partial^2 f_i}{\partial t^2} = -\omega^2\delta f_i,\\
&2u\frac{\partial^2 f_i}{\partial t\partial x} = 2\omega uk\delta f_i, \qquad u^2\frac{\partial^2 f_i}{\partial x^2} = -u^2k^2\delta f_i,\\
&2F_i\frac{\partial^2 f_i}{\partial u\partial t} = 0, \qquad \frac{\partial F_i}{\partial t}\frac{\partial f_i}{\partial u} = -\frac{e}{m_i}\omega k\varphi\frac{\partial f_{0i}}{\partial u},\\
&F_i\frac{\partial f_i}{\partial x} = 0, \qquad u\frac{\partial f_i}{\partial u}\frac{\partial F_i}{\partial x} - \frac{e}{m_i}k^2u\varphi\frac{\partial f_{0i}}{\partial u},\\
&F_i^2\frac{\partial^2 f_i}{\partial u^2} = 0, \qquad \frac{\partial^2 f_i}{\partial u\partial x}2uF_i = 0,
\end{aligned} \quad \text{(A6.9)}$$

which when substituted into Eq. (A6.8) yield

$$\mathrm{i}(ku-\omega)\langle\delta f_i\rangle - \mathrm{i}\frac{e}{m_i}k\langle\varphi\rangle\frac{\partial f_{0i}}{\partial u}$$
$$-(ku-\omega)\tau_i\left\{-(ku-\omega)\langle\delta f_i\rangle + \langle\varphi\rangle\frac{ek}{m_i}\frac{\partial f_{0i}}{\partial u}\right\} = 0, \quad \text{(A6.10)}$$

giving the ion density fluctuation

$$\langle\delta n_i\rangle = -\frac{e}{m_i}\langle\varphi\rangle k\int\frac{\partial f_{0i}/\partial u}{\omega-ku}\,\mathrm{d}u \quad \text{(A6.11)}$$

and the electron density fluctuation

$$\langle\delta n_e\rangle = \frac{e}{m_e}\langle\varphi\rangle k\int\frac{\partial f_{0e}/\partial u}{\omega-ku}\,\mathrm{d}u. \quad \text{(A6.12)}$$

Eqs. (A6.11) and (A6.12) are identical to their BE analogues. Substituting Eqs. (A6.11) and (A6.12) into the Poisson equation

$$\varepsilon_0 k^2\varphi = e(\delta n_i - \delta n_e) \quad \text{(A6.13)}$$

we arrive at the classical dispersion relation (see, for instance, Artsimovich and Sagdeev, 1979)

$$1 = -\frac{e^2}{\varepsilon_0 k}\left\{\frac{1}{m_e}\int_{-\infty}^{+\infty}\frac{\partial f_{0e}/\partial u}{\omega - ku}\,\mathrm{d}u + \frac{1}{m_i}\int_{-\infty}^{+\infty}\frac{\partial f_{0i}/\partial u}{\omega - ku}\,\mathrm{d}u\right\}. \tag{A6.14}$$

Although Eqs. (A6.11) and (A6.12) are a consequence of the general statement that in the absence of the integral collision term the relation

$$\frac{Df_\alpha}{Dt} = 0 \tag{A6.15}$$

(the Vlasov equation) is the solution of the equation

$$\frac{Df_\alpha}{Dt} - \frac{D}{Dt}\left(\tau_\alpha \frac{Df_\alpha}{Dt}\right) = 0, \tag{A6.16}$$

the above argument shows that the GBE can produce correct and expected results, when treated perturbatively.

References*

Abakumov, A.I., Vinogradov, V.S. (1981). Elastic scattering of high-energetic electrons. *Izv. Vyssh. Uchebn. Zaved. Fiz.* **1**, 8–12 (in Russian).

Alexeev, B.V. (1967). *Boundary Layer with Chemical Reactions*. Computer Center of USSR Academy of Sciences, Moscow (in Russian, see also translation realized by Lockheed missiles and space company, 1968).

Alekseev, B.V. (1982). *Matematicheskaya Kinetika Reagiruyushchikh Gazov (Mathematical Theory of Reacting Gases)*. Nauka, Moscow.

Alekseev, B.V. (1987). Hydrodynamic equations in kinetic theory of reacting gases. *J. Comput. Math. Math. Phys.* **27**, 730–740.

Alexeev, B.V. (1988). Generalized Boltzmann equation. In: *Book of Abstracts*. Conference on Mechanics of Media with Chemical Reactions. Krasnoyarsk, pp. 2–4.

Alekseev, B.V. (1990a). Investigation of sound propagation in the frame of generalized Navier–Stokes equations. *Dokl. Akad. Nauk SSSR* **313**, 1078. *Soviet Phys. Dokl.* **35** (1990) 748.

Alekseev, B.V. (1990b). Investigation of sound propagation in the frame of generalized Navier–Stokes equations. In: Alekseev, B.V. (Ed.), *Mekhanika i Elektrodinamika Sploshnykh Sred (Mechanics and Electrodynamics of Continua)*. Izd. Mosk. Univ., Moscow, pp. 20–24.

Alexeev, B.V. (1992). Generalized hydrodynamical equations. In: *Proceedings: IUTAM Symposium Aerothermochemistiry of Spacecraft and Associated Hypersonic Flows*. Universite de Provence (IUSTI/MHEQ), pp. 85–91.

Alekseev, B.V. (1993). To the theory of generalized Boltzmann equation. *High Temp.* **31** (4), 624–633.

Alexeev, B.V. (1994). The generalized Boltzmann equation, generalized hydrodynamic equations and their applications. *Philos. Trans. Roy. Soc. London* **349**, 417–443.

Alekseev, B.V. (1995a). An investigation of the charged particle distribution function in electromagnetic field involving the use of the generalized Boltzmann equation. *High Temp.* **33** (6), 834–845.

Alekseev, B.V. (1995b). An investigation of the charged particle distribution function in electromagnetic field involving the use of the generalized Boltzmann equation. *Teplofiz. Vys. Temp.* **33** (6), 838. English translation: *High Temp.* **33** (6), 834–845.

Alexeev, B.V. (1995c). *Physica A* **216**, 459.

Alekseev, B.V. (1997). Generalized Boltzmann physical kinetics. *Teplofiz. Vys. Temp.* **35** (1), 129. *High Temp.* **35** (1), 125.

Alekseev, B.V. (2000a). *Teplofiz. Vys. Temp.* **38**, 374. *High Temp.* **38** (2000) 351.

Alexeev, B.V. (2000b). *Phys. Usp.* **43** (6), 601.

Alekseev, B.V. (2001). *Teplofiz. Vys. Temp.* **39**, 693. *High Temp.* **39** (2001) 641.

Alekseev, B.V. (2003). *Phys. Usp.* **46** (2), 139.

Alexeev, B.V., Abakumov, A.I. (1982). About one method of solution of Schrödinger equation. *Dokl. Akad. Nauk SSSR* **262**, 5 (in Russian).

Alexeev, B.V., Chikhaoui, A.A. (1993). Application of the generalized hydrodynamic equations for shock wave structure calculations. Shock waves. In: *Proceedings of the 19th International Symposium on Shock Waves, Vol. 4*. Marselle, France, pp. 39–43.

Alekseev, B.V., Grushin, I.T. (1994). *Protsessy Perenosa v Reagiruyushchikh Gazakh i Plazme (Transport Processes in Reacting Gases and Plasma)*. Energoatomizdat, Moscow.

*The author's name in references is given in forms it appeared in cited publications. The translated forms of the author's name vary due to different translators and time of publication, and may differ from the correct translated form "Alexeev, B.V.".

Alekseev, B.V., Lebedev, Yu.A., Michailov, V.V. (1997). Investigation of the generalized Boltzmann equation for electron energy distribution in a constant electric field with due regard for inelastic collisions. *High Temp.* **35** (2), 207–213.

Alexeev, B.V., Mikhailov, V.V. (1999). An investigation of vortex flows of compressible gas involving the use of generalized hydrodynamic equations. *High Temp.* **37** (2), 250–259.

Alexeev, B.V., Mikhailov, V.V. (2001). Generalized hydrodynamic equations: application to the investigation of cas flow in channels with a step. *High Temp.* **39** (4), 567–577.

Alexeev, B.V., Mikhailov, V.V. (2003). Vortex flow past a plate in a channel. *High Temp.* **41** (2), 401–414.

Alekseev, B.V., Poddoskin, A.B., Mikhailov, V.V. (1990). Investigation of sound propagation in the frame of generalized hydrodynamic equations. In: Alekseev, B.V. (Ed.), *Mekhanika i Elektrodinamika Sploshnykh Sred (Mechanics and Electrodynamics of Continua)*, Izd. Mosk. Univ., Moscow. p. 25.

Alekseev, B.V., Polev, V.V. (1990). Calculation of the shock wave structure with the help of the high-accuracy hydrodynamic equations. *Teplofiz. Vys. Temp.* **28** (6), 614–616. *High Temp.* (*USSR*) **28** (1990) 896.

Alexander, F.J., Garcia, A.L., Alder, B.J. (1995). A consistent Boltzmann algorithm. *Phys. Rev. Lett.* **74** (26), 5212–5215.

Aoki, K., Takata, S., Nakanishi, T. (2002). Poiseuille-type flow of rarefied gas between two parallel plates driven by a uniform external force. *Phys. Rev. E* **65**, 026315.

Arnal, M., Laurel, O., Liek, Z., Reric, M. (1992). In: Devill, M., Le, T.-H., Morchois, N.Yu. (Eds.), *Notes on Numerical Fluid Mechanics, Vol. 36*. A GAMM Workshop. Vieweg, Baunsch, p. 13.

Artsimovich, L.A., Sagdeev, R.Z. (1979). *Fizika Plazmy dlya Fizikov (Plasma Physics for Physicists)*. Atomizdat, Moscow.

Bakai, A.S., Sigov, Yu.S. (1996). Mnogolikaya turbulentnost (Multifacial turbulence). In: Makarov, I.M. (Ed.), *Novoe v Sinergetike. Zagadki Mira Neravnovesnykh Struktur (New Concepts in Synergy. Puzzles in the World of Nonequilibrium Systems)*. Nauka, Moscow, p. 10.

Balescu, R. (1960). *Phys. Fluids* **3**, 52.

Balescu, R. (1975). *Equilibrium and Nonequilibrium Statistical Mechanics, Vols. 1, 2*. Wiley, New York. Russian translation: Mir, Moscow, 1978.

Belotserkovski, O.M., Oparin, A.M. (2000). *Numerical Experiment in Turbulence*. Nauka, Moscow (in Russian).

Bhatnagar, P.L., Gross, E.P., Krook, M.A. (1954). A model for collision processes in gases, I. Small amplitude processes in charged and neutral systems. *Phys. Rev.* **511**, 94.

Bird, G.A. (1994). *Molecular Gas Dynamics and Direct Simulation of Gas Flows*. Clarendon, Oxford.

Bogolyubov, N.N. (1946). *Problemy Dinamicheskoi Teorii v Statisticheskoi Fizike (Dynamic Theory Problems in Statistical Physics)*. Gostekhizdat, Moscow. English translation: *The Dynamical Theory in Statistical Physics*. Delhi Hindustan, 1965.

Boltzmann, L. (1872). Weitere Studien über das Wärmegleichgewicht unter Gasmolekulen. *Sitz. Ber. Kaiserl. Akad. Wiss.* **66** (2), 275.

Boltzmann, L. (1912). *Vorlesungen über Gastheorie*. Zweiter unveränderten Abdruck. Johann Barth, Leipzig. 2 Teile.

Born, M., Green, H.S. (1946). A general kinetic theory of liquids 1. The molecular distribution function. *Proc. Roy. Soc.* **188** (1012), 10.

Bourgat, J.F., Le Tallec, P., Malinger, F., Tidriri, D., Qiu, Y. (1992). Numerical coupling of Boltzmann and Navier–Stokes. In: *Aerothermochemistry of Spacecraft and Associated Hypersonic Flows*. Proceedings of IUTAM Symposium. Marseille, France, pp. 60–66.

Braginski, S.I. (1963). In: Leontovich, M.A. (Ed.), *Voprosy Teorii Plazmy (Problems in Plasma Theory), Vol. 1*. Gosatomizdat, Moscow, p. 183. English translation: Leontovich, M.A. (Ed.), *Reviews of Plasma Physics, Vol. 1*. Consultants Bureau, New York, p. 205.

Buchel'nikova, N.S., Matochkin, E.P. (1977a). Preprint No. 77-15, The Budker Institute of Nuclear Physics of the Siberian Branch of the USSR Acad. of Sci., Novosibirsk.

Buchel'nikova, N.S., Matochkin, E.P. (1977b). Preprint No. 77-39, The Budker Institute of Nuclear Physics of the Siberian Branch of the USSR Acad. of Sci., Novosibirsk.

Buchel'nikova, N.S., Matochkin, E.P. (1979). Preprint No. 79-112, The Budker Institute of Nuclear Physics of the Siberian Branch of the USSR Acad. of Sci., Novosibirsk.

Buchel'nikova, N.S., Matochkin, E.P. (1980). *Fiz. Plazmy* **6**, 1097. *Soviet J. Plasma Phys.* **6** (1980) 603.

Cattaneo, C. (1948). *Atti Sem. Mat. Fis. Un. Modena* **3**, 3–21.

Cattaneo, C. (1958). Sur une forme de l'equation de la chaleur eliminant le paradoxe d'une propagation instantanee. *C. R. Acad. Sci. Paris* **247**, 431–433.

Cercignani, C. (1975). *Theory and Application of the Boltzmann Equation*. Scottish Academic Press, Edinburgh and London.

Chapman, S. (1916). On the law of distribution of velocities and on the theory of viscosity and thermal conduction in a non-uniform simple monoatomic gas. *Philos. Trans. Roy. Soc. Ser. A* **216**, 279.

Chapman, S. (1917). On the theory of a gas, part 2. A composite monoatomic gas, diffusion, viscosity and thermal conduction. *Philos. Trans. Roy. Soc. A* **217**, 115.

Chapman, S., Cowling, T.G. (1952). *The Mathematical Theory of Non-Uniform Gases*. University Press, Cambridge.

Chen, F.F. (1984). *Introduction to Plasma Physics and Controlled Fusion*, 2nd edn. Plenum Press, New York. Russian translation: Mir, Moscow, 1987.

Chetverushkin, B.N. (1999). *Kineticheski-Soglasovannye Skhemy v Gazovoi Dinamike: Novaya Model Vyazkogo Gaza, Algoritmy, Parallel'naya Realizatsiya, Prilozheniya (Kinetically Consistent Schemes in Gas Dynamics: New Viscid Gas Model, Algorithms, Parallel Realization, and Applications)*. MGU, Moscow.

Clemmow, P.C., Dougherty, J.P. (1990). *Electrodynamics of Particles and Plasmas*. Addison-Wesley, Reading, MA. Russian translation: Mir, Moscow, 1996.

Compton, K.T. (1916a). *Phys. Rev.* **7** (4), 489.

Compton, K.T. (1916b). *Phys. Rev.* **7** (5), 509.

Cox, H.L., Bonham, R.A. (1967). Elastic electron scattering amplitudes for neutral atoms calculated using the partial wave method at 10, 40, 70 and 100 keV for Z-1 to Z-54. *J. Chem. Phys.* **47** (8), 2599.

Davydov, B.I. (1935). *Dokl. Akad. Nauk SSSR* **2** (7), 474.

Davydov, B.I. (1936). About velocity distribution of electrons moving in electric field. *JETP* **6** (5), 463–480.

Dewan, E.M. (1961). *Phys. Fluids* **4** (6).

Doering, Ch.R., Gibbon, J.D. (1995). *Applied Analysis of the Navier–Stokes Equations*. Cambridge University Press.

Druyvesteyn, M.J. (1930a). *Physica* **10**, 61.

Druyvesteyn, M.J. (1930b). *Z. Phys.* **64** (11–12), 781.

Eaton, J.K., Johnston, A. (1981). Review of research on subsonic turbulent flow reattachment. *AIAA J.* **19** (9), 1093–1100.

Eggert (1919). *J. Phys. Z.* **20**, 570.

Ehrenfest, P. (1972). *Otnositelnost', Kvanty, Statistika (Relativity, Quantum, Statistics)*. Nauka, Moscow.

Ehrenfest, P. (1979). *Collected Scientific Papers*. Klein, M. (Ed.), North-Holland, Amsterdam.

Elizarova, T.G. (1992). Simulation of spacial flow problems on multiprocessor systems. In: *IUTAM Symposium Aerothermochemistry of Spacecraft and Associated Hypersonic Flows*, Marselle, France, pp. 211–213.

Elizarova, T.G., Chetverushkin, B.N. (1986). *The Use of Kinetic Models for Calculation of Gas Dynamic Flows. Mathmatical Modeling*. Nauka, Moscow, pp. 261–278.

Elizarova, T.G., Chetverushkin, B.N. (1989). Kinetical-consistent finite-difference gasdynamic schemes. In: *Intern. Sympos. on Comp. Fluid Dynamics*, Nagoja, Japan, pp. 501–506.

Elsholz, L.E. (1955). *Qualitative Methods in Mathematical Analysis*. Gos. Izd. Tekhnico-Teoret. Literat., Moscow, p. 300 (in Russian).

Enskog, D. (1917). *The Kinetic Theory of Phenomena in Fairly Rare Gases*. Upsala.

Enskog, D. (1921a). The numerical calculation of phenomena in fairly rare gases. *Svensk. Vet. Akad. Arkiv. f. Math. Ast. och Fys.* **16**, 1.

Enskog, D. (1921b). *Kungl. Svenska Vetenskaps Akad. Handl.* **63** (4).

Favre, A.J.A. (1992). Formulation of the statistical equations of turbulent flows with variable density. In: Gatski, Th.B., Sarkar, S., Speziale, Ch.G. (Eds.), *Studies in Turbulence*. Springer-Verlag, New York, pp. 324–341.

Fedoseyev, A.I., Alexeev, B.V. (1998a). Mathematical model for viscous incompressible fluid flow using Alexeev equations and comparison with experimental data. In: Dey, S.K., Ziebarth, G., Ferrandiz (Eds.), *Proceedings of Advances in Scientific Computing Modelling (Special Proceedings of IMACS '98)*. Alicante, Spain, pp. 158–163.

Fedoseyev, A.I., Alexeev, B.V. (1998b). Higher order continuum model for incompressible viscous flow and application to the modelling of thermal vibrational convection. In: Inan, E., Markov, K.Z. (Eds.), *Continuum Models and Discrete Systems (CMDS9)* Proceedings of the 9th International Symposium. Istanbul, Turkey, pp. 130–137.

Fizicheskaya Entsiklopediya. T. 2. Zhidkost' (Physical Encyclopedia. Vol. 2. Liquid). Sovetskaya Entsiklopediya, Moscow, 1990.

Fletcher, C.A.J. (1987). *Computational Techniques for Fluid Dynamics I, II*. Springer-Verlag.

Fowler, R.H., Guggenheim, E.A. (1939). *Statistical Thermodynamics*. Hamburg.

Franklin, R.N., Hamberger, S.M., Smith, G. (1972). *J. Phys. Rev. Lett.* **29**, 914.

Frenkel', Ya.I. (1945). *Kineticheskaya Teoriya Zhidkostei (Kinetic Theory of Liquids)*. Izd. AN SSSR, Moscow–Leningrad.

Gal'tsev, V.E., Dem'yanov, A.V., Kochetov, I.V., Pevgov, V.G., Sharkov, V.F. (1979). Mechanism of excitation of gas discharge lasers. Preprint of the Institute of Atomic Energy, Moscow, no. 3156.

Gibbs, J.W. (1934). *The Collected Works of J. Willard Gibbs, Vol. 2*. Longmans Green, New York. Russian translation: Gostekhizdat, Moscow, 1946.

Ginzburg, V.L., Gurevich, A.V. (1960a). Non-linear effects in plasma placed in alterating electromagnetic field. *Phys. Usp.* **70** (2), 202.

Ginzburg, V.L., Gurevich, A.V. (1960b). Non-linear effects in plasma placed in alterating electromagnetic field. *Phys. Usp.* **70** (3), 393.

Glasstone, S., Laidler, K.J., Eyring, H. (1941). *The Theory of Rate Processes. The Kinetics of Chemical Reactions, Viscosity, Diffusion and Electrochemical Phenomena*. New York and London.

Golant, V.E. (1959). About connection between characteristics surehigh frequency and direct current in the gas. *Izv. Akad. Nauk SSSR Ser. Fis.* **23**, 958–961.

Golant, V.E., Zhilinski, A.P., Sakharov, I.E. (1977). *Basements of Plasma Physics*. Atomizdat, Moscow (in Russian).

Goldman, E., Sirovich, L. Equations for gas mixtures. *Phys. Fluids* **10** (9), 1928–1940.

Grad, H. (1949). *Comm. Pure Appl. Math.* **2** (4), 38.

Green, H.S. (1952). *The Molecular Theory of Fluids*. Amsterdam.

Greenspan, M. (1956). Propagation of sound in five monoatomic gases. *J. Acoust. Soc. Am.* **28** (4), 644–648.

Gryzinski, M., Kunc, J., Zgorzelski (1973). *J. Phys. B* **6**, 2292.

Heisenberg, W. (1924). *Ann. Phys.* (Leipzig) **74** (5), 577.

Heisenberg, W. (1969). In: *Life of Physics; Evening Lectures at the International Centre of Theoretical Physics in Triest*. IAIA, Wiena.

Hildebrand, J.H. (1977). *Viscosity and Diffusivity*. Wiley, New York.

Hirschfelder, I.O., Curtiss, Ch.F., Bird, R.B. (1954). *Molecular Theory of Gases and Liquids*. Wiley, New York, Chapman and Hall, London.

Holway, L.H. (1964). *Phys. Fluids* **7** (6), 141.

Ivanov, Yu.A., Lebedev, Yu.A., Polak, L.S. (1981). *Methods of Contact Diagnostics in Non-Equilibrium Plasma Chemistry*. Nauka, Moscow (in Russian).

Jarzebski, A.B., Thulli, J.W. (1986). A stable highly accurate ADI method for hyperbolic heat conduction equation. *J. Comput. Phys.* **63**, 236–239.

Karoulina, E.B., Lebedev, Y.A. (1988). Computer simulation of microwave and dc-plasma. *J. Phys. D* **21**, 411.

Karoulina, E.B., Lebedev, Y.A. (1992). Influence of cross-sectional shift on electron energy distributions in microwave and dc-plasma. *J. Phys. D* **25**, 401.

Kirkwood, J.G. (1947). The statistical mechanical theory of transports processes II. Transport in gases. *J. Chem. Phys.* **15** (1), 72.

Klimontovich, Yu.L. (1964). *Statisticheskaya Teoriya Neravnovesnykh Protsessov v Plazme (The Statistical Theory of Non-Equilibrium Processes in Plasma)*. MGU, Moscow. English translation: MIT Press, Cambridge, MA, 1967.

Klimontovich, Yu.L. (1975). *Kineticheskaya Teoriya Neideal'nogo Gaza i Neideal'noy Plazmy (Kinetic Theory of Nonideal Gases and Nonideal Plasmas)*. Nauka, Moscow. English translation: Pergamon Press, Oxford, 1982.

Klimontovich, Yu.L. (1992). About necessity and possibility of unified description of hydrodynamic processes. *Theoret. and Math. Phys.* **92** (2), 312.

Klimontovich, Yu.L. (1995). *Statisticheskaya Teoriya Otkrytykh Sistem (Statistical Theory of Open Systems).* Yanus-K, Moscow. English translation: Kluwer Academic, Dordrecht, 1995.

Klimontovich, Yu.L. (1997). *Phys. Usp.* **167** (1), 23.

Koremans, J.M.J., Beenaker, J.M.J. (1960). The influence density on the viscosity coefficients of gases. *Physica* **26**, 653.

Kozeff, J.R., Street, R. (1984a). Vizualization studies of a shear driven three-dimensional recirculating flow. *J. Fluids Engrg.* **106**, 21–29.

Kozeff, J.R., Street, R.L. (1984b). The lid-driven cavity flow: synthesis of qualitative and quantitative observations. *J. Trans. ASME* **106**, 390–398.

Kozeff, J.R., Street, R.L.J. (1984c). On end wall effects in a lid-driven cavity flow. *J. Fluids Engrg.* **106**, 385–389.

Kramers, H.A. (1940). *Physica* **7** (4), 284.

Krieger, V.A., Lukin, V.Yu., Semenov, A.M. (1986). Calculation of the Na thermodynamic parameters from the "first principles". *Teplofiz. Vys. Temp.* **24** (3), 474–482.

Labusov, A.N., Lapin, Yu.V. (1996). Four-parameter two-slice algebraic model of transition boundary layer on the flat plate. *High Temp.* **34** (6), 942.

Landau, L.D. (1941a). *J. Phys. USSR* **5**, 71. *Phys. Usp.* **93** (1967) 527.

Landau, L.D. (1941b). *JETP* **11**, 592.

Landau, L.D. (1946a). *J. Phys. USSR* **10**, 25.

Landau, L.D. (1946b). *Zh. Eksp. Teor. Fiz.* **16**, 574.

Landau, L.D. (1947). *J. Physics* **11**, 91.

Landau, L.D., Lifshitz, E.M. (1988). *Gidrodinamika (Kurs Teoreticheskoi Fiziki, T. 6) (Hydrodynamics Course of Theoretical Physics, Vol. 6).* Nauka, Moscow. English translation: Pergamon Press, Oxford, 1990.

Lebedev, Yu.A., Epstein, I.L. (1995). Simulation of microwave plasma in hydrogen. *J. Moscow Phys. Soc.* **5**, 103–120.

Lenard, A. (1960). *Ann. Phys. (New York)* **10**, 390.

Longo, E., Preziosi, L., Bellomo, N. (1992). The semicontinuous Boltzmann equation: towards a model for fluid dynamics applications. In: *Aerothemochemistry of Spacecraft and Associated Hypersonic Flows.* Proceedings of IUTAM Symposium. Marselle, France, pp. 41–46.

Looney, D.H., Brown, S.C. (1954). *Phys. Rev.* **93**, 965.

Lukovnikov, A.I., Fetisov, I.V. (1969). The electron distribution function in molecular plasma. In: *Fizika Gazorazryadnoi Plazmy, Vol. 2 (Physics of Gas Discharge Plasma).* Atomizdat, Moscow.

Malek, M.M., Baras, F., Garcia, A.L. (1997). *Physica A* **240**, 255.

Maxwell, J.C. (1860). Illustrations of the dynamical theory of gases, I. On the motion and collisions of perfectly elastic spheres; II. On the process of diffusion of two or more kinds of moving particles among one another; III. On the collision of perfectly elastic bodies of any form. *Philos. Mag.* **19**, 19; **20**, 21; **20**, 33.

Meshcherskii, I.V. (1897). *Dinamika Tochki Peremennoi Massy (Dynamics of a Variable Mass Point).* St Petersburg.

Meyer, E., Sessler, G. (1957). On the propagation of sound in monoatomic gases. *Z. Phys.* **149** (15), 15–39.

Moliere, G. (1947). Theorie der Streuung schneller geladener Teilchen. Einzelstreuung am abgeschirmten Coulomb-Field. *Z. Naturforsch.* **2A** (3), 133–145.

Morozov, A.N. (1996). *Zh. Eksp. Teor. Fiz.* **109** (4), 1304. *JETP* **82** (1996) 703.

Morse, T.F. (1963). Exchange between nonequipartition gases. *Phys. Fluids* **6** (10), 1420–1427.

Nayfeh, A.H. (1972). *Perturbation Methods.* Wiley–Interscience, New York.

Nikolis, G., Prigogine, I. (1977). *Self-organization in Nonequilibrium Systems.* Wiley–Interscience, New York.

Pekeris, C.L., Alterman, Z., Finkelstein, L. (1960). In: *Sumposium on the Numerical Treatment of Ordinary Differential Equations, Integral and Integro-Differential Equation of P.I.C.C.* Birkhäuser, Basel, pp. 338–398.

Pekeris, C.L., Alterman, Z., Finkelstein, L., Frankowski, K. (1962). *Phys. Fluids* **5**, 1608–1616.

Pidduck, F.B. (1913). *Proc. Roy. Soc. London Ser. A* **88** (603), 296.

Prigogine, I. (1962). *Introduction to Thermodynamics of Irreversible Processes*. Second ed. Wiley–Interscience, New York.

Reitz, R.D. (1981). One-dimensional compressible gas dynamics calculations using the Boltzmann equation. *J. Comput. Phys.* **42** (1), 103–123.

Reshetnyak, S.A., Shelepin, L.A. (1996). *Kvazistatsionarnye Raspredeleniya v Kinetike (Quasi-Stationary Distributions in Kinetics)*. IPO "Avtor", Moscow.

Reynolds, O. (1894). On the dynamical theory of incompressible viscous fluids and the determination of the criterion. *Philos. Trans. Roy. Soc. Ser. A* **CLXXVI**, 123.

Rezibois, P., De Lener, M. (1980). *Klassicheskaya Kineticheskaya Teoriya Zhidkostei i Gazov (Classical Kinetic Theory of Fluids)*. Mir, Moscow (Russian translation).

Richtmyer, R.D., Morton, K.W. (1967). *Difference Methods for Initial-Value Problems*. Interscience.

Rosenau, Ph. (1993). Random walker and the telegrapher's equation: a paradigm of a generalized hydrodynamics. *Am. Phys. Soc.* **48** (7), 655–657.

Saha, M.N. (1920). *Philos. Mag.* **40**, 472.

Schlichting, H. (1964). *Grenzschicht – Theory*. Karlsruhe, Braun.

Schmidt, B. (1969). Electron beam density measurments in shock waves in argon. *J. Fluid Mech.* **39** (2), 361–373.

Semenov, A.M. (1984). Calculation of the Na thermodynamic parameters from the "first principles". *Dokl. Akad. Nauk SSSR* **278**, 866–869.

Shakhov, E.M. (1974). *Method of Rarefied Gas Investigation*. Nauka, Moscow.

Shaposhnikov, I.G. (1951). *Zh. Eksp. Teor. Fiz.* **21**, 1310.

Shkarovsky, I., Johnstone, T., Bachinsky, M. (1969). *Kinetika Chastits Plazmy (Kinetics of Plasma Particles)*. Atomizdat, Moscow.

Silin, V.P. (1971). *Introduction in Kinetic Theory of Gases*. Nauka, Moscow (in Russian).

Sirovich, L., Thurber, J.K. (1965). Propagation of forced sound waves in rarefied gasdynamics. *J. Acoust. Soc. Am.* **37** (2), 329–339.

Slezkin, N.A. (1951a). *Dokl. Akad. Nauk SSSR* **77**, 205.

Slezkin, N.A. (1951b). *Dokl. Akad. Nauk SSSR* **79**, 33.

Smirnov, B.M. (1985). *Physics of Weakly Ionized Gas*. Nauka, Moscow.

Sobolev, S.L. (1993). Two-temperature discrete model for nonlocal heat conduction. *J. Phys. France* **3**, 2261–2269.

Spencer, L.V. (1955). Theory of electron penitration. *Phys. Rev.* **98** (6), 1597–1615.

Terletskii, Ya.P. (1952). *Zh. Eksp. Teor. Fiz.* **22**, 506.

Terletskii, Ya.P. (1994). *Statisticheskaya Fizika (Statistical Physics)*. Vyssh. Shkola, Moscow. English translation: North-Holland, Amsterdam, 1971.

Tij, M., Santos, A. (1994). *J. Statist. Phys.* **76**, 1399.

Trubnikov, B.A. (1963). In: Leontovich, M.A. (Ed.), *Voprosy Teorii Plazmy (Problems in Plasma Theory), Vol. 1*. Gosatomizdat, Moscow, p. 98. English translation: Leontovich, M.A. (Ed.), *Reviews of Plasma Physics, Vol. 1*. Consultants Bureau, New York, 1965, p. 105.

Uhlenbeck, G. (1971). *Phys. Usp.* **103** (2), 275.

Uribe, F.J., Garcia, A.L. (1999). Burnett description for plane poiseuille flow. *Phys. Rev. E* **60** (4), 4063–4078.

Vallander, S.V. (1951). *Dokl. Akad. Nauk SSSR* **78**, 25.

Vernotte, P. (1958a). La veritable equation de la chaleur. *C. R. Acad. Sci. Paris* **251**, 2103–2105.

Vernotte, P. (1958b). Les paradoxes de la theorie continue de l'equation de la chaleur. *C. R. Acad. Sci. Paris* **246**, 3154–3155.

Verstappen, R.W.C.P., Veldman, A.E.P. (1996). In: Dosidori, J.A., et al. (Eds.), *Proc. of the Third ECCOMAS Computational Fluid Dynamics Conf.*, Paris, France, 9–13 September 1996. Wiley, New York, p. 1073.

Vlasov, A.A. (1978). *Nelokal'naya Statisticheskaya Mekhanika (Nonlocal Statistical Mechanics)*. Nauka, Fizmatlit, Moscow.

Woods, L.C. (1990). On the role of pressure gradients in kinetic theory. In: *Book of Abstracts Aachen, Vol. 2*. 17th Internat. Symp. RGD, pp. 442–444.

Woods, L.C. (1993). *An Introduction to the Kinetic Theory of Gases and Magnetoplasmas*. Oxford Univ. Press, Oxford.

Yvon, J. (1935). *La Theorie Statistique des Fluide et l'Equation d'Etat*. Paris.

Zhdanov, V.M., Roldugin, V.I. (2002). *Zh. Eksp. Teor. Fiz.* **122**, 789. *JETP* **95** (2002) 682.

Subject Index

H-theorem, 61
λ and L scales, 44

alternating electric field, 176
"apocalyptic" points, 74
approximation against the course of time, 52
asymptotic expansion, 25
attachment behind the step, 261
attenuation of sound, 226
attenuation rate, 234

Bernoulli integral, 157
BGK approximation, 58, 89
Blasius formula, 281
Bogolyubov chain, 100
Bogolyubov–Born–Green–Kirkwood–Yvon kinetic equations (BBGKY-equations), 38
Boltzmann collision integral for the multi-component gas, 51
bottom secondary eddies, 248, 251
boundary and initial conditions, 256, 267
boundary layers for solutions of equations, 101
bracket expressions, 200

Cauchy integral, 157
cell scale, 330
"central" eddy, 248
"classical" principle of indeterminacy, 90
"collective" viscosity, 325
complex conductivity, 180
conditions of normalization of the DF, 42
conductivity, 185
correlation functions, 83, 84
Coulomb logarithm, 288
criterion of stability, 147
cyclotron resonance condition, 181

Debye–Hueckel radius, 303
diada, 93
Dieterici's equation, 327
differential cross-section of inelastic collision, 69
diffusive fluxes, 208
diffusive velocity, 91
discrete spectrum of plasma wawes, 311
dispersion equation, 233, 234
drag of a plate, 281
drift velocity, 163
Druyvesteyn distribution function, 175
dynamical viscosity, 115

easy-start method, 318
effective potential energy, 156
electron
– diffusion coefficient, 223
– distribution function, 223
energy temperatures, 164
entropy, 70
Euler number, 244
excitation of plasma oscillations, 314
explicit form of differential part of GBE, 98
external forces, 97

"first" diffusive coefficient, 211
"first" viscosity coefficient, 212
flicker noise, 90
fluctuation of the Boltzmann H-function, 153
free-molecular limit, 56
Frenkel's model, 324
frequency
– of elastic collisions, 221
– of inelastic collisions, 221

general boundary conditions, 153
generalized energy equation, 106
generalized Euler approach with gravitation, 123
generalized Euler energy equation, 119, 122
generalized Euler hydrodynamic equations for one-dimensional motion, 124
generalized Euler momentum equation, 119
generalized Maxwell equations, 299
generalized momentum equation, 101, 105
"ghost" particles, 90

Hall constant, 184
Hildebrand experiments, 324

independent fluctuations, 153
integral of bimolecular collisions, 69
internal energy of particle, 93

kinetically consistent difference schemes, 77
Knudsen layer, 130
Knudsen numbers, 188
Kolmogorov fluctuation scale, 135
Kolmogorov scale length, 138
Kolmogorov–Obukhov law, 134
Kovalevskaya theorem, 147

Landau-damping solution, 315
Langmuir ion frequency, 317
Langmuir waves, 318
linear Fredholm equations, 190
local drag coefficient, 281
local Maxwellian DF, 120
Lorentz gas, 166

mass center velocity, 204
Maxwellian gas, 161
mean
– mass-velocity, 91
– thermal velocity, 91
– time between collisions, 297
– velocity of molecules, 91
method of many scales, 21
mobility, 177
– of electrons, 223
modified Landau formula, 319
molecular markers, 92

non-local integral term, 90
non-local kinetic equations, 83
non-relativistic Hamiltonian, 154

particle correlations, 41
plasma frequency, 299
Poiseuille flow, 255
polarization effects, 85
pressure, 93
– tensor, 210
Prigogine's principle, 63
probability density
– for canonical distribution, 70
– of elastic collisions, 95

quantum liquid, 336
quantum mechanics of open systems, 160

rarefied gases, 42
return flow, 259
Reynolds number, 244
"rigid" boundary condition, 278

scales involved in the gas kinetics problems, 44
Schrödinger equation, 153
second Dieterici equation, 328
self-consistent force, 84
self-consistent forces, 42
"simplest" version, 89
six-dimensional Jacobian, 75
slip velocity, 131
"soft" boundary condition, 278
Sonine expansion, 201
statistical weights, 69
substantial derivatives, 85
super-fluidity, 336

thermal conduction, 211
thermal flux, 212
thermal velocity, 91
time delay effects, 83
turbulent "spot", 262

upper secondary eddy, 250

vector potential, 157
velocity of sound, 226
vortex zones, 260

wave
– damping regime, 319
– number, 229

"zero" diffusive coefficient, 211